SCIENCE matters

A Complete Course for Junior Certificate

Hilary Dorgan
Declan Kennedy
Siobhan Scott

FOLENS

Editor
Alastair Hall

Picture Research
Alastair Hall

Design/Layout/Cover
Karen Hoey

Artwork
Michael Phillips, Gary Dermody, David Benham and Steph Dix of Graham Cameron Illustration, Aileen Caffrey

Photography
Declan Corrigan

ISBN 978-1-84131-812-7

1565

ACKNOWLEDGEMENTS

The writing of this textbook has been a very challenging task and we are extremely grateful to the many people who have given us so much support and encouragement. We wish to thank those who have helped us in so many ways.

We would also like to thank David Benham and Steph Dix of Graham Cameron Illustration, Aileen Caffrey, Michael Phillips and Gary Dermody for their excellent artwork.

We wish to express our sincere thanks and appreciation to Alastair Hall for his excellent work as Editor of this book. He has worked extremely hard to ensure that this textbook is of the highest possible standard. We also wish to thank the other members of the staff at Folens with whom it has been a pleasure to work – particularly Karen Hoey, who designed and set the book, Gary Dermody and, of course, John O'Connor, Managing Director.

Siobhan Scott
I would like to thank my family, especially Michael, for their support and encouragement during the writing of this book.

Declan Kennedy
Sincere thanks to Brian, Esther, Alan and Karen for all their help and support. Thanks are extended to the following for their very useful suggestions on the typescript: Sean Finn, Pat Hanratty, Alan Kennedy, Karen Kennedy, Eoin O'Dwyer, Ted O'Keeffe.

Hilary Dorgan
Also, thanks to my wife, Mary, and family, who provided all the support and encouragement I needed throughout the writing.

The authors and Publisher would like to thank the following who have supplied photographs, appeared in photographs, allowed photographs to be taken on their premises or supplied equipment which appeared in photographs: Cork County Council; Lennox Laboratory Supplies; Cobh Town Council; Ben Coffey; Marian College, Lansdowne Road; Declan Corrigan, Declan Corrigan Photography; John Crowley; Sinead Cunningham; Julie Dorgan; Barry Dorgan; Sean Downey; Inniscarra Water Treatment Plant; Colm McCarthy; Pat O'Connell; Murty O'Dowd; Schering Plough and Department of Chemistry, UCC. The authors and Publisher would also like to thank the following for permission to reproduce photographs: Alamy Images; Science Photo Library; Corbis; DK Images; Getty Images Ireland Inc; London's Transport Museum; GE Infra, Energy; Alan Thomas; Camlab Ltd and Novartis Ringaskiddy Ltd.

TABLE OF CONTENTS

INTRODUCTION

Science Matters is specifically written for the Junior Certificate science syllabus introduced in September 2003 and first examined in June 2006. All of the material for Higher and Ordinary level students is included in this textbook. All Higher Level material is indicated by a red vertical line drawn down the left-hand side of the text, or in some cases specific words, terms or diagrams appear in red lettering.

In presenting the material, great care has been taken to explain the various concepts as clearly as possible and to the depth required by the syllabus. The language used has been kept as simple as possible to cater for students of all abilities. The use of hundreds of colourful diagrams and photographs will also assist in the learning process. Also, the inclusion of sample answers to questions from past examination papers will help to show students the type of answer required to obtain full marks.

The material is presented in a logical teaching sequence, taking into account the demands of various topics. For example, in the chemistry section, topics which can be conceptually demanding for students have been placed towards the end of the course as these topics are best covered in third year. In addition, due to the low number of Mandatory Experiments in these chapters, time is available for the specified *Coursework B* investigations in third year. Since these investigations are worth 25% of the overall marks on the Junior Certificate exam paper, *Chapter 47* has been devoted to helping students carry out this part of the coursework.

Science is a very important and exciting subject! It is not confined to laboratories but is all around us and plays an important part in our lives. Throughout the book we have tried to emphasise the Science Technology Society (STS) component of the syllabus by showing the important part that science plays in our everyday lives.

All of the Mandatory Experiments on the syllabus are included in this textbook. Full step-by-step instructions for each experiment are given in the textbook. In keeping with the aims of the syllabus, the experiments in the *Student Laboratory Notebook* are presented using an investigative approach. We hope that the questions accompanying each experiment in the *Student Laboratory Notebook* will help the student to interpret the results of the various experiments and draw the correct conclusions. A full list of Mandatory Experiments in the order in which they appear in the textbook is given on the following page.

Throughout the textbook, 'Test Yourself' questions give signposts to questions in the textbook and *Workbook* (questions in the *Workbook* are prefixed by the letter *W*). These questions help students to review their work and test their knowledge and understanding of the topic being studied. It is strongly recommended that students use the *Workbook* in conjunction with the textbook, as the questions in the textbook and *Workbook* play an important role in testing the student and in preparing him/her for the Junior Certificate examination. We do not recommend that students write the answers to the completion-type questions into the textbook itself. Rather, we suggest that students write all questions and answers into their exercise copybooks.

To help the students in their study, key definitions are placed in boxes and words are highlighted in bold throughout the textbook. A summary of key points is given at the end of each chapter. In addition, a glossary of key words is given in the accompanying *Workbook*. The *Workbook* also contains a crossword on each chapter. We hope that these will help with the literacy and spelling problems of some students.

The permission of the Department of Education to reproduce questions from past Junior Certificate examination papers is acknowledged.

We hope that this book will help students to become really interested in science and that they will enjoy using this book.

Hilary Dorgan Declan Kennedy Siobhan Scott

LIST OF MANDATORY EXPERIMENTS

Chapter 1 What is Science?

Scientists ask themselves questions like: why do objects fall to the ground? What is in the air that we breathe? What keeps our heart beating? Why does dry hair sometimes crackle when you comb it? What is in vinegar that gives it a sharp taste? In trying to answer questions such as these, many great discoveries have been made about the world around us. That is the reason why science is said to be the study of God's handiwork.

Science is divided into **physical science** and **biological science**. Physical science is that branch of science that deals with inanimate (non-living) things. It is subdivided into subjects like physics, chemistry, astronomy, geology and computer science. Biological science is that branch of science which deals with living things. It is subdivided into botany (the study of plants) and zoology (the study of animals).

Physics is mainly concerned with the **physical properties** of substances. Physical properties are those properties that may be observed without changing the substance into another substance. Examples of the physical properties of an object would be its weight and its volume.

Chemistry is the branch of science that describes the substances that make up the world. Chemists spend a lot of time studying substances around us and making new substances. Without chemistry we would have no colourful clothes to wear, no soap to keep us clean – or no deodorant or perfume or aftershave to help us smell nice!

Biology is the study of life. This includes both plant life and animal life. We study biology in order to learn more about ourselves and the world in which we live. Biologists are concerned with topics such as health, agriculture, food processing, genetics, etc.

When performing experiments, we use various pieces of laboratory equipment. *Fig. 1.1*. Before performing any experiment, the laboratory Safety Rules (inside front cover) should be studied. You must also learn the meaning of the symbols for the labelling of hazardous material.

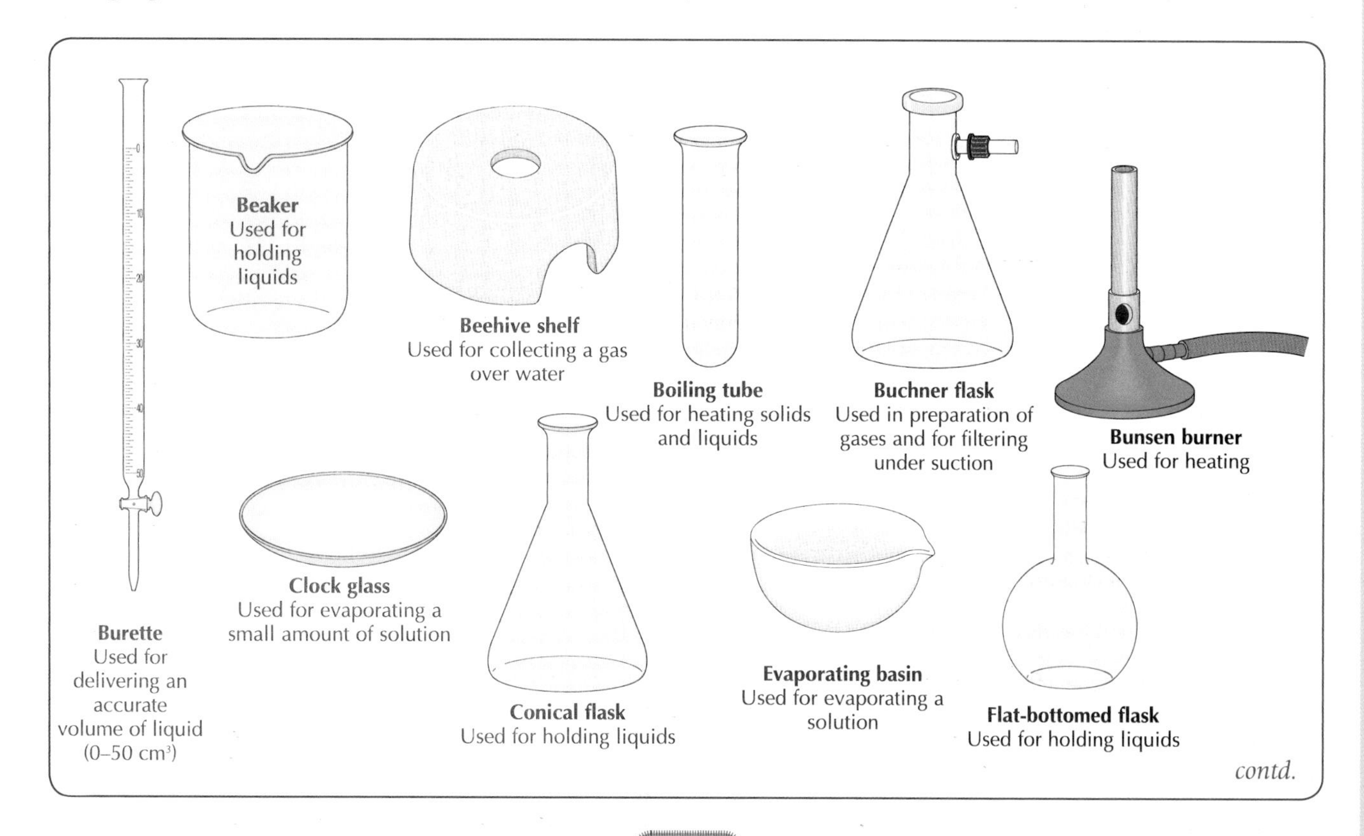

contd.

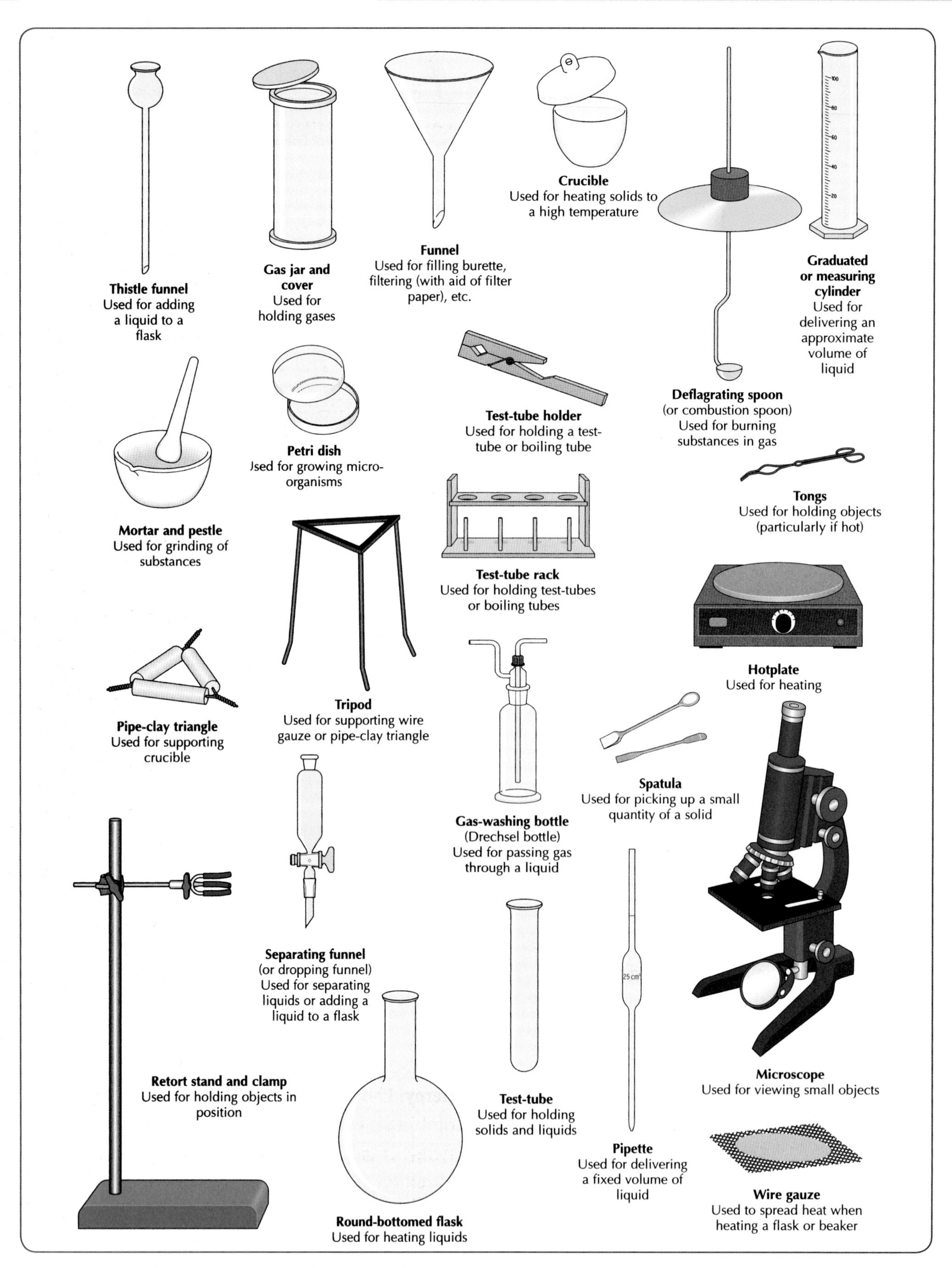

Fig. 1.1 Some laboratory equipment and its uses. Laboratory equipment is commonly called ***apparatus****.*

TEST YOURSELF:

Now attempt question *W1.1*.

Chapter 2 Living Things

2.1 Introduction – What is Biology?

There are millions of different types of living thing in the world. **Biology is the study of living things** like plants, animals, fungi and bacteria. In biology you will learn the following:

- What living things are **made of**.
- How they **function** (work).
- How they are able to **survive** in their environment.

We call living things **organisms**.

An organism is another name for a living thing.

The most important feature that all organisms share is that they are made of tiny units called **cells**. Cells are sometimes called the 'building blocks' of life since they form the basic structure of all living things. You will learn more about plant and animal cells in *Chapter 3*.

Living things are made of cells.

2.2 The Characteristics of Living Things

As well as being made of cells, living things have special features that make them 'alive'. Anything that is living has these life processes and **characteristics** in common. These special characteristics distinguish living things from non-living things like a stone or a box. There are in fact seven characteristics common to all living things. These are:

(1) nutrition
(2) respiration
(3) excretion
(4) growth
(5) reproduction
(6) movement
(7) response

(1) Nutrition

Nutrition is the way in which **living things get and use their food**. Plants can make their own food. Animals cannot make their own food. They eat plants or other animals, or the remains of dead organisms, e.g. caterpillars eat leaves, spiders eat flies and earthworms eat dead leaves.

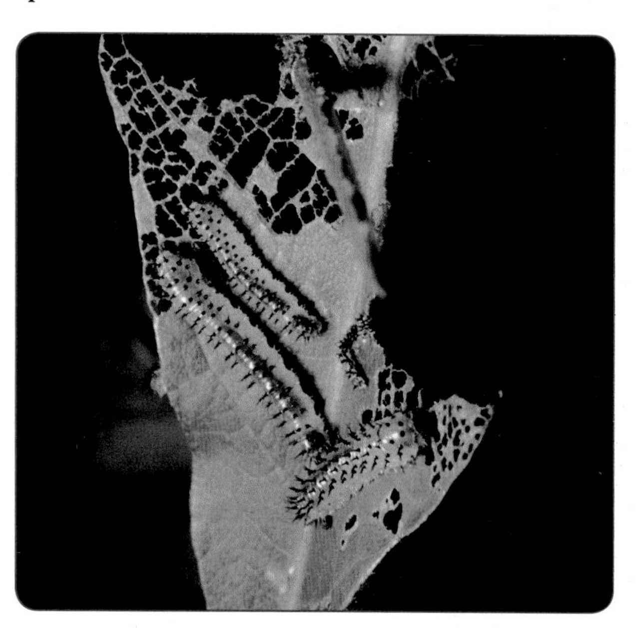

Fig. 2.1 Caterpillars eating leaves – an example of nutrition.

Nutrition is the way in which living things get and use their food.

(2) Respiration

Respiration is the way **living things get energy**. Living things take in oxygen gas and combine it with food in their cells to make energy. Organisms need energy for growth and movement and all their daily activities.

food + oxygen ⟶ energy + carbon dioxide + water

Respiration is the way living things get energy.

(3) Excretion

All living things produce wastes within their cells. Excretion is the getting rid of the **wastes that are made in the cells**. Animals excrete carbon dioxide, water and salts. Plants also produce wastes which they store in their leaves. In the autumn when the leaves fall, the wastes are removed.

Excretion is the getting rid of the wastes that are made in cells.

(4) Growth

All living things grow. Think of how you have grown since you were a baby!, *Fig. 2.2(a)*. Seeds such as acorns grow into mighty oak trees, *Fig. 2.2(b)*. Living things grow because their **cells are able to divide** and make copies of themselves. Growth is a gradual process in which there is an increase in the size of the organism.

Fig. 2.2(a) Growth results in an increase in the size of the organism.

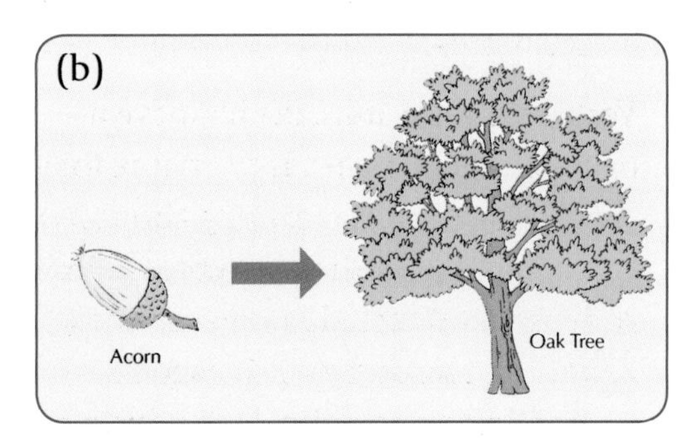

Fig. 2.2(b) Acorns grow into oak trees.

Growth results from cells dividing.

(5) Reproduction

Reproduction is being able to **produce new individuals** of the same kind. If organisms did not reproduce then that type of organism would become extinct, like the dinosaurs.

Fig. 2.3 A mother cat and her kittens.

Reproduction is being able to produce new individuals of the same kind.

(6) Movement

All **living things can move**. Movement in animals is usually easy to see. When animals move, their whole body moves from place to place, e.g. they swim, crawl, run or fly, *Fig. 2.4(a)*. Plants on the other hand stay rooted in one place and move only parts of their bodies, e.g. their petals open and close, *Fig. 2.4(b)*.

Fig. 2.4(a) Hares running.

Fig. 2.4(b) Flowers open and close their petals.

(7) Response

All living things need to be aware **of changes in their surroundings** and be able to react or respond to these changes. For example when you see the bus coming, you put out your hand to get the driver to stop. Animals use their senses to detect changes.

Plants respond more slowly than animals. Plants will grow towards light and water and their petals open in the daytime and close at night.

Response is the ability to react to changes in the environment.

If we say something is alive, it must show all seven characteristics.

TEST YOURSELF: Now attempt questions *2.1–2.3* and *W2.1–W2.2.*

2.3 The Variety of Living Things

There are more than three million different types of organism in the world. But in spite of this huge variety we can sort living things into groups because many share features in common. This is just like the way the CDs in a music store are arranged by the singer or the group or the type, like 'rock', 'pop', 'jazz', 'classical' etc. Being able to sort things into groups on the basis of similarities is called **classification**.

Classifying Plants and Animals

Two major groups of organisms are (1) **plants** and (2) **animals**. The most accurate way to classify an organism as a plant or an animal is to look at the structure of their cells. This we will do in the next chapter, but for the moment it is enough for us to look at the more obvious differences.

(1) Plants

In general, plants are organisms that have leaves, a stem and roots. In addition some plants also have flowers and fruits, seeds and cones. We can use some of these features to distinguish between different types of plants. If you look at *Fig. 2.5*, you can see how we can tell the difference between a horse chestnut and an oak tree by the look of their leaves and the type of fruit they produce.

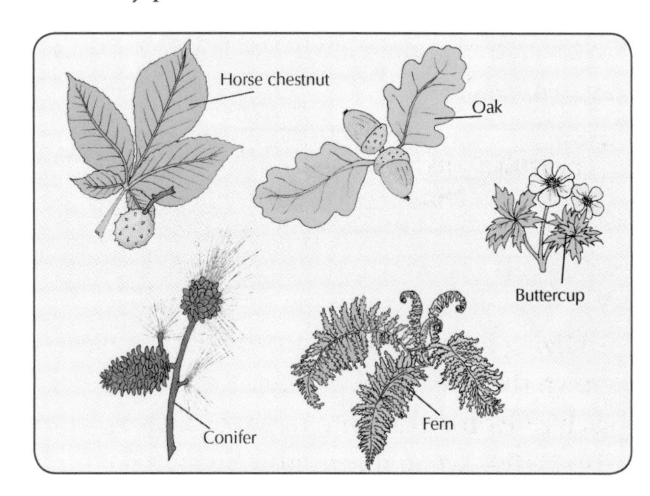

Fig. 2.5 Examples of different plants.

Uses of Plants

Plants are very important on the planet.

- They produce **food** for themselves and for other living things.
- They provide **shelter** for animals.
- They provide **materials** useful to humans such as linen, cotton, medicines and dyes.
- Plants also take in carbon dioxide from the air which helps **reduce global warming**.

(2) Animals

Animals can have eyes, ears, hair, fur, legs, wings, a shell, fins, gills and so on. But not all animals have all of these features. In addition animals can be divided into two big groups: **vertebrates** and **invertebrates**.

- **Vertebrates** are animals that have a **backbone**. A vertebra is the name for a bone in the backbone. Only about 3 per cent of all animals are vertebrates. Examples of vertebrates include humans, frogs, fish, lizards and birds, *Fig. 2.6*, page 6.
- **Invertebrates** are animals that have **no backbone**. Examples of invertebrates include earthworms, insects, spiders, snails, jellyfish and crabs, *Fig. 2.7*, page 6.

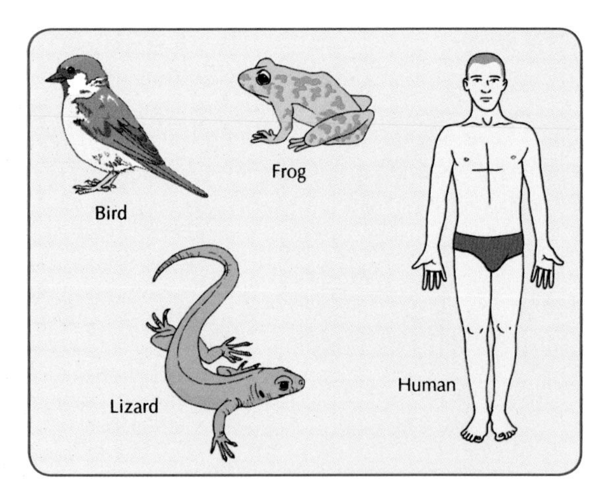

Fig. 2.6 Examples of vertebrates.

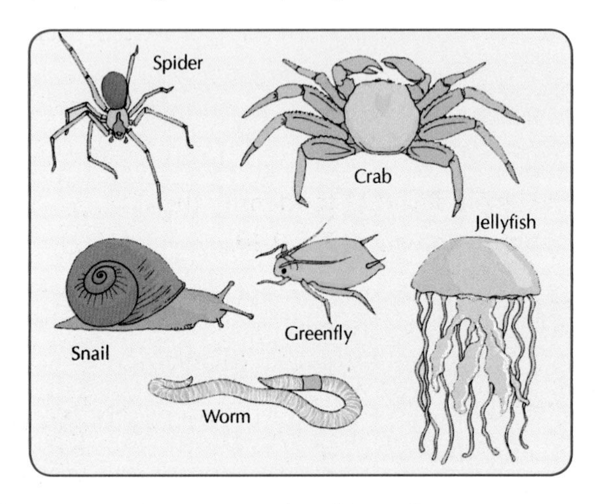

Fig. 2.7 Examples of invertebrates.

Vertebrates are animals that have a backbone.

Invertebrates are animals that do not have a backbone.

Uses of Animals

Like plants, animals are also important.

- They provide **food** for humans and other animals.
- They provide **materials** we need like wool and leather.
- We keep animals as pets which can help to **reduce stress**.
- Many **sports** depend on them, e.g. horse racing.

2.4 How do we Identify Plants and Animals?

Often we don't know the names of the plants and animals we find. One way we can identify them is by comparing what they look like with pictures in a book. Another way is to use what is known as a **key**. As the name suggests, a key opens doors, in this case it opens the door to information about the identity of a plant or an animal, *Fig. 2.8*.

A key asks a series of questions about the features of the organism you want to identify. For plants these could be the shape of the leaf, the colour of the flower, the number of petals etc. By answering the questions you are led to the name of the organism. The best way to understand how a key works is to practice using one yourself.

Fig. 2.8 Keys and books can be used to help identify an organism.

A Simple Plant Key

The following key can be used to identify the leaves of common trees, shown in *Fig. 2.9*. To use the key, start at question 1. Your answer to question 1 will send you to another question. Your answer to this question will either tell you the name of the leaf or send you to another question and so on until all the leaves are identified.

1.	Are the leaves single (not divided into leaflets)?	Go to 2
	Are the leaves divided into leaflets?	Go to 4
2.	Are the leaves divided into many lobes (rounded parts)?	Oak
	Are the leaves not divided into many lobes?	Go to 3
3.	Is the edge of the leaf smooth?	Beech
	Is the edge of the leaf spiked?	Holly
4.	Are the leaves arranged like the fingers of a hand?	Horse chestnut
	Are the leaves arranged in pairs?	Ash

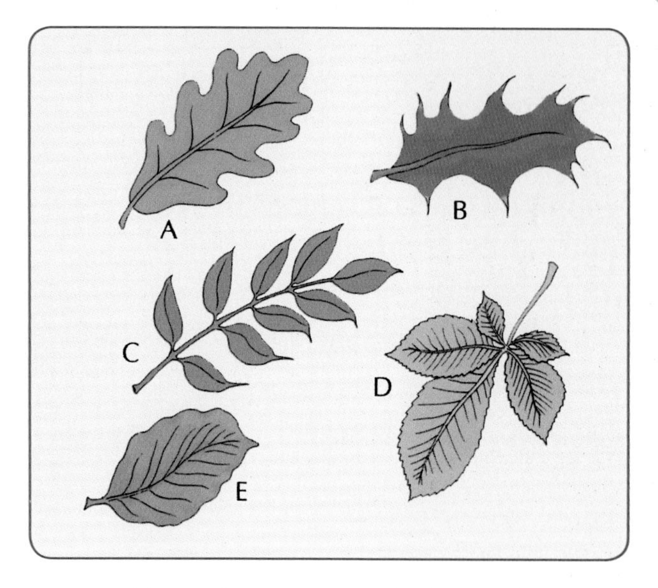

Fig. 2.9 Leaves of trees to be identified.

A Simple Animal Key

Use the key in *Fig. 2.11* to identify the invertebrates shown in *Fig. 2.10*.

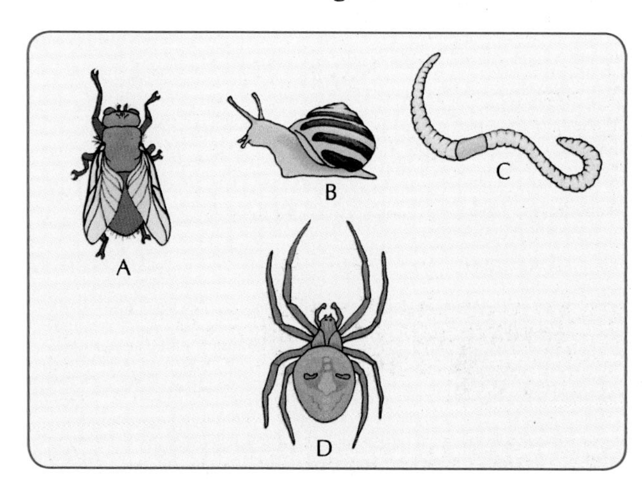

Fig. 2.10 Invertebrates to be identified.

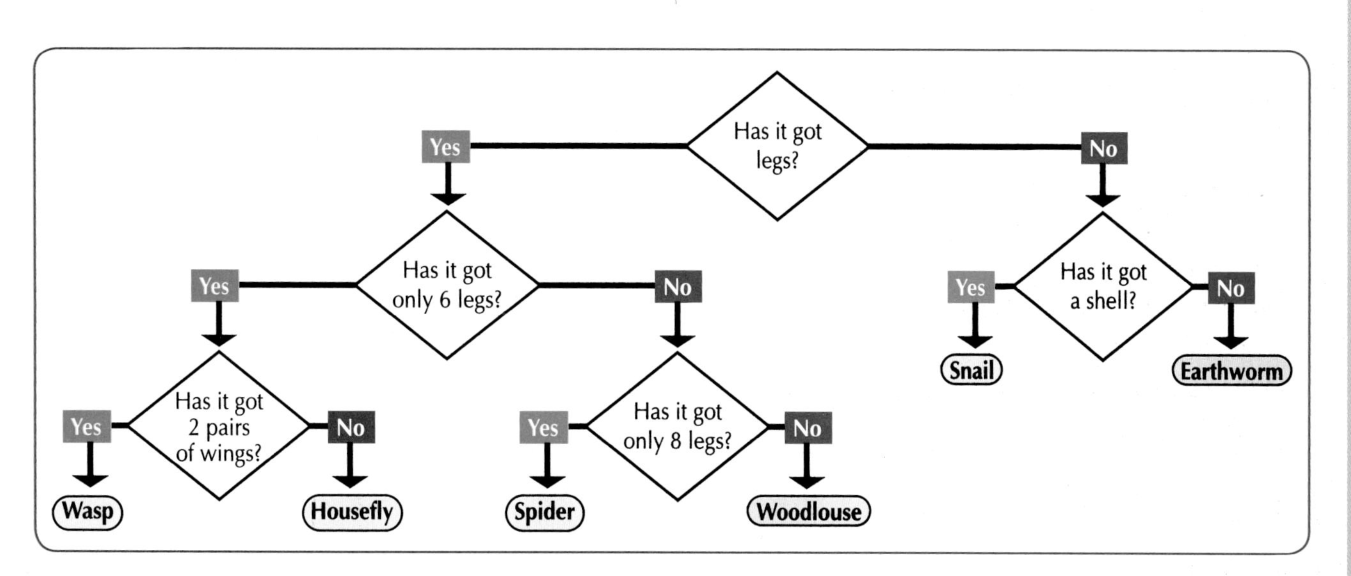

Fig. 2.11 Key to identify invertebrates.

Mandatory experiment 2.1

To investigate the variety of living things by direct observation of plants and animals in their environment. To classify living things as plants or animals, and animals as vertebrates or invertebrates

Materials required: plant and animal identification books; plant and animal keys; note book and pencil

Method

1. Pay careful attention to the instructions of your teacher.
2. Select an area, such as the school grounds or a nearby field, or hedge.
3. Look for as many examples of plants and animals as you can find.
4. Using the identification books and keys, identify and name three plants and three animals within the area.
5. Note which animals are vertebrates and which are invertebrates.
6. Record the information in your note book.

TEST YOURSELF:
Now attempt questions *2.4–2.9* and *W2.3.*

What I should know

- Biology is the study of living things.
- The term organism refers to any living thing.
- All organisms are made of cells.
- There are seven characteristics of living things, these are: nutrition; respiration; excretion; growth; reproduction; movement and response.
- Two main groups of living things are plants and animals.
- Plants usually have leaves, roots and a stem, e.g. roses, grass, ferns.
- Animals may be invertebrates or vertebrates.
- Invertebrates do not have a backbone, e.g. earthworms, spiders.
- Vertebrates have a backbone, e.g. cats, hens and fish.
- Living things can be identified using a key.

QUESTIONS

Write the answers to the following questions into your copybook

2.1 (a) What is biology?

(b) Give another name for a living thing.

(c) What are all living things made up of?

(d) Name four living things.

(e) Name four non-living things.

2.2 (a) List the characteristics of living things.

(b) Explain each characteristic you have listed in (a).

(c) If your hand touched a hot cooker and you pulled your hand away, which characteristic of living things would you be showing?

QUESTIONS

(d) Plants do not need to get rid of wastes. Is this statement true or false?

Give a reason for your answer.

(e) Why do living things need to be able to reproduce?

2.3 Look at the diagram in *Fig. 2.12*, which shows a number of things, some are living and others are non-living.

(a) Name two things in the diagram that are living, other than the girl.

(b) Name two things in the diagram that are non-living, other than the MP3 player.

(c) Give two differences, in each case, to explain how you know the girl is living and the MP3 player is not.

Fig. 2.12

2.4 (a) Living organisms can be sorted into groups. This is known as ….. .

(b) Two main groups of living things are ….. and ….. .

(c) Three features that would allow you to classify a buttercup as a plant are the ….., the ….. and the ….. .

(d) Animals differ from ….. in the way that they ….. .

2.5 (a) Animals can be sorted into ….. or ….., depending upon whether or not they have a ….. .

(b) An example of a vertebrate animal is a ….. .

(c) An example of an invertebrate animal is a ….. .

2.6 (a) What are the general features of a plant?

(b) Name three plants that you know.

(c) Give two ways in which plants are important to humans.

2.7 (a) What are the general features of animals?

(b) Name three vertebrate animals.

(c) Name three invertebrate animals.

QUESTIONS

(d) Explain the difference between a vertebrate and an invertebrate animal.

(e) Give two ways in which animals are important to humans.

2.8 Use the key below to identify the four seashore animals shown in *Fig. 2.13*.

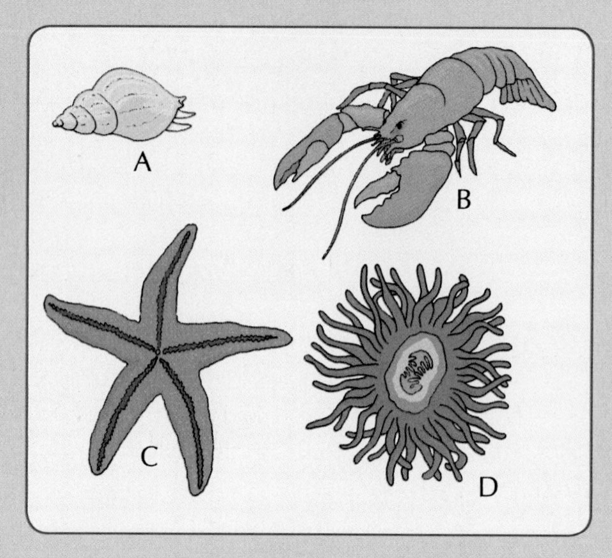

Fig. 2.13

1.	Body divided into segments (partitions)	Go to 2
	Body not divided into segments	Go to 3
2.	Has large claws	Lobster
	Does not have large claws	Shrimp
3.	Has a shell	Go to 4
	Has no shell	Go to 5
4.	Shell has one part	Whelk
	Shell has two parts	Mussel
5.	Body has five arms	Starfish
	Body has more than five arms	Sea anemone

2.9 In each of the following lists, choose the *odd one out*. Give a reason for your choice.

(a) plants, bacteria, bicycle, fungi, animals, humans

(b) spider, greenfly, earthworm, mouse, snail, slug

(c) oak, sycamore, beech, ash, bench, pine

(d) primrose, grass, buttercup, robin, daisy, dandelion

(e) nutrition, movement, reproduction, respiration, talking, excretion

Chapter 3 The Cell

3.1 Introduction

All living things are made of tiny units called **cells**. Some organisms, such as bacteria, are made of only one cell. But most organisms are made up of many hundreds of cells, for example humans have 100 million, million cells.

3.2 Cell Structure

Cells are made up of three main parts:

- **Cell Membrane**. A skin that surrounds the cell. The cell membrane **protects** the cell and it **controls** the things that enter and leave the cell.
- **Cytoplasm**. A watery substance containing many dissolved substances such as proteins and sugars. Other structures can be found suspended in the cytoplasm. The cytoplasm is where the chemical reactions necessary for life take place.
- **Nucleus**. The nucleus acts as the **control centre** of the cell. It contains a special chemical called **DNA**. DNA forms a set of instructions for how the cell will work. The nucleus controls the type of chemicals the cell will make and it is also involved in how the cell divides to make more cells.

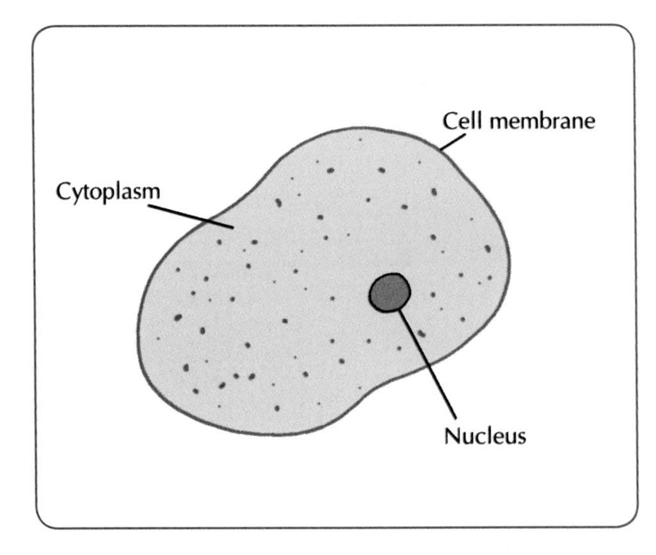

Fig. 3.1 An animal cell.

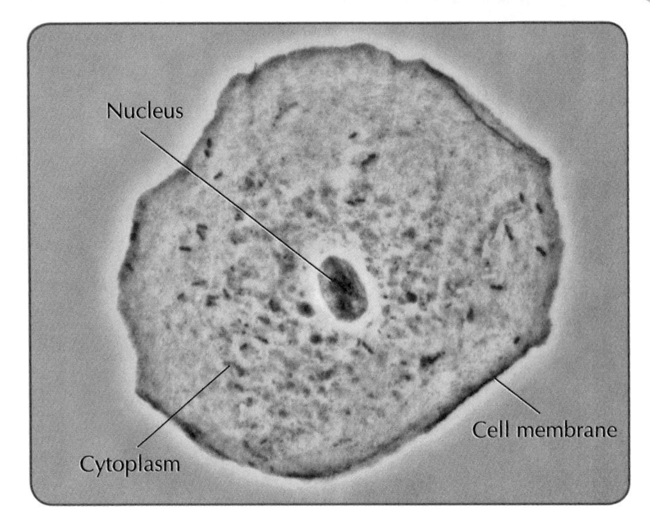

Fig. 3.2 A human cheek cell.

3.3 Differences between Plant and Animal Cells

Table 3.1 summarises the main differences between how the cells of animals differ from those of the cells of plants.

PLANT CELLS	ANIMAL CELLS
1. Plant cells have a **cell wall** outside the cell membrane. The cell wall gives shape and support.	No cell wall
2. Many plant cells have **chloroplasts**. Chloroplasts hold chlorophyll, the green colour that plants use to make food.	No chloroplasts
3. Plant cells have large **vacuoles**. Vacuoles are spaces in the cytoplasm which store things and give support.	Small vacuoles

Table 3.1 Differences between plant and animal cells.

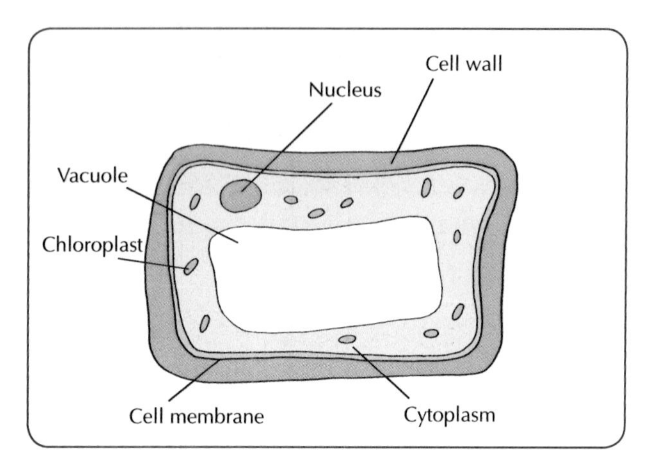

Fig. 3.3 A plant cell.

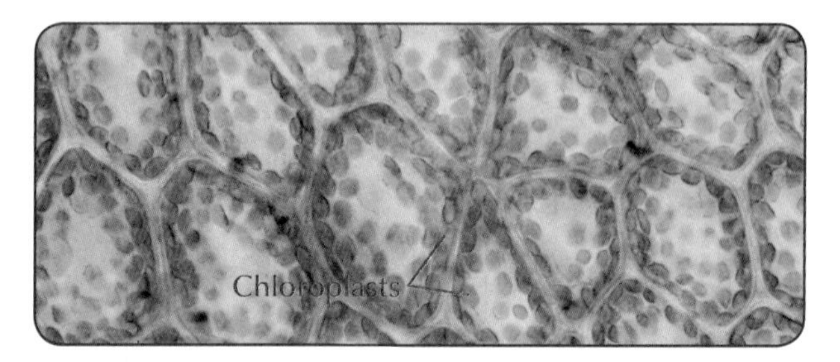

Fig. 3.4 Plants cells.

TEST YOURSELF:
Now attempt questions *3.1–3.5* and *W3.1–W3.3*

3.4 The Microscope

Generally speaking most cells are too small to be seen with our eyes alone. They have to be **magnified** using an instrument called a **microscope**.

A microscope consists of a number of lenses that magnify the cells we are looking at. The earliest microscopes were very simple, often consisting of only one lens.

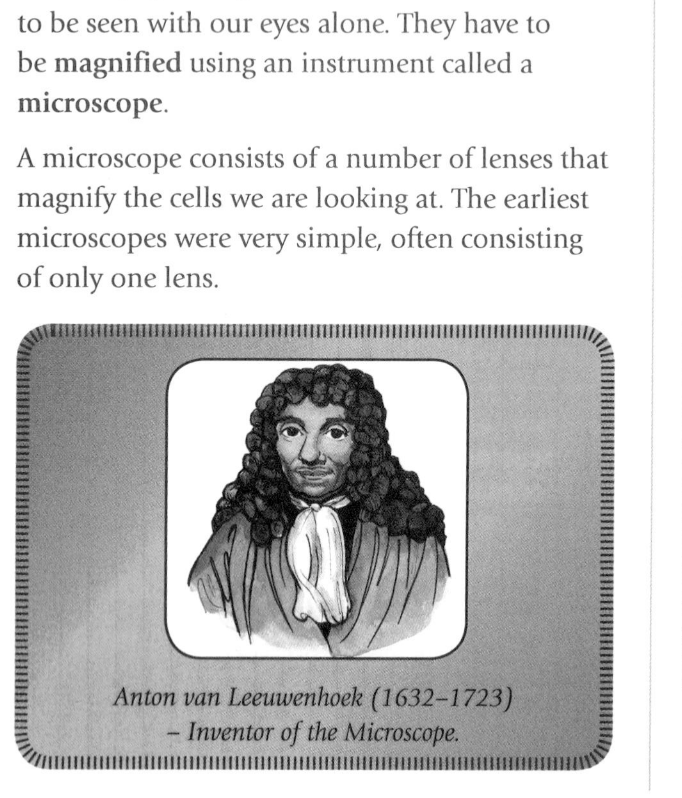

Anton van Leeuwenhoek (1632–1723) – Inventor of the Microscope.

The microscope you will use in school is known as a **light microscope** and it usually has four lenses. Each of the lenses has a different magnification.

- The **eyepiece lens** usually magnifies by ten (x10).
- There are three **objective lenses**: low power (x4), medium power (x10) and high power (x40).

The parts of a light microscope are shown in *Fig. 3.5*, and the functions of the parts are given in *Table 3.2*.

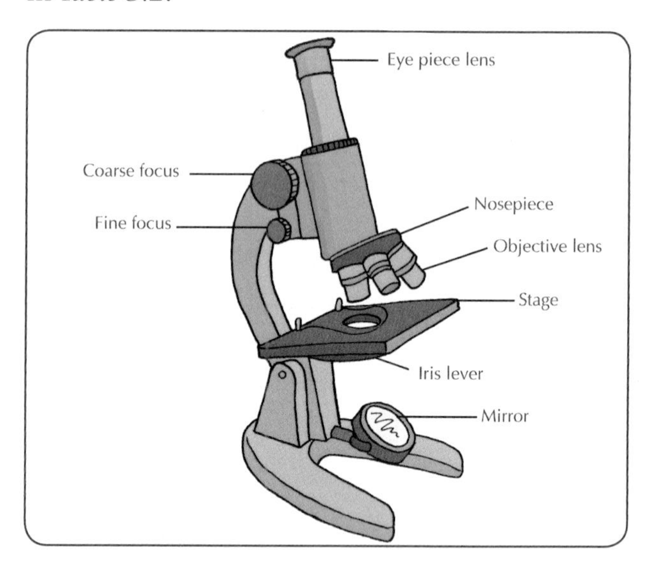

Fig. 3.5 A Light Microscope.

PART	FUNCTION
Eyepiece (lens)	Magnifies the object (usually x10).
Nosepiece	Holds the objective lenses.
Objective lenses	Magnifies the object by different amounts (x4, x10, x40).
Coarse focus wheel	Brings the object into 'rough' focus.
Fine focus wheel	Brings the object into 'sharp' focus.
Stage	Holds the slide.
Mirror/light source	The source of light.

Table 3.2 The functions of the parts of a microscope.

How to Use a Microscope

1. Turn on or plug in the light source.
2. Place a glass slide under the clips on the stage of the microscope.
3. Position the **low power magnification lens** over the slide.
4. Watch from the side and turn the coarse focus wheel so that the objective lens is as close to the stage as possible.
5. Now look down the microscope.
6. Adjust the amount of light by turning the **iris lever**.
7. Gently turn the coarse focus wheel the **opposite** way.
8. The specimen on the slide should become visible, but it may still be a bit fuzzy.
9. Turn the fine focus wheel until the image becomes clear.
10. The specimen is now said to be '**in focus**'.
11. Sometimes a dye or stain is added to the specimen. This makes it easier to see.

The main points to remember when using a microscope are:

- **Turn on the light source.**
- **Place the slide on the stage.**
- **Always use the low power objective lens first.**
- **Use the coarse focus wheel first and then the fine focus wheel.**

Mandatory experiment 3.1

To prepare a slide from plant tissue and sketch the cells under magnification

Apparatus required: onion; cutting board; sharp knife; forceps; glass slide; cover slip; microscope; mounted needle; filter paper or tissue; droppers; water and iodine stain

Method

1. Use the forceps to take a small piece of the inner skin of an onion, *Fig. 3.6(2)*.

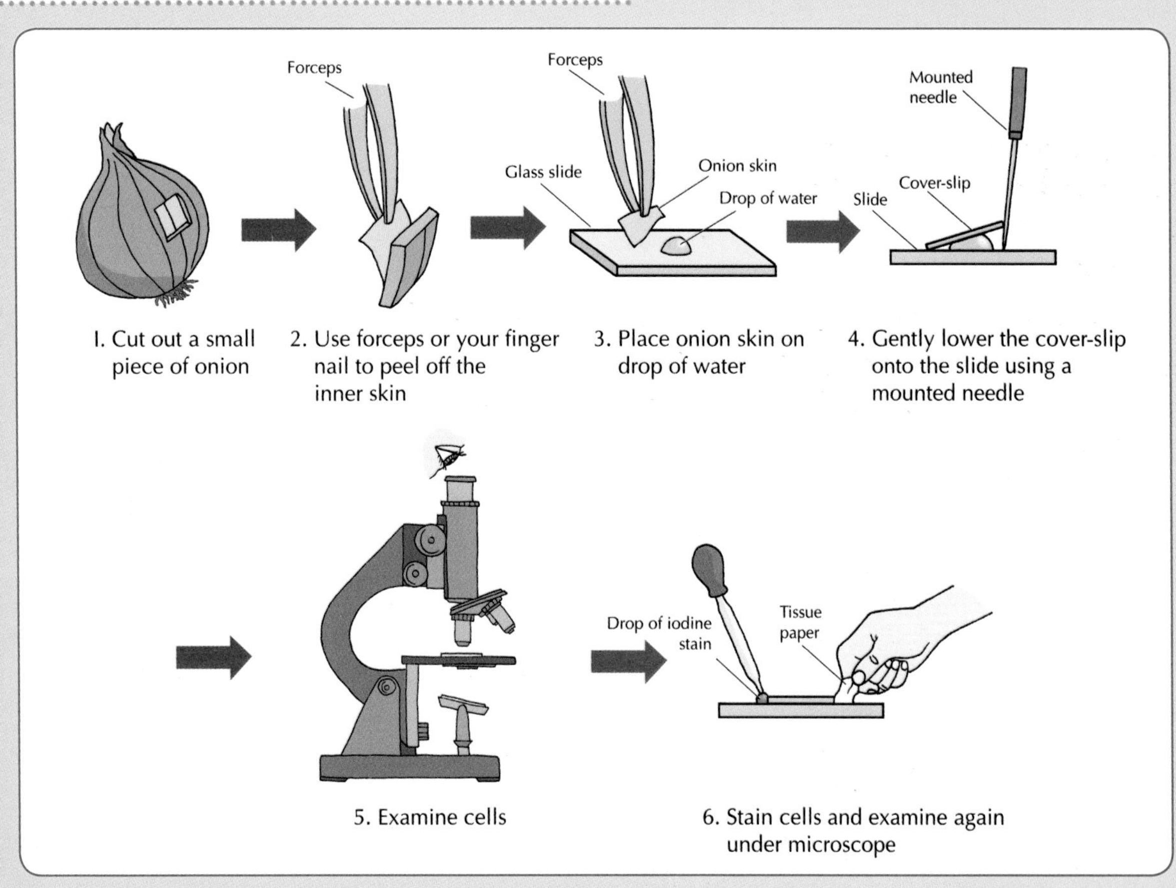

Fig. 3.6 Preparing and staining plant cells.

2. Place a drop of water onto the centre of a clean glass slide. Place your piece of onion skin into the drop of water on the slide, *Fig. 3.6(3)*.
3. Carefully lower the cover slip over the specimen, as shown in *Fig. 3.6(4)*. The cover slip protects the microscope lens from damage.
4. Put the slide onto the microscope stage, under the clips, *Fig. 3.6(5)*.
5. View the slide as described above.
6. Draw and label a diagram of a few of the cells under low power and one cell under high power.

To view the same cells with a stain

1. Remove the slide from the microscope stage.
2. Place one drop of **iodine stain** to the right of the cover slip.
3. Use a piece of tissue paper to draw the stain under the cover slip as shown in *Fig. 3.6(6)*.
4. The iodine will stain the cytoplasm a yellow/brown colour and the nucleus an orange/brown colour.
5. Use the microscope to examine the cells under the low and high power magnifications.
6. Draw a labelled diagram of one cell at each magnification.

TEST YOURSELF:
Now attempt questions *3.6–3.9* and *W3.4–W3.5*.

3.5 Cells, Tissues, Organs and Systems

Tissues

The cells of plants and animals are not all identical. Cells have different shapes and sizes because they have different jobs to do, *Fig. 3.7*.

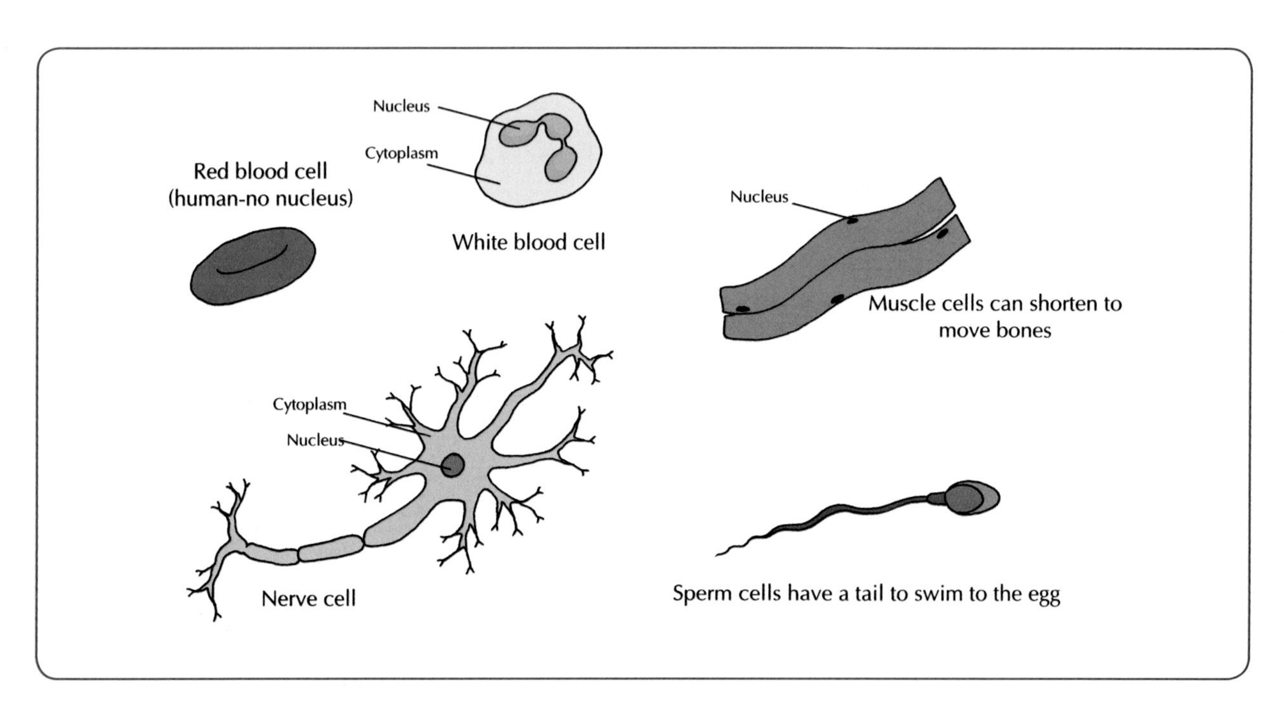

Fig. 3.7 Different animal cell types.

Groups of similar cells with the same function (job) are known as **tissues**.

- Human tissues include **muscle tissue**, **nervous tissue**, **bone tissue** and **skin tissue.**
- Plant tissues include **transport tissue**, **storage tissue** and **growth tissue**.

A tissue is a group of similar cells with the same function, e.g. muscle.

Organs

An **organ** consists of several tissues working together.

- The **heart**, for example, is made of muscle and it has a good blood and nerve supply.
- The **brain** and the **liver** are other organs in the human body.
- In plants, the **leaf**, **flower**, **stem** and **root** are all examples of organs.

An organ is made of many different tissues working together, e.g. the heart.

Systems

When a group of organs work together, they form a **system** of the body.

- The human **digestive system** is made up of the mouth, food pipe, stomach and intestines.
- Other examples include the **circulatory system** and the **skeletal system**.
- In plants, the organs above the ground, the stem, leaves and flowers form the **shoot system**.

The systems work together to form the **whole organism**. *Fig. 3.8* shows how cells, tissues and organs form a system of the body.

A system is a group of organs working together, e.g. the circulatory system.

3.6 Growth and Cell Division

We are made of thousands of cells, but we started life as only one cell – a fertilised egg. This cell then started to divide, first into two cells, then into four, then eight and so on. At a certain stage some of the cells became muscle, blood, nerves etc. until a complete human being was made.

We grow because **cells can divide** to make new cells. Each new cell then increases in size until it is ready to divide again. In this way growth takes place.

Growth results from cells dividing.

TEST YOURSELF:
Now attempt questions *3.10–3.12* and *W3.6–W3.7*.

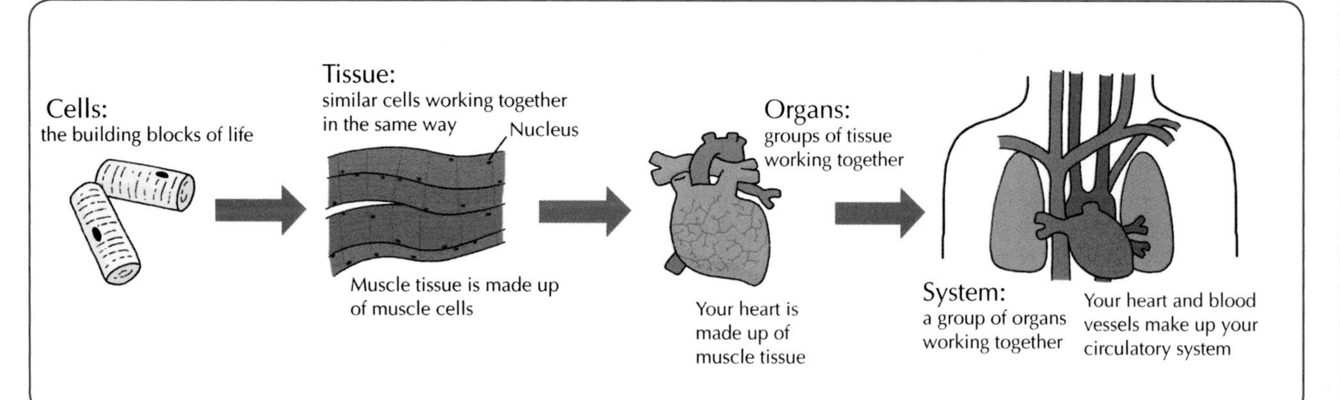

Fig. 3.8 The organisation of muscle cells into tissue, organs and system.

What I should know

- All living things are made of cells.
- Some organisms are made of just one cell. Others are made of thousands of cells.
- Animal cells consist of the following parts: cell membrane, cytoplasm and nucleus.
- In addition, plant cells have a cell wall.
- Plant and animal cells share many features but there are differences, as shown in *Table 3.3* below.

	PLANT CELLS	ANIMAL CELLS
FEATURES THEY HAVE IN COMMON	A nucleus. A cell membrane. Cytoplasm.	A nucleus. A cell membrane. Cytoplasm.
DIFFERENCES	Have a cell wall. Have chloroplasts. Have a large vacuole.	No cell wall. No chloroplasts. No large vacuole.

Table 3.3 Common and different features of plant and animal cells.

- Most cells can only be seen with a microscope.
- Microscopes have different lenses which magnify the image.
- The parts and functions of a light microscope are shown in *Table 3.4* below.

PART	FUNCTION
Eyepiece (lens)	Magnifies the object (usually x10).
Nosepiece	Holds the objective lenses.
Objective lenses	Magnifies the object by different amounts (x4, x10, x40).
Coarse focus wheel	Brings the object into 'rough' focus.
Fine focus wheel	Brings the object into 'sharp' focus.
Stage	Holds the slide.
Mirror/light source	The source of light.

Table 3.4 Parts and functions of a light microscope.

- Stains are used to make cells easier to see. Iodine stain is used for plant cells.
- Cells differ in size, shape and function.
- A tissue is a group of similar cells with the same function, e.g. muscle.
- An organ is made of many different tissues working together, e.g. the heart.
- A system is a group of organs working together, e.g. the circulatory system.
- Cells → Tissues → Organs → Systems → Whole Organism.
- Growth results from cell division.

QUESTIONS

Write the answers to the following questions into your copybook

3.1 Name the tiny units that all living things are made from.

3.2 (a) Which part of a cell controls the activities of the cell?

(b) Which part of a cell controls what gets in and out of the cell?

(c) What is the function of the cytoplasm of a cell?

(d) Where in a cell would you find DNA?

3.3 Which of the following do all cells have in common?

(a) cell wall, chloroplast, cytoplasm

(b) cell membrane, chloroplast, cytoplasm

(c) cell wall, nucleus, cytoplasm

(d) cell membrane, nucleus, cytoplasm

3.4 Draw a labelled diagram to show the structure of a plant cell.

3.5 Name two differences between plant cells and animal cells.

3.6 Name an instrument you would use to look at cells in the laboratory.

3.7 Name the parts *A*, *B*, *C*, *D* and E of the microscope shown in *Fig. 3.9*.

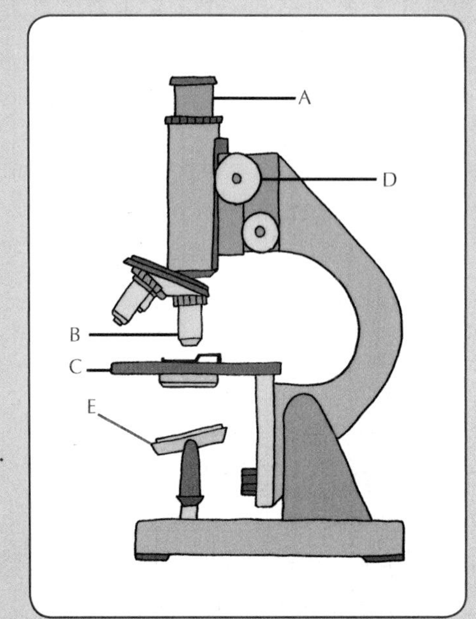

Fig. 3.9

3.8 Copy and complete the following table:

PART OF MICROSCOPE	FUNCTION
Stage	
Eyepiece lens	
	Brings the object into 'rough' focus.
Fine focus wheel	

3.9 (a) Describe how to prepare some plant cells and view them under the microscope.

(b) Draw a labelled diagram of one cell seen under the low power magnification.

3.10 (a) A tissue is a group of ..... cells that have the same ..... .

(b) An example of a plant tissue is ..... and an example of an animal tissue is ..... .

(c) An organ is made of different ..... working ..... .

(d) Two animal organs are the ..... and the ..... .

(e) A ..... is a group of organs ..... together.

3.11 (a) Explain the following terms: cell, tissue, organ, system.

(b) Name one animal and one plant tissue.

(c) Name one animal and one plant organ.

(d) Name one animal and one plant system.

3.12 What causes living things to grow in size?

Chapter 4 Food and Diet

4.1 Introduction

All living things need food.

- **Plants** can make their own food. They use light energy from the sun to combine carbon dioxide and water to form sugar. This process is called **photosynthesis**.
- **Animals**, on the other hand, cannot make their own food. Some eat plants, e.g. sheep eat grass; others eat the animals that have eaten plants, e.g. ladybirds eat greenfly.

Fig. 4.1 Greenfly eating a plant. Some animals eat other animals for food. Some animals eat plants.

Food is needed for the following:

- As a **source of energy**.
- To provide **material for growth**.

4.2 Food Nutrients

The five major **food nutrients** are **carbohydrates**, **proteins**, **fats**, **vitamins** and **minerals**. We need to eat the right amounts of these nutrients regularly in order to stay healthy. Each nutrient has a different function in the body, see *Table 4.1*. A **balanced diet** consists of the correct amount of each of these nutrients plus water.

Water is essential for living things.

- It is an excellent **solvent**.
- It also allows substances to be **transported**.

Fig. 4.2 Good sources of sugars.

Fig. 4.3 Good sources of starch.

Fig. 4.4 Good sources of protein.

Fig. 4.5 Good sources of fats.

NUTRIENT	FUNCTION	SOURCE (FOOD IT IS FOUND IN)
Carbohydrates		
– Sugar	To provide energy.	Fruit, honey, milk, soft drinks.
– Starch	To provide energy.	Potatoes, pasta, bread.
– Cellulose (fibre)	Helps prevent constipation.	Green vegetables and cereals.
Proteins	For growth and repair of cells, to make muscle and hair.	Lean meat, egg white, soya, nuts, fish.
Fats	To provide a store of energy. For insulation.	Butter, milk, cheese, oils.
Vitamins*		
– Vitamin C	For healthy skin and gums. Lack of Vitamin C causes Scurvy.	Citrus fruits (oranges, kiwis, blackcurrants).
– Vitamin D	To build strong bones and teeth. Lack of Vitamin D causes rickets.	Cheese, milk, yogurt.
Minerals*		
– Iron (Fe)	Needed to make red blood cells. Lack of iron causes anaemia.	Red meat, liver, cabbage.
– Calcium	For healthy bones and teeth.	Milk, cheese, tinned salmon.

Table 4.1

*The body needs many different vitamins and minerals. *Table 4.1* only includes two of each.

Fig. 4.6 Good sources of Vitamin C.

TEST YOURSELF:
Now attempt questions *4.1–4.6* and *W4.1–W4.5*.

4.3 Food Tests

Mandatory experiment 4.1

(a) To test for the presence of starch

Apparatus required: 2 test-tubes; droppers

Chemicals required: iodine solution; starch solution

Method

1. Place 2 cm^3 starch solution in a test-tube.
2. Place the same volume of water in the second test-tube.
3. Add two to three drops of iodine solution to both test-tubes.

4. Gently shake the test-tubes and note any colour change.
5. Record your results.
6. Iodine is a blue colour when starch is present and a brown/yellow colour when starch is not present.

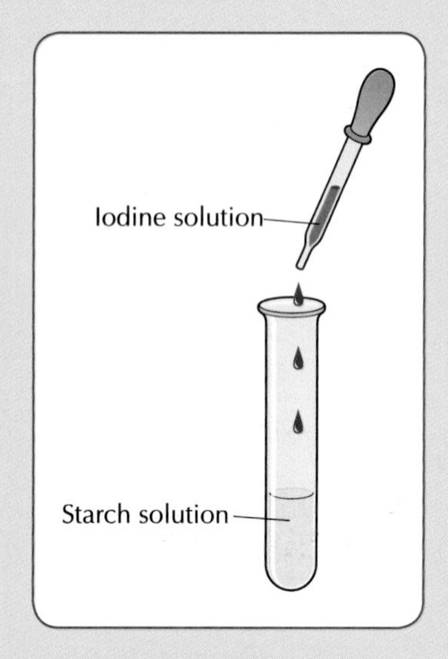

Fig. 4.7 Testing for starch.

(b) To test for the presence of a reducing sugar (glucose)

Note: Sugars like glucose are known as 'reducing sugars'. When heated, a reducing sugar turns Benedict's (or Fehling's) solution a brick-red colour. Not all sugars are reducing sugars, e.g. sucrose (table sugar) is not.

Apparatus required: 2 test- tubes; droppers; water bath (or 400 ml beaker; tripod; gauze; Bunsen burner); tongs

Chemicals required: glucose solution; Benedict's solution (or Fehling's solution)

Method

1. Set up the apparatus as shown in *Fig. 4.8*.
2. Stand the test-tubes into the boiling water bath for about three minutes.
3. Record any colour changes.

Results

The tube with the glucose solution turns a brick-red colour. The tube with the water stays blue.

Conclusion

The appearance of a brick-red colour indicates that a reducing sugar is present.

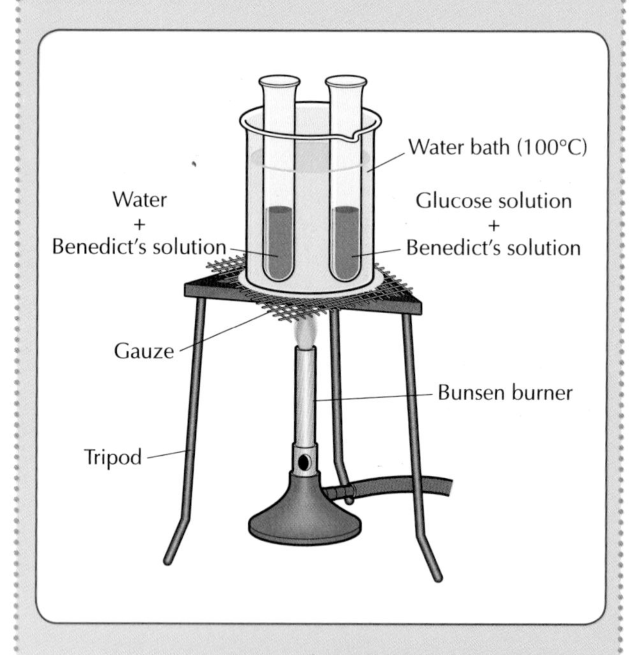

Fig. 4.8 Testing for reducing sugars.

(c) To test for the presence of protein

Apparatus required: 2 test- tubes; droppers

Chemicals required: sodium hydroxide solution; copper sulfate solution; protein solution, e.g. milk

Method

1. Set up two test-tubes as shown in *Fig. 4.9*.
2. Add three to four drops of copper sulfate solution to each test-tube. Swirl the test-tubes gently to mix the contents.

Result

The mixture with the milk turns a purple/violet colour. The second test-tube containing only water remains blue.

Conclusion

Protein is present in the mixture containing milk. Protein is not present in the second test-tube containing only water.

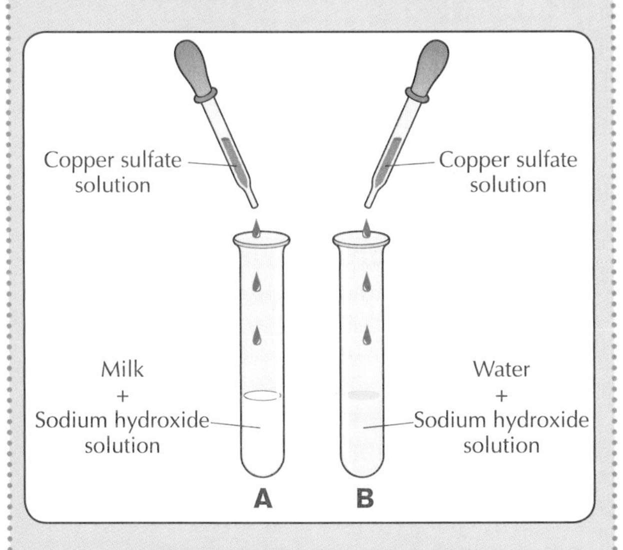

Fig. 4.9 Testing for protein.

(d) To test for the presence of fats

Materials required: brown paper; butter

Method

1. Label two pieces of brown paper *A* and *B*.
2. Rub some butter on *A*.
3. Put a drop of water on *B* and rub it in.
4. Allow both pieces of paper to dry.
5. Hold each piece of paper up to the light. Compare the amount of light that passes through.

Results

The place on the paper where the butter was rubbed will allow light to pass through. This is known as a translucent spot.

Conclusion

Fat is present on the paper rubbed with butter. There is no change in the paper rubbed with water. No light passes through it; therefore, fat is not present.

Fig. 4.10 The brown paper test for fats.

TEST YOURSELF: Now attempt questions *4.7–4.8* and *W4.6.*

4.4 Food and Energy

Different food nutrients provide different amounts of energy. For example, fats produce twice as much energy as carbohydrate or protein. The amount of energy that can be got from a food is known as its **energy value**. This is measured in **kilo joules** (kJ) per gram.

- The energy value for **fats** is 38kJ/g, i.e. 38 kilo joules per gram.
- The energy value for both **carbohydrates** and **protein** is 17 kJ/g.
- The energy value for **vitamins** is 0 kJ/g.

Energy Value

Most foods have their **energy value** on the label. The energy values on the food product labels in *Fig. 4.11*, page 22, are given in **kJ** per 100 grams of the food. Each ingredient is listed in order of decreasing mass. This means that the higher up the list an ingredient is, the more of it there is in the food. In *Fig. 4.11(b)* sugar is listed first. This means that sugar is present in the greatest amount.

The recommended daily energy requirement for teenagers is as follows:

Boys Age 12–15: 11,700 kJ.

Girls Age 12–15: 9,600 kJ.

(a)

NUTRITION INFORMATION

TYPICAL VALUES	PER 100g	PER 208g SERVING (approx 1/3 can)
Energy	392kJ 93kcal	815kJ 193kcal
Protein	3.2g	6.7g
Carbohydrate	15.7g	32.7g
(of which sugars)	8.2g	17.1g
Fat	1.9g	4.0g
(of which saturates)	1.1g	2.3g
Fibre	Trace g	0.1g
Sodium	0.1g	0.2g
Calcium	100mg (12%RDA)	208mg (25%RDA)

ALLERGY ADVICE: Contains Milk

(b)

Ingredients

Sugar, Wheat Flour, Whole Egg, Humectant (Glycerol), **Whey Powder, Powdered Egg White, Salt, Raising Agents** (Disodium Diphosphate, Sodium Bicarbonate), **Flavouring, Preservative** (Potassium Sorbate).

Nutrition

Typical Composition	Each trifle sponge (20g) provides	100g (3½oz) provide
Energy	**284kJ 67kcal**	**1418kJ 335kcal**
Protein	**1.7g**	**8.3g**
Carbohydrate	**13.5g**	**67.7g**
of which sugars	9.5g	47.4g
Fat	**0.7g**	**3.4g**
of which saturates	0.2g	1.0g
mono-unsaturates	0.3g	1.7g
polyunsaturates	0.1g	0.6g
Fibre	**0.2g**	**1.1g**
Sodium	**trace**	**0.2g**

This pack contains **8** trifle sponges.

(c)

INGREDIENTS

Wheat, Glucose-Fructose Syrup, Sugar, Honey (3%), Glucose Syrup, Molasses, Niacin, Iron, Riboflavin (B2), Thiamin (B1).

NUTRITION INFORMATION

Typical Values	per 100g	per 30g serving	% GDA*
Energy	1608kJ (379 kcal)	482kJ (114 kcal)	- (5.7%)
Protein	5.3g	1.6g	-
Carbohydrate	85.8g	25.7g	-
(of which sugars)	35.0g	10.6g	10.6%
Fat	1.6g	0.5g	0.7%
(of which saturates)	0.2g	0.1g	0.5%
Fibre	3.7g	1.1g	4.6%
Salt	trace	trace	trace
Sodium	trace	trace	-

*GDA = Guideline Daily Amount (See back of pack)

Vitamins & Mineral

	per 100g	per 30g serving
Thiamin (B1)	1.0mg/71% RDA*	0.3mg/21%RDA*
Riboflavin (B2)	1.0mg/63% RDA*	0.3mg/19%RDA*
Niacin	10.0mg/56% RDA*	3.0mg/17%RDA*
Iron	8.0mg/57% RDA*	2.4mg/17%RDA*

*RDA = Recommended Daily Allowance

Fig. 4.11 Nutritional information and ingredients on various food labels.

Use the information in this chapter and that shown in the labels in *Fig. 4.11* to answer the following questions:

(i) Which food (*a*, *b* or *c*) would provide the most calcium?

(ii) Which food would provide the most energy?

(iii) Which food would supply the most carbohydrate and the most energy?

(iv) Which food has the highest sugar content?

(v) Which food would best prevent constipation? Explain your answer.

Burning Food to Show Energy

We can show that food has energy by burning it. As the food burns it gives out heat energy. This energy can be used to heat some water. The hotter the water gets, the more energy there is in the food. We can show the energy in a cream cracker or a crisp by carrying out *Experiment 4.2*.

Mandatory experiment 4.2

To investigate the conversion of chemical energy in food into heat energy

Apparatus required: Bunsen burner; mounted needle; thermometer; boiling tube; graduated cylinder; retort stand; balance

Also required: cream cracker or a crisp

Method

1. Place 20cm³ water in a test-tube.
2. Record the temperature of the water using a thermometer.
3. Stick the mounted needle into the cracker.
4. Hold the piece of cracker in the Bunsen flame until it catches fire.

5. Now transfer the burning cracker under the water in the boiling tube, as in *Fig. 4.12*.
6. When the cracker has burnt out take the temperature of the water again.
7. Record the new temperature.
8. Compare the temperature of the water at the start and at the end.

Result

The temperature of the water will have risen.

Conclusion

The chemical energy in the cream cracker is converted into heat energy. Food contains energy.

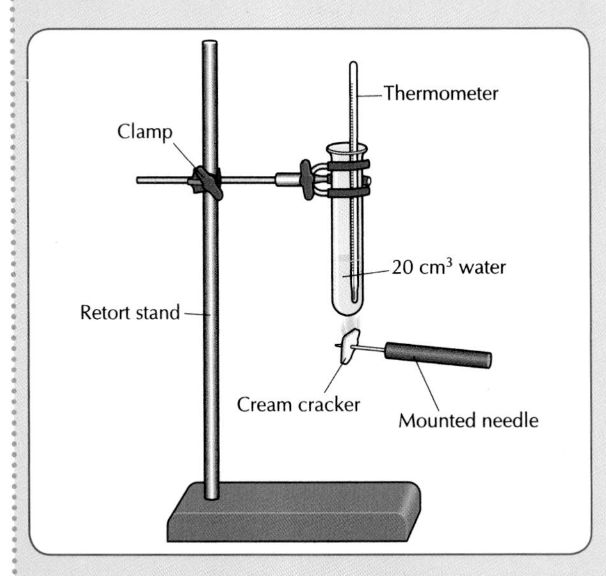

Fig. 4.12 Chemical energy in the cracker is converted into heat energy.

4.5 A Balanced Diet

A **balanced diet** consists of the **correct amounts** of the following: carbohydrates, proteins, fats, vitamins, minerals, water.

Look at *Fig. 4.13*. Which food is a good source of (a) protein? (b) minerals?

Fig. 4.13 A balanced meal.

Each constituent of the diet has a different function, as *Table 4.1* on page 19 shows. In general having a balanced diet depends on the following:

- your age
- your sex (being male or female)
- physical activity – how much exercise you take

According to food scientists if a person does not have a balanced diet they may suffer from **malnutrition**. This can mean either too little food (starvation, **anorexia**) or too much food (**obesity**).

The Food Pyramid

As mentioned above, a balanced diet involves eating the correct amounts of the right types of food. The **food pyramid**, shown in *Fig. 4.14*, page 24, is a guide to the types and the portions of each type of food needed every day. By examining the pyramid we can see that for a healthy diet we should eat the following:

- At least six helpings of cereals, bread or pasta each day. This is the food group we should eat most of.
- At least three helpings each of fruit and green vegetables.
- Two to three helpings of dairy products, such as cheese, milk and yoghurt.
- Two to three helpings of meat, fish, nuts or eggs.
- Very little of the food types at the top of the pyramid. These include sweets, biscuits, fats and oils.

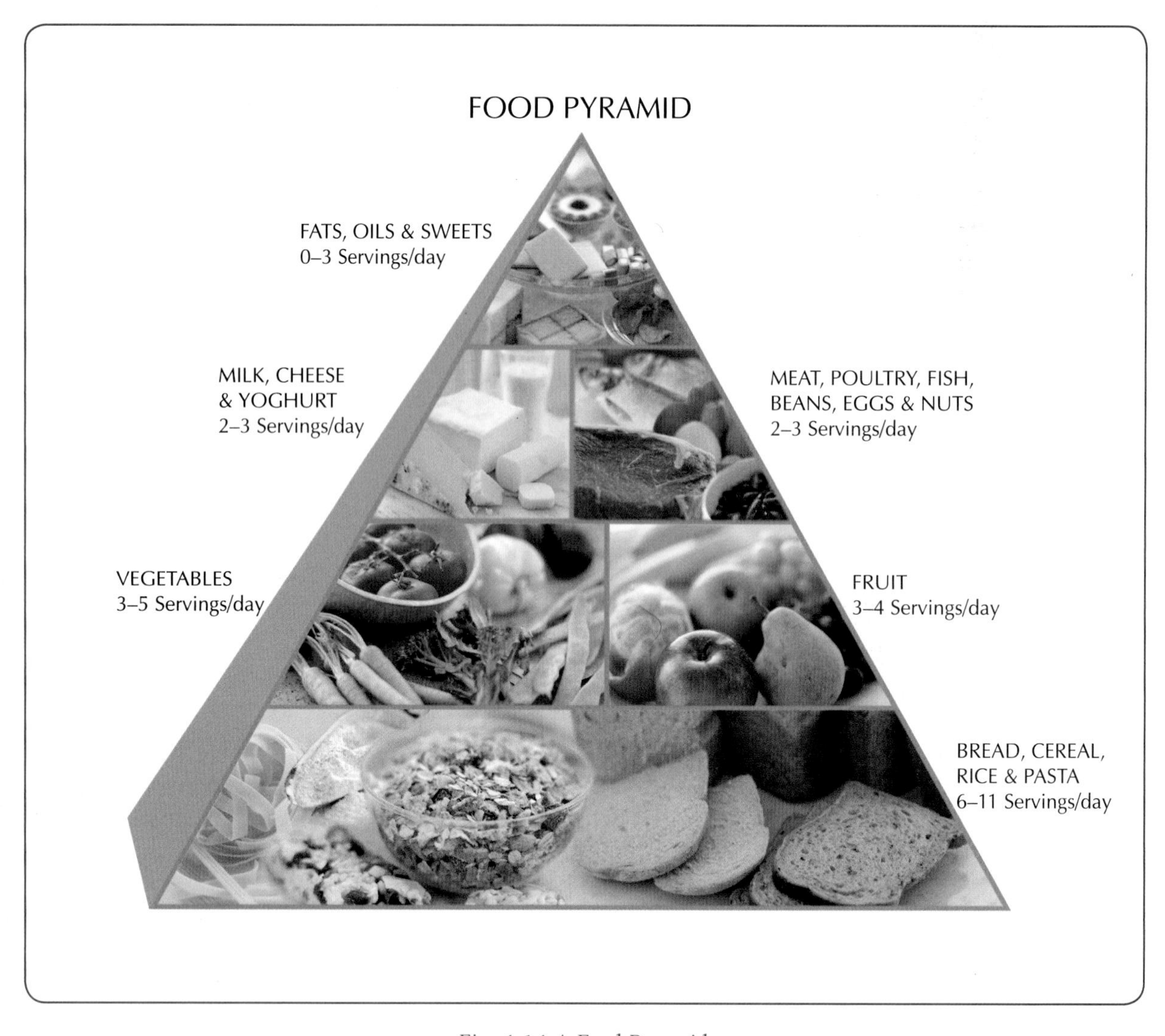

Fig. 4.14 A Food Pyramid.

TEST YOURSELF:

Now attempt questions *4.9–4.13* and *W4.7–W4.10.*

What I should know

- Food is needed for energy and as a growth material for the body.
- Food is made up of chemicals called nutrients.
- The main nutrients are: (a) carbohydrate, (b) protein, (c) fat, (d) vitamins, (e) minerals.
- The nutrients have different functions, see *Table 4.1* on page 19.
- The nutrients can be tested for in different ways, as shown in *Table 4.2* below.

TO TEST FOR	CHEMICAL USED	COLOUR CHANGE IF FOOD IS PRESENT
Starch	Iodine solution	Brown/yellow ⟶ blue-black
Glucose (Reducing Sugar)	Benedict's solution (+ heat)	Blue ⟶ brick-red
Protein	Sodium hydroxide solution + copper sulfate solution	Blue ⟶ purple
Fat	Rub butter on brown paper	Translucent spot appears

Table 4.2 Food Tests Summary.

What I should know

- Water is needed as a solvent and to transport substances.
- The energy in food is measured in kJ (kilojoules).
- Food gives out energy when it is burned.
- A balanced diet is one that contains a carbohydrates (including fibre), fats, proteins, vitamins, minerals and water.
- A balanced diet depends upon the person's age, sex and activity level.
- The food pyramid is a guideline showing the types and portions of food needed each day.

QUESTIONS

Write the answers to the following questions into your copybook.

4.1 (a) Food is needed to provide ..... and raw materials for ..... .

(b) Carbohydrates and fats are nutrients. Name two other nutrients in the diet.

(c) Name a nutrient found in potatoes.

(d) Name a mineral needed in the diet to make red blood cells.

(e) Fibre is needed in the diet to help prevent ..... . A good source of fibre in the diet is ..... .

4.2 Kiwis are a good sources of sugar and vitamins.

(a) Name one important vitamin found in kiwis.

(b) Give one use for this vitamin in the human body.

(c) Name another food rich in sugars.

(d) What is the main function of sugar in the diet?

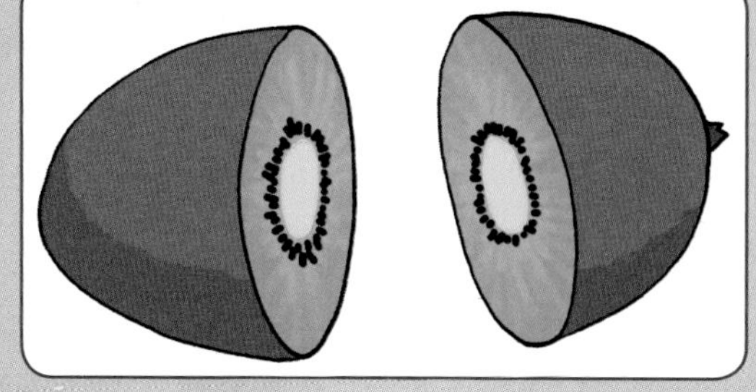

Fig. 4.15

4.3 Which of the following foods contain large amounts of fat?

(i) jam, (ii) grapes, (iii) butter, (iv) lettuce, (v) tomatoes, (vi) chocolate, (vii) sunflower oil

4.4 Name two minerals needed for healthy growth.

(a) Give a function for each named mineral.

(b) Name a good source of each mineral in the diet.

4.5 (a) What type of nutrient is starch?

(b) What is the function of starch in the diet?

(c) Name a food that contains a lot of starch.

4.6 Name your favourite food.

(a) Of which nutrient is it mainly made?

(b) Is it a healthy food? Give a reason for your answer.

4.7 (a) Describe a test to show the presence of starch in some bread.

(b) Name the chemicals used to test for (i) protein, (ii) reducing sugar and (iii) starch.

(c) Describe a test for fats.

QUESTIONS

4.8 Copy and complete the following table about food tests:

	FAT	REDUCING SUGAR	STARCH	PROTEIN
CHEMICALS USED				
IS HEAT NEEDED?				
COLOUR CHANGE FOR A POSITIVE RESULT				

4.9 Draw a labelled diagram of the apparatus you could set up to show that there is chemical energy in a cornflake.

4.10 The information shown in the chart *Fig. 4.16*, was printed on a box of breakfast cereal.

(a) How much protein is there in 100 g of the cereal?

(b) Name the unit used to measure the amount of energy in food.

(c) How much energy is there in 50 g of the cereal?

(d) This cereal is rich in fibre. Why is fibre needed in the diet?

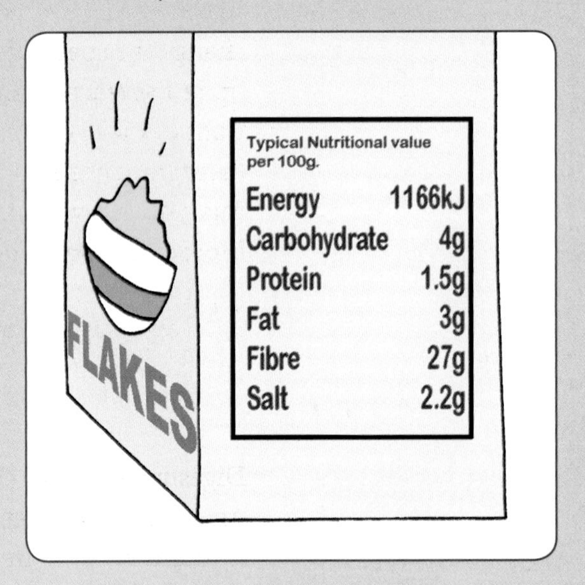

Fig. 4.16

4.11 What is a balanced diet?

4.12 Plan the menu for breakfast, lunch and dinner for a teenager that will provide a balanced diet. Use the information in the food pyramid shown in *Fig. 4.14*, page 24 to help you.

4.13 Give a reason for each of the following:

(a) Eating less sugar in our diet.

(b) Increasing the amount of fibre we eat in our diet.

(c) Eating plenty of fruit and vegetables regularly.

(d) Eating less salt.

Chapter 5 The Digestive System and Enzymes

5.1 Introduction

The food we eat is **too large** to get into our cells for the body to use. It must first be broken down into smaller, soluble pieces. This happens in the **digestive system** or **gut**. After the food is digested it is absorbed into the bloodstream. The bloodstream then carries the digested food to the cells where it is needed.

Digestion is the breakdown of large, insoluble food molecules into small soluble food molecules.

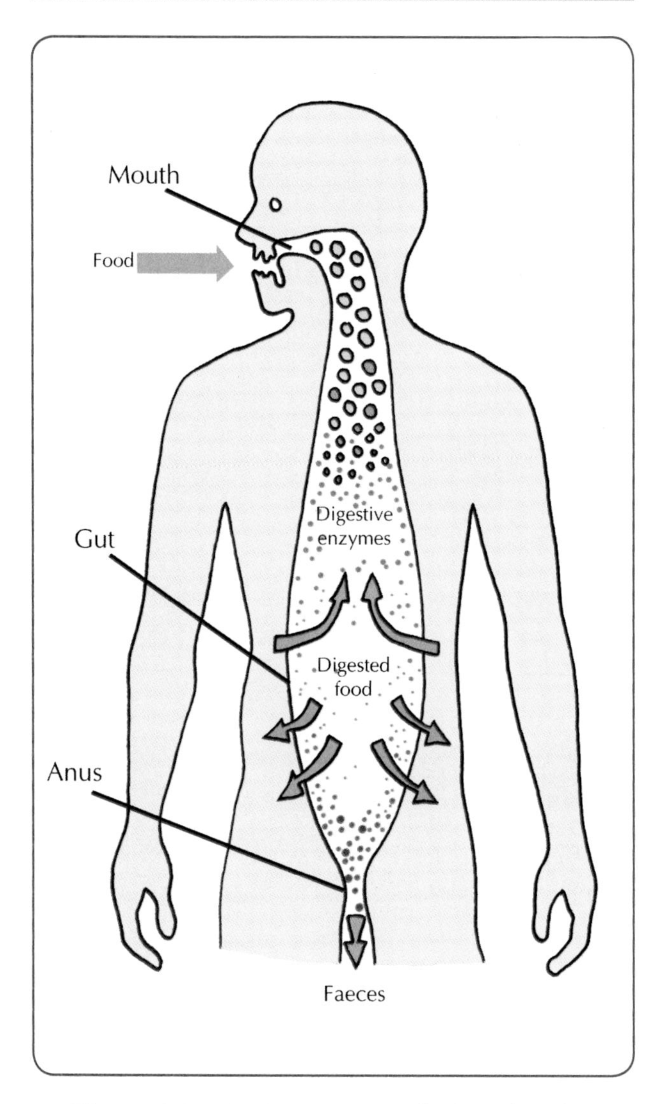

Fig. 5.1 Digestion is necessary so food can be taken into the bloodstream.

There are two types of digestion:

(1) **Physical Digestion**. Where the food is broken down by the teeth and by the churning action of the muscles in the digestive system.

(2) **Chemical Digestion**. Where the food is broken down by special chemicals called **enzymes**.

5.2 Enzymes

Enzymes are chemicals that are found in every cell in the body. Their function is to speed up chemical reactions in cells. Without enzymes the chemical reactions in cells would be too slow.

In *Chapter 25*, we learn that chemicals that change the speed of a reaction but are not used up in the reaction are called **catalysts**. If you want to make oxygen gas in the laboratory, the catalyst manganese dioxide is used. Enzymes are catalysts that work in cells. They are often called **biological catalysts**.

Enzymes are biological catalysts.

Amylase

Digestive enzymes **chemically** break down food. An example of an enzyme is **amylase**. Amylase is found in saliva in the mouth and in the small intestine. Amylase breaks down **starch** into smaller units called **maltose**. There are hundreds of different enzymes working in the body. Each one acts on a different substance.

- The substance on which an enzyme acts is called the **substrate** of the enzyme.
- The substance that is formed as a result of the reaction is called the **product**.

In the case of amylase, starch is the substrate and maltose is the product, see *Fig. 5.2*, page 28. Maltose is a reducing sugar.

starch + amylase → maltose + amylase
(substrate) (enzyme) (product) (enzyme unchanged)

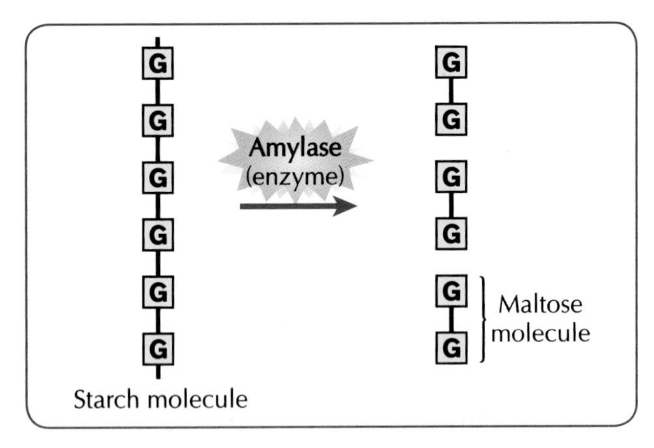

Fig. 5.2 Enzyme action.

Mandatory experiment 5.1

To investigate the action of amylase on starch

Apparatus required: droppers; graduated cylinder; Bunsen burner; tripod; gauze; 400 ml beaker; thermometer (or electric water bath); 4 test-tubes; test-tube rack; tongs; labels/marker pen

Chemicals required: starch solution; amylase solution; iodine solution; Benedict's solution

In this experiment we want to see if amylase will break down starch. In order for the enzyme to work, it must be kept at the correct temperature, in this case human body temperature. We do this by using a water bath kept at 37°C.

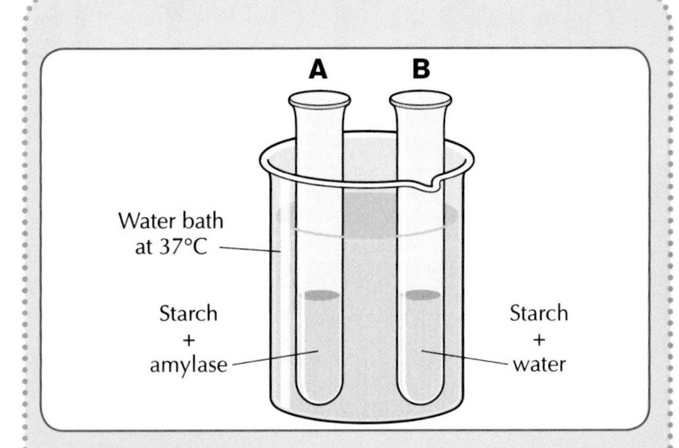

Fig. 5.3 To show the action of amylase on starch.

Method

1. Set up the apparatus as in *Fig. 5.3*.
2. Leave both test-tubes in the water bath for five minutes.
3. Remove the test-tubes and place in a test-tube rack.
4. Place half the contents of tube *A* into test-tube labelled *C*.
5. Test the contents of tube *A* for starch by adding a few drops of iodine solution. Record the result.
6. Test the contents of tube *C* for reducing sugars using Benedict's solution. Record the result.
7. Place half the contents of tube *B* into test-tube labelled *D*.
8. Repeat the starch and reducing sugar tests on the contents of tubes *B* and *D*. Record the results.

TEST-TUBE	CONTENTS AT START	TESTED FOR AT END	RESULT	CONCLUSION
A	Starch + amylase	Starch	No starch present	Starch broken down by amylase
B	Starch + water	Starch	Starch present	Starch not broken down
C	Starch + amylase	Reducing sugar	Reducing sugar present	Starch has been broken down to reducing sugar
D	Starch + water	Reducing sugar	Reducing sugar absent	Starch not broken down

Table 5.1

Result

It is only in the test-tube with the amylase that the starch has been broken down.

Conclusion

The enzyme amylase breaks down starch to reducing sugar.

TEST YOURSELF:
Now attempt questions *5.1–5.4* and *W5.1–W5.2.*

5.3 The Digestive System

Stages in Nutrition (feeding)

Human nutrition involves four stages:

1. **Eating**. First the food is taken into the mouth.
2. **Digestion**. The food is broken down into smaller, soluble molecules.
3. **Absorption**. The digested food is taken into the bloodstream and carried to the cells where it is used.
4. Getting rid of undigested waste as faeces.

The digestive system or gut is a long tube that stretches from mouth to anus, *Fig. 5.4*. It consists of the following parts:

(1) mouth
(2) oesophagus
(3) stomach
(4) small intestine
(5) large intestine

(1) Mouth

The mouth is where digestion begins.

- **Enzymes** in saliva (amylase) begin to chemically break down the starch in our food. The starch is broken down into maltose sugar.
- At the same time, the **teeth** physically break the food into smaller pieces.

Teeth

The teeth cut and crush the food we eat. They make the food smaller, so that it is easier for the enzymes to work on it. There are four different types of tooth. *Fig. 5.5*, page 30, shows the arrangement of the teeth in the mouth.

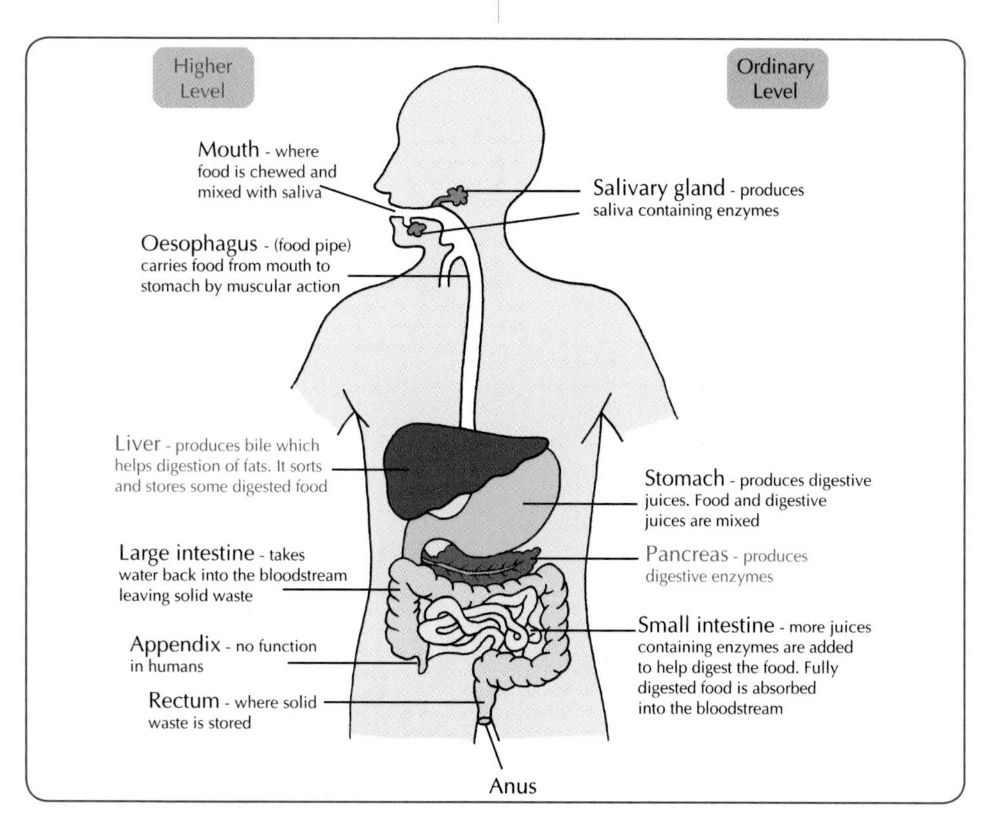

Fig. 5.4 The Digestive System: its parts and functions.

BIOLOGY

Different animals have different numbers of teeth depending on the food they eat. An adult human has 32 teeth.

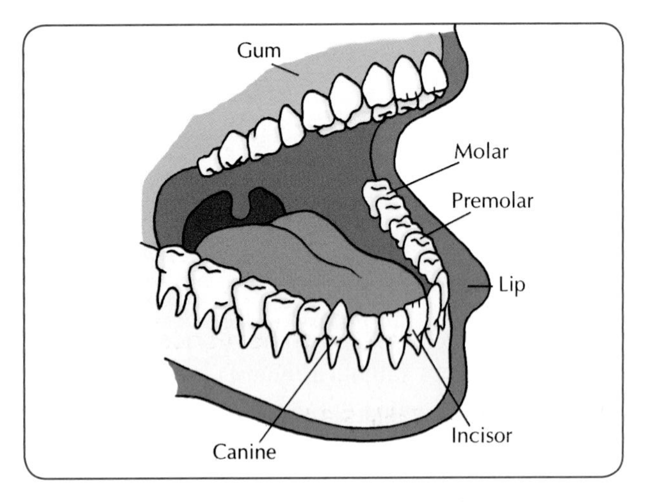

Fig. 5.5 Arrangement of the teeth in the mouth.

TOOTH TYPE	SHAPE OF TOOTH	FUNCTION
Incisor		Cutting and biting food
Canine		Tearing food
Premolar		Crushing and grinding food
Molar		Crushing and grinding food

Table 5.2 Tooth types and functions.

(2) Oesophagus

The oesophagus is a muscular tube which leads to the stomach. The food is '**shunted**' down the oesophagus by muscular contractions of the wall.

(3) Stomach

The stomach is a muscular bag which holds the food for three to four hours.

- It produces **enzymes** and other digestive juices which chemically digest the food.
- The stomach muscle **churns** and mixes the food.
- **Hydrochloric acid** produced by the stomach kills bacteria and other germs.

(4) Small Intestine

In the small intestine, more enzymes are added to complete the break down of the food.

- The **digested food** now passes from the small intestine into the bloodstream. This is known as **absorption**.
- The leftovers are water and the **undigested food**. This is pushed into the **large intestine**.

Liver and Pancreas

In addition to the long tube of the gut there are two other organs associated with digestion: these are the **liver** and the **pancreas**.

- The **liver** has many functions including the production of **bile**. Bile passes from the liver to the small intestine where it helps break up fats into smaller pieces. This makes it easier for enzymes to digest fats.
- The **pancreas** produces **enzymes**. These enzymes pass into the small intestine where they break down food.

(5) Large Intestine

The main function of the large intestine is to **take water back** into the bloodstream. This helps prevent the body becoming dehydrated. The remaining material, called **faeces**, is stored in the **rectum** and released from the body through the **anus**.

TEST YOURSELF:

Now attempt questions *5.5–5.8* and *W5.3–5.5.*

What I should know

- Enzymes are biological catalysts.
- The substance an enzyme works on is called the substrate.
- The substance(s) produced by the enzyme is called the product.
- Amylase is an enzyme.
- starch + amylase ⟶ maltose
  (substrate) (enzyme) (product)
- The stages in human nutrition are: 1. Eating, 2. Digestion, 3. Absorption, 4. Removal of undigested waste.
- Digestion is the break down of food. It takes place in the gut.
- Digestion is needed so that the food molecules are small enough to pass into the bloodstream.
- Muscle action shunts food through the gut.
- The parts and functions of the digestive system (gut) are shown in *Table 5.3* below.

PART	FUNCTION
Mouth	Physical digestion by teeth. Chemical digestion by enzymes.
Oesophagus	Carries food from mouth to stomach.
Stomach	Food is mixed with digestive enzymes and churned.
Liver	Produces bile.
Pancreas	Produces digestive enzymes.
Small Intestine	Completes digestion. Digested food absorbed into bloodstream.
Large Intestine	Takes water back into bloodstream.

Table 5.3 Parts and functions of the digestive system (gut).

- The four types of tooth are shown in *Table 5.4* below.

TOOTH TYPE	FUNCTION
Incisor	Cutting and biting food
Canine	Tearing food
Pre molar	Crushing and grinding food
Molar	Crushing and grinding food

Table 5.4 The four types of tooth and their functions.

QUESTIONS

Write the answers to the following questions into your copybook.

5.1 (a) What is meant by digestion?

(b) Why is digestion of food necessary?

(c) Name two types of digestion.

(d) Which type of digestion involves enzymes?

5.2 (a) An enzyme is a ..... .

(b) enzyme + substrate ⟶ ..... + ...

(c) The function of an enzyme in a reaction is to ..... .

5.3 Enzymes are catalysts that work in cells.

(a) What is a catalyst?

(b) Name a catalyst that is an enzyme.

(c) Name a catalyst that is not an enzyme.

5.4 Describe fully how you could show the action of amylase on starch.

5.5 Rewrite the following parts of the digestive system in the correct order through which food passes.

stomach, large intestine, oesophagus, anus, small intestine, mouth, rectum

5.6 (a) There are ..... types of tooth.

(b) The incisors ..... the food and the ..... tear the food.

(c) The teeth at the back of the mouth are the ..... . Their function is to ..... and ..... the food.

(d) Teeth help to ..... break down the food.

5.7 *Fig. 5.6* shows the human digestive system.

(a) Name the parts labelled *P, Q, R, S, T, U* and *V*.

(b) What is the main function of *T*?

(c) How is food moved through the digestive system?

(d) What is the function of the part labelled Q?

(e) Give one function of the part labelled *R*.

(f) A boy swallows a piece of potato. List the names of the parts of the digestive system that the potato will pass through, in the correct order, until it is fully digested.

5.8 (a) What is bile?

(b) Where is bile made?

(c) What is the function of bile?

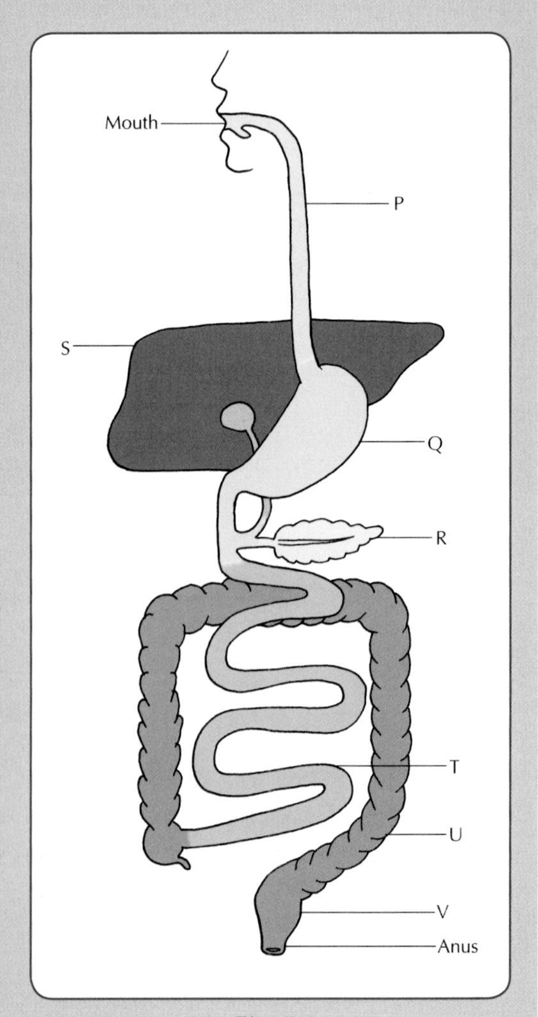

Fig. 5.6

Chapter 6 Respiration

6.1 Introduction

All living things need energy for growth and for movement. They get this energy from food in a process called **respiration**.

Respiration is the release of energy from food.

Fig. 6.1 Football players need a lot of energy.

- **Green plants** make their own food. They use this food to get energy for all their activities.
- **Animals** then either eat the plants or eat other animals that have eaten plants. Caterpillars eat leaves and then robins eat the caterpillars. In this way energy from the sun is passed into all living things.

Don't forget, plants need energy too. They use it to grow, to open and close their petals and to make seeds.

During respiration, food is broken down in the presence of oxygen, producing energy, carbon dioxide and water. The carbon dioxide and water are waste products, *Fig. 6.2*.

Aerobic Respiration

The process of **aerobic respiration** can be summarised by the following word equation:

glucose (food) + oxygen → energy + carbon dioxide + water

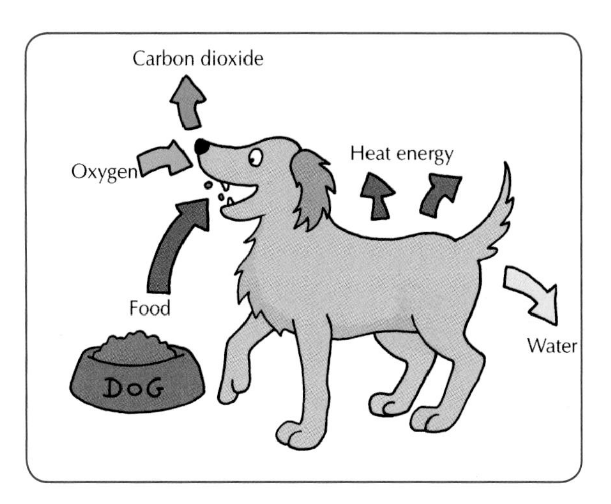

Fig. 6.2 Respiration produces energy and wastes.

When we burn a sample of food, e.g. the cream cracker in *Mandatory experiment 4.2*, the energy in the food is converted into heat energy. In order for the cream cracker to burn, oxygen must be present.

A similar reaction takes place in every living cell and is known as **cellular respiration**. Because oxygen is required for respiration by most cells, the process is known as **aerobic respiration**. Aerobic respiration requires oxygen to break down food. The release of energy is carefully controlled by enzymes.

TEST YOURSELF:
Now attempt questions *6.1–6.4* and *W6.1–W6.2*.

6.2 Experiments on Respiration

If living organisms are respiring aerobically it should be possible to show that they produce (1) carbon dioxide, (2) energy and (3) water.

Experiment 6.1

To show that respiring animals produce carbon dioxide

Apparatus required: 2 boiling tubes; wire gauzes to fit the boiling tubes; 2 rubber stoppers; graduated cylinder

Chemical required: lime water

Also required: 3–4 woodlice

Note: A chemical called **limewater** turns a **milky** colour when **carbon dioxide** is passed through it. This fact is used to test for the presence of carbon dioxide.

Method

1. Set up the apparatus as shown in *Fig. 6.3*.

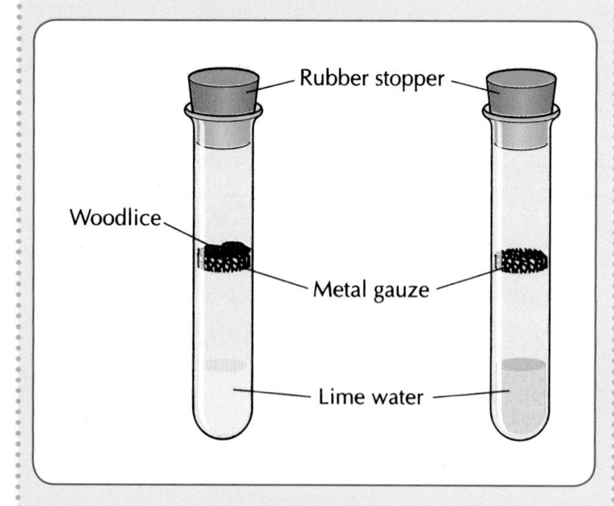

Fig. 6.3 To show woodlice produce carbon dioxide.

Result

The limewater in the tube with the woodlice turns milky. There is no change in the limewater in the other tube.

Conclusion

- The limewater turns milky because carbon dioxide has passed in to it. This only happens in the **tube with the woodlice**. Therefore the woodlice must be producing carbon dioxide.
- The **second tube** acts as a control in this experiment. It shows that only the woodlice produce the carbon dioxide.

Experiment 6.2

To show that respiring plants produce carbon dioxide

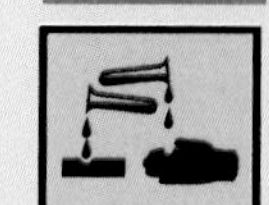

Apparatus required: set up as in *Fig. 6.4*

Chemicals required: limewater; sodium hydroxide solution

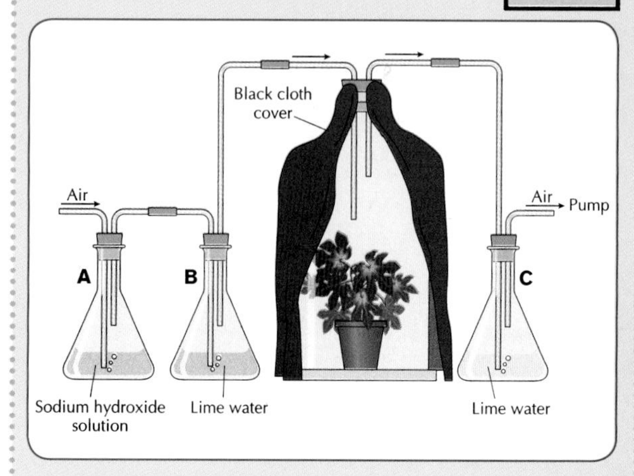

Fig. 6.4 To show plants produce carbon dioxide.

Method

1. Set up the apparatus as shown in *Fig. 6.4*.

   Note: the **sodium hydroxide** solution **removes carbon dioxide** from the air entering the apparatus.

2. The bell jar is covered with a black cloth to prevent photosynthesis happening in the plant. During photosynthesis the plant would use up the carbon dioxide it makes in respiration and it would not be released.
3. When the pump is turned on, air is pulled through the apparatus.
4. Observe what happens to the limewater in flasks *B* and *C*.

Result

After a while the limewater in flask *C* turns milky. The limewater in flask *B* stays clear.

Conclusion

- Sodium hydroxide solution removes carbon dioxide from the air reaching the plant. We show that the carbon dioxide has been removed by passing the air through the limewater in **flask *B***. It stays clear.
- The limewater in ***C*** turns milky. The only source of carbon dioxide to do this must be the plant. We can conclude therefore, that the plant has produced carbon dioxide.

A **control** for this experiment would be to use the same apparatus, but with no plant in the bell jar. Can you think of any other suitable control for this experiment?

Experiment 6.3

To show that respiring organisms produce heat energy

In this experiment we are using germinating pea seeds as the organism. The thermos flasks are used because they have a special lining which (a) does not allow any heat produced by the peas to escape and (b) does not allow any heat from the room outside to get into the flasks.

Apparatus required: 2 thermos flasks; cotton wool; 2 thermometers; labels

Also required: a beaker-full of pea seeds (soaked over night in water); a beaker-full of dead pea seeds

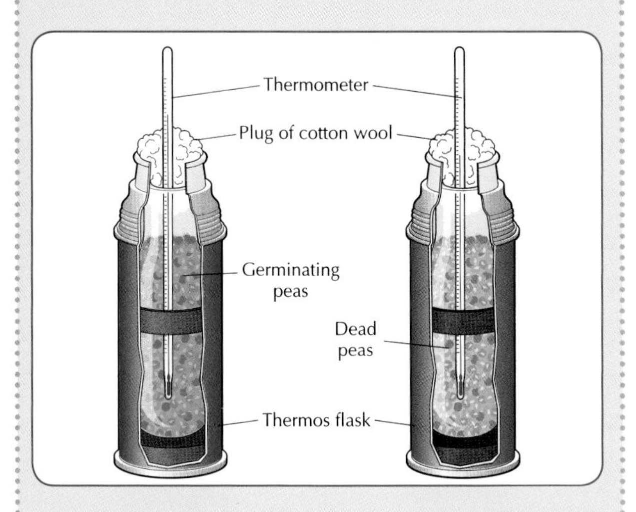

Fig. 6.5 To show respiring seeds produce heat energy.

Method

1. Set up the flasks as shown in *Fig. 6.5*.
2. Read and record the temperature in both flasks.
3. Note the temperature readings each day for three days.
4. Plot a graph of temperature (°C) on the vertical axis (y-axis), against time (number of days) on the horizontal axis (x-axis).

Results

The temperature in the flask with the live pea seeds will increase. There will be no change in the flask with the dead seeds.

Conclusion

Dead seeds do not respire. It is only the living seeds that produce the heat energy. Respiring seeds produce heat energy.

Experiment 6.4

To show that respiring organisms produce water

Chemical required: blue cobalt chloride paper

Note: The test for water. Water turns blue cobalt chloride paper pink.

Fig. 6.6 We breathe out water vapour.

Method

1. Hold a piece of blue cobalt chloride paper in your hand, as in *Fig. 6.6*, page 35.
2. Breathe in and out onto the piece of paper a number of times.
3. Observe any change in the colour of the cobalt chloride paper.

Result

The blue cobalt chloride paper turns pink.

Conclusion

The pink colour of the cobalt chloride paper shows that we breathe out water vapour. This water is a waste product of respiration in our cells. So we can conclude that living things produce water.

TEST YOURSELF:

Now attempt questions *6.5–6.9* and *W6.3–W6.4.*

What I should know

- All living things need energy. Energy is needed for movement, growth and chemical reactions in cells.
- Respiration is the release of energy from digested food.
- Respiration takes place in every living cell.
- Most organisms need oxygen for respiration.
- Aerobic respiration requires the presence of oxygen.
- The word equation for respiration is:

  food (glucose) + oxygen ⟶ energy + carbon dioxide + water
- The products of aerobic respiration are energy, carbon dioxide and water.
- During respiration, chemical energy is converted to heat energy.

Questions

Write the answers to the following questions into your copybook.

6.1 (a) Living things need ..... for ..... and ..... .

(b) The process in which plants and animals make energy is called ..... .

(c) Plants need energy to ..... and ..... .

(d) Plants make food using energy from the ..... .

(e) Animals get their food from eating either ..... or ..... .

6.2 Name three human activities that need energy.

6.3 Name three plant activities that need energy.

6.4 (a) Write down the word equation for respiration.

(b) Name the gas normally needed for respiration to take place.

(c) What is the main product of respiration?

(d) Name the two waste products of respiration.

(e) What is aerobic respiration?

QUESTIONS

6.5 (a) Name a chemical that tests for the presence of carbon dioxide.

(b) State the colour of this chemical when carbon dioxide is present.

(c) State the colour of this chemical when carbon dioxide is not present.

(d) Name a chemical that absorbs carbon dioxide.

(e) Name a chemical that tests for the presence of water.

6.6 Describe an experiment to show that living things produce heat energy.

6.7 Draw a labelled diagram of the apparatus you could set up to show that respiring plants produce carbon dioxide.

6.8 Look at the apparatus used in an experiment, shown below in *Fig. 6.7*.

(a) Suggest a title for this experiment.

(b) The tubes were left for a number of hours. Predict what results you would expect.

(c) What was the purpose of tube *B*?

(d) What is the purpose of the gauze?

(e) Would it make any difference to the results if the tubes were covered with a black cloth or aluminium foil during the experiment? Give a reason for your answer.

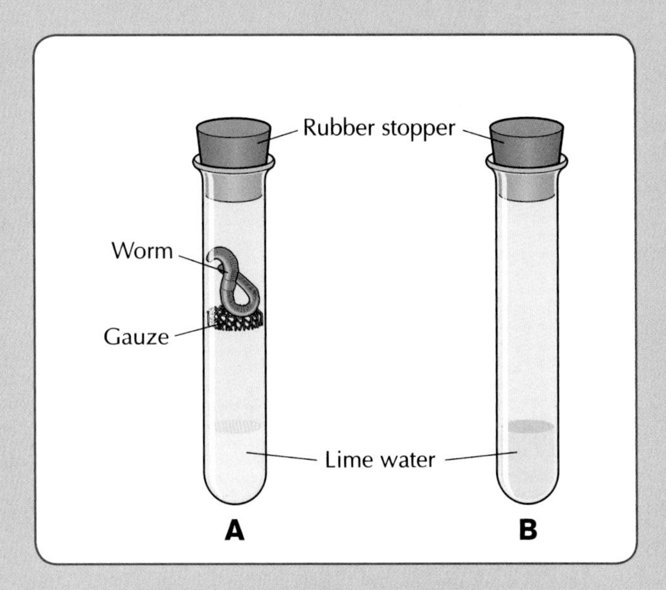

Fig. 6.7

6.9 State which of the following statements about respiration is true or false. Give a reason for any false statement.

(a) It takes place only in animal cells.

(b) It produces oxygen.

(c) It only takes place at night.

(d) It uses up energy.

(e) It takes place all the time.

(f) It releases energy.

(g) It produces carbon dioxide.

Chapter 7 The Breathing System

7.1 Introduction

We need **oxygen** to get **energy** from the food we eat. The breathing system lets us take in this oxygen.

- Breathing in is called **inhaling**.
- Breathing out is called **exhaling**.

What Happens during Breathing

1. We breathe **oxygen** into our **lungs**.
2. The **oxygen** then passes from our lungs into the **bloodstream**.
3. The blood carries the **oxygen** to the **cells** where it is used in respiration.
4. **Carbon dioxide** and **water** are waste products of **respiration**.
5. They pass from the cells, into the **bloodstream** and are carried back to the lungs.
6. At the **lungs**, **carbon dioxide** and **water** vapour move from the blood into the lungs and get breathed out.

7.2 The Breathing System

Fig. 7.1 shows the parts of the human breathing system and their functions.

The functions of the breathing system are the following:

(1) To **take in oxygen** gas for respiration.

(2) To **remove** waste **carbon dioxide** gas.

(3) To **remove** waste **water** vapour.

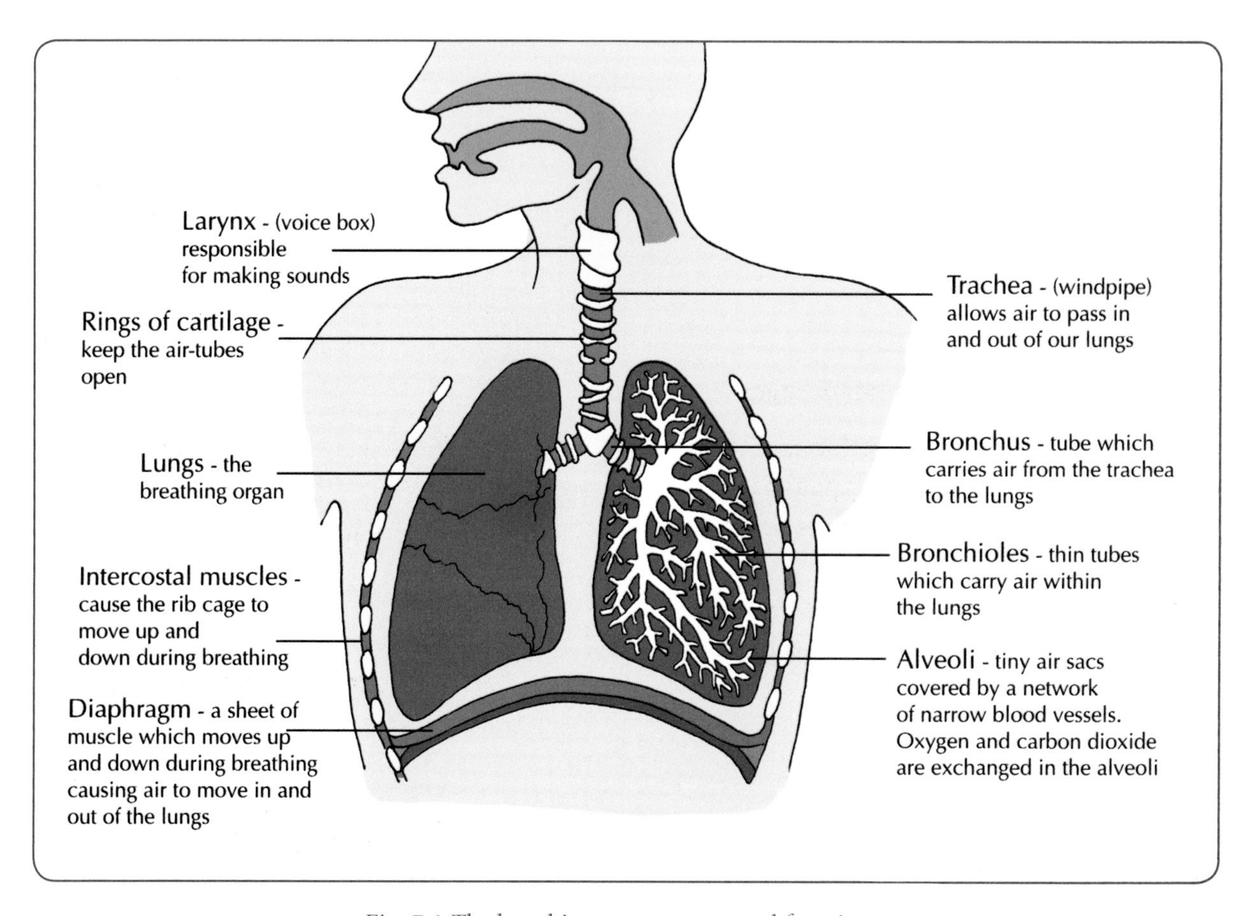

Fig. 7.1 The breathing system, parts and functions.

7.3 How Breathing Happens

We can demonstrate how the diaphragm works using the 'bell-jar' model. Look carefully at the model in *Fig. 7.2*. Note which parts of the model represent which parts of the breathing system.

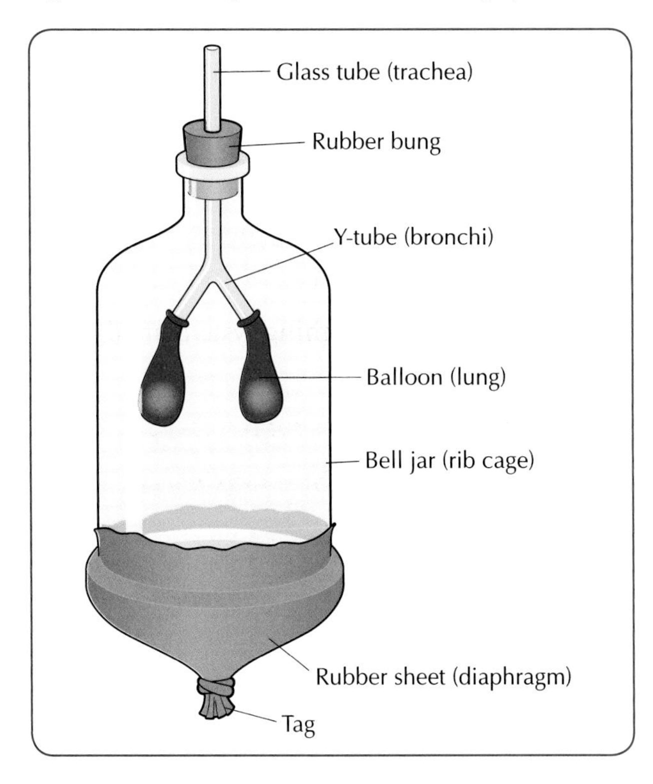

Fig. 7.2 The bell-jar model.

Now watch closely as the tag is pulled gently downwards and then quickly pushed back up again. Watch what happens to the balloons as these movements are repeated.

Diaphragm

The rubber sheet represents the **diaphragm**.

- When it moves down, the balloons fill with air. This is similar to what happens to the lungs when the diaphragm moves down when we breathe in.
- When the rubber sheet moves up, air is pushed out of the balloons. This is the same as when we breathe out.

Mandatory experiment 7.1

To compare the carbon dioxide levels of inhaled and exhaled air

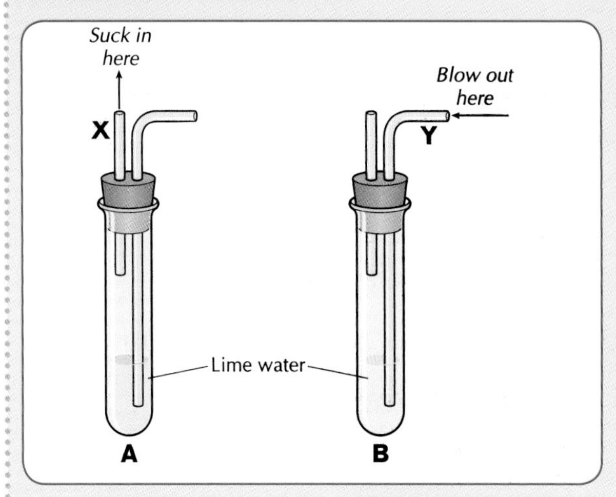

Fig. 7.3 To compare the carbon dioxide levels of inhaled and exhaled air.

Chemical required: limewater

***Note:* Limewater** turns a **milky** colour when **carbon dioxide** is passed through it.

Method

1. Set up the apparatus as in *Fig. 7.3*.
2. Hold tube *A* in one hand and tube *B* in the other.
3. Suck in through *X* of test-tube *A* (this represents inhaled air).
4. Hold this breath and then breathe out through *Y* of test-tube *B* (this represents exhaled air).
5. Repeat steps 3 and 4 until the limewater in one of the tubes goes milky.
6. Record in which tube the limewater goes milky first.

Result

The limewater in test-tube *B* turns milky quicker than the limewater in *A*.

Conclusion

The level of carbon dioxide in exhaled air is greater than the level of carbon dioxide in inhaled air.

BIOLOGY

Examine *Table 7.1*, which compares the percentages of gases in inhaled and exhaled air.

GAS	AIR BREATHED IN	AIR BREATHED OUT
Nitrogen	78%	78%
Oxygen	21%	16%
Carbon dioxide	0.03%	4.0%

Table 7.1 Comparison of the percentages of gases in inhaled and exhaled air.

What do you think the figures for nitrogen gas tell us?

TEST YOURSELF:
Now attempt questions *7.1–7.4* and *W7.1–W7.3*.

7.4 Gaseous Exchange

Look at *Fig. 7.4*, it shows a group of alveoli. The lungs are made up of millions of tiny air sacs called **alveoli**. The alveoli lie at the end of the bronchioles. They are covered by a network of tiny tubes that carry blood.

It is at the alveoli that oxygen and carbon dioxide gases are exchanged during breathing.

The walls of the alveoli are very thin and this allows the gases to **diffuse** through quickly.

Gaseous exchange occurs when oxygen from the air is swapped with carbon dioxide from the body.

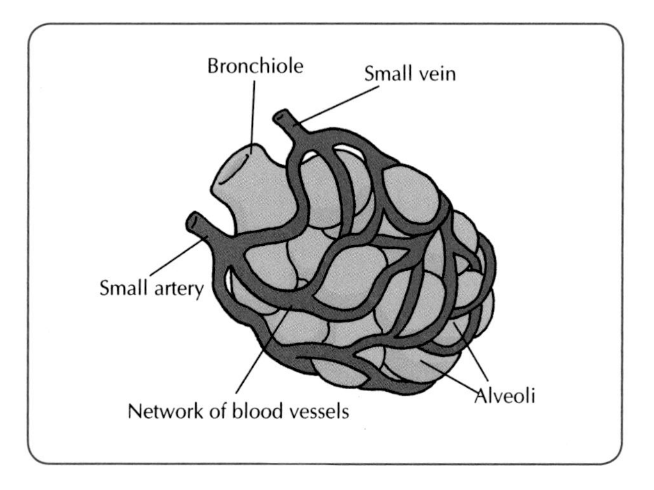

Fig. 7.4 Alveoli with blood supply.

- During **gaseous exchange**, **oxygen** in the air we breathe in, diffuses across the wall of the alveolus into the bloodstream.
- **Carbon dioxide** diffuses into the lungs from the bloodstream and is breathed out.

Fig. 7.5 illustrates how oxygen and carbon dioxide gases are swapped at the alveoli.

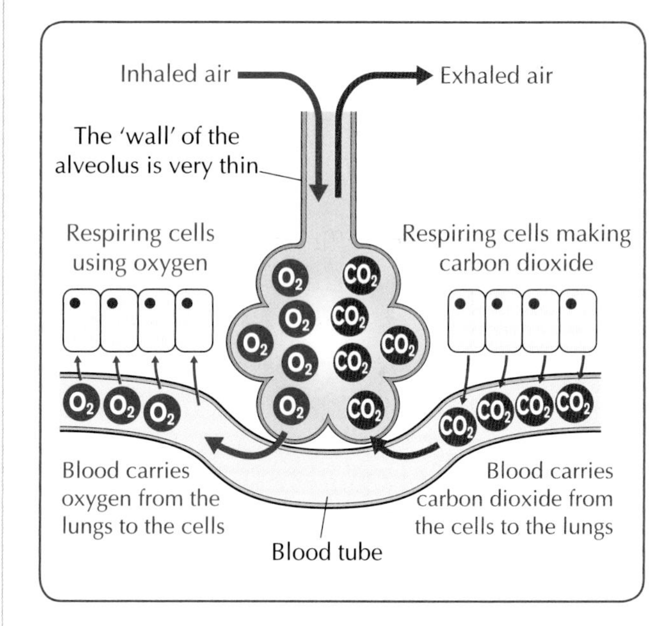

Fig. 7.5 Gaseous exchange takes place at the alveoli.

7.5 Smoking and Breathing

Certain things can interfere with the proper exchange of gases in the lungs. One such thing is **smoking**.

Smoking is bad for you. It is a **dangerous**, **harmful** habit. You only have to look at the Government health warnings on the sides of cigarette packets to see that smoking is not a good idea, *Fig. 7.6*.

Fig. 7.6 Government warnings on cigarette packets.

Dangers of Smoking

Smoking damages your breathing system and your general health in the following ways:

- **Tar** in tobacco smoke irritates and damages the lining of the bronchioles. This causes the production of too much mucous which cannot be easily cleared. This can lead to bronchitis.
- Nicotine is just one of many hundreds of **harmful chemicals** in cigarette smoke. Nicotine damages the heart and blood vessels.
- Smoking prevents efficient gas exchange. **Carbon monoxide** is a harmful gas in cigarette smoke. Carbon monoxide replaces oxygen in the blood. This means not enough oxygen gets to our cells. This may cause the smoker to breathe faster which can cause heart strain.
- Smoking during **pregnancy** prevents the baby in the womb from getting enough oxygen. This can cause the stunted growth of the baby.
- Smoking causes **lung cancer**. A cancer happens when cells start to grow out of control. Sometimes this produces a lump of cells called a tumour which prevents the body's organs from working properly.

The best advice you can take is **NEVER SMOKE**.

Experiment 7.2

To show the effect of exercise and rest on the breathing rate

Breathing rate is a measure of how many times you inhale in one minute. On average, the human **breathing rate**, at rest, is **17 breathes per minute**.

Apparatus required: stopwatch/watch

Method

1. Draw up a table like the one shown in *Table 7.2*.
2. After a short period of rest, count how many times you inhale in one minute.
3. Record this figure in your table.
4. Repeat twice.
5. Calculate the average. This is your breathing rate at rest.
6. Now run on the spot for a minute.
7. Straight away take your breathing rate.
8. Record this figure.
9. Take your breathing rate again.
10. Record this figure.
11. Take your breathing rate one last time.
12. Record this figure.
13. Draw a graph of breathing rate against time before and after exercise.

Result

The breathing rate increases with exercise. When we exercise, we need more energy in our muscles. This means we need to take in more oxygen for respiration and get rid of more waste carbon dioxide.

TIME	NUMBER OF BREATHS PER MINUTE
At rest (first time)	
At rest (second time)	
At rest (third time)	
Average at rest	
After 1 minute of exercise	
1 min later	
1 min later again	

Table 7.2

TEST YOURSELF:
Now attempt questions *7.5–7.9* and *W7.4–W7.6.*

BIOLOGY

What I should know

- The human breathing system consists of the nose and mouth, the trachea, bronchi, bronchioles, lungs and the alveoli. See *Fig. 7.1* on page 38 for the functions of each part.
- The breathing system allows us to take in oxygen and get rid of carbon dioxide.
- Gas exchange occurs at the alveoli.
- The alveoli have a large blood supply and thin walls. This allows efficient gas exchange.
- Oxygen gas passes from the alveoli into the bloodstream.
- Carbon dioxide passes from the bloodstream into the alveoli.
- Exhaled air contains more carbon dioxide than inhaled air.
- Carbon dioxide turns lime water milky.
- Smoking can damage your breathing system, it prevents efficient gas exchange.
- Exercise increases a person's breathing rate. This brings more oxygen to the cells and removes more waste carbon dioxide.

QUESTIONS

Write the answers to the following questions into your copybook.

7.1 (a) Name six parts of the human breathing system.
(b) Give the function of each part you have named in (a).
(c) List the functions of the breathing system.

7.2 (a) The ..... are the human breathing organs.
(b) The windpipe is also known as the ..... .
(c) Rings of ..... on the trachea help to keep the tube open.
(d) The ..... connect the trachea to the bronchioles.
(e) The ..... are the tiny air sacs at the end of the bronchioles.
(f) The ..... moves ..... when we breathe in.

7.3 Name the parts of the breathing system shown in *Fig. 7.7*. Give the function of parts *A*, *B* and *E*.

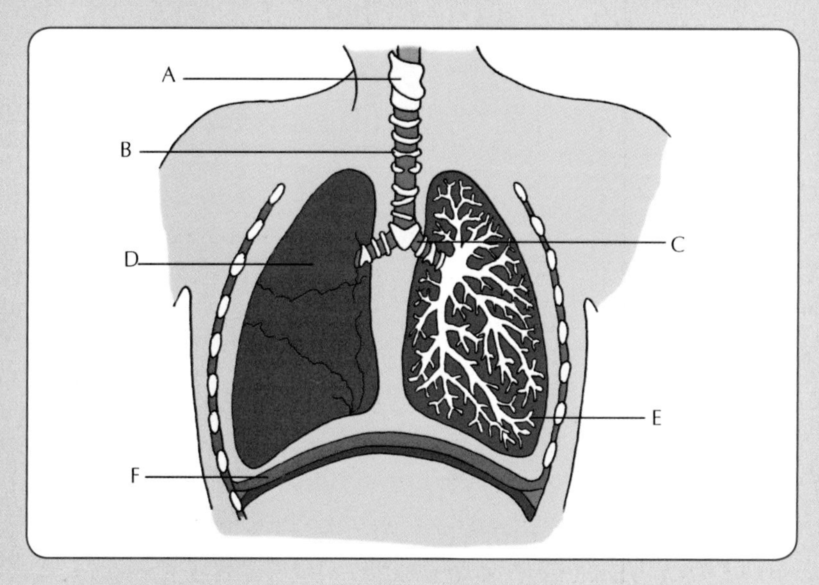

Fig. 7.7

QUESTIONS

7.4 *Fig. 7.8* shows the apparatus you could use to investigate that the air we breathe out contains carbon dioxide.

(a) Name the chemical you would use at *X*.

(b) Describe the colour of this chemical <u>before</u> and <u>after</u> you breathe into the test-tube.

(c) Describe, using a labelled diagram, a control experiment that would show that we breathe in less carbon dioxide than we breathe out.

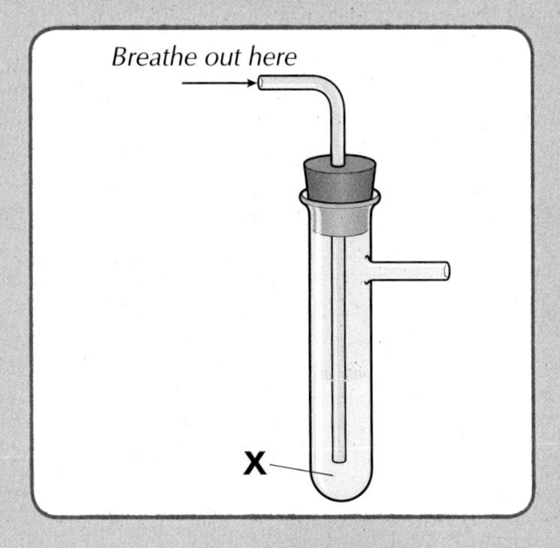

Fig. 7.8

7.5 (a) What is gas exchange?

(b) Where does gas exchange take place?

(c) Draw a labelled diagram to show how gas exchange happens.

7.6 List three harmful effects of smoking on the breathing system.

7.7 (a) What is the average breathing rate of a person at rest?

(b) How is a person's breathing rate measured?

7.8 Describe an investigation you could carry out to show that the breathing rate increases with exercise.

7.9 The graph in *Fig. 7.9* shows the average rate of breathing of a 14 year old girl.

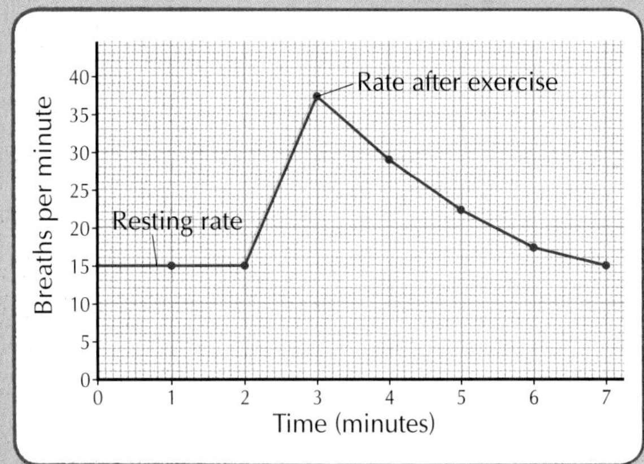

Fig. 7.9

(a) How is the average breathing rate calculated?

(b) Use the graph to find out

(i) The girl's breathing rate at rest.

(ii) The breathing rate immediately after exercise.

(iii) What was the difference in breathing rate between the rate at rest and the rate immediately after exercise?

(iv) How long did it take for the breathing rate to return to the resting rate after exercise?

(c) Name the muscles used in breathing.

(d) Why does breathing rate increase with exercise?

(e) State two things that could slow down a person's breathing rate.

(f) State two things, other than exercise, that could speed up a person's breathing rate.

Chapter 8 The Circulatory System

8.1 Introduction

Living things need a **transport system** to carry substances around their bodies. The human transport system is also known as the **circulatory system**. Our circulatory system consists of the **blood**, the **blood vessels** and the **heart**, *Fig. 8.1*.

Blood is a liquid which carries substances to and from our cells. Blood travels in blood vessels (blood tubes) called **arteries**, **veins** and **capillaries**. The heart pumps the blood around the body.

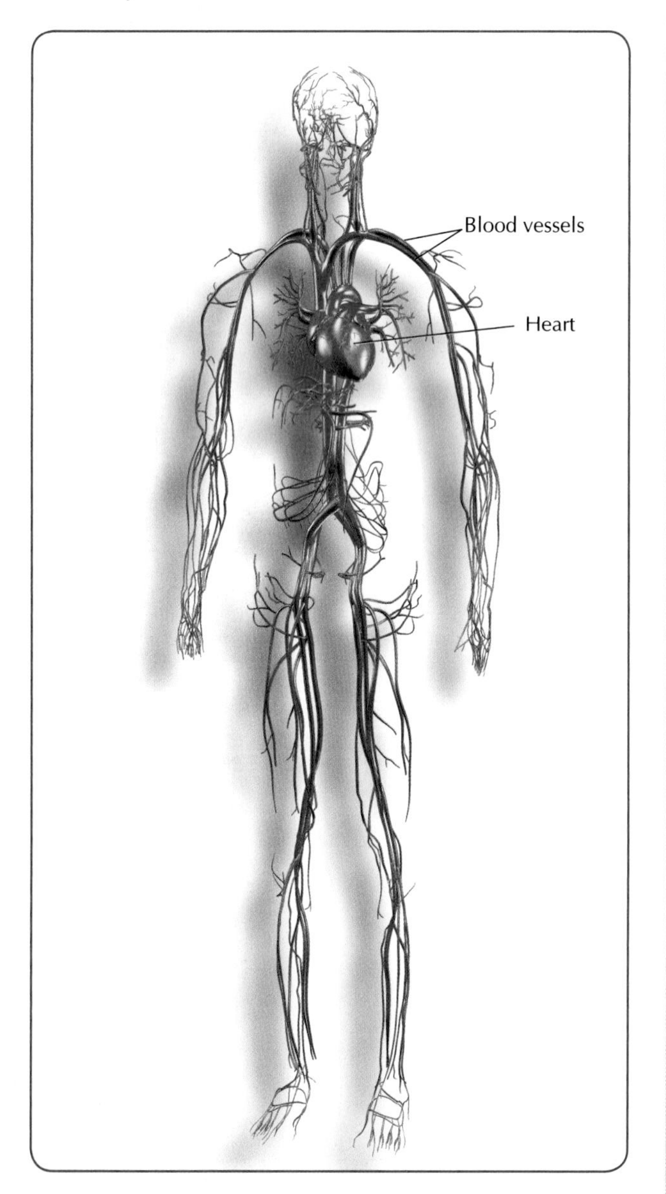

Fig. 8.1 The human circulatory system.

8.2 The Functions of Blood

(1) **Transport**. Blood transports many substances. For example oxygen, digested foods, waste materials and hormones are all carried by the blood.

(2) **Defence against Disease**.

- The blood **destroys** harmful **bacteria and viruses**. In this way it helps protect us against infection.
- The blood can **clot** when we cut ourselves. This prevents loss of blood and keeps out harmful bacteria or viruses.

(3) **Constant Body Temperature**. The blood also plays a part in keeping our **body temperature constant**. Chemical reactions that take place in our cells produce heat energy. The blood carries the heat all around the body.

Normal **human body temperature is 37°C**. A change in body temperature can be a sign that a person is ill.

Fig. 8.2 Being ill can raise your body temperature.

The functions of the blood are:

- **transport**
- **defence against disease**
- **keep body temperature constant**

8.3 The Composition of Blood

An average adult has five litres of blood circulating in their blood vessels. Blood is made up of the following:

- A liquid called **plasma**.
- Three types of blood cell: the **red blood cells**, the **white blood cells** and the **platelets**, see *Fig. 8.3* and *Fig. 8.4*. The blood cells are made in the bone marrow.

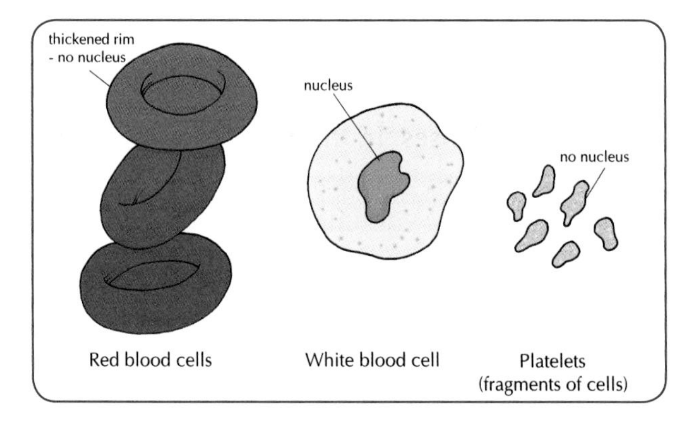

Fig. 8.3 Blood is made of three types of cell.

(1) Blood Plasma

Blood plasma is a straw-coloured liquid. It is mostly made up of water and **dissolved substances**. The blood cells float in the plasma.

- The plasma **transports** dissolved substances. These include digested foods, e.g. glucose; and gases, e.g. carbon dioxide. The plasma also carries wastes, e.g. urea; and hormones such as insulin.
- The plasma **carries heat** around the body. This helps to keep our body temperature at a constant 37°C.

The function of the plasma is to transport substances and heat.

(2) Red Blood Cells

Red blood cells get their name from the red pigment called **haemoglobin** found inside the cell. Haemoglobin picks up oxygen. In this way red blood cells **transport oxygen**.

The function of the red blood cells is to transport oxygen.

(3) White Blood Cells

White blood cells fight disease. One type of white blood cell produces **antibodies** to fight harmful bacteria, another type 'gobbles up' harmful bacteria and viruses, see *Fig. 8.5*.

The function of white blood cells is to fight disease.

(4) Platelets

Platelets are **tiny fragments** of larger cells.The platelets help the blood to **clot**. This helps seal wounds when we cut ourselves.

The function of platelets is to clot the blood.

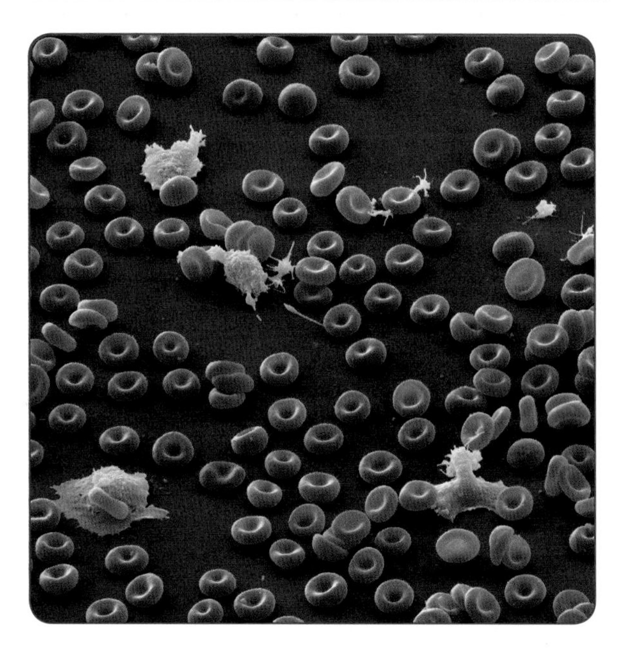

Fig. 8.4 Red and white blood cells seen under an electron microscope.

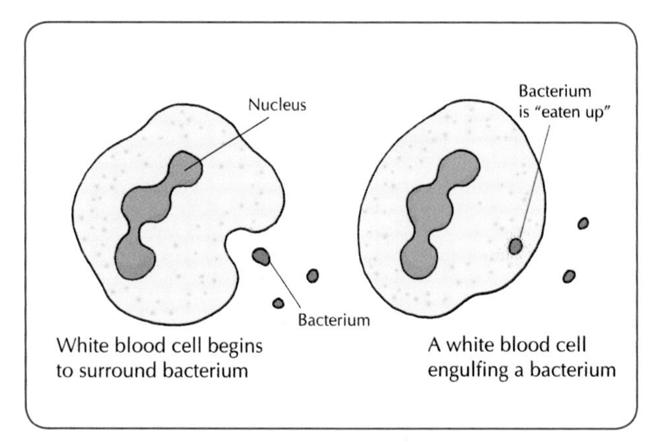

Fig. 8.5(a) A white blood cell engulfing harmful bacteria.

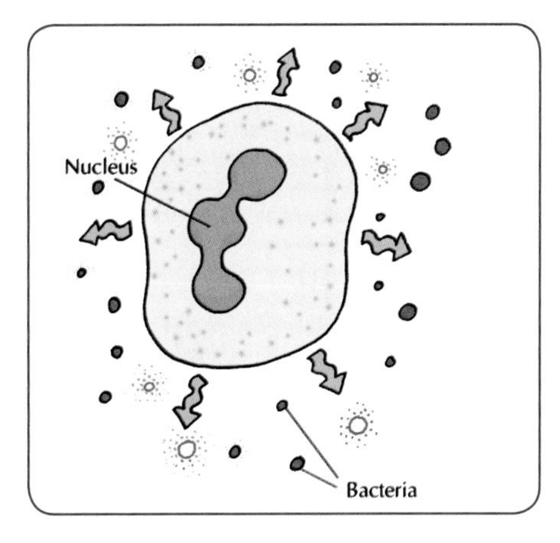

Fig. 8.5(b) A white blood cell producing antibodies to destroy bacteria.

TEST YOURSELF:
Now attempt questions *8.1–8.4* and *W8.1–W8.2*.

8.4 Blood Vessels

The three main types of blood vessel are:
(1) **arteries**, (2) **veins** and (3) **capillaries**.

(1) **Arteries** carry blood **away** from the heart.

(2) **Veins** carry blood **to** the heart.

(3) **Capillaries** are tiny blood vessels that **link** the arteries and veins, *Fig. 8.6*.

The walls of the capillaries are very thin. This allows substances such as oxygen, and glucose to pass into the cells. In turn, wastes made in the cells can pass back into the capillaries.

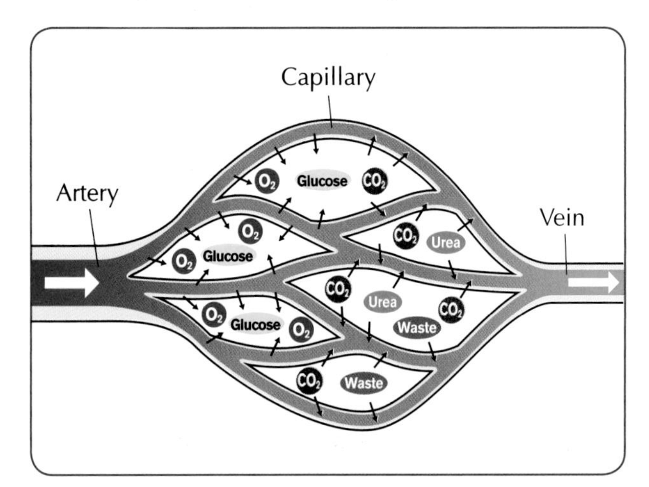

Fig. 8.6 Arteries are connected to veins by capillaries.

The differences between an artery, vein and capillary are listed in *Table 8.1*.

ARTERY	VEIN	CAPILLARY
Thick wall and small lumen (the space down which blood flows)	Thin wall with large lumen	Very thin wall (only one cell thick), tiny lumen
Blood carried under high pressure from the heart	Blood flow not under pressure	Blood flow under low pressure
No valves	Have valves to prevent backflow of blood	No valves
Carry blood **away** from the heart	Carry blood **to** the heart	**Connect** the arteries to the veins
Usually contain oxygen-rich blood	Usually contain oxygen-poor blood	

Table 8.1 Differences between arteries, veins and capillaries.

TEST YOURSELF:
Now attempt questions *8.5–8.6* and *W8.3–W8.4*.

William Harvey (1578–1657) was an English doctor. He discovered that blood flows around the body, i.e. it circulates.

8.5 The Heart

Blood is carried around the body in the arteries and veins. The **heart** is a **pump** which pushes blood into the arteries.

The heart is about the size of your clenched fist.

It lies between the lungs in the chest. The walls of the heart are made of cardiac muscle. This type of muscle never tires; it keeps working right throughout our lives.

The function of the heart is to pump blood around the body.

The Structure of the Heart

The photograph in *Fig. 8.7* shows a sheep's heart from the outside. The creamy-white material on the surface of the heart is fat. Too much fat on a heart can prevent it from working properly.

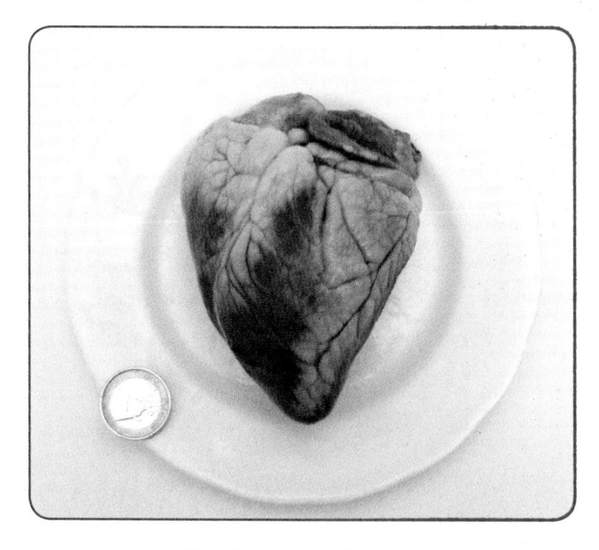

Fig. 8.7 A sheep's heart.

If the heart is cut open, its structure can be seen and the parts identified, see *Fig. 8.8*. Important points to note are:

- The heart consists of **four chambers**. Two **atria** (one atrium) at the top, one left and one right; and two **ventricles** at the bottom, one left and one right.
- The **ventricles are larger** than the atria.
- The **left ventricle wall is thicker than the right ventricle wall**. This is because blood from the left ventricle has to be pushed all around the body. Blood in the right ventricle only has to be pushed a short distance to the lungs.
- The left side of the heart is separated from the right side by a muscular wall called the **septum**.
- **Valves** in the heart stop the backflow of blood.

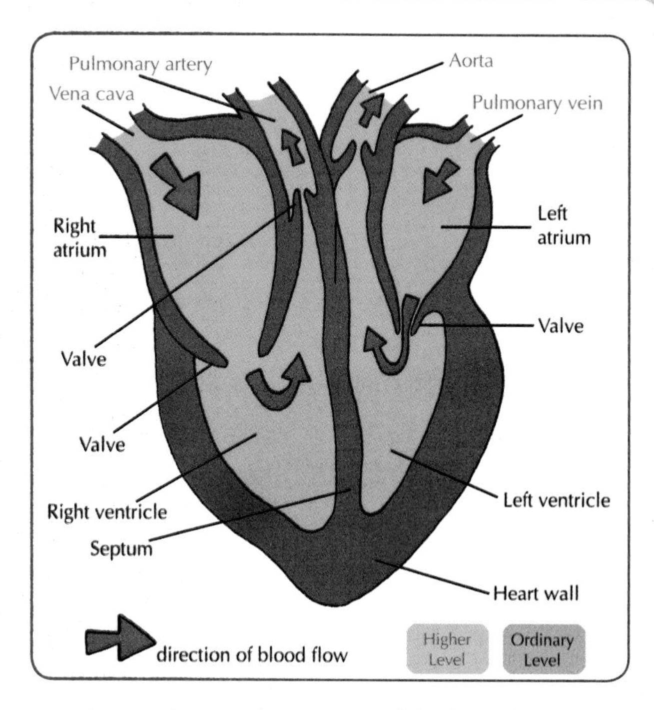

Fig. 8.8 The structure of the heart.

The main arteries and veins attached to the heart are:

- The **aorta** which carries oxygenated blood from the left side of the heart to the main organs of the body.
- The **pulmonary artery**. This vessel carries deoxygenated blood from the right side of the heart to the lungs.
- The **vena cava**. This vessel carries deoxygenated blood to the right side of the heart from the body organs.
- The **pulmonary vein**. This vessel carries oxygenated blood from the lungs to the left side of the heart.

TEST YOURSELF:
Now attempt questions *8.7–8.9* and *W8.5–W8.6.*

8.6 How the Heart Pumps the Blood

1. Blood flows through the **veins** (vena cava and pulmonary vein) into both atria at the same time.

2. Blood from the head, arms, legs and other parts of the body enters the **right atrium** through the **vena cava**. This blood is low in oxygen.

3. The muscles in the wall of the atrium contract and push the blood down into **the right ventricle**. The **valve** shuts. This stops the blood flowing back into the atrium.
4. The muscle wall of the right ventricle contracts and pushes the blood out of the heart through the **pulmonary artery**. The blood passes to the lungs where carbon dioxide is removed and oxygen is collected.
5. This oxygen-rich blood passes back to the left atrium through the **pulmonary vein**.
6. The muscles in the wall of the atrium contract and push blood down into the **left ventricle**. The **valve** shuts.
7. Finally the thick muscle wall of the left ventricle contracts and pushes the blood out of the heart, through the **aorta** to all parts of the body. *Fig. 8.9* shows the pathway of blood through the heart and the main body organs.

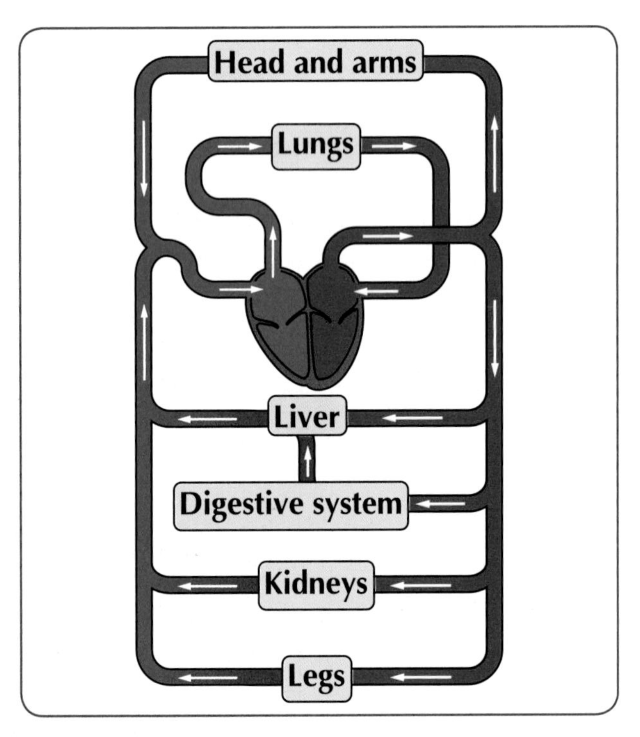

Fig. 8.9 Passage of blood through the heart and the main body organs.

The Pulse Rate

The average adult heart beats about **70 times per minute at rest**. Every time the heart beats, it forces blood into the main arteries. The pressure of this blood against the walls of arteries produces a throbbing sensation called **a pulse**. A pulse can be found at the wrist, temple and neck where arteries lie close to the surface of the skin.

The pulse rate tells us if the heart is beating at its normal rate. **Pulse rates vary** depending upon a person's age, sex, health and how fit they are. In general people who are fit have a lower pulse rate.

The Effect of Exercise and Rest on the Pulse Rate

- When we exercise, our muscles need more oxygen and food. The blood brings these things to the cells. At the same time, the cells produce more waste carbon dioxide. As a result, the heart beats faster and the pulse rate increases.
- Exercise strengthens the muscle of the heart and makes it pump more efficiently.
- After exercise the pulse rate drops back down. This happens because the need for oxygen and food by the cells decreases.
- A balance between exercise and rest is good for our health.

Experiment 8.1

To show the effect of exercise and rest on your pulse rate

To find your pulse, place the index, middle and ring fingers of one hand on the inside of your other wrist (or on your neck or temple). You should feel the pulse as a repeated throbbing. Another method is to use an electronic pulse meter, with the sensor attached to your earlobe.

Apparatus required: stopwatch/watch

Fig. 8.10 Exercise increases your pulse rate.

Method

1. Draw up a table like the one shown in *Table 8.2*.
2. Find your pulse and count the number of pulses you feel in one minute.
3. Record this figure in your table.
4. Repeat twice.
5. Calculate the average. This is your pulse rate at rest.
6. Now run on the spot for two minutes.
7. Straight away take your pulse rate.
8. Record this figure.
9. Take your pulse rate again.
10. Record this figure.
11. Take your pulse rate one last time.
12. Record this figure.
13. Draw a graph of your pulse rate against time.

TIME	PULSE RATE
At rest (first time)	
At rest (second time)	
At rest (third time)	
Average at rest	
After 1 minute of exercise	
1 min later	
1 min later again	

Table 8.2

Conclusion

Exercise causes the pulse rate to increase.

Exam Question

Junior Certificate Higher Level Examination Question

The diagram shows a person's pulse rate being taken.

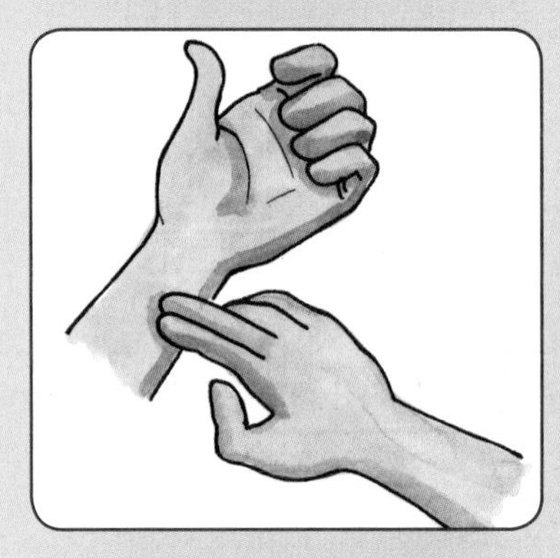

Fig. 8.11

(i) What causes a person's pulse?

A: **A person's pulse is caused by changes in blood pressure in an artery. It results from the heart pumping the blood.**

(ii) How is a person's pulse rate measured using this method?

A: **Count the number of heart beats (pulses) in one minute.**

(iii) An athlete's resting pulse rate is 58. After 10 minutes strenuous exercise their pulse rate was 120. After resting for 5 minutes their pulse rate reduced to 63. Clearly account for the rise and fall in pulse rate experienced by the athlete.

A: **The rise in pulse rate is due to the athlete needing more oxygen in their cells (to give them more energy during exercise).**

The fall in pulse rate is due to the athlete no longer needing as much oxygen (less energy needed).

TEST YOURSELF: Now attempt questions *8.10–8.11* and *W8.7–W8.8*.

What I should know

- The circulatory system is composed of the blood, the blood vessels and the heart.
- Blood is composed of a liquid called plasma and three types of cell: red blood cells, white blood cells and platelets.
- The functions of blood are transport, defence against disease and to keep body temperature constant.
- The functions of the parts of the blood are:
  (1) Plasma transports dissolved substances and carries heat around the body.
  (2) Red blood cells transport oxygen to all the cells.
  (3) White blood cells help fight disease.
  (4) Platelets clot the blood.
- Normal human body temperature is 37°C. Illness may cause a change in body temperature.
- Blood travels around the body in arteries, veins and capillaries.
- The differences between arteries, veins and capillaries can be found in *Table 8.1* on page 46.
- Valves prevent the backflow of blood.
- Capillaries link the arteries and the veins.
- The function of the capillaries is to allow materials to pass between the blood and the cells.
- The function of the heart is to pump blood around the body.
- The heart is made of four chambers: the left and right atria, and the left and right ventricles.
- Blood flows into the atria and out of the ventricles.
- The wall of the left ventricle is much thicker than the wall of the right ventricle. This is because the left ventricle pushes blood further.
- Blood enters the heart through the vena cava and the pulmonary vein.
- Blood leaves the heart through the aorta and the pulmonary artery.
- The pulse is caused by the blood being pushed along arteries near the surface of the body.
- The average pulse rate for an adult at rest is 70 beats per minute (bpm).
- The pulse and breathing rates increase during exercise because the cells of the body need more oxygen and food.
- Pulse rate can be affected by diet, age, sex, level of exercise and stress.
- A balance between exercise and rest is important for good health.

QUESTIONS

Write the answers to the following questions into your copybook.

8.1 (a) What is a circulatory system?

(b) Why do we need a circulatory system?

(c) Name the three main parts of the circulatory system.

8.2 Blood is composed of a liquid called ….., red blood cells which carry ….., ….. cells which help fight disease; and platelets which help the blood to ….. . The main functions of blood are (i) ….. , (ii) ….. and (iii) ….. . Normal human body temperature is …..°C.

8.3 (a) Name the type of blood cell labelled *A*, *B* and *C* in *Fig. 8.12*.

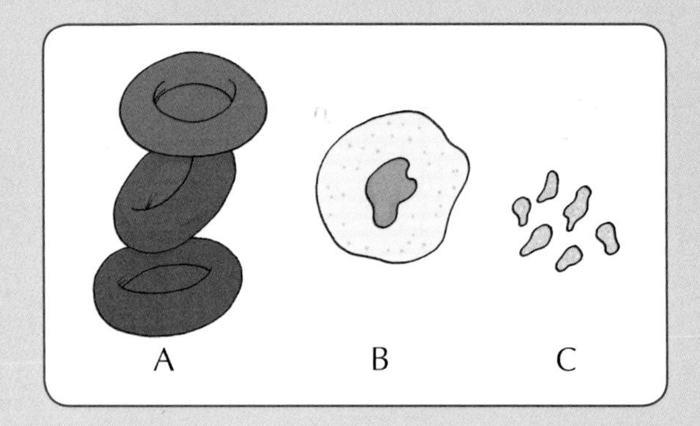

Fig. 8.12

(b) What substance gives the cells labelled *A* their red colour?

(c) List two differences between cells *A* and *B* that can be seen in the diagram.

(d) Which of the cell types is responsible for (i) helping the blood to clot, (ii) protecting us against disease and (iii) carrying oxygen to our cells?

(e) State what you think would happen to a person who had very few (i) type *A* cells and (ii) type *B* cells?

(f) (i) Name the liquid part of the blood in which these cells are found.
(ii) Name two things, other than cells, that this liquid transports.

8.4 (a) Name the liquid part of the blood.

(b) Name (i) a gas transported by the blood, (ii) a waste material transported by the blood.

(c) Name the red pigment found in the blood.

(d) Which of the following are responsible for blood clotting?

red blood cells, plasma, platelets, white blood cells

8.5 (a) What are capillaries?

(b) Where are capillaries found?

(c) Name the type of blood vessel that carries blood away from the heart?

(d) All arteries have valves. Is this true or false?

(e) What is the function of the valves in blood vessels?

8.6 (a) Name two types of blood vessel.

(b) State one way in which the types of blood vessel you named in (a) differ in structure.

(c) State one way these types of blood vessel differ in function.

QUESTIONS

8.7 Name the parts of the heart labelled A–G on the diagram of the heart shown in *Fig. 8.13.*

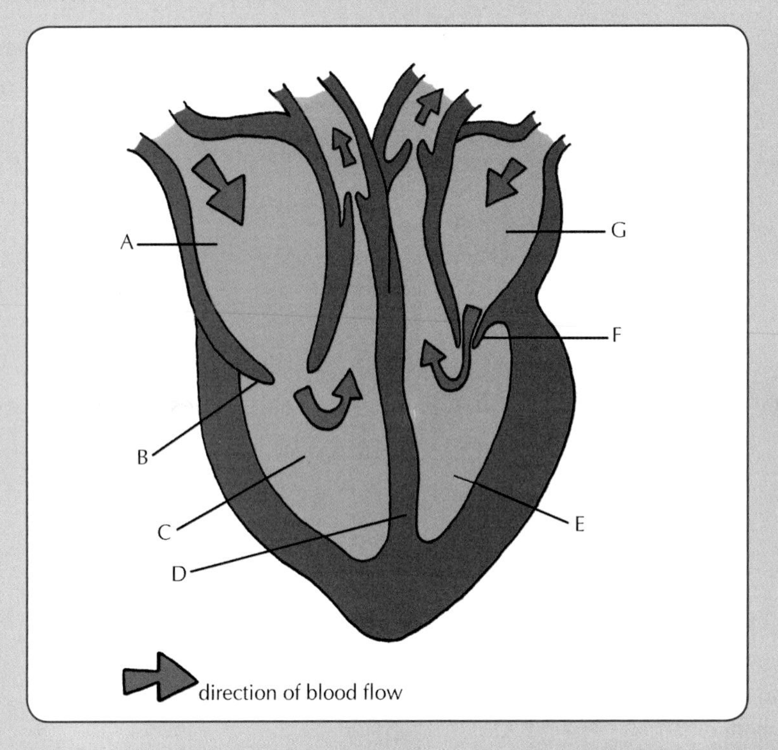

Fig. 8.13

8.8 (a) How many chambers are there in the heart? Name each one.

(b) Which chamber of the heart has the thickest wall? Give a reason for this.

(c) Name the structure that separates (i) the left side of the heart from the right side and (ii) the right atrium from the right ventricle.

(d) Where does the blood go to when it leaves the right ventricle?

8.9 There are ..... main blood vessels entering and leaving the heart. The vena cava brings blood from the ..... to the right ..... . Blood passes into the left atrium from the lungs through the ..... . The ..... is the main artery of the body. It carries blood from the .... ventricle all over the body.

8.10 Describe the passage of blood through the heart and lungs. You can use a labelled diagram as part of your answer.

8.11 (a) What causes a person's pulse?

(b) What is the normal pulse rate of an adult at rest?

(c) How is the pulse rate measured?

(d) A boy's pulse rate was 84 beats per minute at rest. He ran around the school building three times and then took his pulse rate again. Which of the following is most likely to be his new pulse rate?

(i) 40 bpm, (ii) 84 bpm, (iii) 120 bpm, (iv) 260 bpm

(e) Why does the pulse rate increase during exercise?

(f) List three things, other than exercise, that can increase the pulse rate.

Chapter 9 The Excretory System

9.1 Introduction

There are many chemical reactions going on in cells. These reactions produce wastes which could be harmful. The excretory organs remove harmful wastes from the body. The main wastes made in the body are **carbon dioxide**, **urea** and **water**.

Excretion is the removal of wastes that are made in the body cells.

The Excretory Organs

Table 9.1 summarises the main excretory organs and their waste products.

EXCRETORY ORGAN	WASTE PRODUCTS
Lungs	Carbon dioxide and water vapour
Kidneys	Urea, water and salts
Skin	Water and salts

Table 9.1 The excretory organs and products.

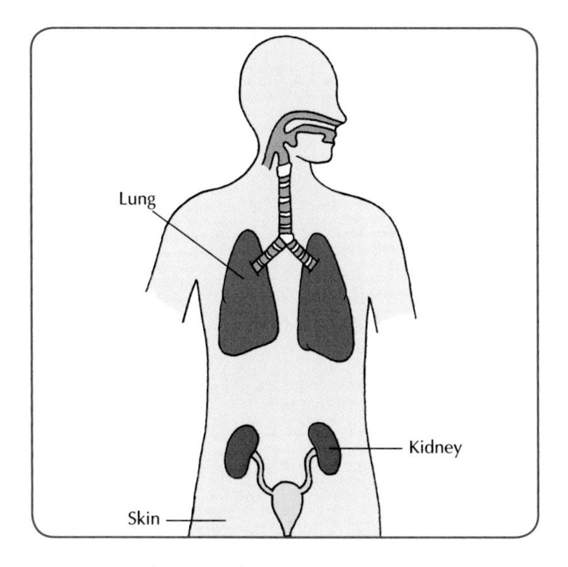

Fig. 9.1 The excretory organs.

- **Lungs**. Carbon dioxide and water are waste products of respiration in cells. The lungs get rid of carbon dioxide and water vapour. We can test for the presence of these substances by carrying out *Experiments 6.1 and 6.4*.
- **Kidneys**. The kidneys make urine. Urine is made up of urea, water and salts. The protein we eat in our diet is used to build and repair cells. But if more protein is eaten than is needed, we cannot store it. The excess **protein** is broken down in the **liver** to form urea. Urea is a waste chemical.
- **Skin**. The skin excretes excess water and salts. These wastes are excreted from the skin as sweat.

TEST YOURSELF:
Now attempt questions *9.1–9.2* and *W9.1*.

9.2 The Urinary System

The human urinary system (*Fig. 9.2*) is made up of the following parts:

- the kidneys
- the renal arteries and the renal veins
- the ureters
- the bladder
- the urethra

The function of the urinary system is to get rid of urea, excess water and salts.

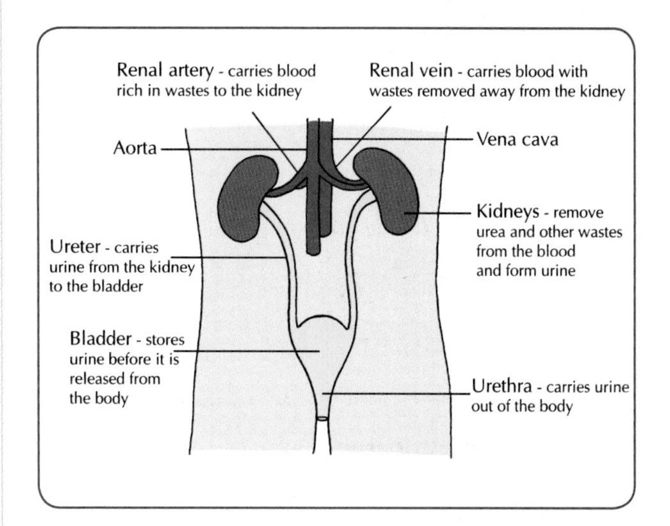

Fig. 9.2 The urinary system, its parts and functions.

BIOLOGY

9.3 How the Kidneys Make Urine

Blood enters the kidneys through the renal arteries. This blood contains many useful substances such as glucose, in addition to the wastes like urea.

The body wants to get rid of the wastes but keep the useful substances. So, the main function of the kidneys is to **filter** the blood. In this way the urea, excess water and salts are converted into urine. Glucose and other useful substances are taken back into the bloodstream and are kept in the body.

Urine is made up of urea, water and salts.

The urine travels from the kidneys, through the ureters to the bladder. The bladder stores the urine before it is released from the body.

TEST YOURSELF:
Now attempt questions 9.3–9.6 and *W9.2–W9.4.*

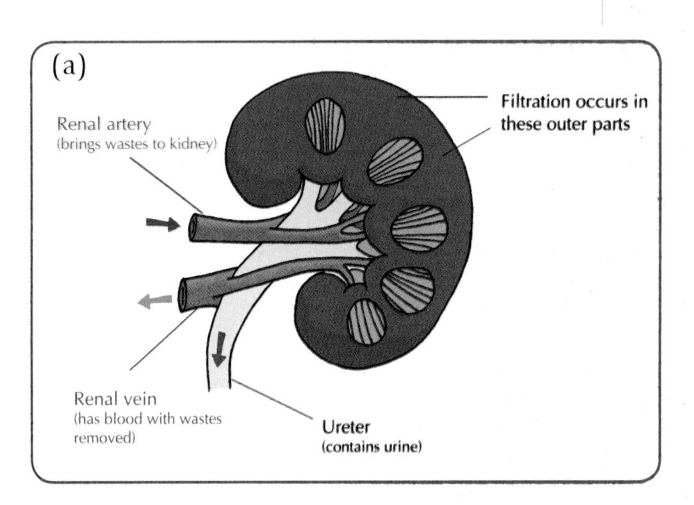

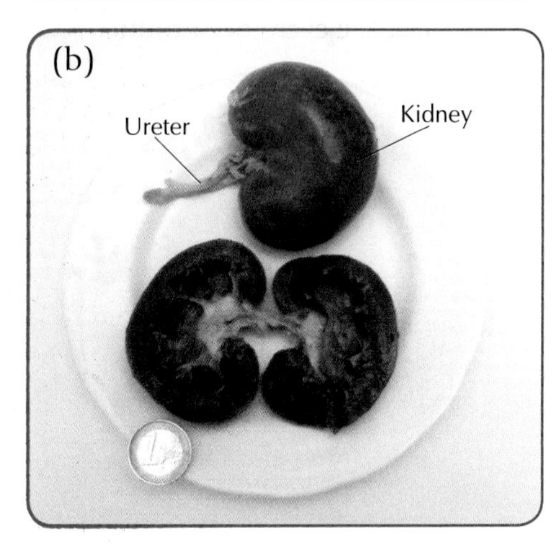

Fig. 9.3(a) The kidney filters the blood to make urine. (b) A lamb's kidney cut lengthways (L.S.).

What I should know

- Excretion is the removal of the wastes that are made in the body's cells.
- The excretory organs are the lungs, the kidneys and the skin.
- The excretory products of the lungs are carbon dioxide and water vapour.
- The excretory products of the kidneys are urea, water and salts.
- The excretory products of the skin are excess water and salts.
- The urinary system consists of the kidneys, the ureters, the bladder, the urethra and the renal arteries and veins.
- The kidneys form urine by filtering the blood.
- Urine is made of urea, water and salts.
- Urine is stored in the bladder.

Questions

Write the answers to the following questions into your copybook.

9.1 (a) Excretion is getting rid of ..... made by the ..... reactions in the ..... of the body.

(b) Our waste products are removed from the body by the ..... system.

(c) Three of the excretory organs are the ....., the ..... and the ..... .

QUESTIONS

(d) The excretory products of the kidneys are ….., ….. and ….. .

(e) Carbon dioxide and ….. are excreted through the ….. .

9.2 (a) Name the substances present in urine.

(b) Name a chemical you could use to test for the presence of water vapour in the breath you breathe out?

(c) Which of the following are not a waste product of humans?

(i) carbon dioxide, (ii) water, (iii) oxygen, (iv) salts

(d) Name three substances excreted by the human body.

(e) What is urea?

(f) List the excretory product(s) common to the kidneys, lungs and skin.

9.3 Look at *Fig. 9.4* which shows the urinary system. There are parts labelled *A*, *B*, *C* and *D*.

(a) Which labelled part is the kidney?

(b) Which labelled part is the bladder?

(c) Which labelled part brings blood to the kidney?

(d) Name the part of the diagram that connects the kidney to the bladder.

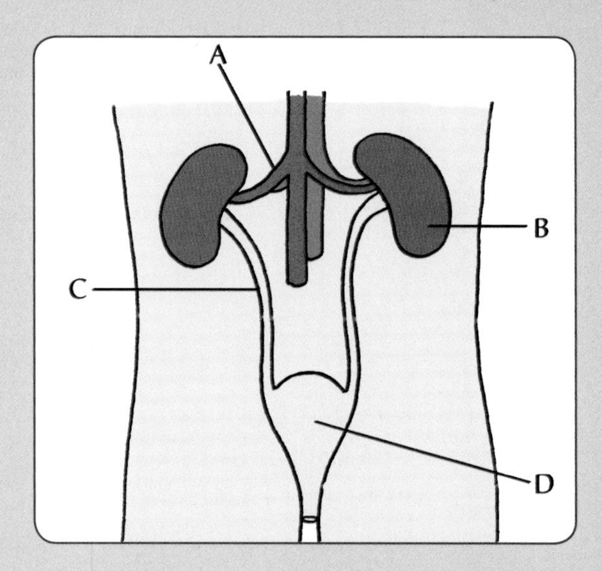

Fig. 9.4

9.4 (a) Urea is made in the ….. . It is carried to the kidney in the ….., where it combines with water and salts to form ….. .

(b) The function of the renal artery is to ….. wastes to the ….. .

(c) The function of the kidney is ….. .

(d) The bladder stores ….. .

(e) Urine is made in the ….. and passes out of the body through the ….. .

(f) Waste products are removed from the ….. by ….. in the kidneys.

(g) The ….. transport urine from the kidney to the ….. .

9.5 Describe how the kidneys form urine.

9.6 Distinguish between the ureter and the urethra.

Chapter 10 The Skeletal and Muscular Systems

10.1 The Skeleton

The skeleton is made of bones. In fact there are more than **200 bones** in the human skeleton. The main bones are shown in *Fig. 10.1*.

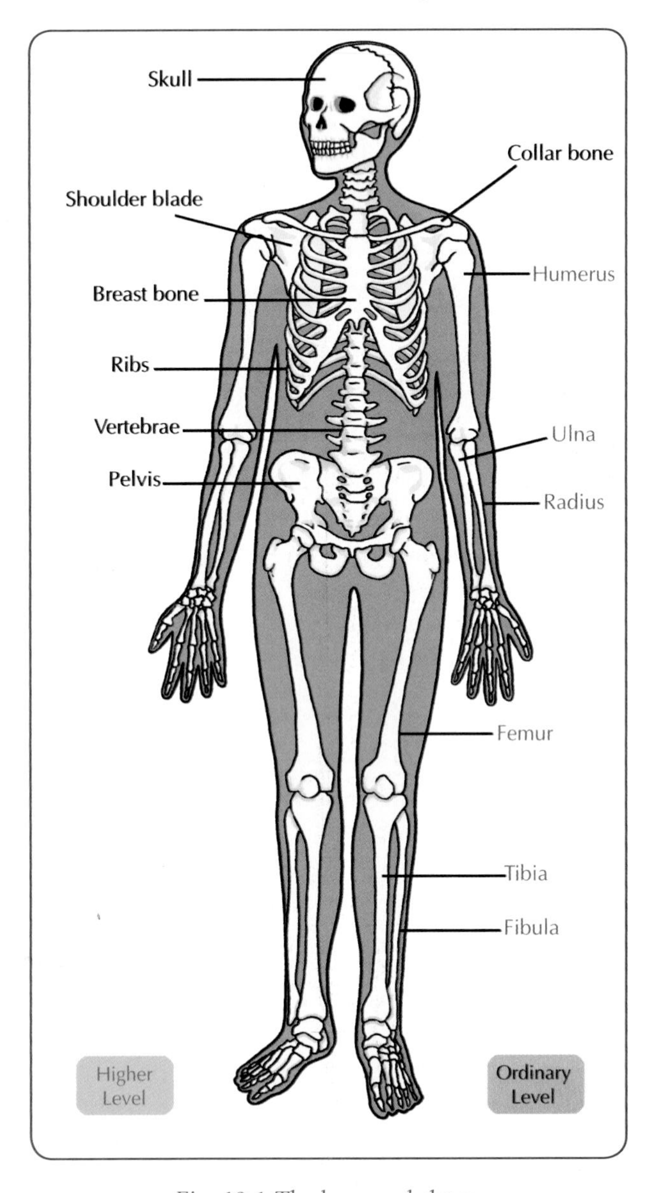

Fig. 10.1 The human skeleton.

10.2 The Functions of the Skeleton

The three main functions of the skeleton are: (1) **Support**, (2) **Movement**, (3) **Protection**.

(1) Support

The skeleton supports the body and gives it shape. If the skeleton were removed, the human body would collapse, see *Fig. 10.2*.

Fig. 10.2 The body without bones.

(2) Movement

Our bones together with our muscles allow us to move about. Muscles are attached to our bones. When these muscles contract (shorten), they pull on the bones causing movement.

(3) Protection

Many of the bones act as a protection for the soft parts of the body.

- The **skull** protects the brain; the **rib cage** protects the heart and lungs.
- The bones of the **backbone** protect the spinal cord.

TEST YOURSELF: Now attempt questions *10.1–10.3* and *W10.1–W10.2*.

10.3 Bone

Bone is a living tissue. It contains special bone cells, a blood supply and nerves. Because bone is living, it can grow and repair itself.

Bone also contains the mineral calcium. **Calcium makes bone very hard** which enables it to be so strong. This is why it is important to include calcium-rich foods such as milk and dairy products in our diet.

Fig. 10.3 Milk is an excellent source of calcium for our bones.

TEST YOURSELF:
Now attempt question *10.4* and *W10.3.*

10.4 Joints and Movement

Bones are connected to one another by structures called **joints**. Muscles attached across a joint allow the bones at the joint to move. Without the muscles, movement would not be possible.

> **A joint is formed where two or more bones meet.**
>
> **In general the function of joints is to allow movement.**

Three joints are described here:

(1) Fused Joints

A fused joint occurs when there is **no movement** between the bones that meet.

- For example, the bones of **the skull** are 'locked' together and they do not move in relation to each other, *Fig. 10.4*.

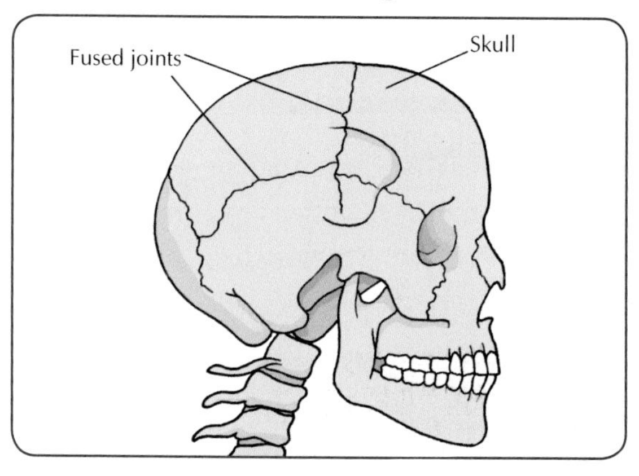

Fig. 10.4 The skull bones have fused joints.

(2) Ball and Socket Joint

A ball and socket joint is a moveable joint. It allows movement of the bones in **many directions**. Because the joint is cushioned by synovial fluid, it is also known as a synovial joint.

- A ball and socket joint is found at the **hip**, between the hip bone (socket) and the top of the femur (ball), *Fig. 10.5*.
- The joint at the **shoulder**, between the shoulder blade and the top of the humerus, is another example of a ball and socket joint.

Fig. 10.5 A ball and socket joint is found at the hip.

(3) Hinged Joint

A hinged joint is also a synovial joint. This type of joint allows movement in only **one direction**, *Fig. 10.6*.

- Hinged joints are found at the **knee**, between the femur and the tibia.
- Another example is at the **elbow** between the end of the humerus and the ulna.

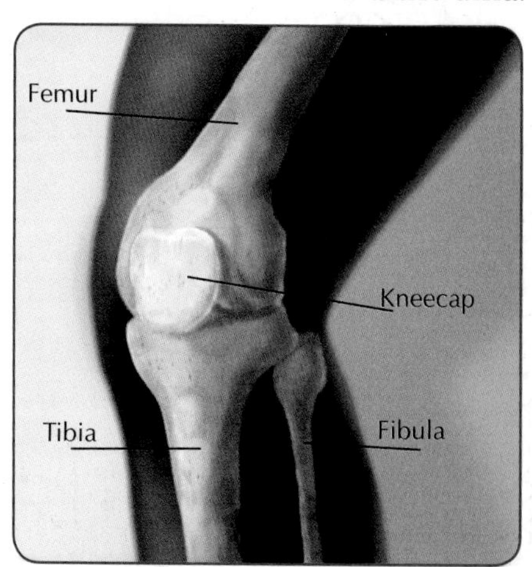

Fig. 10.6 A hinged joint is found at the knee.

10.5 The Structure of a Moveable Joint

The bones at a joint are held together by tough elastic fibres called **ligaments**. The bones are prevented from rubbing off each other in the following two ways:

- A pad of **cartilage**.
- A lubricating fluid called **synovial fluid**.

Together the cartilage and the synovial fluid help in the smooth movement of the joint. The structure of a typical moveable joint can be seen in *Fig. 10.7*.

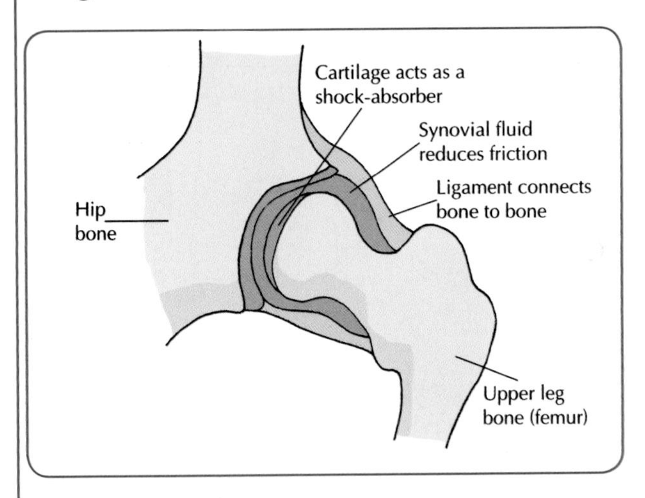

Fig. 10.7 A typical moveable joint.

Ligaments connect bone to bone.

TEST YOURSELF:
Now attempt questions *10.5–10.7* and *W10.4–W10.5*.

10.6 Muscles and Movement

On their own, bones do not cause movement. Muscles are needed to enable bones to move.

Muscles are attached to bones by non-elastic fibres called **tendons**. A tendon is really an extension of a muscle. *Fig. 10.8* shows the calf muscle attached to the heel bone by tendons.

Tendons connect muscle to bone.

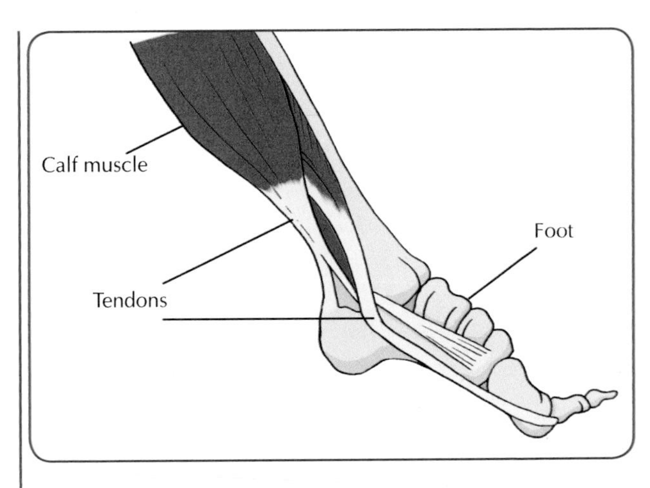

Fig. 10.8 Tendons attach the muscles to the bones.

Antagonistic Muscles

When muscles **contract** they pull on the skeleton and cause the bones at a joint to move. But muscles can only get shorter, they cannot get longer. This means that the muscle which causes a bone to move at a joint cannot straighten that bone again. There has to be another muscle which can pull in the opposite direction.

Pairs of muscles which work opposite one another are called **antagonistic pairs of muscles**.

Biceps and Triceps

An example of an antagonistic pair of muscles is the **biceps** and **triceps** muscles of the upper arm. These muscles raise and lower the lower arm. This can be seen in *Fig. 10.9*.

- When the **biceps contracts**, it pulls on the radius and the forearm (lower arm) is pulled up. At the same time, and the triceps muscle relaxes.
- If the **triceps** now **contracts**, it pulls on the ulna causing the forearm to straighten out. At the same time, the biceps relaxes.

In each case when one muscle is contracting, the other is relaxing.

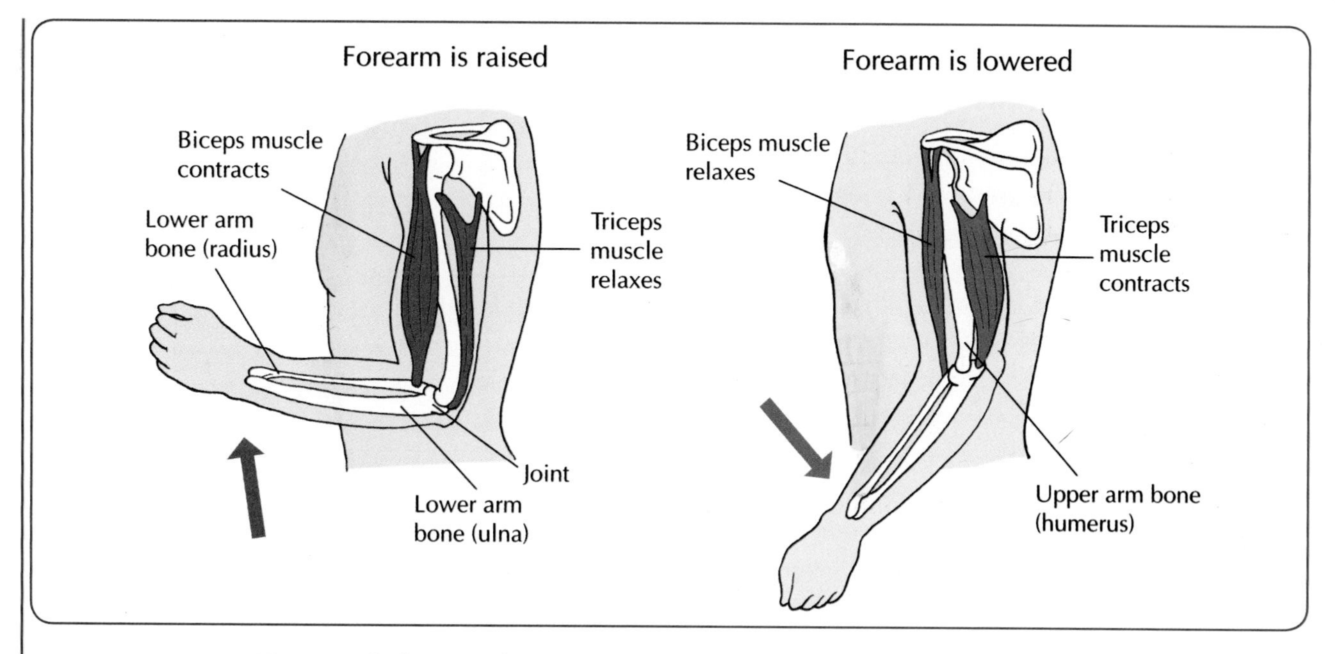

Fig. 10.9 The biceps and triceps muscles pull the lower arm in opposite directions.

TEST YOURSELF:

Now attempt questions *10.8–10.9* and *W10.6–W10.8.*

What I should know

- The functions of the skeleton are support, movement and protection.
- The bones work with the muscles to bring about movement.
- The skull protects the brain, the ribcage protects the lungs and heart, the backbone protects the spinal cord.
- The main bones of the body include the skull, ribs, vertebrae, collarbone, shoulder blade, pelvis, humerus, radius, ulna, femur, tibia and fibula, see *Fig. 10.1* on page 56.
- Bone is a living tissue.
- Calcium is an essential mineral needed to keep bones strong.
- A joint is formed where two or more bones meet.
- Two types of joint are: fused and moveable joints.
- A fused joint does not allow any movement of the bones, e.g. in the skull.
- A moveable joint allows bones to move, e.g. ball and socket joint, hinged joint.
- Ball and socket joints allow movement in many directions.
- Ball and socket joints are found at the shoulder and at the hip.
- Hinged joints allow movement in only one direction.
- Hinged joints are found at the elbow and knee.
- Cartilage prevents the bones from rubbing against each other.
- A fluid lubricates and cushions the joint.
- Ligaments connect one bone to another.
- Muscles contract and relax to bring about movement of bones.
- Tendons connect muscle to bone.
- Antagonistic muscles work opposite each other to bring about movement.
- The biceps and triceps are examples of antagonistic muscles.

QUESTIONS

Write the answers to the following questions into your copybook.

10.1 (a) What is the skeleton made of?

(b) Name three functions of the skeleton.

(c) Name an organ of the body that is protected by the skull.

(d) Name an organ of the body that is protected by the ribcage.

10.2 Name the bones of the skeleton labelled *A–I* in *Fig. 10.10*.

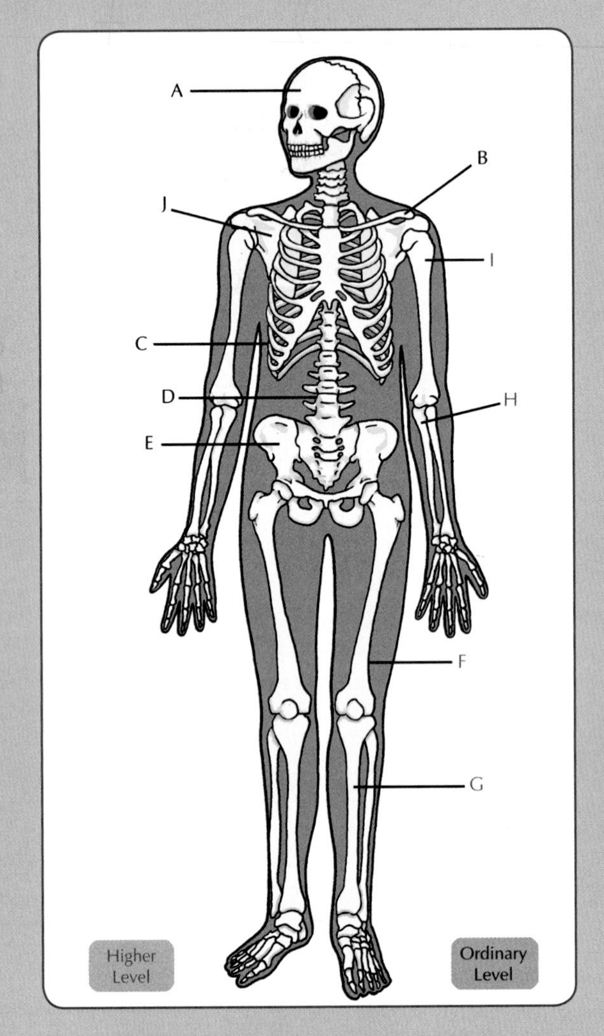

Fig. 10.10

10.3 (a) Give another name for the upper arm bone.

(b) How many bones are there in the arm?

(c) What name is given to the bone that lies on the thumb side of the lower arm.

(d) What is the name of the upper leg bone?

(e) Where in the body would you find the tibia?

(f) Give another name for the hips.

10.4 (a) Bone is a ….. tissue.

(b) A tissue is a group of similar ….. that have the same …… .

QUESTIONS

(c) An important mineral needed for making bones is ….. .

(d) ….. is a good source of calcium which is needed for making strong bones.

10.5 (a) What is a joint?

(b) Name two types of joint in the human body.

(c) Give one difference between the types of joint you named in (b).

(d) Name the type of joint found at the elbow.

10.6 (a) Muscles are needed to help bones to ….. .

(b) A ….. connects bone to bone.

(c) ….. at a joint stops the bones rubbing off each other.

(d) A ….. joint is found at the knee. It allows movement in ….. direction.

(e) A ….. and ….. joint is found at the shoulder. It lies between the ….. blade and the ….. .

10.7 Label the parts *A–D* of the moveable joint shown in *Fig. 10.11*. Describe the function of each labelled part.

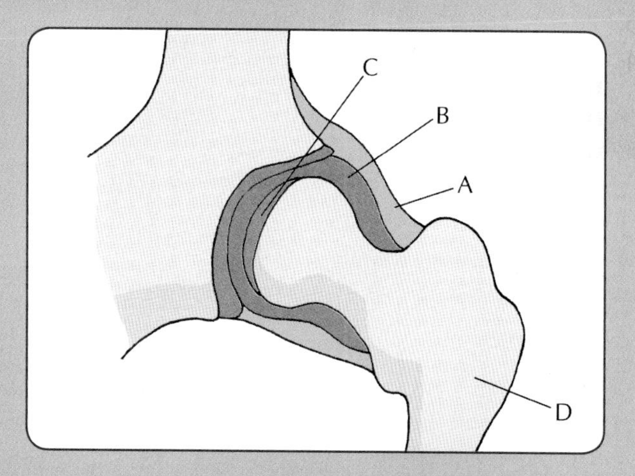

Fig. 10.11

10.8 (a) What is a tendon?

(b) What is a ligament?

(c) What do antagonistic muscles do?

(d) Name a pair of antagonistic muscles.

10.9 (a) Muscles can only get shorter. True or false?

(b) Name the muscles of the upper arm and describe how they work to raise and lower the forearm.

BIOLOGY

Chapter 11 The Sensory and Nervous Systems

11.1 Introduction

Animals must be able to notice changes in their surroundings. This will allow them to avoid danger or to find food. For example, if you accidentally touch a hot cooker, you immediately pull your hand away.

- The change to which the body responds is called the **stimulus**, i.e. the heat.
- The reaction is the **response**, i.e. taking your hand away.

Many other examples illustrate the importance of being able to respond to stimuli, such as the urge to drink water when you are thirsty, or feeling hungry when you smell nice food. Many animals hibernate as the days shorten in autumn and birds fly south to warmer climates.

11.2 The Senses

Humans use their sense organs to gather information from their surroundings.

- The five main sense organs are the **ears**, **eyes**, **nose**, **skin** and **tongue**.
- The five senses are **sound**, **sight**, **smell**, **touch** and **taste**. The information we receive through our sense organs is passed to the brain via nerves. The brain sorts out the information.

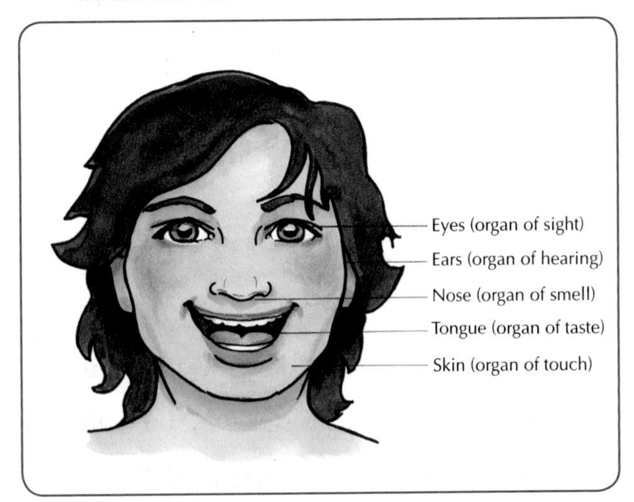

Fig. 11.1 Our five sense organs.

TEST YOURSELF:
Now attempt questions *11.1–11.2* and *W11.1–W11.2.*

11.3 The Eye

We are going to learn more about the eye as an example of one of our sense organs. The eyes allow us to detect light and to distinguish colours. To do this, the eyes have special **sensory cells** which are found on a part of the eye called the **retina**.

1. The sensory cells change light energy into **nerve impulses**.
2. These nerve impulses pass from the retina to the brain along the **optic nerve**.
3. The brain **sorts out** the impulses so that we see things.

Structure of the Eye

When you look at a person's eye, you really only see a small part of the front of it, see *Fig. 11.3*. Most of the eye is hidden, protected by a bony socket of the skull. The structure of the eye can be seen in *Fig. 11.2*.

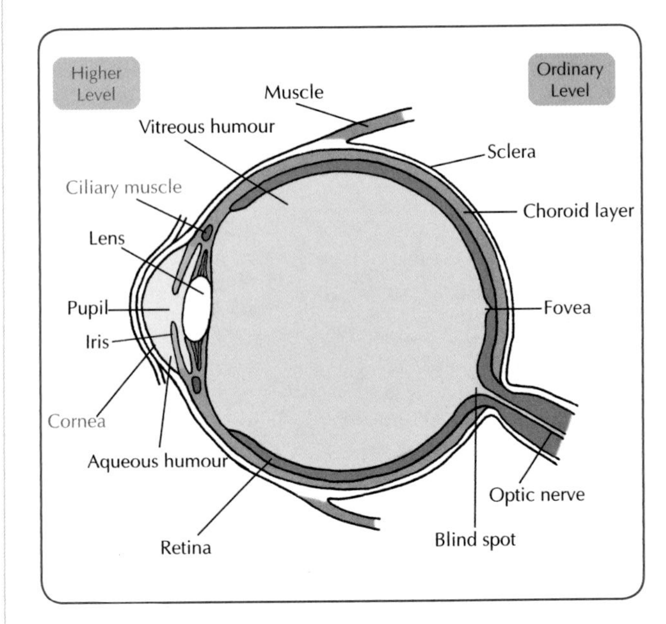

Fig. 11.2 The structure of the eye.

The eye is made up of three main layers:
(1) the **sclera**, (2) the **choroid**, (3) the **retina**.

(1) The Sclera

The **sclera** is the outermost layer of the eye. It is white in colour, except at the very front where it becomes transparent. The sclera has a protective function.

- The transparent front of the eye is called the **cornea**. The cornea allows light into the eye and it also bends (focuses) the light.

(2) The Choroid

The **choroid** is the middle layer of the eye. It has a dark colour and has the blood vessels to nourish the eye. The choroid stops light bouncing around inside the eye.

- At the front of the choroid there is a ring of muscle called the **iris**. The iris is the coloured part of the eye. It controls the amount of light getting into the eye.
- In the middle of the iris there is a hole called the **pupil**. Light passes into the eye through the pupil. *Fig. 11.3* shows how the pupil changes size in bright and dim light.

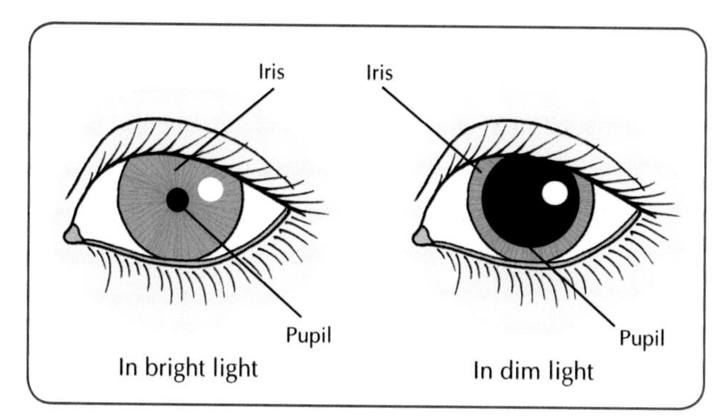

Fig. 11.3 The iris changes the size of the pupil.

- Behind the iris lies the **lens**. The lens focuses light onto the retina.
- The lens is held in place by the **ciliary muscles**. These muscles change the shape of the lens and allow us to see things either close-up or far away.

(3) The Retina

The **retina** is the innermost layer of the eye. It contains the light receptor cells and the nerves.

- The receptor cells detect light. They pass the information to the brain along the **optic nerve**.
- There are no light receptor cells on the part of the retina where the optic nerve leaves the eye. This means that when light falls on this area no messages are sent to the brain. For this reason it is called the **blind spot**.

Inside the Eyeball

Inside the eyeball there are two liquids:

- The **aqueous humour** in the front of the eye.
- The **vitreous humour** at the back.

These liquids help focus the light entering the eye and give shape to the eyeball.

Table 11.1 summarises the main parts of the eye and their functions.

PART OF THE EYE	FUNCTION
Iris	Controls the amount of light entering the eye.
Lens	Focuses light onto the retina.
Optic nerve	Carries nerve impulses from the eye to the brain.
Pupil	Allows light to enter the eye.
Retina	Has the light receptor cells/ detects light.
Ciliary Muscle	Changes the shape of the lens.
Cornea	Allows light into the eye and bends the light rays.

Table 11.1 Functions of parts of the eye.

TEST YOURSELF:
Now attempt questions *11.3–11.5* and *W11.3–W11.4.*

11.4 The Nervous System

Our nervous system receives information from the senses and organises the body's responses to it.

The nervous system consists of the following:

- The **central nervous system** (CNS).
- The **nerves** that lead to and from the CNS.

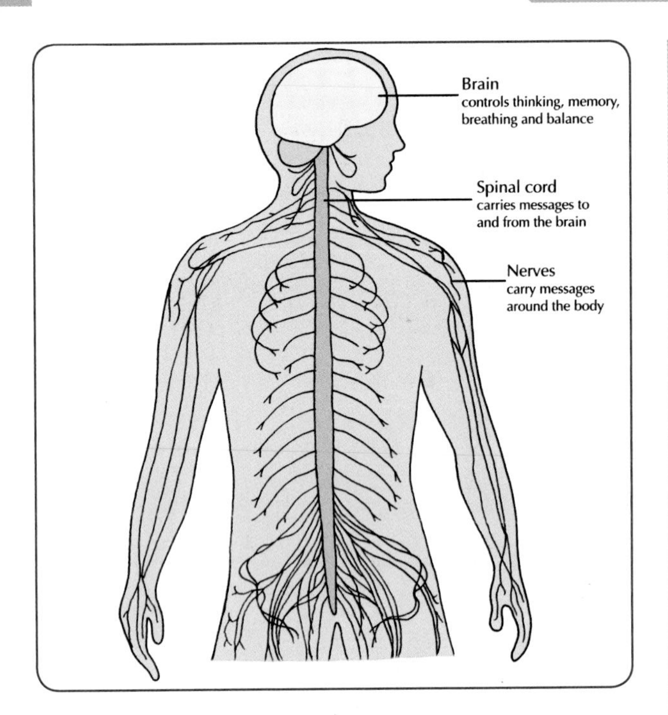

Fig. 11.4 The nervous system.

The Central Nervous System

The central nervous system consists of the **brain** and the **spinal cord**, *Fig. 11.5*. The brain is protected by the skull and the spinal cord is protected by the vertebrae of the backbone.

Messages from the senses are brought to the CNS for interpretation. The brain is the most complex organ of the body. It is made up of millions of special nerve cells.

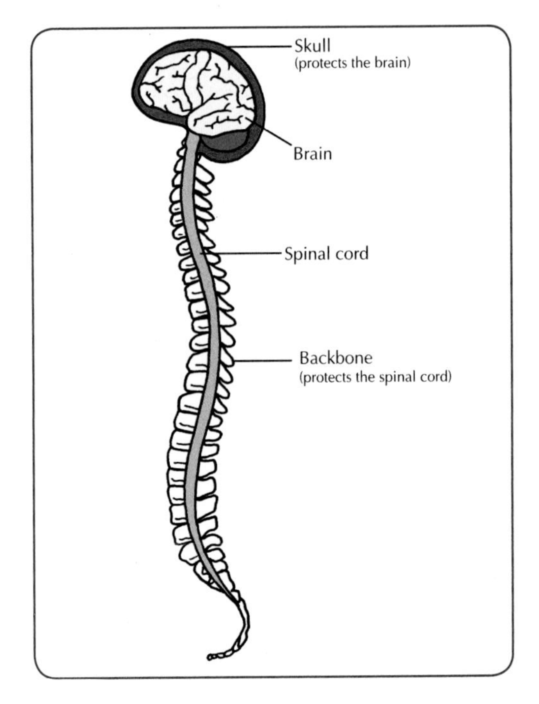

Fig. 11.5 The central nervous system is made up of the brain and the spinal cord.

The Role of the CNS

The role of the brain:

- To **receive information** from the sense organs.
- To **send a response** to the muscles and glands.
- To be responsible for **memory**, **intelligence** and our **emotions**.

The role of the spinal cord:

- To **bring information** from the sense organs to the brain.
- To **relay information** from the brain to the muscles and glands of the body.

The Nerves

A nerve is a structure made up of bundles of special nerve cells called **neurons**. Electrical signals or impulses pass along these cells to and from the brain. The function of our nerves is to carry messages to and from the CNS.

There are two different types of nerve – (1) **sensory nerves** and (2) **motor nerves**.

(1) Sensory nerves carry impulses **to the brain** and **spinal cord**. The optic nerve in the eye is an example of a sensory nerve.

(2) Motor nerves carry impulses **from the brain** and **spinal cord** to the muscles and glands.

The sensory and motor nerves co-operate to bring about a nervous response.

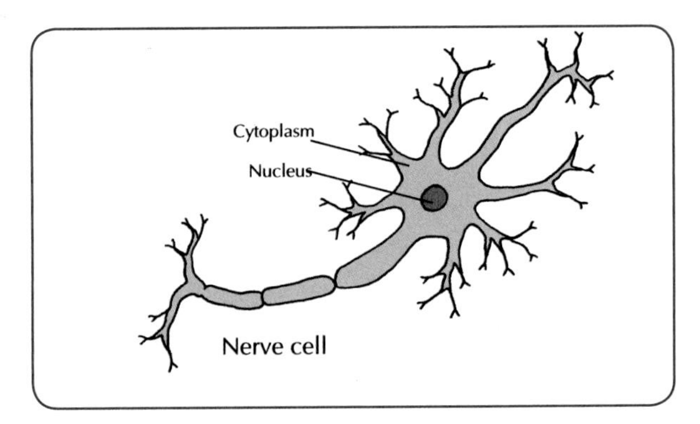

Fig. 11.6 A nerve cell.

TEST YOURSELF:
Now attempt questions *11.6–11.8* and *W11.5–W11.7.*

What I should know

- In order to survive, animals must be sensitive to changes in their surroundings.
- Our sense organs help us to be aware of and respond to changes in our surroundings.
- The five senses are: the ears, the eyes, the nose, the skin and the tongue.
- Our senses allow us to gather information from our surroundings, as shown in *Table 11.2* below.

SENSE ORGAN	STIMULUS TO WHICH IT RESPONDS	SENSE
Ear	Vibrations	Hearing
Eye	Light	Vision
Nose	Chemicals in solution	Smell
Skin	Heat, pain, pressure	Touch
Tongue	Chemicals in solution	Taste

Table 11.2 The sense organs and their functions.

- The eye is made up of three layers: the sclera, the choroid and the retina.
- The parts and function of the eye are shown in *Fig. 11.2* on page 62 and *Table 11.1* on page 63.
- The nervous system consists of the central nervous system (CNS) and the nerves.
- The CNS consists of the brain and the spinal cord.
- The role of the brain includes: receiving and responding to information from the senses; being responsible for memory, intelligence, emotions and making decisions.
- The role of the spinal cord is to relay information to and from the brain.
- Nerves branch out from the brain and spinal cord.
- There are two types of nerve: sensory nerves and motor nerves.
- The function of the sensory nerves is to carry messages **to** the CNS.
- The function of the motor nerves is to carry messages **from** the CNS.

QUESTIONS

Write the answers to the following questions into your copybook.

11.1 (a) Why is it important that organisms be able to respond?

(b) Describe how animals gather information from their surroundings.

(c) The little boy jumped when he saw the monster. Which is the (i) stimulus and (ii) the response?

11.2 (a) Name the five senses.

(b) Name the sense organ that detects sound.

(c) Name the sense organ that detects heat.

(d) Name the sense organ that detects light.

(e) Name the sense organ that detects curry flavoured chips.

(f) Name the sense organ that detects different smells.

11.3 Name the parts of the eye labelled *A–K* in *Fig. 11.7*.

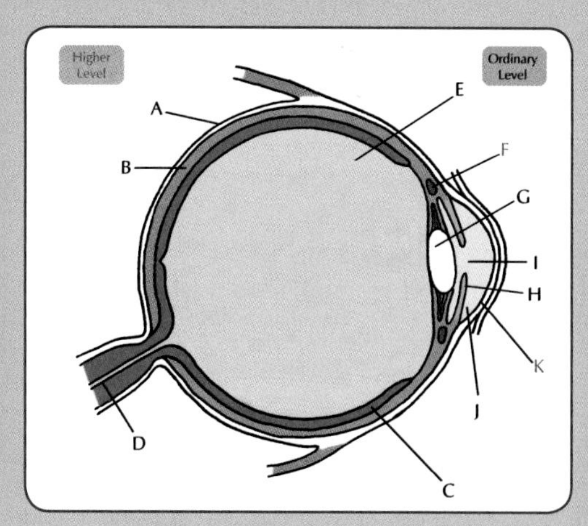

Fig. 11.7

QUESTIONS

11.4 (a) What is the coloured part of the eye called?

(b) Name the part of the eye which carries messages to the brain.

(c) Name the part of the eye which focuses light onto the retina.

(d) Which layer of the eye detects light?

(e) What is the pupil of the eye?

(f) What is the function of the pupil?

(g) What is the function of the cornea?

(h) State the role of the ciliary muscle.

11.5 (a) Why is the cornea of the eye transparent?

(b) What part of the eye controls the amount of light that enters the eye.

(c) Explain how the eyeball can keep its shape.

(d) How does the ciliary muscle bring light to focus on the retina?

11.6 Use the following terms to complete the sentences below:

central nervous system, skull, interpret, senses, nerves, brain, gathers, spinal cord

(a) The nervous system ..... information from the ..... .

(b) The nervous system is made up of the ..... and the ..... .

(c) The CNS is made up of the ..... and the ..... .

(d) The role of the CNS is to ..... messages from the senses.

(e) The brain is protected by the ..... .

11.7 (a) Label the parts *A* and *B* on *Fig. 11.8*.

(b) Give two functions of the part labelled *A*.

(c) Give two functions of the part labelled *B*.

(d) Name the part of the skeleton that protects the part labelled *B*.

(e) Name the part labelled *C*.

(f) Which two parts of the diagram represent the central nervous system?

11.8 The brain, spinal cord and the nerves make up the nervous system.

(a) Name the bone that protects the brain.

(b) Name the bones that protect the spinal cord.

(c) What do the letters CNS stand for?

(d) Which letters on *Fig. 11.8* represent the CNS?

(e) Describe the role of the CNS.

(f) What is a nerve?

(g) Give one difference between the function of a sensory nerve and a motor nerve.

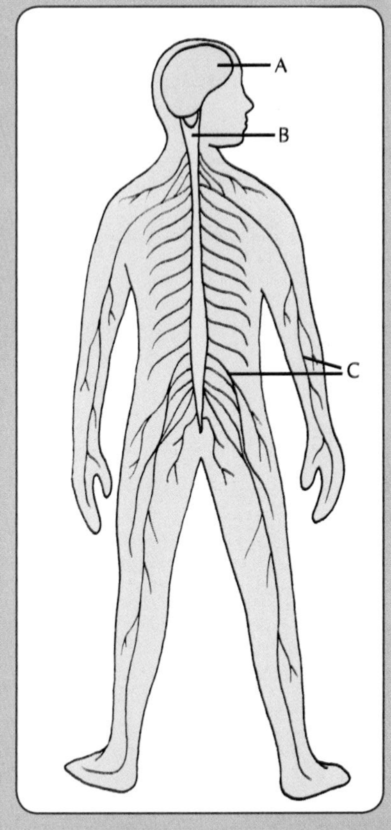

Fig. 11.8

Chapter 12 The Reproductive System and Genetics

12.1 Introduction

Reproduction is one of the characteristics of all living things. It means being able to produce **new individuals**. Reproduction ensures that particular species survive and do not become extinct.

Humans reproduce by a method known as **sexual reproduction**. Sexual reproduction always involves **two parents**. Each parent produces special sex cells called **gametes**.

- The female gamete is called the **egg**.
- The male gametes are the **sperm**.

A gamete is a sex cell.

During reproduction the male and female gametes join together and form a **zygote**. This process is called **fertilisation**. The zygote then grows into a new individual.

Fertilisation is the fusion between male and female gametes to produce a zygote.

12.2 The Male Reproductive System

The human male reproductive system consists of special reproductive organs, which are shown in *Fig. 12.1* and *Fig. 12.2*.

The role of the male reproductive system:

- To produce sperm.
- To deliver the sperm to the female body.

The Parts of the Male Reproductive System

- The **testicles (testes)** produce millions of **sperm** cells. The testicles also produce chemicals called **hormones**. The male sex hormones control the production of sperm and the development of the reproductive organs.

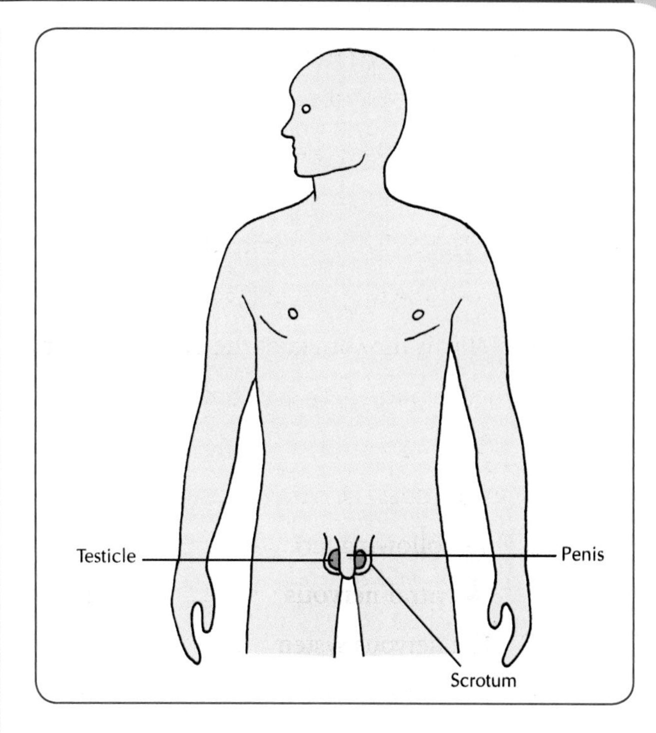

Fig. 12.1 The male reproductive organs.

- The **scrotum (scrotal sac)** is a fold of skin in which the testes lie. The scrotum holds the testes outside the body, at a temperature slightly lower than body temperature. This temperature is more suitable for sperm to be produced.
- The **sperm duct** carries sperm from the testes to the urethra which runs through the penis.
- The **seminal vesicles** and the **prostate gland** secrete **fluids** which nourish the sperm and help them to swim. The mixture of sperm and these fluids is known as **semen**.
- The **penis** is made of a spongy tissue. When blood fills this tissue, the penis becomes hard and erect. This allows the penis to be placed inside the woman's body and transfer semen during sexual intercourse.

A summary of the parts and functions of the male reproductive organs can be found in *Table 12.2* in the summary at the end of this chapter.

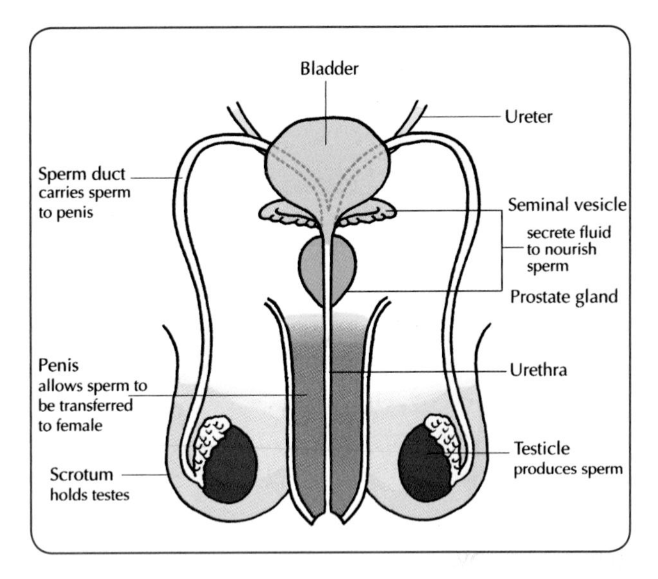

Fig. 12.2 The male reproductive system; its parts and functions.

12.3 The Female Reproductive System

The female reproductive organs are found inside the body, see *Fig. 12.3*.

The role of the female reproductive system:

- To produce eggs.
- To hold and nourish the developing baby during pregnancy.
- To give birth to the baby.

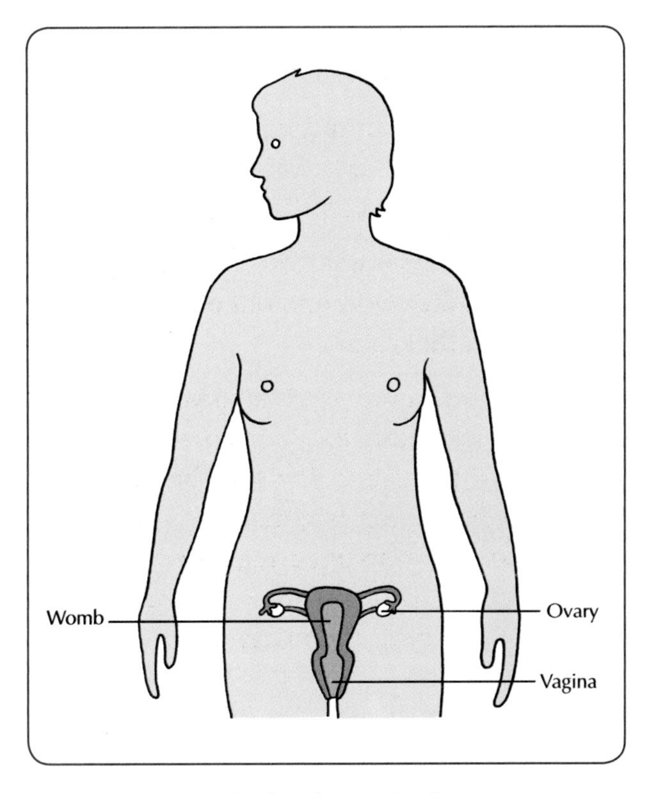

Fig. 12.3 The female reproductive organs.

The Parts of the Female Reproductive System

- There are two **ovaries**. They produce the **eggs**. Usually one egg is produced from alternate ovaries each month. The ovaries also produce female sex **hormones**. These hormones control the female reproductive system.
- The **fallopian tubes** (oviducts) carry the egg from the ovary to the womb. The fallopian tube is the place where **fertilisation** occurs.
- The **womb** (uterus) is the place where the developing baby is held and nourished during pregnancy.
- The **cervix** is the entrance to the womb.
- The **vagina** holds the penis during sexual intercourse. It is also the passage down which the baby passes at birth.

A summary of the parts and functions of the female reproductive organs can be found in *Table 12.3* in the summary at the end of this chapter.

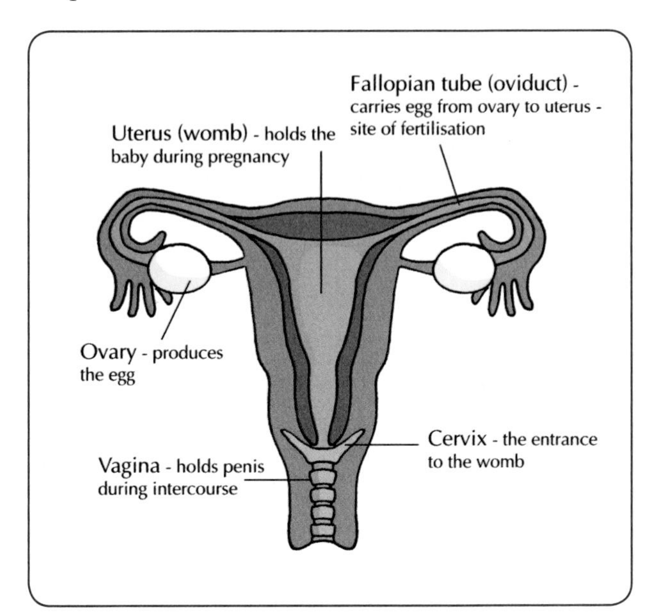

Fig. 12.4 The female reproductive system; its parts and functions.

TEST YOURSELF:
Now attempt questions *12.1–12.4* and *W12.1–W12.4.*

12.4 Puberty

Between the ages of about 10–15 years, the reproductive organs mature and begin to function. This period of development is called **puberty**. In general, puberty is a time when teenagers experience 'growth spurts' when they grow taller and other changes occur. The changes that take place during puberty are brought about by hormones.

Puberty is the time when the sex organs mature and other physical and emotional changes take place.

Puberty in Boys

In boys, puberty begins at about 12–13 years of age, somewhat later than in girls. Some boys reach maturity before this age while others do not reach puberty until 15 or 16.

The sexual features that develop during puberty in boys include the following:

- The testes and penis grow in size.
- The shoulders broaden and there is an increase in body muscle.
- The voice 'breaks', then deepens.
- Hair grows around the sex organs (pubic hair), face, chest and underarms.
- Production of sperm begins.

Puberty in Girls

Girls usually begin puberty one or two years before boys, though again there can be quite a wide variation in this. Some girls reach puberty as early as ten years old whereas others do not reach it until they are 16.

The sexual features that develop during puberty in girls include the following:

- The ovaries and womb grow in size.
- The breasts enlarge and the hips broaden.
- Pubic hair grows.
- Production of eggs begins.
- The menstrual cycle begins.

TEST YOURSELF: Now attempt question *12.5* and *W12.5.*

12.5 The Menstrual Cycle

Girls are born with thousands of unripe (immature) eggs in their ovaries. From puberty onwards, one egg cell ripens every month or so. If the egg is fertilised, then it may grow into a baby in the womb. This means that from age 12–13 onwards, a girl is physically capable of having a baby.

As well as an egg ripening each month other changes occur in a girl's reproductive system. These changes and the production of the egg, are controlled by hormones and are known as the **menstrual cycle**.

The menstrual cycle is a series of changes which take place in the female body to prepare it for pregnancy.

Stages of the Menstrual Cycle

On average, a menstrual cycle is **28 days long**, but this can vary from one person to another and from one month to another.

1. Every month, the **lining of the womb is built up** with blood. This is needed to be able to nourish a fertilised egg.
2. But if the **egg** is not fertilised then it passes through the womb and **out of the body** through the vagina.
3. The lining of the womb then breaks down and is shed from the womb over a period of four to five days. This shedding of the womb lining each month is known as **menstruation** or having a '**period**'. Menstruation occurs at the start of each menstrual cycle, i.e. on day one of the cycle.

Summary of the Menstrual Cycle

The events of the menstrual cycle are summarised in *Table 12.1* and in *Fig. 12.5*.

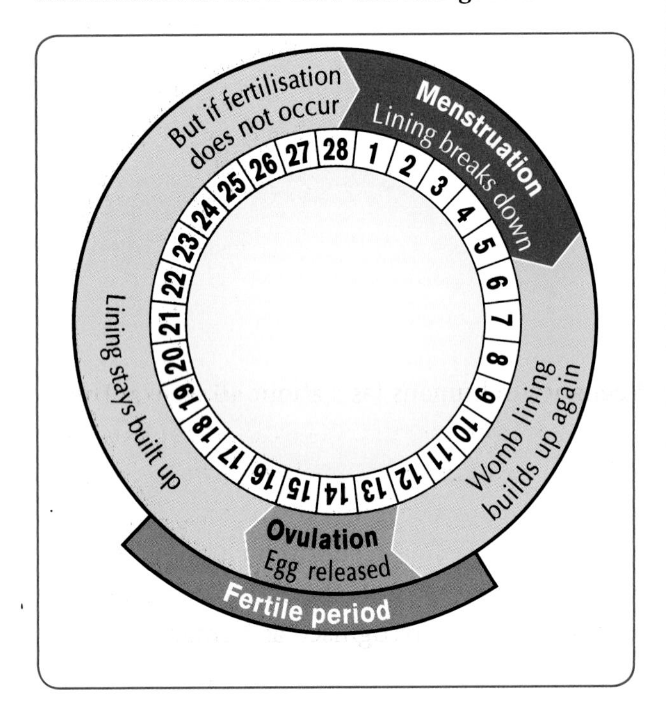

Fig. 12.5 The menstrual cycle and the fertile period.

Unless an egg is fertilised, menstruation occurs once a month throughout a woman's reproductive life. At about age 45–50, a woman stops making eggs and her periods stop. This stage of a woman's life is known as the **menopause**.

The Fertile Period

In order for fertilisation (and pregnancy) to happen, there must be an egg and sperm present at the same time. In a typical cycle of 28 days, the egg is released from the ovary on day 14. Sperm can live in a woman's body for two to three days and the egg for about a day or two. This means that in a typical menstrual cycle, the **fertile period** extends from day 11 to day 17.

The time during the menstrual cycle when fertilisation is most likely to occur is known as the fertile period.

What happens during the fertile period?

1. The ovary **releases the egg**.
2. The **lining of the uterus thickens** with blood to prepare it to receive a fertilised egg.

TEST YOURSELF:

Now attempt questions *12.6–12.8* and *W12.6–W12.7.*

12.6 Sexual Intercourse and Fertilisation

Sexual intercourse and 'making love' are both names used to describe the way sperm are placed in a woman's body.

DAYS OF CYCLE	MAIN EVENT	REASON
1–5	Menstruation – blood lining of the womb is shed.	The egg from the last cycle was not fertilised, the lining is no longer needed.
6–13	Blood lining of womb repairs and builds up again.	Getting new lining ready.
14	Ovulation – an egg is released from the ovary into the fallopian tube.	To make the egg available for fertilisation.
15–28	Egg travels down to the womb. Lining of the womb fills with blood and continues to be built up. But when fertilisation does not occur, the lining begins to break down.	Womb is prepared in case fertilisation occurs.

Table 12.1

Stages of Fertilisation

1. During **intercourse,** the erect penis is placed inside the woman's vagina.
2. This results in **ejaculation**, i.e. the release of semen from the penis. Ejaculation usually occurs at the entrance to the womb, the cervix. Millions of sperm are released per ejaculation, but it only takes **one sperm** to fertilise the egg.
3. Following ejaculation, the sperm swim up through the womb and into the fallopian tubes. If an egg is present in one of the tubes, then **fertilisation** may occur. Successful fertilisation occurs in the fallopian tube.

Fertilisation is the fusion between male and female gametes to produce a zygote.

4. Fertilisation in humans is known as **conception**. Once conception has occurred, the zygote divides many times to form a ball of cells. At the same time as it is dividing, the fertilised egg **travels down the fallopian tube** to the womb. This journey takes about seven days, *Fig. 12.6*.
5. When it reaches the womb, the ball of cells attaches itself and sinks into the lining of the womb wall. This event is called **implantation**. The ball of cells is now known as the **embryo**. The lining of the womb wall has already been prepared for implantation by becoming filled with blood vessels. The blood vessels help to nourish the embryo.

Once implantation is completed the woman is said to be **pregnant**.

Implantation is the attachment of the embryo to the womb lining.

12.7 Pregnancy and Birth

The Foetus

Pregnancy in humans lasts about **40 weeks**. This is counted from the time the woman had her last period. During the first eight weeks, the cells of the embryo continue to divide forming all the tissues and organs of the body, including the brain, heart, limbs etc. At the end of eight weeks the embryo can be recognised as human. It is now called a **foetus**.

Fig. 12.7 Foetus in the womb – 12/16 weeks old.

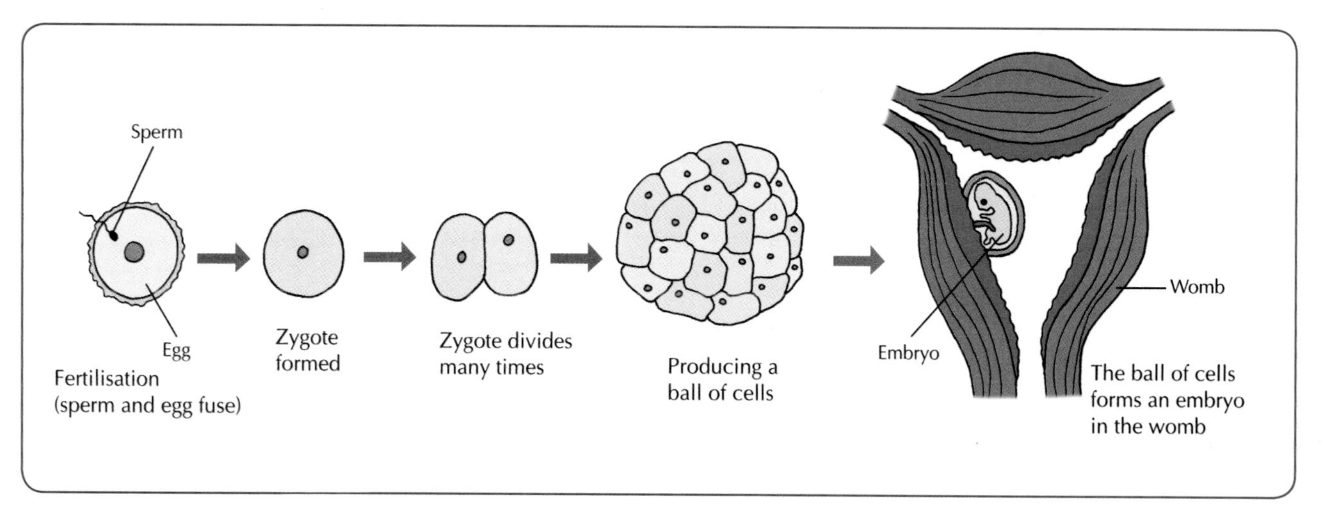

Fig. 12.6 The fertilised egg divides many times to form the embryo in the womb.

Over the next 32 weeks, the foetus develops and matures until it has developed enough to survive outside its mother. While it is in the womb, the foetus is surrounded by a fluid called **amniotic fluid**. This fluid acts as a shock absorber which protects the foetus from physical injury.

The Placenta

During pregnancy, a structure called the **placenta** develops. The placenta is needed to nourish the developing baby and it acts like a 'life support system'. It lies against the womb lining and is connected to the foetus by the **umbilical cord**. The cord attaches to the foetus at the navel (belly button).

The functions of the placenta:

- To pass **digested food and oxygen** from the mother's blood to the baby's blood.
- To **remove wastes and carbon dioxide** from the baby's blood and pass them to the mother's blood for disposal.

Harmful Substances

A pregnant woman must avoid taking or breathing in substances which might harm her unborn baby. Such things include alcohol, certain medicines and other drugs and the chemicals in cigarette smoke. These kinds of substances can pass directly across the placenta and into the baby causing serious harm.

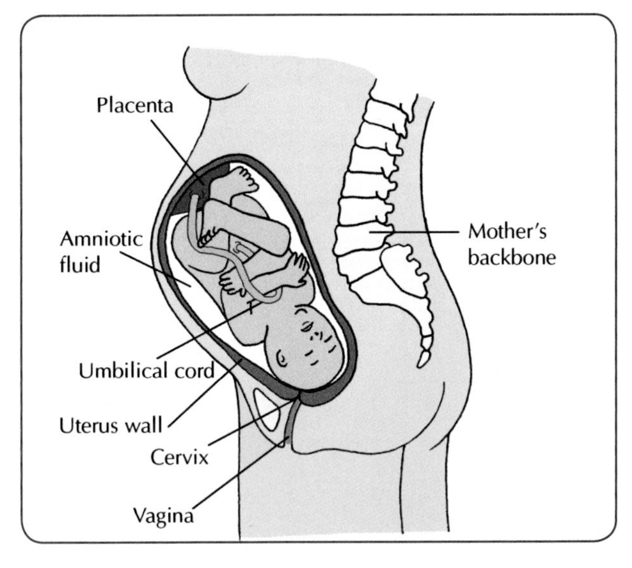

Fig. 12.8 A baby in the womb.

Birth

At the end of pregnancy, the baby is fully formed and ready to be born. Just before birth, the baby's head usually comes to lie against the cervix. The following events take place:

1. Labour

- The bag of **amniotic fluid bursts** (commonly known as the 'breaking of the waters').
- Muscles in the **womb wall** begin to **contract** in a definite rhythmic pattern. This causes the cervix to widen (to allow the baby's head to pass through).

2. Delivery of the baby

- The baby's head is pushed down out of the womb through the vagina and out into the world. The **baby is born**.

3. Delivery of the placenta

- The umbilical cord is tied or clamped and cut. The baby now **breathes on its own** for the first time.
- The placenta comes loose from the womb wall and passes out of the mother's body as the '**after birth**'.

Feeding

Shortly after it is born, the baby will need to be fed. The most natural method of feeding a baby is **breastfeeding**. During pregnancy the mother's breasts have been preparing to produce milk.

- Breast milk contains all the **food, vitamins and salts** that a normal healthy baby requires for the first four to six months of its life.
- It also contains **antibodies** that protect the new born baby from infection.

After this time, solid food can be introduced.

TEST YOURSELF:
Now attempt questions *12.9–12.13* and *W12.8–W12.10.*

12.8 Contraception

Some couples may want to control the number of children that they have, or how close together their children are born. They can do this by using contraception as a method of birth control or family planning.

Contraception is the prevention of fertilisation and implantation.

Examples of Contraception

There are many forms of **contraception**; some prevent eggs being made, others prevent ovulation and fertilisation. Apart from abstaining from intercourse, none of these methods of contraception, mentioned below, are guaranteed to be 100 per cent effective.

Some forms of contraception are listed below:

- **Abstinence from sexual intercourse** means no fertilisation can take place.
- **Natural methods** are based on a woman's fertile period and avoiding intercourse at this time.
- The use of **condoms**, which cover the penis, prevents the sperm entering the vagina and womb and so prevents fertilisation.
- The contraceptive '**pill**' stops eggs being produced. If no eggs are produced, then fertilisation cannot occur.

TEST YOURSELF:
Now attempt questions *12.14* and *W12.11*.

12.9 Genetics

Genetics is the study of the way **characteristics** (traits) are passed on from one generation to the next.

- We are all human beings, therefore we have many **features in common**. These include hair on the body, we walk upright, have two arms and two legs and a large brain.
- But some of us have brown hair, others blond. Some of us have freckles, others do not. Some of us are male, and some female.

In other words, although we are all humans, we all vary. It is these **variations** which make each of us an individual. We inherit these variations from our parents.

Genetics is the study of heredity.

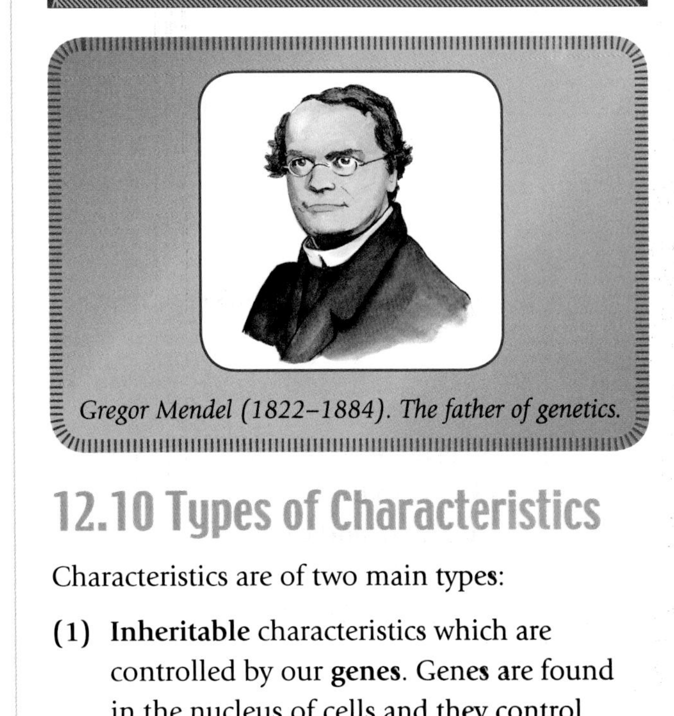

Gregor Mendel (1822–1884). The father of genetics.

12.10 Types of Characteristics

Characteristics are of two main types:

(1) **Inheritable** characteristics which are controlled by our **genes**. Genes are found in the nucleus of cells and they control all the activities of the cell. Inheritable characteristics include eye colour, hair colour, ability to speak, whether you are a girl or a boy, what blood group you have and so on, *Fig. 12.9*. Inheritable characteristics are passed on from parent to child during sexual reproduction.

Inheritable characteristics are controlled by genes.

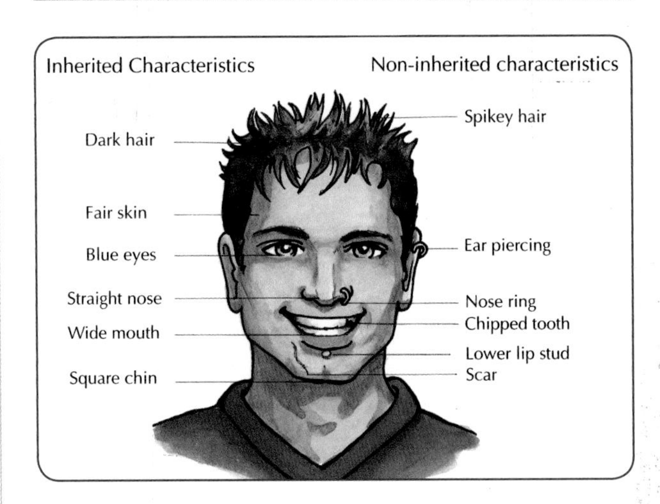

Fig. 12.9 Inheritable and Non-inheritable characteristics.

BIOLOGY

(2) **Non-inhertiable** characteristics are ones that are **learned** throughout our life and can be influenced by the environment. They are not controlled by genes and they cannot be passed on from parent to child. Examples of non-inheritable characteristics include being able to read and write, being able to skateboard, play the guitar, play hockey and having a tan and a different accent. *Fig. 12.10*.

Fig. 12.10 Non-inheritable characteristics.

12.11 Chromosomes and Genes

Chromosomes are thread-like structures found in the nucleus of plant and animal cells. Each chromosome is made up of two chemicals, **DNA** (deoxyribonucleic acid) and **protein**.
Fig. 12.11(a) shows chromosomes and genes in the nucleus of a cell.

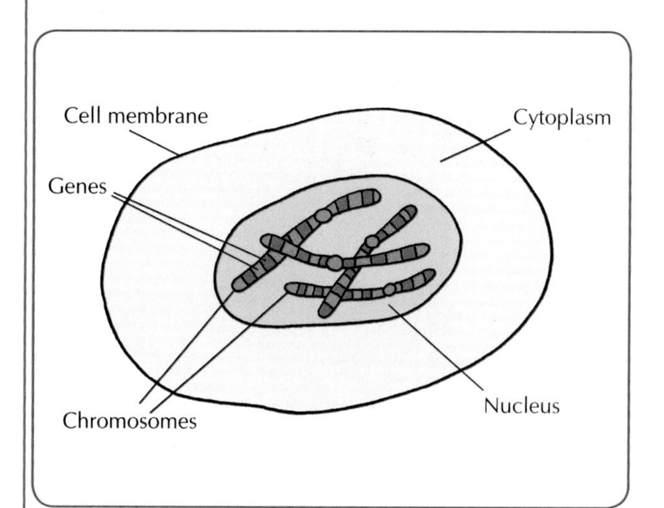

Fig. 12.11(a) A cell with nucleus showing chromosomes and genes.

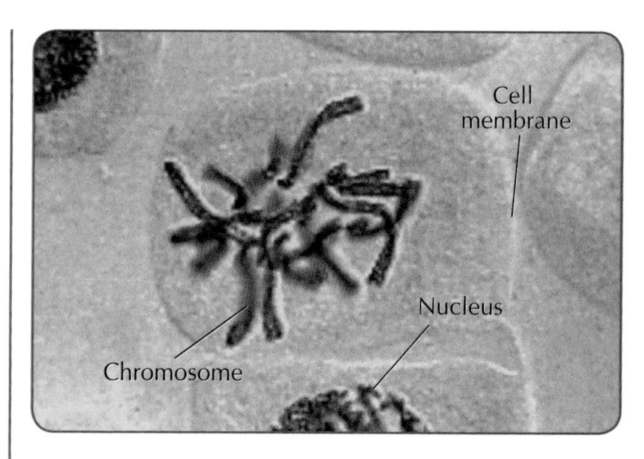

Fig. 12.11(b) If chromosomes are stained, they become visible under the light microscope.

Number of Chromosomes

Humans have **23 *pairs* of chromosomes** in the nucleus of every cell in the body except the gametes. The egg and sperm each have only 23 chromosomes. At fertilisation when the sperm containing 23 chromosomes and the egg also containing 23 chromosomes fuse, the zygote they produce will have **46 chromosomes** (i.e. 23 pairs), *Fig. 12.12*. So in each new individual, half of the chromosomes (and genes) come from the father, and half from the mother. In this way it is easy to see how we usually show characteristics of both of our parents.

Chromosomes are made of DNA and protein.

Genes

- Genes are found along the length of a chromosome, see *Fig. 12.11(a)*. Each gene is **a section of the DNA in a chromosome**.
- Genes are responsible for making the **proteins** in a cell. For example, one gene will instruct the cell to make the haemoglobin in the blood. Another gene will form the colour of your eye and so on.
- It is now known that humans have approximately **25,000 genes** in each body cell.

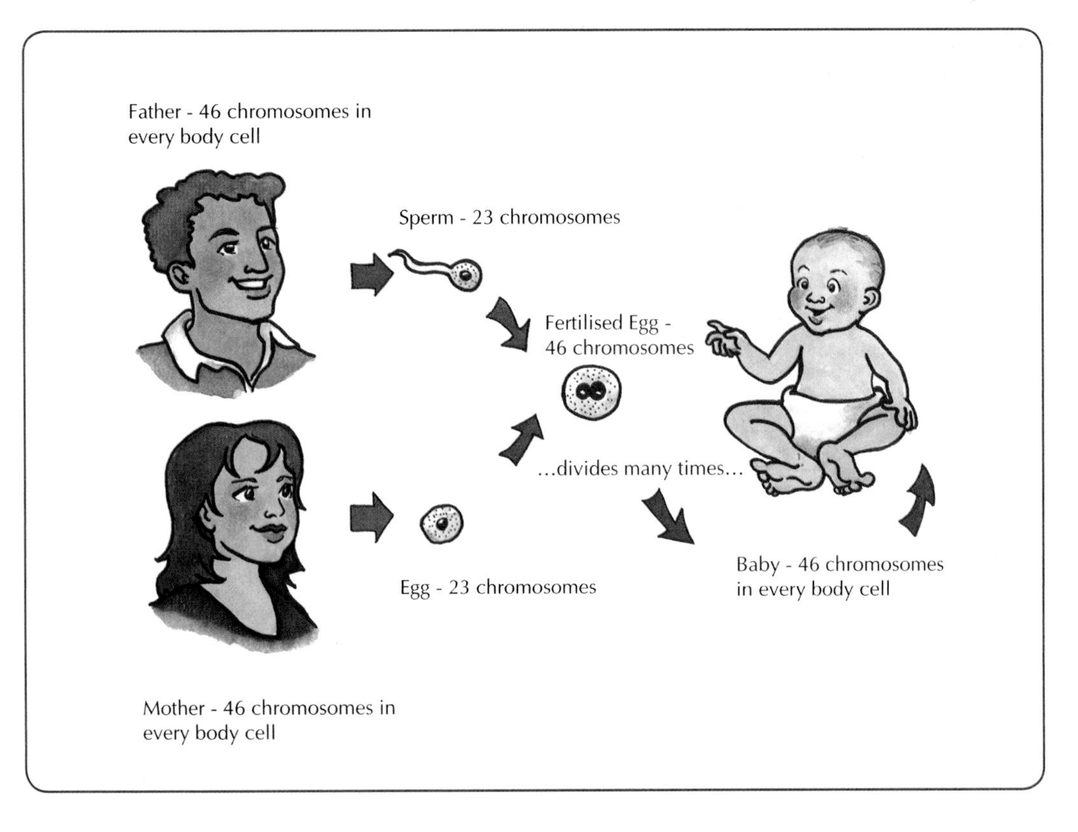

Fig. 12.12 Each new individual results from the fusion of an egg and a sperm, and shares characteristics of their father and their mother.

TEST YOURSELF:
Now attempt questions *12.15–12.20* and *W12.12–W12.14*.

What I should know

- Sexual reproduction involves two parents, one male and the other female.
- Gametes are sex cells.
- The female gamete is the egg, and the male gamete is the sperm.
- Fertilisation is when a male and female gamete fuse to form a new cell called a zygote.
- The zygote divides many times and develops into a new individual.
- The parts of the male reproductive system and their functions are shown in *Table 12.2* below.

PART	FUNCTION
Testicles (testes)	Produce sperm. Produce hormones.
Scrotum	Supports and holds testes outside the body.
Sperm duct	Carries sperm from the testes to the urethra in the penis.
Prostate gland and seminal vesicles	Secretes fluid which protect and nourish the sperm.
Penis	Transfers sperm to the woman's body.

Table 12.2 Parts and functions of the male reproductive organs.

What I should know

- The parts of the female reproductive system and their functions are shown in *Table 12.3* below.

PART	FUNCTION
Ovary	Produces eggs. Produces hormones.
Fallopian tube (oviduct)	Transfers egg to womb. Site of fertilisation.
Uterus (womb)	Site of implantation. Where baby develops during pregnancy.
Cervix	Where sperm are released during sexual intercourse.
Vagina	Holds the penis during sexual intercourse. Birth canal.

Table 12.3 Parts and functions of the female reproductive organs.

- Puberty is the time when the sex organs mature and other physical and emotional changes take place.
- In boys, the sex organs grow in size, there is an increase in body muscle, pubic and facial hair grows and the voice deepens.
- In girls, the sex organs grow in size, the breasts enlarge, the ovaries begin to produce eggs and pubic hair grows.
- The menstrual cycle is a series of changes that take place in the female body.
- The menstrual cycle lasts about 28 days.
- Menstruation (a period) occurs at the start of the menstrual cycle (between days 1–5).
- The purpose of the menstrual cycle is to produce an egg and prepare the womb for pregnancy.
- Ovulation is the release of an egg from the ovary.
- Ovulation usually occurs on day 14 of the menstrual cycle.
- The fertile period is the time during the menstrual cycle when fertilisation is most likely to occur (between day 11–17).
- During the fertile period, the ovary releases the egg and the lining of the uterus thickens with blood.
- During sexual intercourse, the erect penis is placed in the vagina and semen is ejaculated.
- Fertilisation occurs when the egg and sperm fuse together to form the zygote.
- Fertilisation takes place in the fallopian tube.
- Implantation is the attachment of the embryo to the lining of the womb.
- Human pregnancy lasts 40 weeks from the date of the last period.
- After eight weeks, the embryo becomes known as the foetus.
- By the eighth week, all the body organs have been formed.
- Over the next 32 weeks, the foetus matures.
- During pregnancy, the placenta forms to nourish the developing baby and to remove wastes.
- The baby is surrounded and protected by the amniotic fluid.
- The umbilical cord attaches the baby to the placenta.
- Birth of the baby is caused by contraction of the muscles of the womb.
- There are many forms of contraception, e.g. condoms and the 'pill'. Some of these prevent fertilisation.
- Genetics is the study of heredity.
- Inheritable characteristics, e.g. having freckles, are controlled by genes.
- Non-inheritable characteristics e.g. ability to ride a bike, must be learned.
- Chromosomes are made of DNA and protein. They are found in the nucleus of cells.
- There are 23 pairs of chromosomes in human body cells.
- Genes are found along the length of the chromosomes. Genes control the production of proteins.

QUESTIONS

Write the answers to the following questions into your copybook.

12.1 (a) What is reproduction?

(b) Explain why reproduction is important for all organisms.

(c) Name the type of reproduction that is used by humans.

12.2 (a) Give another name for a sex cell.

(b) Name the human male sex cell.

(c) Name the human female sex cell.

(d) What is a zygote?

(e) What is fertilisation?

12.3 *Fig. 12.13* shows the male reproductive system.

(a) Name the parts labelled *A, B, C* and *D*.

(b) Give one function of each of the labelled parts.

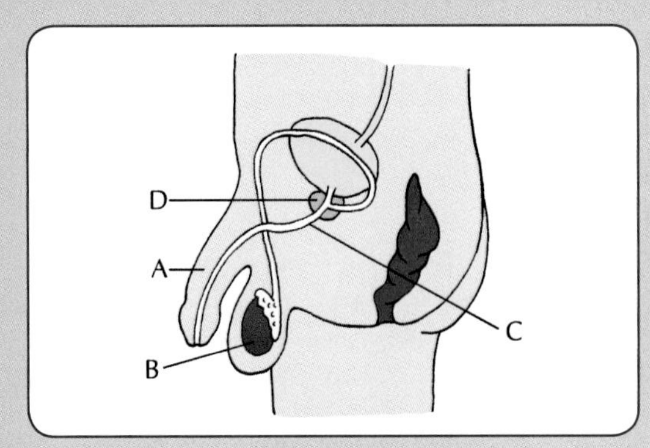

Fig. 12.13

12.4 *Fig. 12.14* shows the female reproductive system.

(a) Name the parts labelled *A, B, C* and *D*.

(b) Give one function of each of the labelled parts.

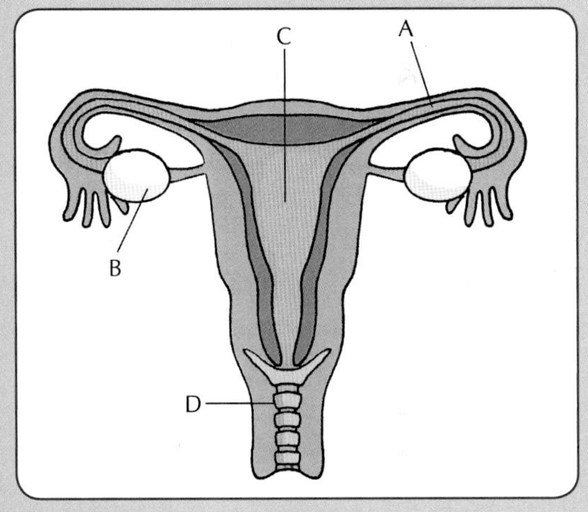

Fig. 12.14

12.5 (a) What does the term puberty mean?

(b) At what age does puberty usually begin in (i) boys and (ii) girls?

(c) Describe three changes that occur in a girl during puberty.

(d) Describe three changes that occur in a boy during puberty.

12.6 (a) What is the menstrual cycle?

(b) What is the average duration (length) of a menstrual cycle?

(c) What is menstruation?

(d) What is the average duration (length) of menstruation?

(e) On what day of the menstrual cycle does menstruation usually begin?

(f) On what day in a typical menstrual cycle does ovulation take place?

(g) What happens to an unfertilised egg?

12.7 (a) What name is given to the time during a menstrual cycle when fertilisation is likely to occur?

(b) If a woman's menstrual cycle occurs every 28 days, on what day(s) of her cycle will the following events be possible?
(i) ovulation
(ii) fertilisation

QUESTIONS

12.8 Match each of the events of the menstrual cycle with the approximate days in the right hand column.

EVENT	DAYS OF THE CYCLE
Fertile period	1-5
Womb lining stays built up	6-13
Womb lining repairs	14
Ovulation	11-17
Menstruation	15-28

12.9 (a) In which part of the female reproductive system does fertilisation usually occur?

(b) How many sperm fertilise an egg?

(c) What other name is given to fertilisation in humans?

(d) What happens to the egg after fertilisation?

(e) What is meant by implantation? Where does it take place?

12.10 Using words from the list below complete the following sentences:

fallopian, organs, eight, womb, placenta, implantation, 40, tissues, foetus

(a) Human pregnancy usually lasts for ..... weeks.

(b) Once the egg is fertilised, it passes down the ..... tube to the ..... .

(c) ..... is when the ball of cells burrows into the lining of the womb wall.

(d) After ..... weeks, the embryo has formed all the ..... and ..... of the body and it is now called the ..... .

(e) The ..... is a structure that helps to nourish the developing baby in the womb.

12.11 (a) Name the tube through which a baby feeds and breathes while in the womb.

(b) Name a waste material that passes from the baby into the mother's bloodstream.

(c) Name two substances that pass from the mother to her baby.

(d) Why are pregnant women advised not to take certain medicines?

(e) Explain how the baby is cushioned and protected while in the womb.

12.12 (a) What is the placenta?

(b) Where is the placenta found?

(c) List two functions of the placenta.

12.13 (a) Describe how a baby is born.

(b) What is the best type of food for very young babies?

12.14 (a) What is meant by the term contraception?

(b) Name two methods of contraception.

(c) Explain how each of the methods named in (b) prevent fertilisation occurring.

QUESTIONS

12.15 (a) Genetics is the study of ….. .

(b) Humans all belong to the same ….., but we also show differences called ….. .

(c) There are ….. types of characteristic: ….., and ….. .

(d) ….. characteristics are controlled by genes.

(e) Having bleached hair is an example of a ….. characteristic.

(f) Having brown eyes is an example of an ….. characteristic.

12.16 (a) Name three characteristics that all human beings share.

(b) Name three characteristics that differ between two human beings.

12.17 (a) Which of the following are not inherited characteristics?

(i) hair colour, (ii) ability to roll your tongue, (iii) ability to swim, (iv) having freckles

(b) How do we develop non-inherited characteristics?

12.18 Decide whether each of the following is an inheritable or a non-inheritable characteristic.

(a) green eyes

(b) long hair

(c) speaking Spanish

(d) square chin

(e) producing saliva

(f) ability to play the guitar

(g) having a scar on your arm

12.19 (a) What are chromosomes?

(b) Where in a cell are chromosomes found?

(c) What are chromosomes made of?

(d) How many chromosomes are there in a human muscle cell?

(e) How many chromosomes are there in a human sperm cell?

12.20 (a) What are genes?

(b) Where in a nucleus are genes found?

(c) What are genes made of?

(d) How many genes are there in a typical human cell?

(e) What is the function of genes in a cell?

Chapter 13 Plant Structure and Transport

13.1 Introduction

Although many different types of flowering plant exist, they have the same basic structure. Typically a flowering plant consists of **roots**, **stem**, **leaves** and **flower(s)**. The main parts of a typical flowering plant are shown in *Fig. 13.1*.

- The roots of a plant usually grow **down** into the ground.
- The stem, leaves and flowers form the shoot which is **above** the ground.

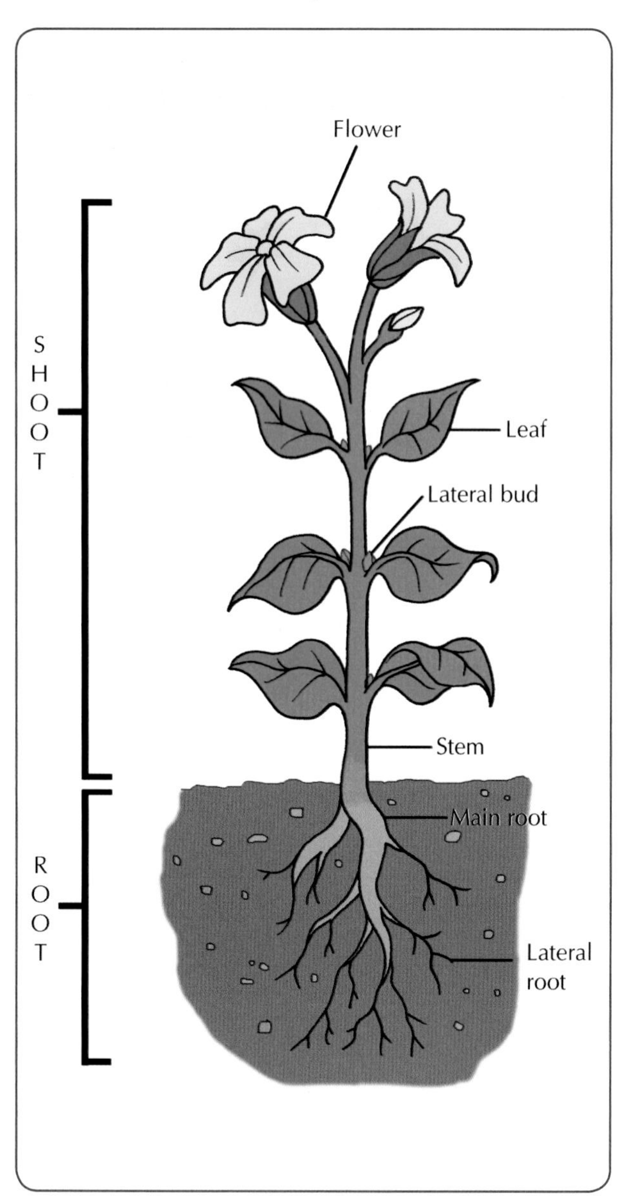

Fig. 13.1 The main parts of a typical flowering plant.

Fig. 13.2 Buttercups.

13.2 The Functions of the Main Parts

(1) The Root

The functions of the root are:

- To **anchor** the plant in the ground and give support.
- To **take in water and minerals** from the soil.
- Some roots can **store food**, e.g. carrots and turnips.

(2) The Stem

The functions of the stem of a plant are:

- To **hold up** the leaves, flowers and fruits.
- To **carry water and minerals** from the roots to the leaves and flowers.
- To **carry food**, made in the leaves, to other parts of the plant.

(3) The Leaves

The functions of the leaves of a plant are:

- To make food and oxygen in the process called **photosynthesis**.
- To allow the **exchange** of oxygen and carbon dioxide **gases**.
- To allow the **loss of water vapour**.

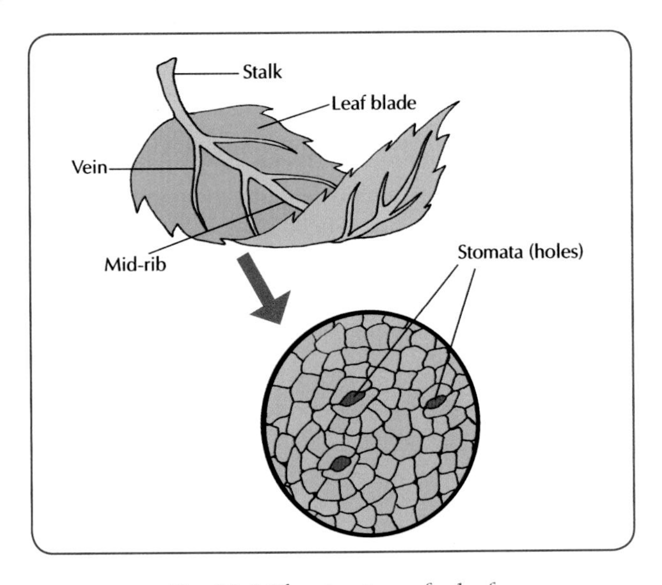

Fig. 13.3 The structure of a leaf.

(4) The Flower

- The main function of the flower is sexual **reproduction**. Flowers produce **seeds**.

You will learn more about the structure of a flower and flowering plant reproduction in *Chapter 15*.

TEST YOURSELF:
Now attempt questions *13.1–13.3* and *W13.1–W13.2*.

13.3 Transport in Plants

If plant cells are to carry out their activities they must be able to get the materials to make food and energy. In addition, the wastes they produce must be removed. This means that plants need a **transport system**.

Transport Tissue

The transport tissue in plants consists of a system of fluid-filled tubes through which substances are transported.

There are two types of transport tissue:

(1) **Xylem**. It carries **water and minerals** from the roots upwards to the leaves and flowers.

(2) **Phloem**. It carries **food** made in the leaves, to all the other parts of the plant.

Xylem transports water and minerals up the plant.

Phloem transports food up and down the plant.

13.4 Passage of Water and Minerals through a Plant

1. Water is taken **into the roots** of a plant from the soil.
2. It travels **up the stem** to the leaves in the water transporting tissue (xylem).
3. In the leaves some of the water is used to make food in **photosynthesis**.
4. The remaining water **passes out of the leaves** into the air.

The blue arrows in *Fig. 13.4* show the pathway of water movement up the plant. Minerals are also taken in at the roots and travel up the plant with the water.

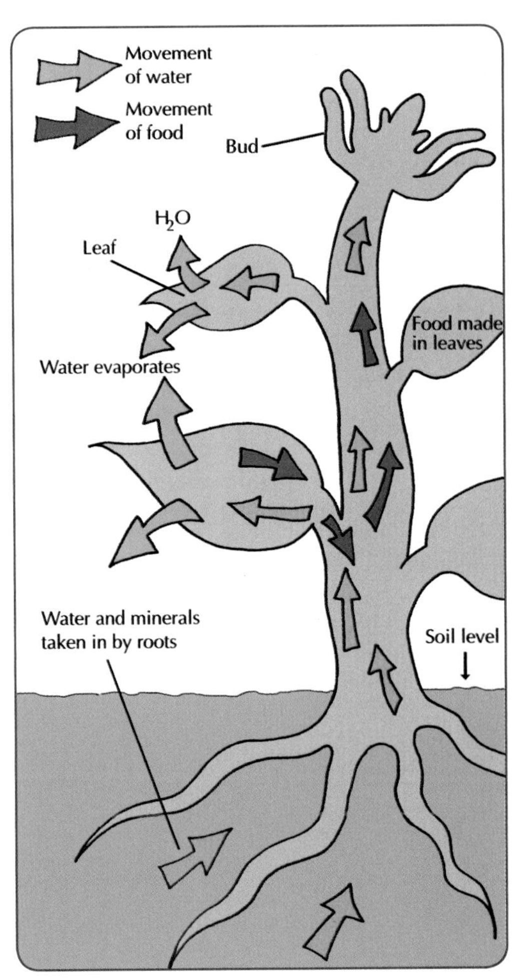

Fig. 13.4 The pathway of water through the plant – the transpiration stream.

Experiment 13.1

To show that water is lost from a plant

Apparatus required: 2 test-tubes; marker

Also required: seedling; oil

Method

1. Set up the apparatus as in *Fig. 13.5*. The layer of oil prevents water evaporating from the tubes.
2. Mark the level of water in each test-tube at the start of the experiment.
3. Set the tubes aside for two to three days.
4. Again mark the level of water in each tube.

Result

The level of water in the tube with the seedling has fallen. The level of water in the other tube stays the same.

Conclusion

The layer of oil in each tube prevented evaporation of water from the tubes. Therefore the fall in water level in the tube with the seedling suggests that the water has passed through the plant and out into the air.

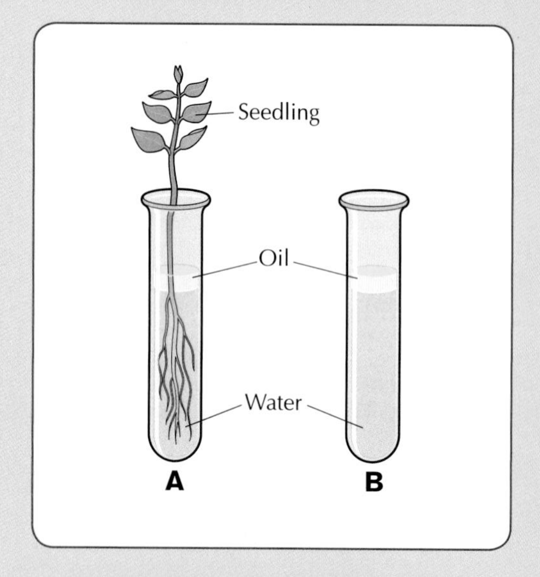

Fig. 13.5 The level of water in tube A drops after a few days. This shows water is lost from the plant.

Experiment 13.2

To show the path of water through plant tissue

Apparatus required: beaker; knife; cutting board

Also required: celery with leaves, red food dye

Method

1. Set up the apparatus as in *Fig. 13.6*.
2. Leave the beaker for one to two days.
3. Remove the piece of celery and rinse off the red dye.
4. Describe the appearance of the leaves and the leaf veins.
5. Place the piece of celery on a cutting board and use the knife to cut across the stem.
6. Look at the cut end of the stem. Can you see where the dye is?

Result

The veins in the leaves of the celery will be a red colour. When cut across, tiny red dots will be found at the edge of the stem. These are the transport tubes that carry the water.

Conclusion

Water travels upwards through the stem and into the leaves. The red dots represent the xylem tubes that carry water up the plant.

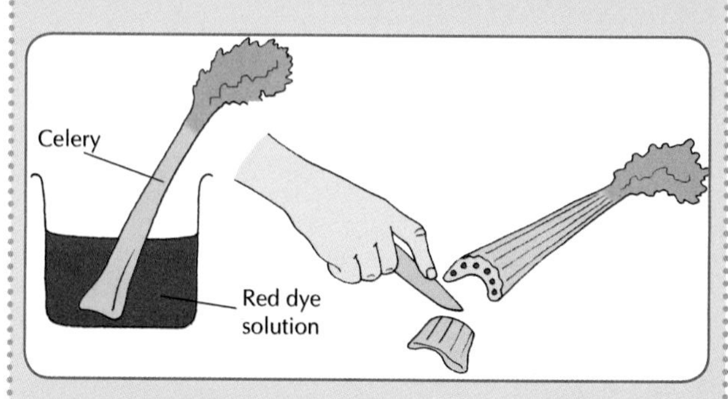

Fig. 13.6 The stem and leaves of the celery become red as the dye is carried up through the plant.

TEST YOURSELF:
Now attempt questions *13.4–13.6* and *W13.3–W13.5.*

13.5 Transpiration

The loss of water vapour from the surface of the plant is known as **transpiration**. This loss of water usually takes place through tiny openings in the leaves called **stomata**, see *Fig. 13.3*, page 81.

Transpiration is the loss of water vapour from the surface of a plant.

As water is lost from the leaves more is drawn up from the roots. It is rather like sucking water through a straw. The movement of water up the plant is called the **transpiration stream**, see *Fig. 13.4*, page 81.

The functions of transpiration are:

- To **bring water** needed for photosynthesis, from the roots to the leaves.
- To **carry minerals** dissolved in the water up the plant.
- To **cool the plant** – in a similar way to that in which perspiring cools us down.

Experiment 13.3

To show that water evaporates from the surface of a leaf by transpiration

Chemical required: blue cobalt chloride paper

Also required: a well-watered pot plant; a well-watered pot plant with no leaves; 2 clear plastic bags; 2 plastic bags to cover the soil

Note: The test for water. Water turns cobalt chloride paper pink.

Method

1. Set up the apparatus as shown in *Fig. 13.7*.
2. Leave the plants in a warm, bright place for several hours.
3. The plastic bags covering the soil in each pot prevent evaporation of water.
4. After several hours, remove the plastic bags covering each plant.
5. Test any liquid that has formed with blue cobalt chloride paper.

Result

Drops of a colourless liquid will have formed on the inside of the plastic bag covering the leaves in plant *(a)*. There are no drops of liquid in the bag on plant *(b)*.

The liquid droplets from plant *(a)* turn blue cobalt chloride paper to a pink colour.

Conclusion

The liquid that forms in the plastic bag around plant *(a)* is water. No water forms from the plant without leaves (the control). Therefore we can conclude that water is lost from the leaves of plants.

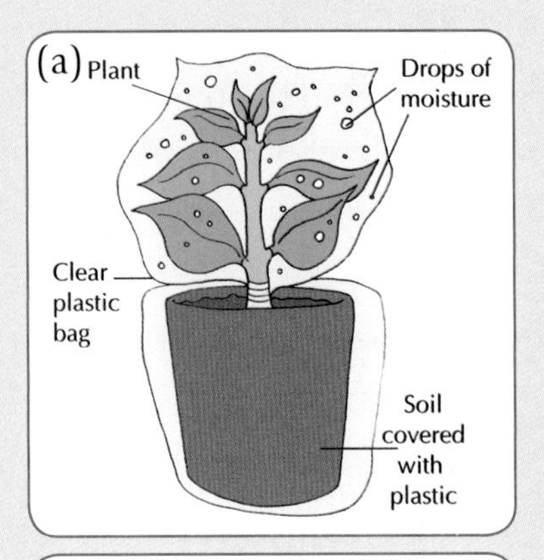

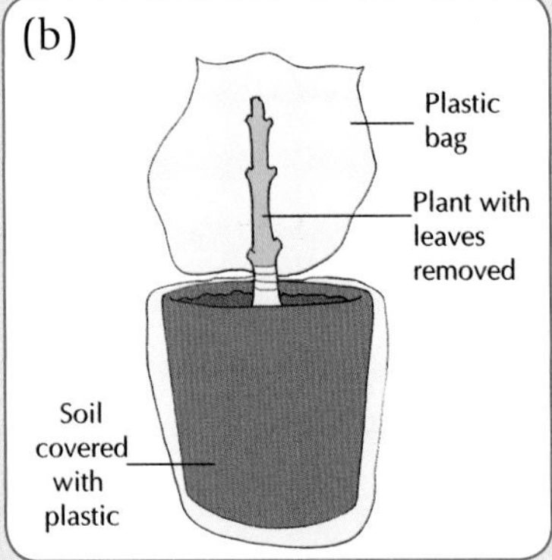

Fig. 13.7 To show transpiration in plants.

TEST YOURSELF: Now attempt questions *13.7–13.9* and *W13.6–W13.7*.

BIOLOGY

What I should know

- Flowering plants consist of four main parts: the root, stem, leaf and flower.
- The root is for anchorage, storage, taking in water and minerals.
- The stem is for transport and support of leaves and flowers.
- The leaves are for making food, gas exchange and for transpiration.
- The flower is for sexual reproduction.
- Plants have transport tissue.
- Xylem transports water and minerals up the plant.
- Phloem transports food up and down the plant.
- Water enters the root, passes into the transport tissue (xylem), travels up the stem to the leaves and is used by the plant or released in transpiration.
- Stomata are pores in the leaves through which water and gases pass out of the plant.
- Transpiration is the release of water vapour from the surface of a plant.
- The functions of transpiration are to bring water and minerals up to the leaves and to cool the plant.

Questions

Write the answers to the following questions into your copybook.

13.1 (a) Name the parts *A–F* of the flowering plant shown in *Fig. 13.8*.

(b) Which labelled part is responsible for making seeds.

(c) Give two functions of the part labelled C.

(d) List the functions of the part labelled *A*.

(e) Name three flowering plants.

13.2 (a) The flowering plant is made of two systems: the ….. below the ground and the ….. above the ground.

(b) The shoot is formed from the stem, ….., and ….. .

(c) Two functions of the root are ….. , and ….. .

(d) The ….. produces seeds during reproduction.

(e) The leaves and flowers of plants are held upright on the ….. .

(f) Leaves make food for the plant in a process called ….. . They also allow the exchange of ….. gas and ….. gas.

13.3 (a) List the functions of the root.

(b) List the functions of the stem.

(c) List the functions of the leaf.

(d) List the function of the flower.

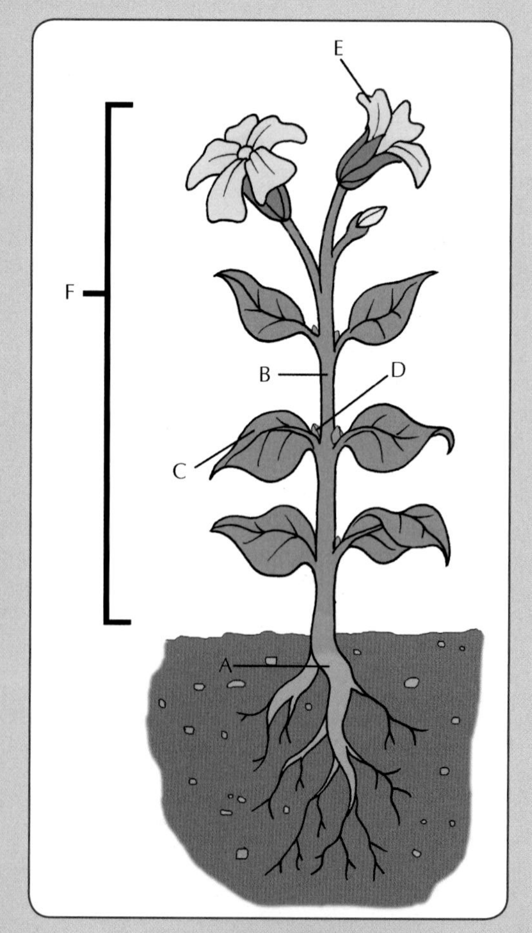

Fig. 13.8

QUESTIONS

13.4 (a) Why is a transport system needed by plants?

(b) Name the type of transport tissue that carries water.

(c) Name the transport tissue that carries food.

(d) Which of the transport tissues also carries minerals?

13.5 You are given a stick of celery and a beaker of blue dye solution. The celery is placed in the beaker of dye and left for 24 hours.

(a) Describe the appearance of the stick of celery after this time.

(b) After 24 hours the stick of celery is cut across with a knife. Draw a diagram of the cut celery to show where the blue dye would be found.

13.6 A plant was set up as shown in the diagram in *Fig. 13.9*.

(a) What will happen to the level of water in the conical flask after a couple of days?

(b) What is the purpose of the layer of oil?

(c) Which of the following conclusions can be drawn from this activity:

(i) The plant is taking in water from the air.
(ii) The plant is taking in water into its roots.
(iii) The plant is losing water through its roots.
(iv) The plant is losing water through its leaves.

(d) Describe in words or draw a labelled diagram of a suitable control apparatus you could set up.

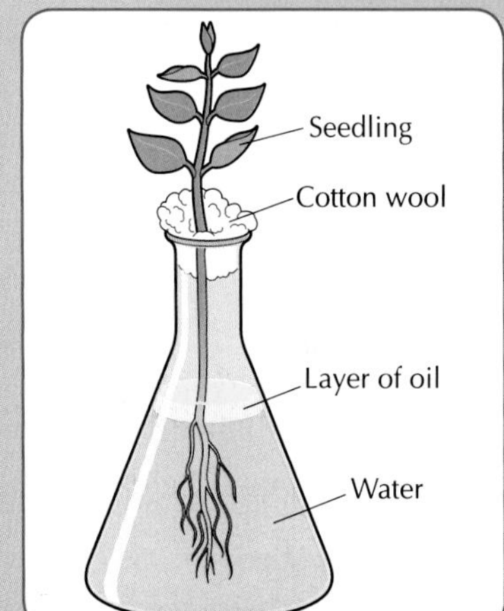

Fig. 13.9

13.7 (a) What term is given to the loss of water vapour from the surface of a plant?

(b) Name the tiny pores in a leaf through which water vapour is lost.

(c) Which of the following would you use to show that a liquid is water:

(i) litmus paper, (ii) cobalt chloride paper, (iii) iodine solution

13.8 (a) What is transpiration?

(b) List the functions of transpiration in a plant.

(c) What are stomata? Where are they found? What is their function in a plant?

13.9 The apparatus shown in *Fig. 13.10* was set up to show that water evaporates from the surface of leaves.

(a) What is the name of this process?

(b) Suggest why the following were carried out:

(i) A pot without a plant was set up?
(ii) Each pot of soil was covered with a plastic bag?
(iii) A Vaseline ® seal was placed around the base of each bell jar.

(c) Describe how you would test that the droplets forming in the bell jar are water.

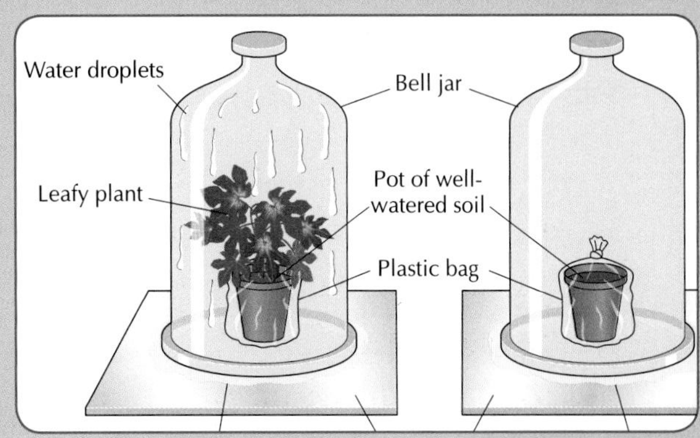

Fig. 13.10

Chapter 14 Photosynthesis and Tropisms

14.1 Introduction to Photosynthesis

Plants make their own food using light energy from the sun. This process is called **photosynthesis**. Photosynthesis takes place in the leaves and other green parts of a plant. The green colour in plants is due to the presence of the pigment **chlorophyll**.

Photosynthesis is the way in which plants make their food.

- During photosynthesis, carbon dioxide and water combine to form **sugars** such as glucose. The energy needed for this comes from the sun, see *Fig. 14.1*.
- As well as food, photosynthesis produces **oxygen gas**. Oxygen is a waste product of photosynthesis. It can be used by the plant in respiration or released out into the air for use by other living things.

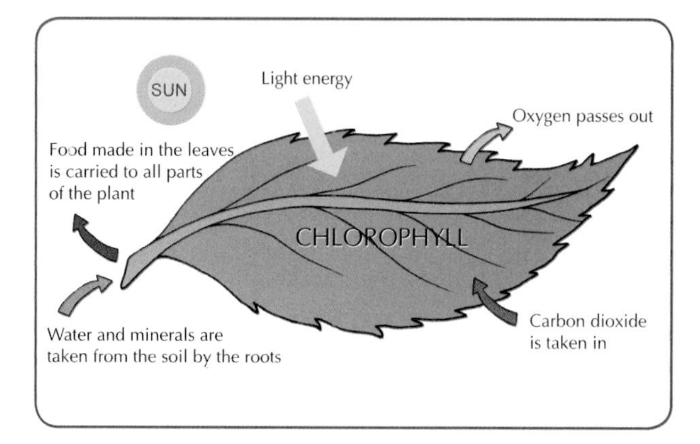

Fig. 14.1 The process of photosynthesis.

14.2 Factors Needed for Photosynthesis

Carbon dioxide, water, chlorophyll and light are needed for photosynthesis.

(1) **Carbon dioxide** comes from the air. It enters the plant through the stomata in the leaves.

(2) **Water** for photosynthesis comes from the soil. It travels from the roots, up the stem to the leaves in the transport tissue (xylem).

(3) **Chlorophyll** is the green pigment found in the leaves. It traps light energy and converts it to chemical energy.

(4) **Light** comes from the sun. It is absorbed by the chlorophyll in the leaves and is the energy needed to make food.

14.3 Products of Photosynthesis and their Uses

The **products** of photosynthesis are **glucose** and **oxygen**.

Glucose

- glucose can be used in **respiration**, to produce energy for the plant.
- glucose can be **stored** as starch in the leaves or **transported** around the plant to the stem, flower and roots.
- glucose can be **converted** into protein.

Oxygen

- oxygen can be used in **respiration**, to produce energy in the plant.
- oxygen can be **released** from the plant and used by other organisms in respiration.

We can summarise the process of photosynthesis in a word equation:

$$\text{carbon dioxide + water} \xrightarrow[\text{chlorophyll}]{\text{light}} \text{glucose + oxygen}$$

The Importance of Photosynthesis

- Photosynthesis **produces food** for plants which in turn provide food for humans and other animals.
- Photosynthesis **produces oxygen** gas which is the oxygen breathed in by humans and other organisms.

- Photosynthesis **removes carbon dioxide** from the air, thus reducing the green house effect, *see Chapter 18*.

Mandatory experiment 14.1

To show that starch is produced by a photosynthesising plant

Apparatus required: beaker; Bunsen burner; tripod; gauze; forceps; test-tube; test-tube holder; white tile

Chemicals required: alcohol; iodine solution

Also required: pot plant

Note: If we want to show that a plant produces starch during photosynthesis, we must use a plant that has no starch in it to begin with. We do this by placing the plant in the dark for 48 hours. This is known as **de-starching** the plant. Without light, the plant cannot make any food and it uses up its store of starch.

Method

1. Place the pot plant in the dark for 48 hours. This is necessary to de-starch the plant.
2. Cover part of some of the leaves with aluminium foil as shown in *Fig. 14.2(a)*.
3. Leave the plant in bright light for four to six hours. (This allows the plant to photosynthesise.)
4. Set up a water bath as shown in *Fig. 14.2(b)* and bring the water to the boil.
5. Remove one of the leaves with a foil strip and draw a sketch of it to show the position of the foil strip.
6. Remove the foil strip. Drop the leaf into the boiling water for one minute. (This kills the leaf.)
7. Turn off the Bunsen burner.
8. Half-fill a test-tube with alcohol.
9. Using the forceps, remove the leaf from the water. Gently push the leaf into the test-tube of alcohol.
10. Stand the test-tube in the warm water for about ten minutes, *Fig. 14.2(c)*. (The warm alcohol removes the chlorophyll from the leaf. This makes it easier to see the reaction of starch with the iodine solution.)
11. Use the forceps to remove the leaf from the test-tube. (The leaf will be creamy/white in colour and it will be very brittle.)
12. Dip the leaf into the warm water in the water bath, *Fig. 14.2(d)*. (This softens the leaf.)
13. Carefully spread the leaf out onto a white tile and cover with iodine solution. *Fig. 14.2(e)*. (Iodine solution tests for the presence of starch.)
14. Draw a new diagram of the leaf showing where starch is present and compare it with your first diagram.

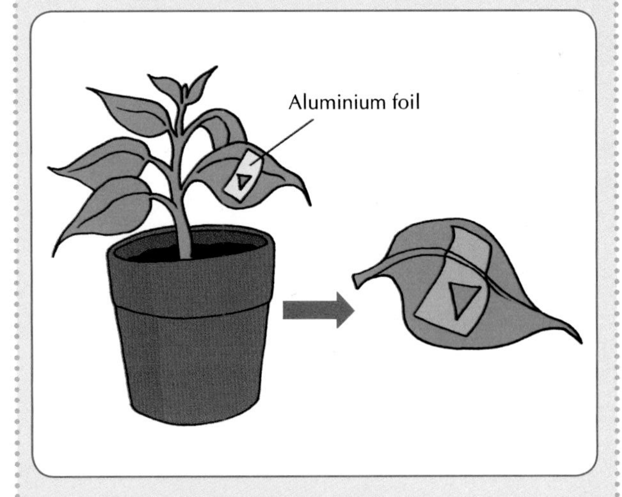

Fig. 14.2(a) Cover part of de-starched leaf with aluminium foil.

BIOLOGY

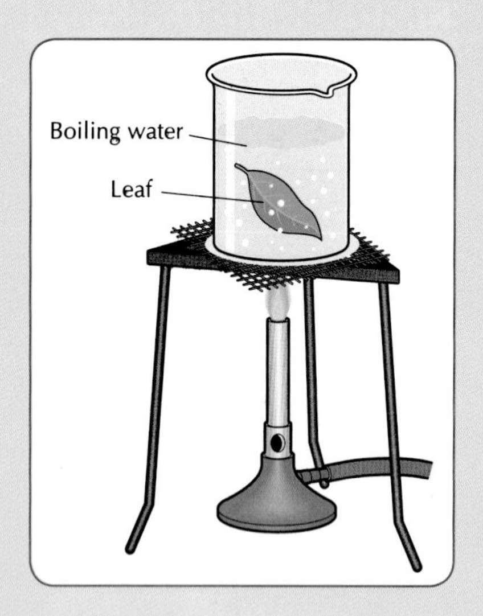

Fig. 14.2(b) Boil leaf in water – to kill leaf.

Fig. 14.2(c) Place in warm alcohol – to remove the chlorophyll.

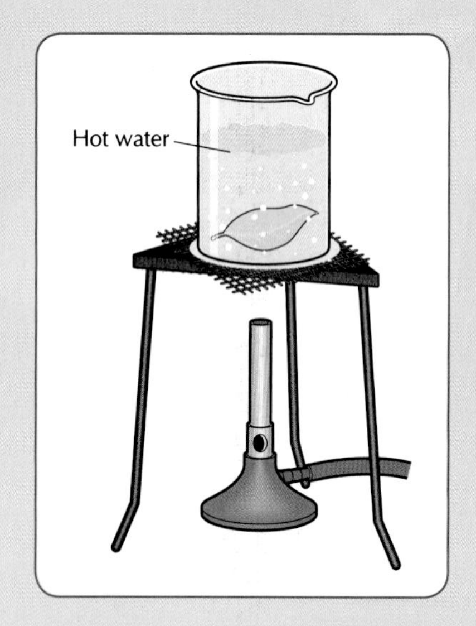

Fig. 14.2(d) Dip in warm water – to soften the leaf.

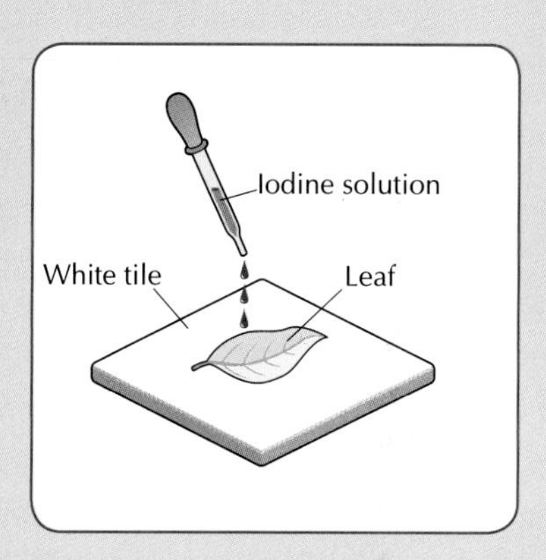

Fig. 14.2(e) Add iodine solution – to test for starch.

Results

Part of leaf in the light: the leaf turns a blue-black colour with iodine solution. This shows starch is present.

Part of leaf that was covered by foil: the leaf stays brown-yellow (the colour of iodine solution). This shows starch is not present.

Conclusion

Starch is only found in the part of the leaf that was exposed to light. Therefore starch is produced by photosynthesis.

TEST YOURSELF:
Now attempt questions *14.1–14.5* and *W14.1–W14.2*.

14.4 Tropisms

In order to survive, plants must be able to respond to changes in their surroundings. Plants respond to stimuli such as light, the force of gravity, water and some plants respond to touch. For example, plants grow in the direction of light to be able to make food. Roots grow down to anchor and support a plant.

The growth of a plant in response to a stimulus is called a **tropism**. Two tropisms that occur in plants are: **phototropism** and **geotropism**.

A tropism is the growth response of a plant to a stimulus.

Phototropism

Phototropism is the growth response of a plant to light. The shoots of plants grow towards light, see *Fig. 14.3*. Phototropism allows the plant to get as much light as possible. As a result the plant makes more food.

Fig. 14.3 Plant shoots grow towards the light – an example of phototropism.

Experiment 14.2

To investigate the growth response of plants to light – phototropism

Apparatus required: 3 petri dishes; cotton wool; scissors or knife

Also required: cress seeds; shoe box with lid; cardboard to make room 'dividers'; sellotape®

Method

1. Build a *seed house* as shown in *Fig. 14.4*.
2. Set up three Petri dishes as follows: fill each dish with moist cotton wool; place ten seeds on top.
3. Place one dish in each *room* of the *seed house*. Place the lid on the *seed house*.
4. Leave on a window ledge or in a bright place for three to four days. Take care that the cotton wool does not dry out.
5. After three to four days record the appearance of the seedlings in each dish.

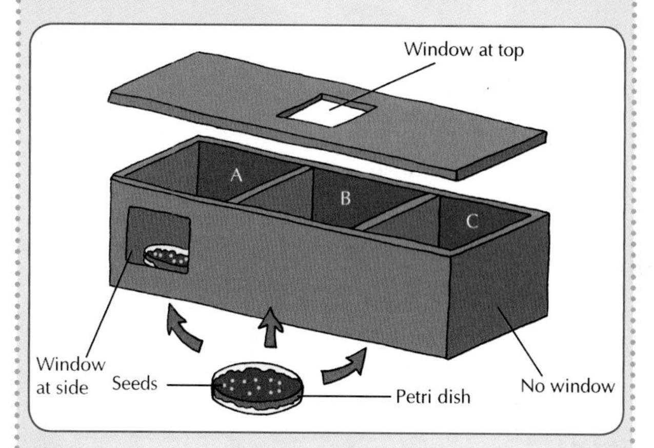

Fig. 14.4 To investigate phototropism.

Results

Dish from *Room A*: the seedlings are green and have grown towards the window (light).

Dish from *Room B*: the seedlings are green and have grown straight up to the window (light).

Dish from *Room C*: the seedlings are creamy/ yellow and they have a straggly growth.

Conclusion

The seedlings have grown towards the light. This experiment demonstrates phototropism.

Phototropism is the growth response of a plant to light.

Geotropism

Geotropism is the growth response of a plant to the force of gravity. The roots of plants grow towards the force of gravity whereas the shoots grow away from it, see *Fig. 14.5*, page 90. Geotropism makes sure that the roots find water and minerals in the soil and that the plant is well anchored.

BIOLOGY

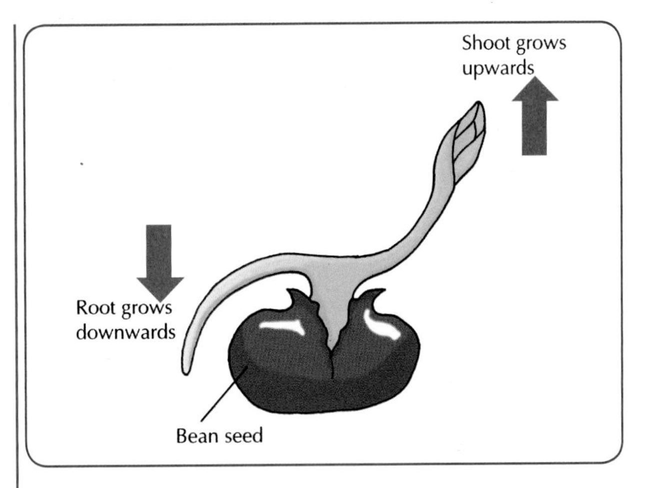

Fig. 14.5 Roots grow down towards the force of gravity; shoots grow up away from the force of gravity.

Experiment 14.3

To investigate the growth response of plants to gravity – geotropism

Apparatus required: a jam jar or a beaker

Also required: blotting paper; potting compost; 3 soaked broad bean or pea seeds

Method

1. Set up the apparatus as shown in *Fig. 14.6*.
2. Make sure the seeds are each facing in a different direction.
3. Leave the jar in a dark place for one week. Take care to check every few days to see that the compost does not dry out.

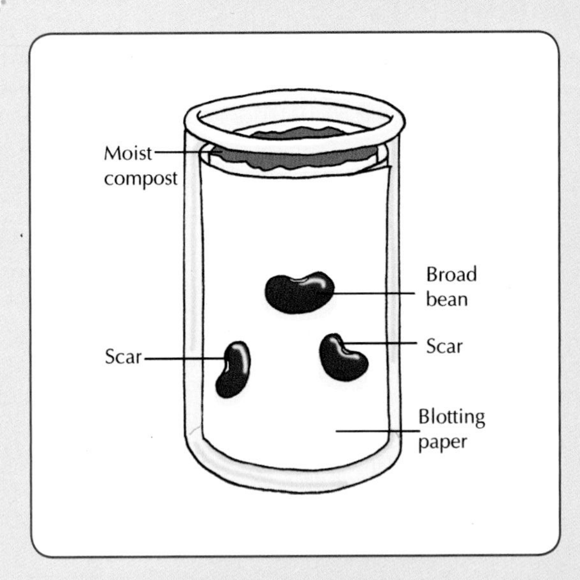

Fig. 14.6 To investigate geotropism – bean seeds are placed facing in different directions.

Results

The seeds show the shoots growing up and the roots growing down regardless of the position of the seed.

Conclusion

The roots of plants grow in the direction of the force of gravity. This demonstrates geotropism.

Geotropism is the growth response of a plant to the force of gravity.

TEST YOURSELF:

Now attempt questions *14.6–14.9* and *W14.3–W14.6*.

What I should know

- Photosynthesis is the way that plants make food.
- Photosynthesis takes place in the leaves and other green parts of plants.
- Energy from the sun is used to make glucose in photosynthesis.
- Carbon dioxide, water, light and chlorophyll are needed for photosynthesis to occur.
- Glucose and oxygen are produced during photosynthesis.
- Glucose can be stored in a plant in the form of starch.
- The energy conversion in photosynthesis is:
  light energy ⟶ chemical energy

What I should know

- The word equation for photosynthesis is:

$$\text{carbon dioxide + water} \xrightarrow[\text{chlorophyll}]{\text{light}} \text{glucose + oxygen}$$

- Plants respond to changes in their surroundings.
- A tropism is a growth response of a plant to a stimulus.
- Phototropism is the growth response of a plant to light.
- Plant shoots grow towards light; roots grow away from light.
- Geotropism is the growth response of a plant to the force of gravity.
- Plant roots grow towards gravity; shoots grow away from gravity.

QUESTIONS

Write the answers to the following questions into your copybook.

14.1 (a) Green plants can make their own food by ..... .

(b) The green pigment in leaves is ..... . It traps the sun's ..... .

(c) Plants need ..... from the air and ..... from the soil, to be able to make food.

(d) During photosynthesis, ... energy is converted into ..... energy.

(e) The sugars made in the leaves can be stored as ..... .

(f) The gas ..... is a waste product of photosynthesis.

14.2 (a) What is photosynthesis?

(b) Write down the word equation for photosynthesis.

(c) Where does photosynthesis take place?

(d) Why is photosynthesis so important to humans?

(e) State the energy conversion that takes place during photosynthesis.

14.3 (a) Name the four factors that plants need for photosynthesis.

(b) State where the plant gets each of these factors.

14.4 Describe an experiment to show that starch is produced by a photosynthesising plant.

14.5 Decide whether you think the following statements are true or false. If false, give a reason for your choice.

(a) Photosynthesis only occurs in green plants.

(b) The green colour in plants is chlorophyll.

(c) Photosynthesis produces carbon dioxide gas.

(d) During photosynthesis, light energy is converted to chemical energy.

(e) Plants do not need oxygen.

(f) Animals need photosynthesis.

(g) Plants are de-starched by placing them in the light for 48 hours.

QUESTIONS

14.6 (a) What is a tropism?

(b) Name a tropism.

(c) What advantage does this tropism have for a plant?

14.7 Look at the diagram in *Fig. 14.7* and answer the following:

(a) What is the stimulus?

(b) What is the response of the plant to the stimulus?

(c) What is the advantage of this response to the plant?

(d) What is the name of this response?

Fig. 14.7

14.8 Describe an experiment you could carry out to demonstrate phototropism.

14.9 (a) What is geotropism?

(b) What advantage does geotropism have for plants?

(c) Describe an experiment to demonstrate geotropism. Include a labelled diagram of the apparatus you would set up.

(d) When a gardener goes out to sow seeds, why does she not have to worry which way up the seeds land in the soil?

Chapter 15 Plant Reproduction

15.1 Introduction

Reproduction is being able to produce new individuals of the same kind. This ensures that a particular species does not become extinct.

Types of Reproduction

There are two types of reproduction: **asexual** and **sexual**.

- **Asexual reproduction** involves only **one parent**. It does not involve gametes and no fertilisation takes place. The new individuals are genetically identical to each other and to the parent. This type of reproduction is quite common in plants, fungi and bacteria.
- **Sexual reproduction** involves **two parents**. Each parent produces sex cells called **gametes**. During reproduction, male and female gametes come together and fuse to form a zygote. This is known as **fertilisation**. The zygote then divides many times and grows into the new individual.

Fertilisation is the fusion between male and female gametes to produce a zygote.

15.2 Asexual Reproduction in Plants

An example of a plant that reproduces asexually is the strawberry. The strawberry plant produces a new stem at the base of the parent plant that grows across the surface of the ground. This new stem is called a **runner**, see *Fig. 15.1*.

At certain places where the runner touches the ground it produces new roots and a new shoot. In this way new strawberry plants are formed. Daffodils and tulips reproduce asexually using bulbs.

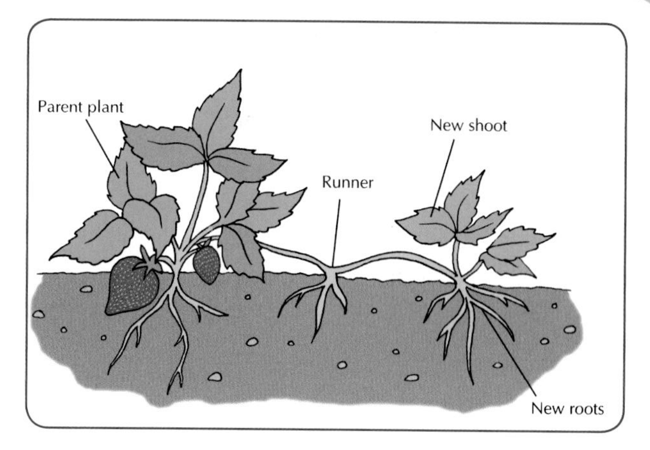

Fig. 15.1 Strawberry runners are an example of asexual reproduction.

TEST YOURSELF:
Now attempt questions *15.1–15.2* and *W15.1*.

15.3 Sexual Reproduction in Plants

The Structure of a Flower

The flower of a plant contains the reproductive organs.

- The male organs are the **stamens**.
- The female organs are the **carpels**.

The structure of a typical flower is shown in *Fig. 15.2*.

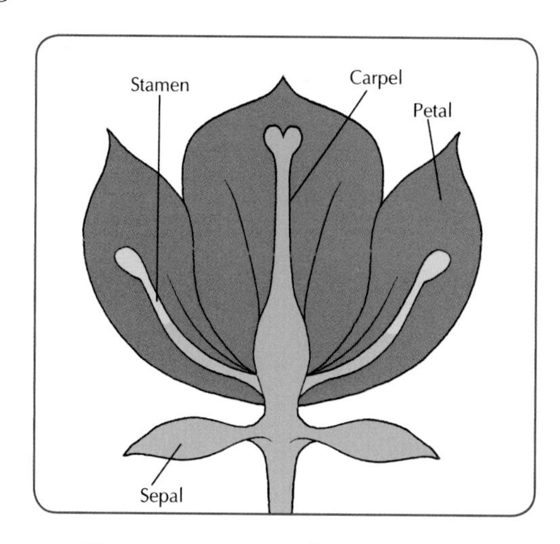

Fig. 15.2 Structure of a typical flower.

BIOLOGY

The Functions of the Parts of the Flower

- **Sepals** protect the flower before it blooms.
- **Petals** attract insects to the flower.
- **Stamens** produce the pollen grains which contain the male gamete.
- **Carpels** produce the eggs (in the ovary).

Different types of flowers vary greatly in the number and arrangement of their parts, but they all have the same basic structure.

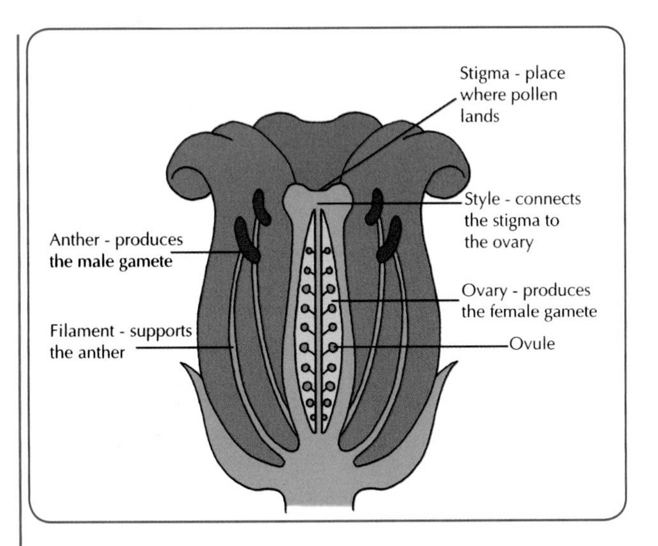

Fig. 15.3 Wallflower flower.

The Parts and Functions of the Stamen and Carpel

Stamens

This is the male part of the flower. Stamens consist of a filament and an anther. Most flowers have lots of stamens.

- The **filament** is the stalk of the stamen, it supports the anther.
- The **anther** produces the pollen grains which contain the male gamete.

Carpels

This is the female part of the flower. Carpels consist of three parts: the stigma, the style and the ovary. Some flowers have a single carpel, others have many.

- The **stigma** is the place where the pollen lands.
- The **style** connects the stigma to the ovary.
- The **ovary** contains ovules which produce the female gamete, the egg.

Stages of Sexual Reproduction in Plants

Sexual reproduction in flowering plants involves the following stages:

1. Pollination
2. Fertilisation
3. Seed (and fruit) formation
4. Seed (and fruit) dispersal
5. Germination

1. Pollination

In order for fertilisation to happen, the male gamete (in the pollen grain) must get to the female gamete (the egg) in the ovary. This is made possible by pollination.

Pollination is the transfer of pollen from the stamen to the carpel.

Pollination is brought about by one of two agents: **wind** or **insect**. Flowers differ in the arrangement of their parts, depending on whether they are wind or insect pollinated. *Figs. 15.4* and *15.5* show the different structure of wind and insect pollinated flowers.

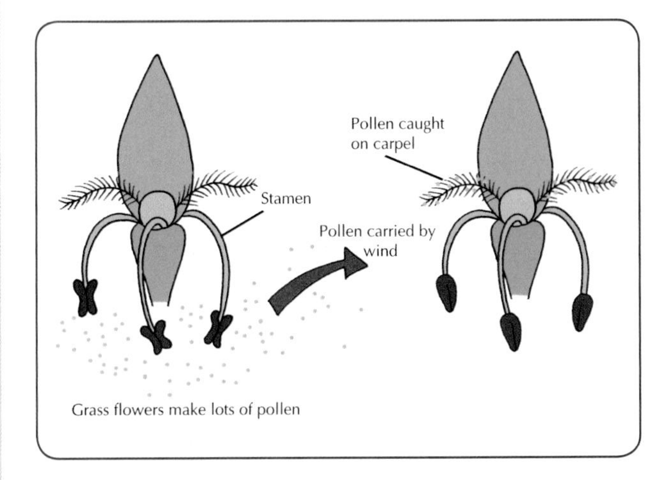

Fig. 15.4 Flower of grass, a typical wind pollinated flower.

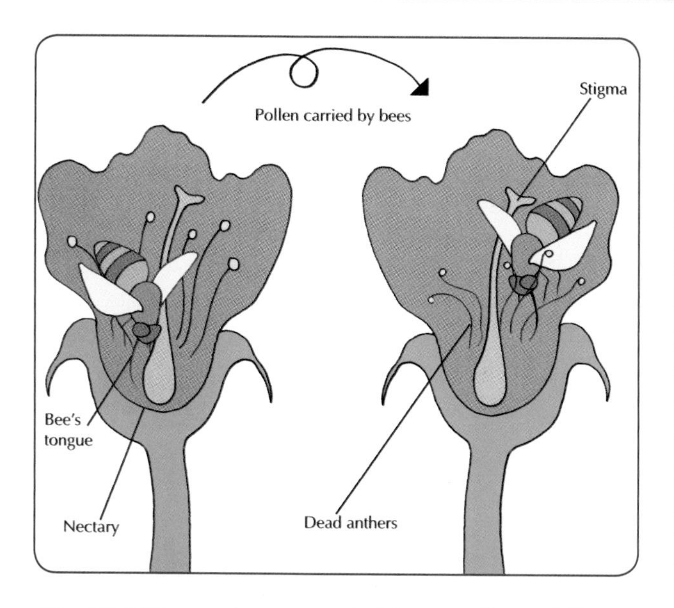

Fig. 15.5 Flower of rose, a typical insect pollinated flower.

Fig. 15.6 The bee transfers pollen as it feeds on the flower.

2. Fertilisation

1. Once the pollen has landed on the carpel, a **tube grows out** of the pollen grain.
2. This tube, seen in *Fig. 15.7*, grows down to the ovule in the ovary. The ovule contains the egg cell.
3. The male **gamete passes** from the pollen tube **into the ovule** and fuses with the egg.
4. A single cell called the **zygote** is formed.
5. The zygote develops into the **new plant**.

Fertilisation is the fusion between male and female gametes to produce a zygote.

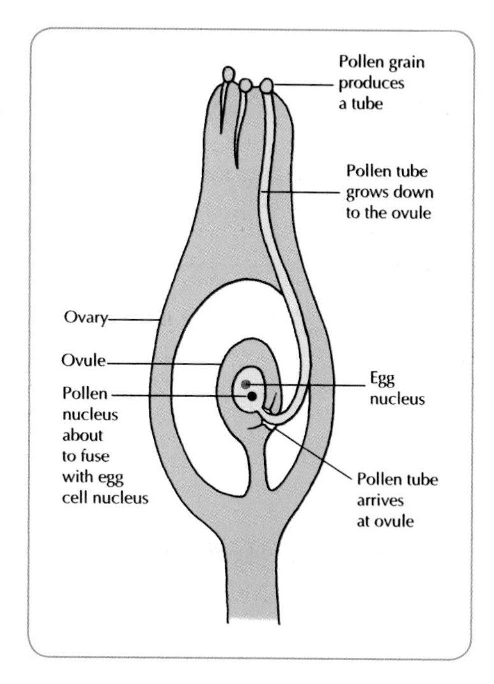

Fig. 15.7 A carpel just before Fertilisation.

TEST YOURSELF:
Now attempt questions *15.3–15.7* and *W15.2–W15.4*.

3. Seed (and Fruit) Formation

- Once fertilisation has occurred, the fertilised ovule becomes a **seed**.
- At the same time, the **sepals, stamens and petals fall off** the flower.
- The ovary swells to become the **fruit**. The fruit protects the seeds.

Some fruits are dry and hard, e.g. the shell around nuts such as hazelnuts and almonds. Other fruits are juicy and good to eat such as raspberries, grapes and oranges, *Fig. 15.8*.

Fig. 15.8 Fruits contain the seeds.

Seeds consist of the following:

- A protective outer coat called the **testa**.
- A baby shoot called the **plumule**.
- A baby root called the **radicle**.
- A **food supply**.

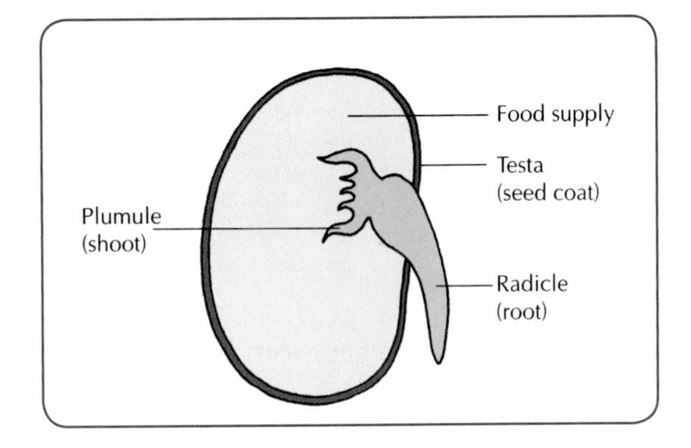

Fig. 15.9 The structure of the broad bean seed.

4. Seed (and Fruit) Dispersal

It is important that seeds and fruits should be **dispersed** (scattered) away from the parent plant. This is to avoid competition between the seedlings and the parent plant. Plants compete for light, water, minerals and space.

Seeds and fruits are dispersed by four main methods: **wind**, **animal**, **water** and **self-dispersal**. *See Table 15.1.*

METHOD	EXAMPLE	SPECIAL FEATURE OF FRUIT/SEED
Wind	Dandelion, sycamore, grasses, *Fig. 15.10*	Seeds/fruits are light, may have wings or hairs
Animal	Raspberry, goose grass (stick weed), *Fig. 15.11*	Fleshy and tasty, may have hooks
Water	Water lily, alder, duck weed, *Fig. 15.12*	Light and buoyant to float
Self	Peas, wallflowers	Have pods which explode to release the seeds

Table 15.1 Methods of seed dispersal.

Fig. 15.10 Wind dispersed fruits and seeds. (a) Dandelion 'parachute'. (b) Sycamore 'helicopter'.

Fig. 15.11 Animal dispersed fruits and seeds – the blackberries are eaten by birds and the seeds deposited away from the bramble bush.

Fig. 15.12 The seeds of the water lily are dispersed by water.

5. Germination

Germination is the growth of a seed into a new plant.

Once the seeds have been dispersed they will germinate, but only if conditions are suitable. Many seeds can not germinate in the winter because it is too cold. Seeds need the following conditions in order to germinate:

- Water
- Oxygen
- Correct temperature

Fig. 15.13 shows the stages in the germination of a broad bean seed.
Mandatory experiment 15.1, investigates the conditions needed for germination.

Mandatory experiment 15.1

To investigate the conditions necessary for germination

Apparatus required: 4 test-tubes; cotton wool

Also required: cress seeds; oil; cool boiled water

Method

1. Set up the four test-tubes as shown in *Fig. 15.14*, page 98.
2. Boiling the water used in test-tube *C* removes the oxygen from it. The layer of oil prevents oxygen getting back into the water.
3. Place test-tubes *A*, *B* and *C* in a warm place for a week.
4. Place test-tube *D* in a fridge for a week.

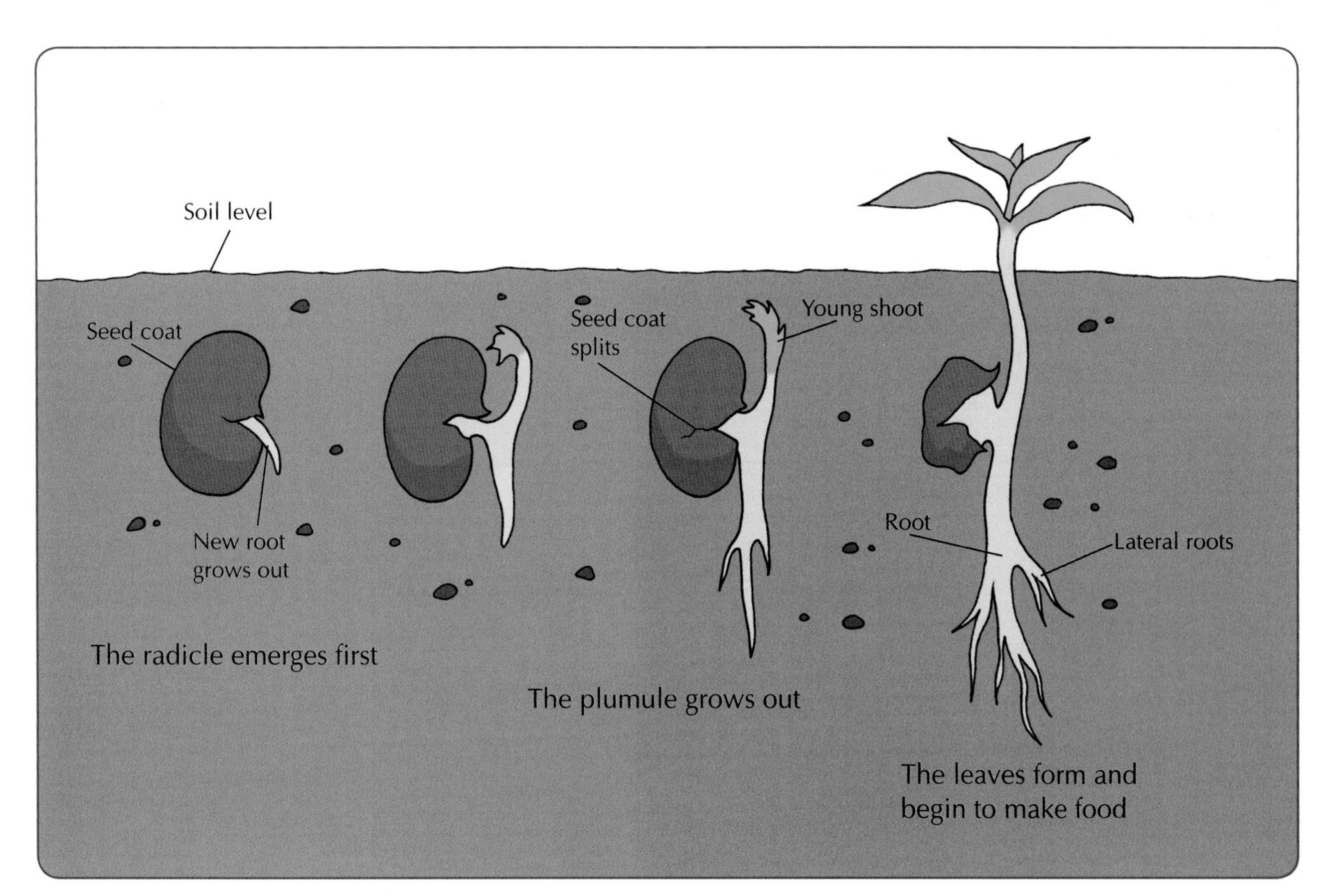

Fig. 15.13 The main events in the germination of the broad bean seed.

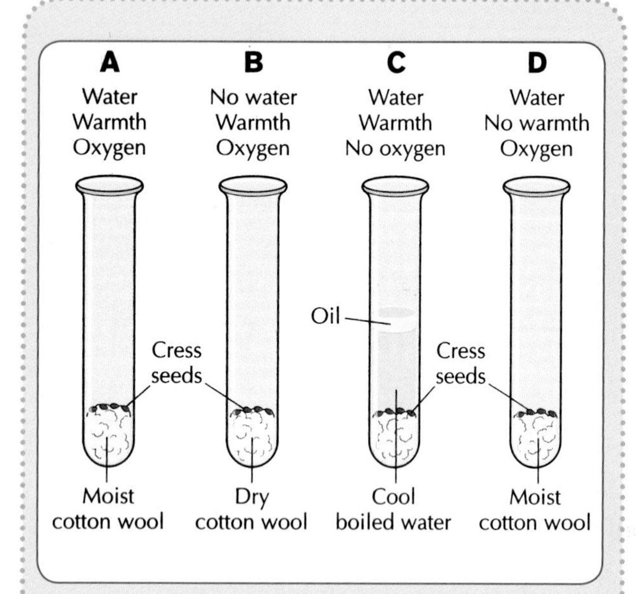

Fig. 15.14 Set up to show the conditions necessary for germination.

Results

Test-tube *A*: The cress seeds germinate. These seeds have all the conditions for germination.

Test-tube *B*: The seeds do not germinate. They have no water.

Test-tube *C*: The seeds do not germinate. They have no oxygen.

Test-tube *D*: The seeds do not germinate. They have no warmth.

Conclusion

Seeds need water, oxygen and warmth to germinate. If any of these factors are missing the seeds do not germinate.

Exam Question

Junior Certificate Ordinary Level Examination Question

A number of cress seeds were set up as shown in the diagram and left for a few days to investigate the conditions necessary for germination. Test-tubes A,B and D were kept in the laboratory at room temperature. Test-tube C was placed in the fridge at 4°C.

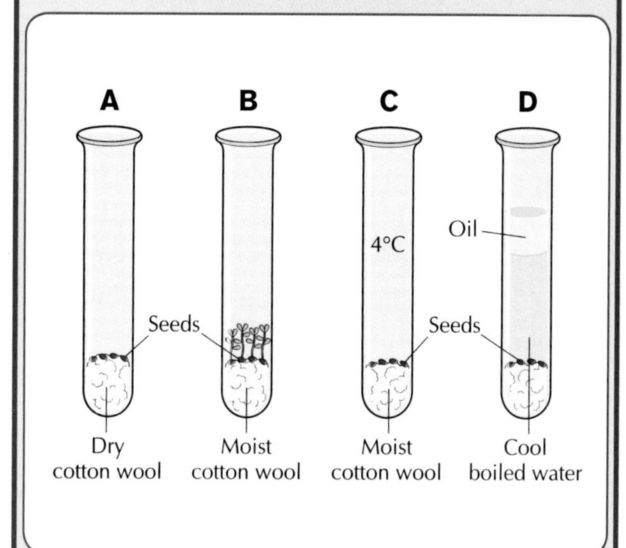

Fig. 15.15

(i) Why do only the seeds in test-tube B germinate?

A: Because these seeds are the only ones to have all the things needed for germination, i.e. they have moisture, oxygen and a suitable temperature.

(ii) Why is the water in test-tube D boiled before use?

A: Boiling the water removes the oxygen from the water.

(iii) Explain why the seeds in test-tube C failed to germinate.

A: The seeds did not have a suitable temperature, it is too cold.

(iv) Why is this investigation considered to be a 'fair test'?

A: Because only one thing changed at a time in each of the tubes.

TEST YOURSELF:

Now attempt questions *15.8–15.12* and *W15.5–W15.8.*

What I should know

- Reproduction is being able to produce new individuals of the same kind.
- Plants can reproduce by sexual and asexual means.
- Sexual reproduction involves two parents and the fusing of male and female gametes to produce a zygote.
- Asexual reproduction involves only one parent. It does not involve gametes.
- Strawberry runners are an example of asexual reproduction.
- The flower is the reproductive organ in flowering plants.
- The parts of the flower:

  The sepals are for protection.
  The petals are to attract insects.
  The stamens are the male part and produce pollen grains which contain the male gamete.
  The carpels are the female part and they produce the egg.

- The stamen is made up of the filament and anther.
- The anther produces the pollen containing the male gamete.
- The carpel is made up of the stigma, style and ovary.
- The ovary produces the egg.
- The stages in flowering plant reproduction are: 1. Pollination, 2. Fertilisation, 3. Seed Formation, 4. Seed Dispersal and 5. Germination.
- Pollination is the transfer of pollen from the stamen to the carpel.
- Flowers may be pollinated by insects or by wind.
- Fertilisation is the fusion of the male and female gametes to produce a zygote.
- Following fertilisation, seeds are formed.
- Seeds are dispersed to avoid competition with the parent plant.
- The methods of seed dispersal:

  Wind (sycamore)
  Animals (blackberries)
  Water (water lily)
  Self-dispersal (peas)

- The parts of a seed are the testa, the radicle, the plumule and a food store.
- The testa protects the seed; the radicle becomes the root; the plumule becomes the shoot and the food store provides food during germination.
- Germination is the growth of the seed into a new plant.
- The conditions necessary for germination are: water, oxygen and warmth.

QUESTIONS

Write the answers to the following questions into your copybook.

15.1 (a) What is reproduction?

(b) Name two types of reproduction.

(c) Give two differences between the types of reproduction you name in (b).

15.2 (a) Name a method of asexual reproduction used by plants.

(b) Describe how asexual reproduction is carried out by the strawberry plant.

(c) Name another plant that reproduces asexually.

15.3 (a) Name the parts of the flower labelled *A–D* in *Fig. 15.16*.

(b) Give one function for each of the parts named in (a).

15.4 (a) What is the name of the female part of the flower?

(b) Which parts of the flower protects the flower before it blooms?

(c) Where are the pollen grains made?

(d) What do the pollen grains contain?

(e) In which part of the flower does fertilisation take place?

(f) What does the carpel produce?

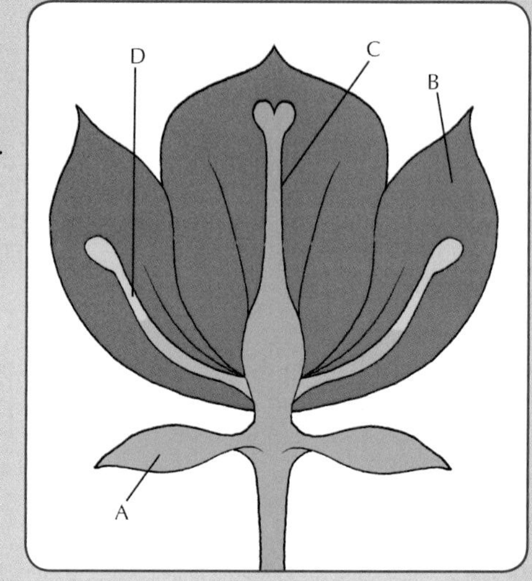

Fig. 15.16

15.5 (a) Name the parts of the carpel and stamen shown in *Fig. 15.17*.

(b) Give the functions of parts *X* and *S*.

15.6 (a) What is pollination?

(b) Name two agents of pollination.

(c) Name the method of pollination used by the flower shown in *Fig. 15.16* above, giving a reason for your answer.

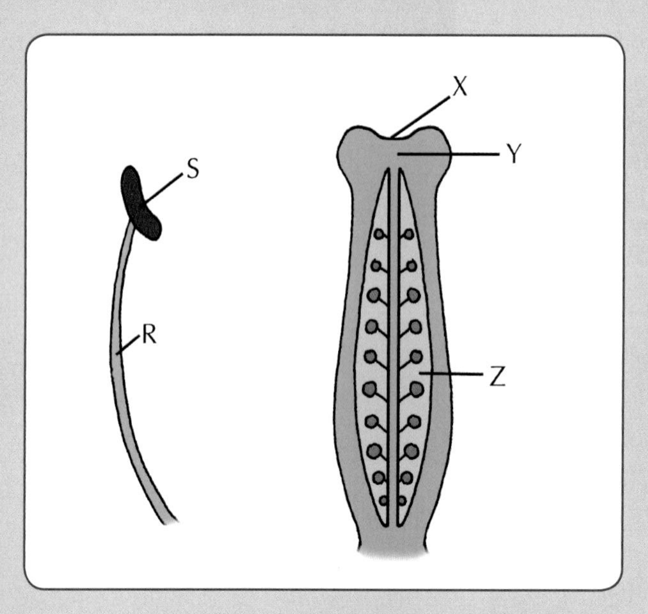

Fig. 15.17

QUESTIONS

15.7 (a) What is meant by fertilisation?

(b) Does fertilisation take place in the anther or the ovary?

(c) Name the structure that the pollen grain produces after it has landed on the stigma.

(d) What does the zygote develop into?

15.8 Following fertilisation, ….. are formed.

15.9 (a) What is the function of the fruit of a plant?

(b) Name three different fruits and for each one say how it is dispersed.

(c) Why is seed and fruit dispersal necessary?

15.10 (a) Name the parts labelled *A–D* of the seed shown in *Fig. 15.18.*

(b) Which labelled part grows into the shoot?

(c) Which labelled part grows into the root?

(d) List the four methods by which seeds are dispersed and give an example in each case.

(e) Humans use a lot of seeds as a source of food, e.g. pumpkin seeds. Name three other types of seeds used by humans for food.

(f) How could the seed be tested to show it has a store of food?

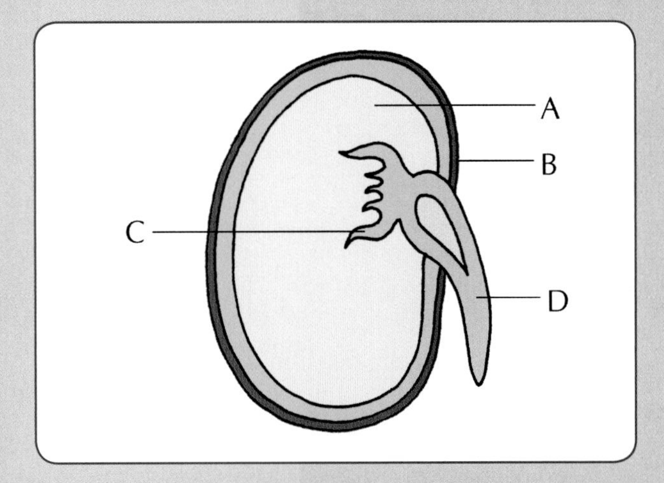

Fig. 15.18

15.11 (a) What is germination?

(b) List the conditions necessary for germination.

(c) Describe an experiment to show these conditions are necessary.

15.12 Describe how you could investigate if the amount of moisture affects how quickly seeds germinate.

Chapter 16 Ecology

16.1 What is Ecology all about?

All living things are affected by their environment and respond to changes that happen in the places around them.

Ecology is the study of how living things interact with each other and with their environment.

Examples of How Organisms Interact

- **Animals** need plants for food and shelter. If food is scarce, animals will move to somewhere else, otherwise they might starve.
- **Plants** rely on animals to transfer pollen and to scatter their seeds.
- When plants and animals die, **bacteria and fungi** rot them down and return minerals to the soil. These are all examples of how organisms interact.

Fig. 16.1 Animals rely on plants for food and shelter, e.g. birds nest in trees.

16.2 Habitats

The place where an organism lives is known as its **habitat**. Examples of habitats include a hedgerow, woodland, meadow, pond, or rocky sea shore, *Fig. 16.2*.

The habitat of an organism is the place where it lives.

Each habitat has its own collection of plants and animals and the conditions they need to survive. For example, earthworms, daisies, green fly and robins are commonly found in garden habitats. You would not expect to find crabs or lobsters in a garden because the conditions they need are not to be found in a garden.

Together the habitat and its community of different plants and animals forms a unit called an **ecosystem**.

Fig. 16.2 A seashore habitat.

Populations in the Habitat

Each habitat is made up of different **populations** of plants and animals, e.g. a hedgerow might have a population of woodlice, a population of nettles and a population of hawthorn trees.

The numbers in a population of plants and animals can vary. These changes depend on things like whether the organism can get enough light or food and whether they are being killed for food by other animals.

16.3 Feeding Relationships in a Habitat

(1) **Producers**. Green plants are known as **producers** because they make their own food. When they photosynthesise, plants produce food for themselves and for animals.

(2) **Consumers**. All other organisms get their food either by eating plants or eating other animals. These are known as **consumers**.

There are three types of consumer:

- **Herbivores** feed on plant material only, e.g. greenfly, limpets, snails.
- **Carnivores** feed on animal material only, e.g. centipedes, starfish, sparrow hawks.
- **Omnivores** feed on both plant and animal material, e.g. badgers, barnacles, humans.

(3) **Decomposers**. Finally there are the **decomposers**, such as bacteria, fungi, woodlice and earthworms. Decomposers break down dead plant and animal material. As a result nutrients are put back into the soil, in other words the nutrients are recycled.

- **Producers make their own food, e.g. green plants.**
- **Consumers cannot make their own food. They feed on other organisms, e.g. herbivores, carnivores and omnivores.**
- **Decomposers break down dead material, e.g. bacteria and fungi.**

Food Chains

One of the most important relationships in a habitat is the **feeding relationship** between different organisms. A **food chain** is the way in which energy and nutrients are passed from one organism to another.

Examples of a Food Chain

- grass → rabbit → fox
- seaweed → limpet → sea gull
- nettle leaves → caterpillar → thrush → sparrow hawk

In each case, the chain begins with a green plant – the producer. The arrows show the direction of the energy and nutrients. The arrow means 'eaten by'.

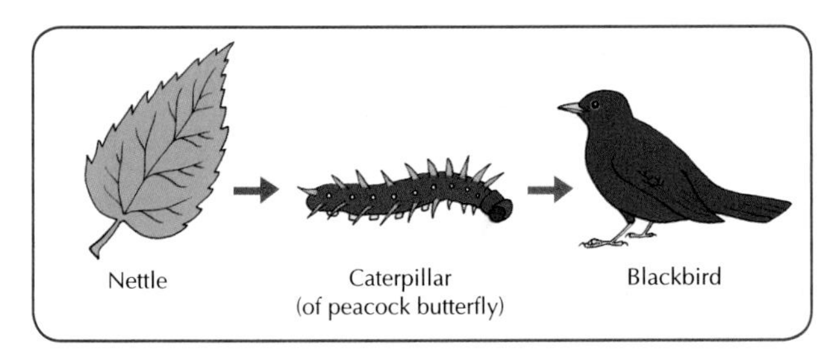

Fig. 16.3 A simple food chain.

Some examples of food chains from a variety of habitats can be seen in *Table 16.1*.

HABITAT	FOOD CHAIN
Grassland	grass → rabbit → fox
Woodland	oak leaves → caterpillars → thrush
Rocky sea shore	plankton → mussels → starfish → sea gull
Hedgerow	hawthorn → greenfly → spider → blackbird

Table 16.1

Energy Flow in a Food Chain

Fig. 16.4 shows a simple food chain. Energy from the sun is used to make food in the leaves. The leaves, in turn, provide energy for the consumers. A huge number of leaves are needed to provide enough energy containing materials (food) for a large number of caterpillars. But these caterpillars can only provide enough food for a small number of thrushes. This is because the consumers in the chain either use up or waste nearly all the materials they take in. As a result, very few are left to be passed on to the next consumer.

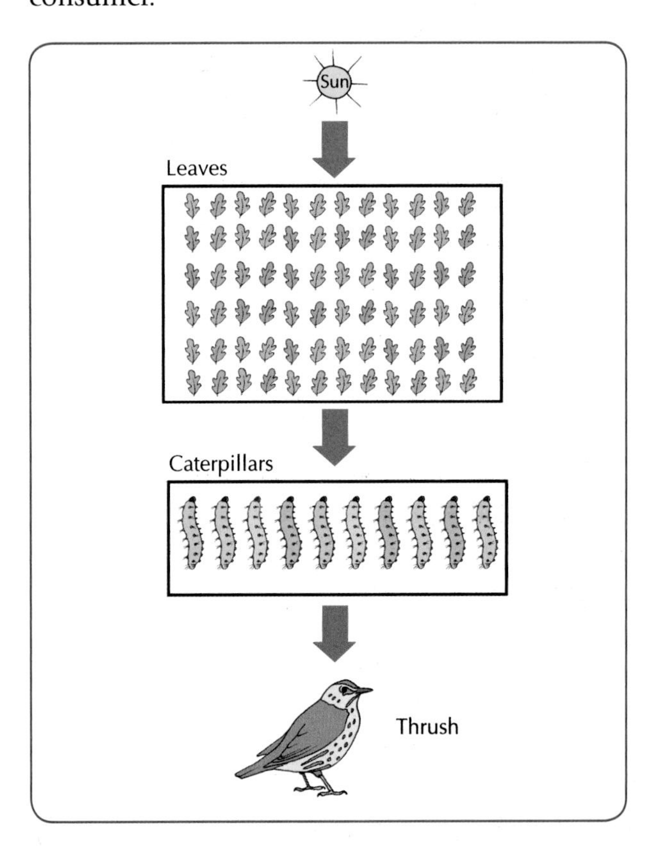

Fig. 16.4 Energy flow in a food chain.

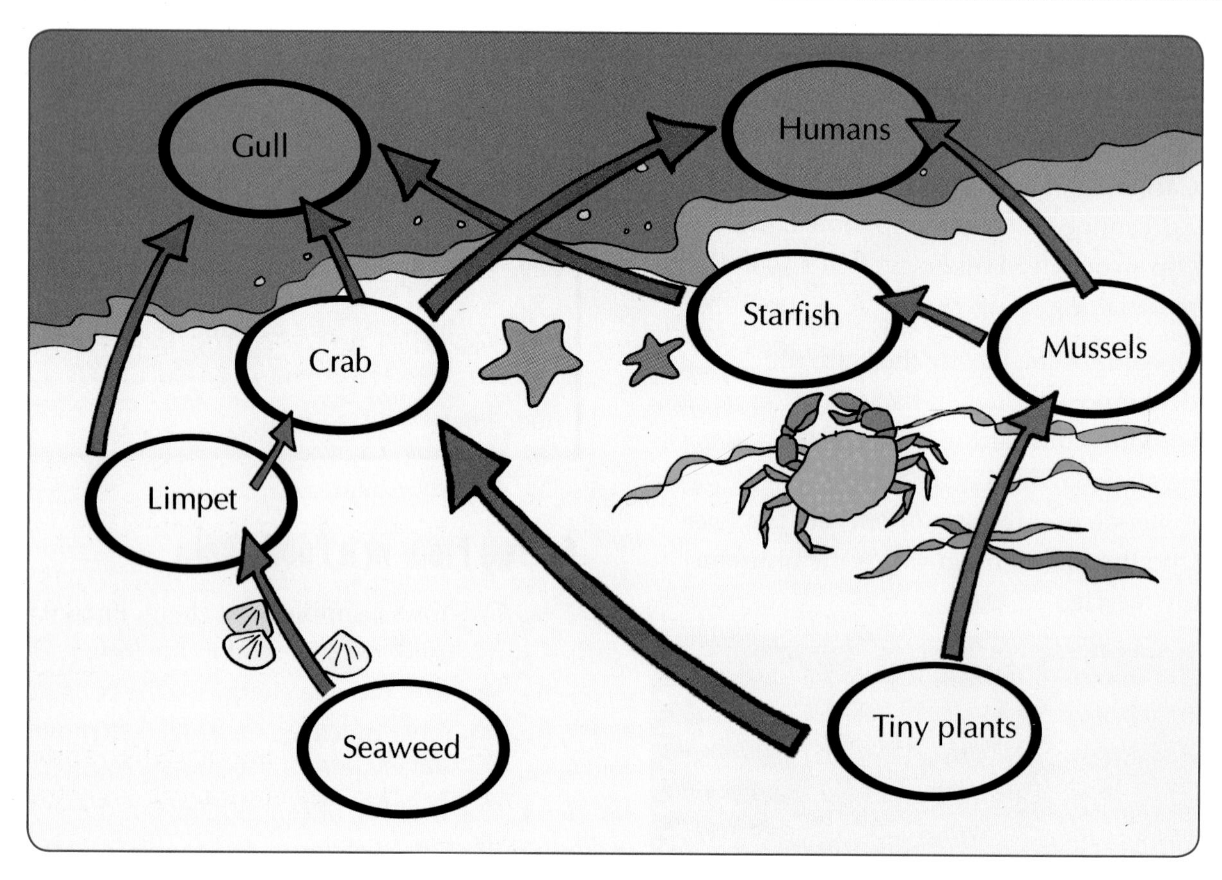

Fig. 16.5 Part of a seashore food web.

Food Webs

In a habitat, the feeding relationships are rarely as simple as the ones described so far. Usually the various food chains are inter-connected, *Fig. 16.5*. When two or more food chains are inter-connected a **food web** is formed. A food web clearly shows that organisms in a habitat depend on one another for food.

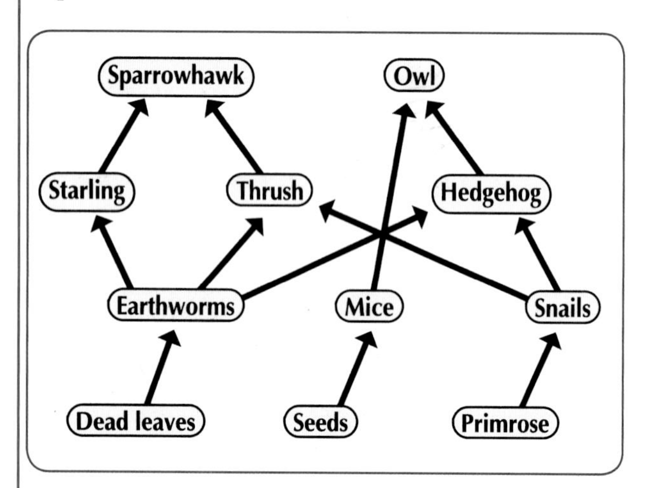

Fig. 16.6 A food web from a meadow.

A food web is a series of inter-connecting food chains.

TEST YOURSELF:

Now attempt questions *16.1–16.6* and *W16.1–W16.5*.

16.4 Adaptation

In order to survive in a habitat, plants and animals must be able to cope with changes that occur. To do this, living things have **adaptations** which are characteristics that enable them to function more effectively and survive longer.

Adaptation is when an organism has characteristics that make it well suited to its environment.

Examples of Adaptations

Ladybirds are brightly coloured to warn predators of their bitter taste, *Fig. 16.7*. This helps them survive. Ladybirds also release a nasty-smelling fluid which wards off birds and humans. *Table 16.2* describes some adaptations in different habitats.

HABITAT	PLANT ADAPTATION AND BENEFIT	ANIMAL ADAPTATION AND BENEFIT
Hedgerow	Nettles have stinging hairs to prevent them being eaten by certain animals.	Peacock butterflies have long tube-like mouthparts that allow them to suck nectar.
Grassland	Buttercups have bright yellow petals to attract insect pollinators.	Earthworms are a dark colour for camouflage.
Woodland	Primroses flower early in the spring and get more light, before the leaves come out on the trees.	Sparrowhawks have large feet with needle-sharp talons for catching their prey.
Rocky sea shore	Seaweeds produce a slimy mucilage that prevents them drying out at low tide.	Limpets have a muscular foot to anchor it to the rocks and prevent them from being swept away.
Pond	Common water-crowfoot has two types of leaf, one for under water, the other for above.	Dragonflies have huge eyes and keen eyesight to identify their insect prey.

Table 16.2 Examples of plant and animal adaptations in different habitats.

Fig. 16.7 Ladybirds show adaptations for survival.

16.5 Competition

Competition is when organisms 'fight' over things that they need.

- **Plants** compete for light, water, space and minerals. For example, daisies and grass compete for light in the lawn.
- **Animals** compete for food, territory, mates and shelter. Sometimes different types of animals compete for the same resource, e.g. thrushes and blackbirds compete with each other for earthworms and insects. In other cases, competition occurs between animals of the same type, e.g. male robins sing to warn off rivals from their territory. Because animals can move around, they are able to overcome the problems of shelter, space and food shortage much more easily than plants.

HABITAT	PLANT COMPETITION	ANIMAL COMPETITION
Hedgerow	Honeysuckle and ivy compete for light.	Blackbirds and thrushes compete for berries.
Rocky seashore	Different brown seaweeds compete for space.	Limpets and periwinkles compete for space.

Table 16.3 Competition in two different habitats.

Competition takes place when organisms struggle for the same resources in the habitat.

16.6 Interdependence

In any habitat, many organisms depend on each other to stay alive.

- **Animals depend on plants** for shelter and for food, eg. grey squirrels nest in beech trees and rabbits eat grass.
- **Plants also depend on animals**, e.g. to transfer pollen (e.g. bees and flies) and to scatter their seeds (e.g. blackbirds).

Sometimes the same plant and animal need each other.

- For example, the grey squirrel needs the beech tree for shelter and at the same time the squirrel disperses the beech nuts. In this case the plant and animal are said to be **interdependent**.

Interdependence is when two different types of organism need each other.

TEST YOURSELF:

Now attempt questions *16.7–16.9* and *W16.6–W16.8.*

What I should know

- Ecology is the study of how living things interact with each other and their environment.
- The habitat of an organism is the place where it lives.
- An ecosystem is the habitat plus the community of organisms.
- A food chain is a feeding relationship between organisms through which energy is transferred, e.g. grass → rabbit → fox.
- A producer makes its own food, e.g. grass.
- A consumer cannot make its own food, it eats plants or animals.
- There are three types of consumer: herbivore, carnivore and omnivore.
- Decomposers break down dead matter; examples are bacteria and fungi.
- A food web is a series of interconnecting food chains.
- Adaptation is when an organism has characteristics that make it well suited to (able to cope with) its environment.
- Competition takes place when organisms struggle for the same resources in the habitat.
- Plants compete for light, water and minerals.
- Animals compete for space, food, mates and shelter.
- Interdependence is when two different types of organism need each other.

Questions

Write the answers to the following questions into your copybook.

16.1 (a) What is ecology?

(b) What is meant by the environment of an organism?

(c) What is a habitat? Name two habitats.

(d) What is an ecosystem?

16.2 Name a common habitat for each of the following organisms:

(i) snail, (ii) dandelion, (iii) salmon, (iv) periwinkle, (v) wheat, (vi) primrose

16.3 (a) Name a habitat you have studied.

(b) Name two animals from the habitat.

(c) Name two plants from the habitat.

16.4 (a) Plants are the ..... in a habitat because they can make their own food.

(b) Two examples of decomposers in a habitat are ..... and ..... .

(c) Animals that can only eat plants are called ..... . An example is a ..... .

(d) An omnivore can eat ..... and ..... .

(e) Rabbits, slugs and greenfly are examples of ..... . They eat only ..... .

QUESTIONS

16.5 (a) What is a food chain?

(b) Give an example of a food chain that has three links.

(c) Give an example of a food chain with four links.

(d) Suggest a reason why most food chains cannot have more than four to five links in the chain.

16.6 (a) What is a food web?

(b) *Fig. 16.8* shows part of a garden food web. Use the information in the diagram to answer the following:

(i) List as many types of animal as you can see.
(ii) What is the most common plant in the diagram?
(iii) Name two producers in the diagram.
(iv) From the food web, write down three different food chains.
(v) Identify one example of competition from the food web.
(vi) If all the rosebushes and cabbages in the garden were killed by some chemical, what effect do you think this would be likely to have on the population of (a) greenfly, (b) ladybirds and (c) snails?

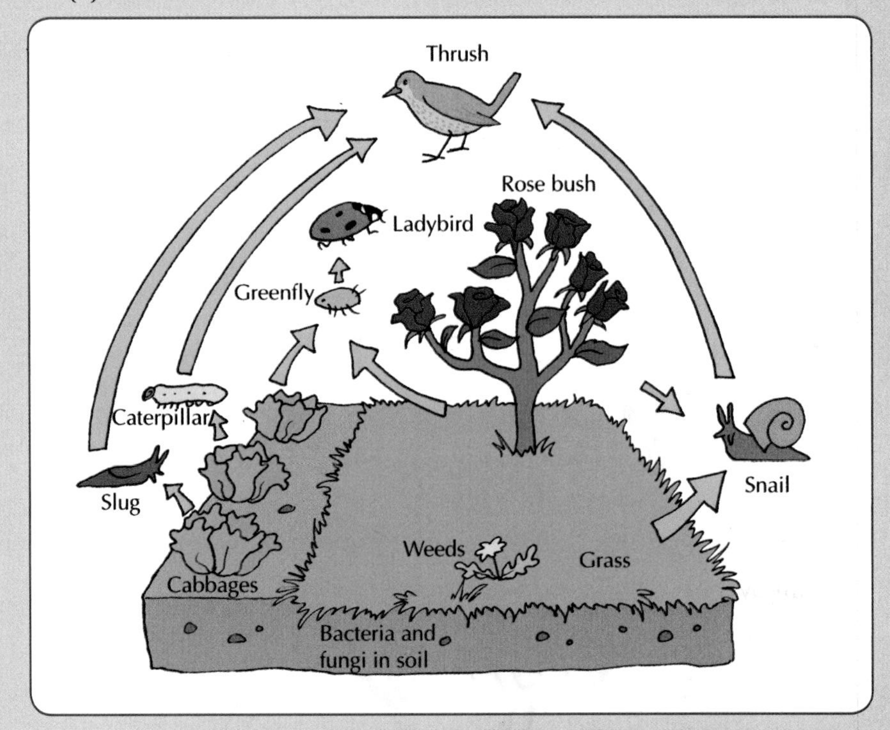

Fig. 16.8

16.7 (a) What is meant by an adaptation in ecology?

(b) Give one example of a plant adaptation that you can see in *Fig. 16.8*.

(c) Give one example an animal adaptation that you can see in *Fig. 16.8*.

16.8 (a) What is meant by competition in a habitat?

(b) Give an example of competition between animals in a named habitat.

(c) Give an example of competition between plants in a named habitat.

16.9 Give an example of interdependence between plants and animals in a named habitat.

Chapter 17 Habitat Study

17.1 Introduction

There are many different habitats that you might choose to study. It could be a meadow, a hedgerow, the rocky sea shore or the school grounds. Whichever it is, studying a habitat involves obeying the **country code**, which your teacher will explain to you. It can be summarised by the quotation:

'Take nothing but photographs
Leave nothing but footprints
Kill nothing but time.'

Steps in Studying a Habitat

1. Make a simple **map** of the habitat to illustrate the main features of the area, e.g. trees, shrubs, paths, ditches etc.
2. Make a note of the **environmental factors** affecting the habitat, e.g. the weather, the type of soil, the height above sea level etc.
3. **Identify** and list the types of plants and animals present.
4. **Estimate the number and distribution** of the organisms in the habitat.
5. Prepare a **report** on your findings.

Fig. 17.1(a) A pond. (b) A hedgerow habitat.

17.2 Environmental Factors Affecting the Habitat

A number of different factors affect the plants and animals living in any habitat. These include:

(1) **Factors relating to the soil**. E.g. soil type, soil pH, moisture content. The type of soil present will determine the type of plant and animal that can live in the habitat.

(2) **Weather factors**. These include light intensity, rainfall, temperature and wind.

(3) **Physical factors**. These include the direction the habitat faces (north, south, east or west), height above sea level, whether the ground is flat or on a slope etc.

(4) **Biotic factors**. These are the living factors such as those caused by competition between plants and animals. Humans can have a major effect on a habitat, e.g. when we pollute an environment.

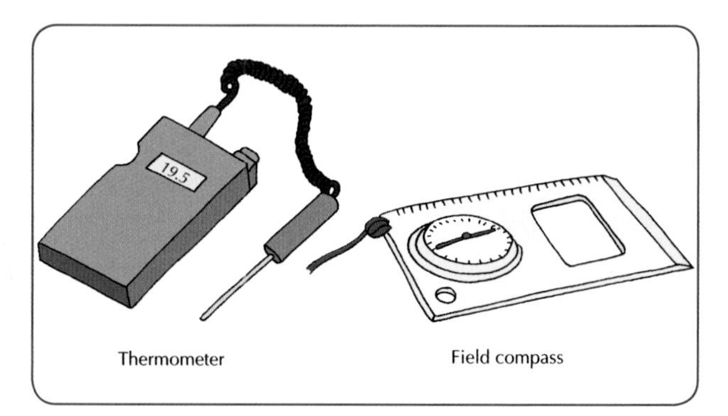

Fig. 17.2 A thermometer and a compass can be used to measure environmental factors affecting a habitat, such as temperature and direction.

The environmental factors can be measured and recorded. Some methods for measuring these factors are given in *Table 17.1*.

ENVIRONMENTAL FACTOR	HOW IT MAY BE MEASURED
Soil pH	Universal indicator paper/pH meter
Rainfall	Rain gauge
Temperature	Thermometer
Light intensity	Light meter
Direction	Compass

Table 17.1 Measuring environmental factors.

TEST YOURSELF:
Now attempt questions *17.1* and *W17.1.*

17.3 Collecting and Identifying Plants and Animals

Plants

Plants do not move around so it is easy enough to see which ones are present.

- Sometimes **a wild flower or a tree-spotter's guide book** can be used, or an identification 'key'. In *Chapter 2* you learned how to use a **key** to identify plants and animals. Remind yourself of how to use a key by revising *Section 2.4* of this chapter.

APPARATUS	HOW IT IS USED	USED TO COLLECT
(1) Pooter	One tube is placed over the organism and you suck through the tube with gauze (the gauze prevents you swallowing the organism).	Small insects (such as greenfly) and spiders.
(2) Beating tray	A white sheet/tray is held under the leaves of a tree or shrub. The branches are shaken or beaten with a stick (this dislodges the animals and they fall onto the sheet).	Insects and small animals that live on the leaves of shrubs and trees.
(3) Pitfall trap	A jar is sunk into the ground (the mouth of the jar must be level with the soil). A flat stone supported by small stones forms a lid (the lid prevents rain getting in).	Animals, e.g. ground beetles, that walk along the surface of the ground. Useful to compare animals that are out at night with those around during the day.

Table 17.2 Methods of collecting animals.

- If you cannot identify a plant, then a leaf, fruit or flower of it can be placed in a plastic bag and **taken back to the laboratory**. Your teacher can then help you identify the unknown specimen. But always remember that rare plants should never be collected.

Animals

Some animals are fairly easy to find, e.g. snails and slugs, which don't move around too fast. Limpets and whelks on the sea shore can be dislodged from rocks using a knife. But because a lot of animals move very quickly or are not visible during the day, special collecting methods are needed.

Three pieces of apparatus used for collecting animals are: a **pooter**; a **beating tray** and a **pitfall trap**. Their appearance, how they are used and what they are used to collect is shown in *Table 17.2*, page 109.

TEST YOURSELF:
Now attempt questions *17.2–17.7* and *W17.2–W17.4*.

17.4 Estimating the Number of Organisms in a Habitat

It would be impossible to count all the plants and animals present in a given habitat. Instead we count only a few of them, i.e. we take a random sample which gives us some idea of the overall numbers.

Two methods of sampling are to use (1) a **quadrat** and (2) a **line transect**.

(1) Quadrat

A quadrat is a square area/frame which is thrown at random in the habitat. Quadrats vary in size depending upon the habitat, but are usually $1 m^2$ or $0.5 m^2$. A quadrat may be made of metal, wood, plastic or string, *Fig. 17.3*.

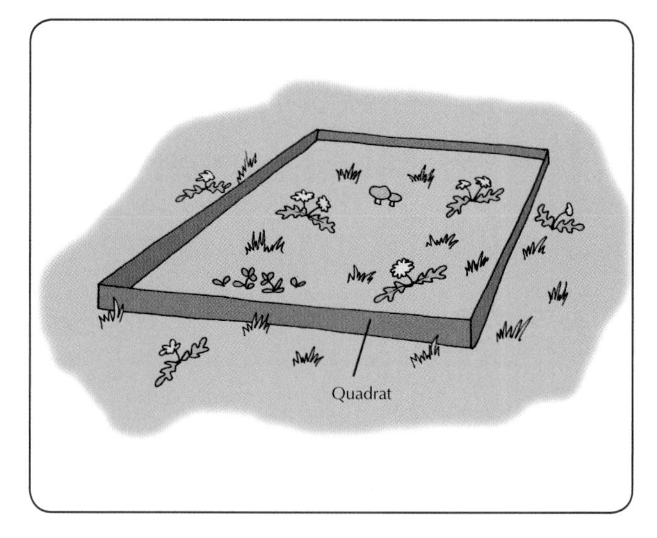

Fig. 17.3 A quadrat.

A quadrat is a square frame used to estimate plant numbers.

(2) Line Transect

The word transect means 'cut across'. A transect shows how the type of plant or animal changes across an area, e.g. from a shaded area under trees out onto a playing field, or across the seashore from the road to the water. A **line transect** is a method of showing the **distribution of organisms** in a habitat.

A line transect consists of a rope marked off with tape at regular intervals. Each mark is called a **station**. The transect is laid across the habitat and pegged down at each end, *Fig. 17.4*, page 112. The types of plant or animal found touching each station are recorded.

A line transect is a rope marked out at regular intervals across a habitat. It is used to investigate the distribution of organisms.

Mandatory investigation 17.1

(a) To study a local habitat

The study of the habitat should be written up as a report to cover the following steps:

1. Make a simple outline **map** of your habitat. Record the position of north.

2. Record the **environmental factors** affecting the habitat.
3. **Collect** and **identify** at least five plants and five animals in the habitat.
4. Use a **quadrat** to examine the **variety of organisms** present.
   See *Mandatory investigation 17.1(b)*.
5. Use a **line transect** to investigate the **distribution of organisms** across contrasting areas of the habitat, e.g. from sunny to shade.
   See *Mandatory investigation 17.1(c)*.
6. Use the information gathered to find examples of the following from the habitat:

   (i) food chains, (ii) a food web (HL only), (ii) competition, (iv) adaptation and (v) interdependence (HL only).

(b) To show the variety of plants in a habitat using a quadrat

Apparatus required: quadrat; clipboard; pen; recoding sheet

Also required: plant identification book/plant key

Method

1. Throw a pencil over your shoulder at random (take care that there is no one behind you).
2. Place the quadrat where the pencil lands.
3. Record the names of the types of plants present inside the quadrat boundary, as in *Table 17.3*. The actual numbers of plants does not matter.
4. The quadrat should be thrown at least ten times. The more times you throw the quadrat, the more accurate your results will be.
5. You can use the information gathered to calculate the percentage frequency of a particular plant. For example, if dandelions are found in six out of ten quadrats, then you would expect 60 dandelions to occur in 100 quadrats. Therefore we can say dandelions have a 60 per cent frequency.
6. Use your results to draw a bar chart. This makes your results easier to understand. It also makes it simpler to compare with the types of plants present in other habitats.

PLANT	QUADRAT NUMBER										TOTAL	PERCENTAGE FREQUENCY
	1	2	3	4	5	6	7	8	9	10		
Dandelion		√	√		√		√	√	√		6	60%

Table 17.3 Results for a quadrat survey.

DAFOR Scale

An alternative method to estimate the occurrence of plants present is to use the DAFOR scale. The letters DAFOR stand for:

D – dominant

A – abundant

F – frequent

O – occasional

R – rare

The problem with the DAFOR scale is that it is subjective, i.e. what one person might consider abundant might appear dominant to another person. This method is not very accurate.

(c) To investigate the distribution of plants using a line transect

Apparatus required: a length of rope marked off at intervals; stakes; clipboard/recording sheet/ pen

Also required: plant identification books and keys

Method

1. Lay the marked rope across an area of change in the habitat, see *Fig. 17.4*. Mark this line on your map of the habitat. Stake the rope at either end.
2. Record the name of any plant which is touching or under the rope at each station (*Table 17.4*).

STATION NUMBER	NAME OF PLANT
1	Hawthorn tree
2	Fern
3	None – path
4	Short grass
5	Dandelion
6	Tall grass

Table 17.4

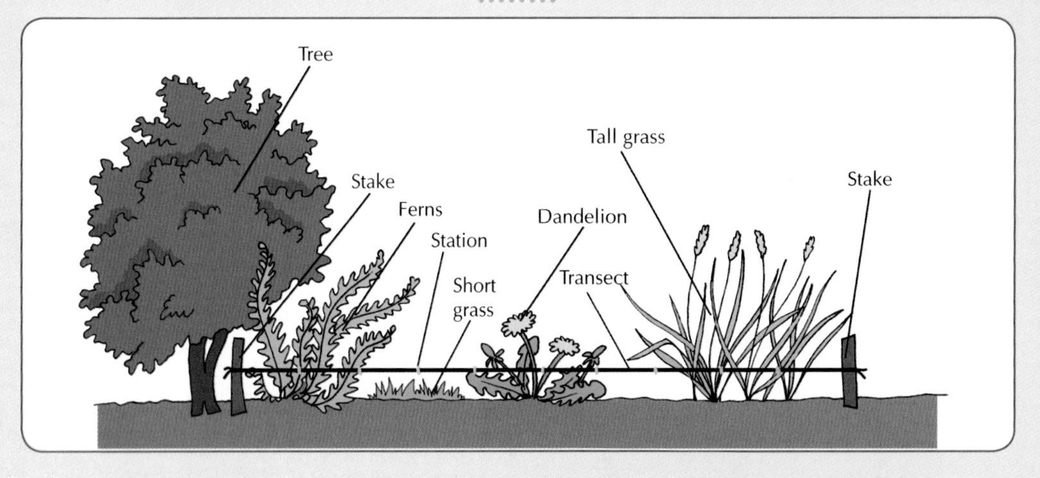

Fig. 17.4 A line transect is used to show the distribution of organisms in a habitat.

Exam Question

Junior Certificate Science Higher Level Examination Question

Many species of plant are protected in National Parks. The manager of one of these parks is asked to measure the frequency with which a protected species occurs in a habitat within a park.

Describe how this might be carried out. Include a diagram of any equipment that might be used.

A: The frequency of a plant means how common it is in the area. In this case it would be impossible for the park manager to count every one of the protected plants. Instead s/he takes a number of samples and from these an estimate of the frequency of the plants can be made.

Method

1. **Throw a quadrat over your shoulder at random.**
2. **Note if the protected species is present.**
3. **Repeat a large number of times in different locations.**
4. **Record the results in a table.**

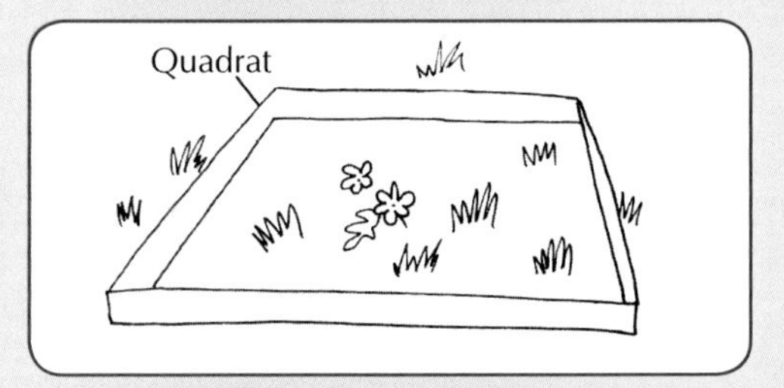

Fig. 17.5 You must give a diagram to get full marks for this question.

TEST YOURSELF:
Now attempt questions *17.8–17.11* and *W17.5–W17.6*

What I should know

- A habitat study involves: (a) mapping the habitat, (b) recording the environmental factors, such as temperature and soil pH, (c) identifying the plants and animals, (d) estimating the number and distribution of the plants and animals, (e) writing a report.
- Plants and animals can be identified using keys and books.
- Pooters, pitfall traps and beating trays are methods of collecting animals, see *Table 17.2* on page 109.
- A quadrat is a square frame used to estimate plant numbers.
- A line transect is a rope marked out at regular intervals across a habitat. It is used to investigate the distribution (spread) of organisms.

QUESTIONS

Write the answers to the following questions into your copybook.

17.1 (a) What is a habitat?

(b) Name three habitats.

(c) Name two weather factors that affect a habitat.

(d) What would you use to measure the light intensity in a habitat?

(e) What would you use to measure the pH of the soil in a habitat?

17.2 (a) List three pieces of apparatus suitable for collecting animals.

(b) Which part(s) of a plant are most suitable to help identify a plant?

(c) What is a key used for in Ecology?

17.3 (a) Name the piece of equipment in *Fig. 17.6*.

(b) Describe how you use it.

(c) What is the function of the part labelled *B*?

(d) Name two animals you would expect to collect using it.

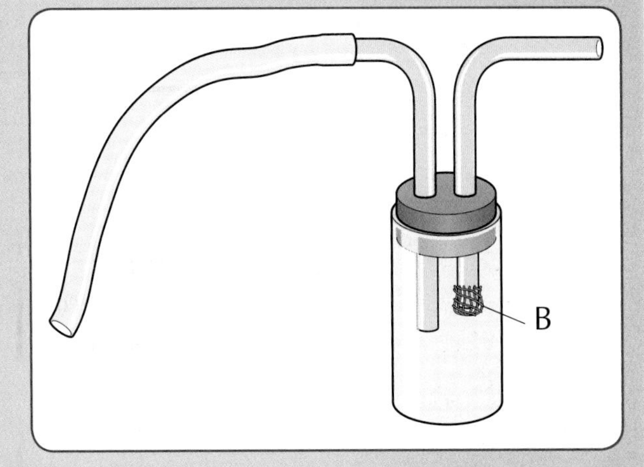

Fig. 17.6

17.4 (a) What is a beating tray?

(b) How is it used?

(c) What types of animals would you expect to collect using it?

17.5 (a) Draw a labelled diagram of a pitfall trap.

(b) Describe how you would use it to collect some ground beetles.

17.6 Which of the following pieces of apparatus would you use to collect the insects that fall into a beating tray?

(i) pitfall trap, (ii) transect, (iii) quadrat, (iv) pooter

17.7 (a) Name the habitat you have studied.

(b) List five animals that you found in the named habitat.

QUESTIONS

(c) State how *one* of the named animals is adapted to this habitat.

(d) List five plants that you found in the named habitat.

(e) State how *one* of the named plants is adapted to this habitat.

17.8 Describe how you would find the percentage frequency of plants in a named habitat.

17.9 The table below shows the results of a quadrat survey of plants in a field. Fifteen quadrats were thrown at random and the types of plants found recorded. Each √ represents the presence of that plant type in that quadrat.

(a) Use the results given to calculate the percentage frequency of each plant type.

(b) Draw a bar graph to represent these results. Put % frequency on the y-axis.

(c) Why was the quadrat thrown at random?

(d) Use the DAFOR scale to suggest which of the plants in the survey you would consider to be

(i) dominant, (ii) frequent and (iii) rare.

(e) Give one reason why the DAFOR scale is not considered a very useful estimate of plant numbers.

PLANT	1	2	3	4	5	6	7	8	9	10	11	12	13	14	15
Clover		√			√			√							√
Grass	√	√	√	√	√	√	√	√	√	√	√	√	√	√	√
Poppies		√			√		√		√			√		√	√
Plantain	√	√									√				
Buttercup					√										
Self heal	√	√									√				

17.10 (a) What is a transect?

(b) Describe how you could use a line transect to determine the distribution of limpets across the sea shore or bluebells from inside a woodland to the outside.

17.11 Rearrange the following list to form the correct order in which you should make a study of a habitat.

(a) Record the environmental conditions.

(b) Write up a report.

(c) Determine the number and distribution of the plants and animals.

(d) Identify and collect the plants and animals.

(e) Draw a map of the habitat.

Chapter 18 Conservation, Pollution and Waste Management

18.1 Conservation

In Ireland we have more than 400 types of birds, over 12,000 different insect types and at least 28 types of mammal. If these and all the hundreds of different types of plants are to survive, we must take care of their habitats.

Over the past hundred years or more, changes in **farming practices**, **drainage of wetlands**, increased **pollution** by humans and major **housing developments** have seen the habitats of many organisms disappear. In addition, many animals have been hunted by humans, some to the point of **extinction**.

- In Ireland, **animals** like the corncrake, pearl mussel and the barn owl are now on the endangered or threatened list. Such animals are in danger of becoming extinct.
- **Plants** like the cornflower and the bog orchid are also threatened.

We have a duty to protect the environment so that the wide variety of organisms (biodiversity) can be maintained.

Conservation is the protection and careful management of our natural resources.

(a) Barn owl

(b) Corncrake

(c) Cornflower

Fig. 18.1 Endangered Irish plants and animals. These organisms are threatened mainly because their habitats have been destroyed by humans.

How Humans are Endangering Nature?

- Agriculture provides us with food crops and livestock (which produce meat and milk). But certain **agricultural practices** including drainage of land, removal of hedgerows and forests, and the use of chemicals to kill pests on crops must be carried out with caution. We always need to have the long term effects of such actions in mind.
- Much of our wildlife can be killed when land is cleared for building motorways and houses. For example, many different types of bird nest in our hedgerows and when these are cut down, their natural **habitat is destroyed**.
- Certain **insecticides** used to kill insect pests on crops can have the effect of killing many useful insects as well. The spray from insecticides can also affect the plants and animals in nearby hedgerows. So the **wise use** of such chemicals is needed if the balance of nature is not to be upset.
- **Over-fishing** means that fishermen and women keep *all* the fish that are caught, both young and adult. If this continues then the young fish do not survive to breed

and the numbers of fish decline. To help conserve our fish stocks **special laws** have to be enforced.

What can we do?

- We have a **duty** to future generations to ensure that the environment and its plants and animals are protected and conserved.
- Governments, industrialists and farmers need to consider the environment before **planning** new roads and developments.
- Zoos and Botanical Gardens around the world are encouraged to take part in **breeding programmes** to prevent the extinction of endangered species.
- We can join local **conservation groups** and take part in conservation activities at a local level such as park, sea shore and river 'clean up' events.
- We can take part in the **annual garden and wild bird surveys** that keep track of our bird populations.
- We can become more aware and more caring of our environment and always obey the **Country Code**.
- Check out the National Parks and Wildlife Service website on **www.npws.ie** and the Environment Information Centre on **www.enfo.ie** for lots of information on wildlife issues.

TEST YOURSELF:
Now attempt question *18.1–18.2* and *W18.1.*

18.2 Pollution

Pollution occurs when humans add **harmful substances** to the environment. Some of the main causes of pollution include: raw sewage, domestic waste, oil, excess fertilisers, harmful gases and litter, *Fig. 18.2.*

Pollution is the addition of harmful substances to the environment.

Fig. 18.2 Litter is a major cause of pollution in Ireland.

The three main types of pollution are: (1) air, (2) water and (3) soil pollution.

(1) Air Pollution

The main cause of air pollution is the **burning** of **fossil fuels**, such as oil, gas and coal. As a result smoke, dust, sulfur dioxide, carbon monoxide and lead are produced – all of which act as air pollutants.

Summary of Air Pollution

A summary of the effects of these air pollutants can be found in *Table 18.1*.

CAUSE	EFFECT
Smoke and dust	Irritate the linings of the bronchii. This can cause bronchitis in humans. Smog – a mixture of smoke and dense fog is equally harmful.
Sulfur dioxide	Prevents plants from being able to photosynthesise. When it combines with rain it forms **'acid rain'** which wears away stone buildings and kills plants.
Carbon monoxide and lead	Produced by burning 'leaded' petrol and diesel. Carbon monoxide and lead are poisonous to humans.
Carbon dioxide and methane gas	Cause the **'greenhouse effect'**, i.e. the warming of the earth. This causes global warming and increased flooding.
CFCs (chlorofluorocarbons) from fridges and aerosols	Reduce the **ozone layer**, which protect us against UV light. Excess UV light causes skin cancer.

Table 18.1 Causes and effects of air pollution.

Fig. 18.3 Air pollution.

(2) Water Pollution

Water pollution is caused by a variety of substances being dumped into our rivers, lakes and seas: These include slurry from silage pits, untreated sewage, excess fertilisers, litter and oil, *Fig. 18.4*.

Algal Bloom

When chemicals with a high mineral content get washed into our waterways, it causes a sudden increase in the growth of microscopic algae. This is known as an '**algal bloom**'. When the algae die, the decomposing bacteria use up the available oxygen and so prevent fish and other animals and plants that live in water from surviving.

Summary of Water Pollution

A summary of the effects of these water pollutants can be found in *Table 18.2*.

CAUSE	EFFECT
Excess fertiliser and slurry	Less oxygen, causing fish to die.
Untreated sewage	Decomposers use up the oxygen, causing fish to die.
Oil	Poisonous to birds and fish. Coats feathers of sea birds preventing flight. Unsightly on our beaches.
Litter, e.g. plastic packaging	Unsightly and can be a health hazard.

Table 18.2 Causes and effects of water pollution.

Fig. 18.4 Water pollution.

(3) Soil Pollution

The soil can be polluted when pesticides and **excess fertilisers** get into the soil. Also chemicals from waste dumps and factories can be the cause of soil pollution.

TEST YOURSELF:

Now attempt questions *18.3–18.5* and *W18.2–W18.3*.

18.3 Waste Management

In Ireland we produce huge quantities of **waste**. This includes waste from households, building sites, factories, hospitals and sewage sludge (faeces and urine). Getting rid of all this waste is a big problem. Not only is the waste unsightly, but it can be dangerous to ourselves and to the environment.

Waste management is the way that we deal with our waste.

It is important that we **manage our waste** in a safe and efficient manner. There are a number of methods of dealing with the waste that is generated.

(1) Waste can be put into **landfill** sites.

(2) Waste can be burned in an **incinerator**.

(3) Waste can be **recycled**.

(1) Landfill Sites

Landfill sites are huge 'holes' in the ground where waste is dumped and buried, *Fig. 18.5*. Before being placed in the landfill, the waste is usually compressed or compacted to reduce the amount of it. This helps to fit more waste into the landfill.

Advantages

- They are **convenient**.
- The **methane gas** that is produced by the rotting rubbish can be used as a fuel.

Disadvantages

- Badly managed landfill sites can be unsightly, produce bad smells, and **attract vermin** (rats).
- In addition, **poisonous chemicals** from the waste can seep into the ground and affect water supplies.
- Also we are **running out of space** for these 'dumps'.

Fig. 18.5 A landfill site.

(2) Incineration

Incineration is a method of reducing waste by **burning** it.

Advantages

- Incineration can deal with **huge volumes** of waste.
- It is a **clean and efficient** method of dealing with waste.
- The waste is burned in huge incinerators that **don't use up much land space**.

Disadvantages

- Incinerators produce large quantities of **ash** which then has to be disposed of.
- The burning of materials produces **poisonous fumes** which have to be dealt with. Many people do not want an incinerator built near their houses because they are afraid of the harmful gas emissions.

(3) Recycling

Recycling, as the name suggests, involves either using something again or using the materials it is made of for another purpose. A good example of **re-use** is to use the same plastic bags over and over to do the shopping.

Composting our garden and vegetable waste reduces the amount of organic waste that goes into landfill sites. Composting uses bacteria and fungi to break down the waste and provide us with natural fertiliser for the garden. These examples illustrate the **three R's** of waste management, which are:

- **Reduce**. Reduce the amount of waste produced in the first place.
- **Re-use**. Use again for a different purpose.
- **Recycle**. Recover for re-use.

Fig. 18.6 Everyone can play their part in reducing waste.

Advantages

This is the method of waste management that we as individuals can use.

- It is **environmentally friendly** and easy to

do. Glass bottles, paper, cardboard, plastics and metal can all be recycled.

- Recycling also **saves money and energy**. For example, it takes 11 months to make a tin can from extraction of the ore to the supermarket shelf, but only one day to recycle and only 5 per cent of the energy costs.

Disadvantages

There are very **few** disadvantages to recycling.

- However, some of the material brought to 'bring banks' **may not be re-usable** and it then has to be put into landfill either here or abroad.

Waste Management Pyramid

Fig. 18.7 shows the waste management pyramid. We should all be trying to prevent the waste in the first place. *Fig. 18.8* shows the length of time it takes for the various items to be broken down by bacteria and fungi.

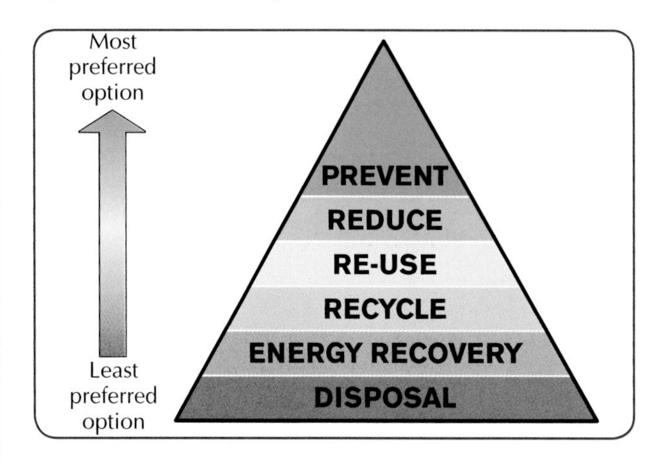

Fig. 18.7 Waste management pyramid.

Biodegradation Timeline

1 month	Paper towel Newspaper	
2 months	Apple core Cardboard box	
6 months	Orange peel	
1 year	Biodegradable nappy Wool gloves	
10 years	Painted wooden stick	
50 years	Tin can Styrofoam cup	
450 years	Plastic bottle Disposable nappy	
Never?	Glass bottles and jars	

Fig. 18.8 The table shows the relative lengths of time it takes the various items to be broken down by micro-organisms.

18.4 How Human Activities Affect the Environment

In learning about conservation, pollution and waste management we can see how the actions of humans affect the environment.

Positive Human Activity

Human activity can affect the environment in a **positive** way. Two examples of this are:

(1) **Protecting** the **habitats** of organisms so that the great variety of living things is maintained for us and for future generations.

(2) Implementing **anti-pollution** and **waste management laws** will help reduce the effects that pollution and increasing waste have on the planet.

Negative Human Activity

Human activity can affect the environment in a **negative** way. Two examples of this are:

(1) **Destroying habitats**.

(2) The continued **burning** of **fossil fuels** contributes to **global warming**, which is a gradual increase in the temperature of the earth. This increase in temperature may give rise to increased flooding (melting of the ice-caps), a rise in sea levels and extreme weather events.

TEST YOURSELF:

Now attempt questions *18.6–18.10* and *W18.4–W18.7.*

What I should know

- Conservation is the protection and careful management of our natural resources.
- Conservation is necessary to prevent habitat destruction and protect the wide variety of organisms on the planet from extinction.
- Threatened species in Ireland include the corncrake and the cornflower.
- Pollution is the addition of harmful substances to the environment.
- Air pollution is mainly caused by burning fossil fuels such as oil and gas.
- When fossil fuels are burned, they produce sulfur dioxide gas.
- Water pollution is caused by the addition of chemicals, oil, and litter to water.
- Soil pollution is caused by the addition of excess fertiliser and other chemicals to the soil.
- Waste management is the way that we deal with our waste.
- Landfill sites, incineration and recycling are methods of waste management.
- Two positive effects of humans on the environment are: reducing pollution and protecting endangered species.
- Two negative effects of humans on the environment are: habitat destruction and increasing pollution.

Questions

Write the answers to the following questions into your copybook.

18.1 (a) What is conservation?

(b) Why is conservation necessary?

(c) Name two Irish animals that are on the 'threatened' list.

(d) Name two Irish plants that are on the 'threatened' list.

(e) Describe any method by which living things can be conserved.

QUESTIONS

18.2 (a) Name two natural resources.

(b) Name two ways in which habitats are being destroyed by humans.

(c) Outline two ways in which humans can help to conserve the variety of plants and animals on our planet.

18.3 (a) What is pollution?

(b) Name two types of pollution.

(c) For each named type of pollution, give one cause and one effect of the pollution on the environment.

18.4 State one cause and one effect of each of the following types of air pollution:

(a) the greenhouse effect

(b) reduction of the ozone layer

(c) smog

(d) acid rain

18.5 (a) Sewage, oil and litter can cause water pollution. State the effect of each on the environment.

(b) Name two causes of soil pollution.

18.6 Explain what is meant by the following:

(a) waste management

(b) landfill

(c) incineration

(d) recycling

18.7 (a) Give two disadvantages of landfill sites.

(b) Give two disadvantages of incinerators.

(c) Give one advantage of incinerators.

(d) Describe two ways that we can reduce the waste we produce.

18.8 (a) Describe two examples of how human activities affect the environment in a positive way.

(b) Describe two examples of how human activities affect the environment in a negative way.

18.9 Use the Biodegradation timeline in *Fig. 18.8* to answer the following:

(a) How long does it take each of the following to break down?

(i) a tin can, (ii) a disposable nappy, (iii) a glass bottle

(b) Which of the following decays fastest? Is it (i) plastic bottles or (ii) plastic bags?

(c) Name two things from the timeline that you would consider 'alright' to throw away? Explain your answer.

18.10 Give one reason why we should:

(a) Have a compost bin in the garden.

(b) Keep and maintain hedgerows in the countryside.

(c) Not burn our rubbish on bonfires.

(d) Take our litter home.

(e) Take our own bags with us to carry things home from the shops.

(f) Buy goods that do not come already packaged, e.g. vegetables and fruit etc.

Chapter 19 Microbiology and Biotechnology

19.1 Introduction

Microbiology is the study of very tiny organisms, hence the name **micro-organisms**. Generally speaking, micro-organisms can only be seen with a microscope. Examples of micro-organisms are (1) **fungi**, (2) **bacteria** and (3) **viruses**.

19.2 Fungi

Fungi are the largest type of micro-organism. They include the **mushrooms** and bread mould which are clearly visible to our eyes. In addition, there are microscopic fungi such as **yeast**. Fungi cannot make their own food. They live off other things.

Fig. 19.1 Various fungi.

Beneficial Fungi

Many fungi are very useful:

- Some **act as decomposers**, living on and breaking down dead things. For example, many fungi breakdown the **dead leaves** in Autumn. Organisms that feed on dead matter are known as **saprophytes**.
- **Yeast** is another useful fungus. It is used in the **baking** and **brewing** industries.

Harmful Fungi

Other fungi cause **disease**, like the **potato blight** fungus that was responsible for the failure of the potato crop in Ireland in the middle of the 19th century. We still have potato blight today, but now we have chemicals that can control the spread of the disease. *Table 19.1* shows some of the ways fungi can be beneficial and harmful.

BENEFICIAL EFFECTS OF FUNGI	HARMFUL EFFECTS OF FUNGI
(1) They act as decomposers.	(1) They cause athlete's foot and ringworm of the scalp.
(2) They are used to produce antibiotics, e.g. penicillin.	(2) They cause food spoilage, e.g. they cause bread to go mouldy.
(3) They are used to make flavoursome cheeses.	(3) They cause dry rot of wood.

Table 19.1 The beneficial and harmful effects of fungi.

19.3 Bacteria

Bacteria are made of only **one cell** and can only be seen with a microscope. They may be very small, but there are thousands of different types of bacteria and they can live almost anywhere. Bacteria are found in soil, water, the air, as well as in and on the bodies of plants, humans and other animals.

Beneficial Bacteria

Most bacteria are very useful to humans. They **act as decomposers**, breaking down dead and rotting plant and animal matter, e.g. dead leaves and birds. Bacteria help to put the minerals from these organisms back into the soil. *Table 19.2* shows some of the many ways in which bacteria are useful.

Harmful Bacteria

However, not all bacteria are good. Many cause disease and illness, and others cause food to go bad, see *Table 19.2*.

Three common illnesses caused by bacteria:

- tonsilitis/strep throat
- food poisoning
- tetanus

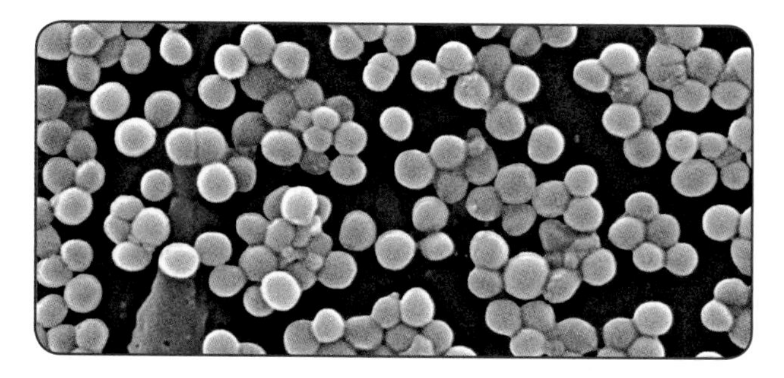

Fig. 19.2 Staphylococcus aureus bacteria, commonly referred to by the acronym MRSA, magnified 9560×.

BENEFICIAL EFFECTS OF BACTERIA ☺	HARMFUL EFFECTS OF BACTERIA ☹
(1) They act as decomposers.	(1) They cause common illnesses, e.g. tonsillitis and food poisoning.
(2) They are used to produce antibiotics.	(2) They cause food spoilage, e.g. they cause milk to go sour.
(3) They are used to make yoghurt.	(3) They cause tooth decay.

Table 19.2 The beneficial and harmful effects of bacteria.

Mandatory experiment 19.1

To investigate the presence of micro-organisms in air and soil

- *Apparatus required:* 3 sterile Petri-dishes, each containing sterile nutrient agar; innoculating loop; Bunsen burner; incubator
- *Also required:* sample of soil; marker pen

Note: Bacteria and fungi grow well if conditions are favourable, i.e. if they have a good supply of food, oxygen, moisture and warmth. In the laboratory, a suitable food for micro-organisms is a jelly-like substance called **nutrient agar**.

Method

1. You are provided with three Petri-dishes containing sterile nutrient agar. Label the dishes *A*, *B* and *C*. (The dishes and agar are sterile which means there are no organisms of any kind present.)
2. Lift the lid off dish *A* and leave it in the laboratory for ten minutes. This exposes the agar to the air. Then put the lid back on.
3. Heat the inoculating loop in the Bunsen flame to sterilise it. Allow to cool. Open the lid of dish *B* just wide enough to sprinkle a sample of soil across the surface of the agar, *Fig. 19.3*. Replace the lid.
4. Leave dish *C* unopened. This dish acts as a control.
5. Place the dishes, upside down, in an incubator at 25°C for 48 hours. The dishes are placed upside down to prevent condensation from blocking our view of the micro-organisms that grow.

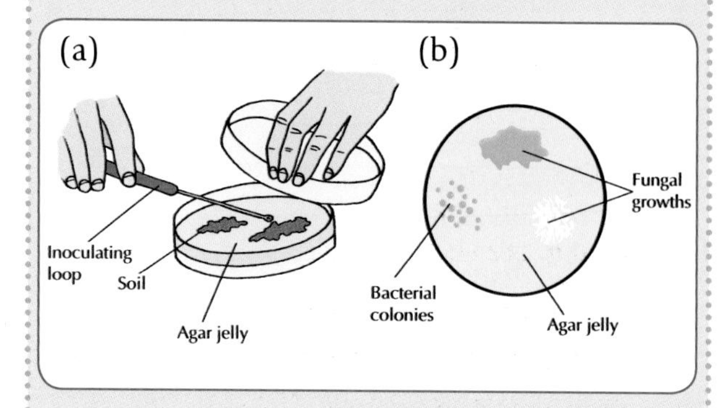

Fig. 19.3(a) Scattering soil on an agar dish. (b) Colonies of bacteria and fungi on agar.

Results

Bacteria grow in groups or **colonies** of many

BIOLOGY

thousands of cells. These can be seen as shiny, circular blobs on the surface of the agar. Fungi tend to appear as fluffy, hairy or mouldy growths.

Dish *A*: bacterial and fungal colonies will be visible.

Dish *B*: bacterial and fungal colonies will be visible around the clumps of soil.

Dish *C*: the dish will have no growths present.

Conclusion

There are micro-organisms present in air and soil.

TEST YOURSELF:
Now attempt questions *19.1–19.8* and *W19.1–W19.4*

19.4 Viruses

Viruses are much smaller than bacteria or fungi. They can only be seen with an electron microscope. All viruses are **parasites**. This means that they must live in or on another living thing.

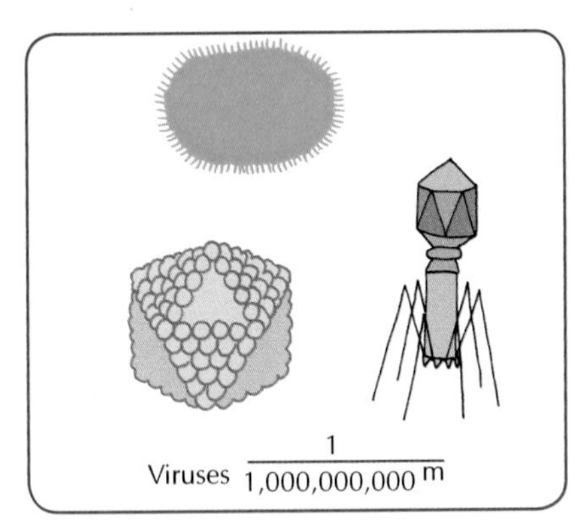

Fig. 19.4 Micro-organisms: viruses.

Many of the **diseases** you may have suffered from are caused by viruses, e.g. the common cold, the 'flu', and chicken pox. In addition, viruses cause SARS, measles, mumps, and German measles.

Three common illnesses caused by viruses:

- Common cold
- Chicken pox
- Influenza

Edward Jenner (1749–1823). Edward Jenner discovered a vaccine against the deadly disease of smallpox.

19.5 Biotechnology

Biotechnology is the use of living things or parts of living things to produce useful substances. For thousands of years, humans have used micro-organisms to make food and drink, e.g. yeast to make bread and wine, and bacteria to make cheeses and yoghurt, *Fig. 19.5*.

Fig. 19.5 Traditional products of biotechnology.

Uses of Biotechnology

More recently, micro-organisms have been used to manufacture a wide range of things, such as antibiotics, vaccines, foods, hormones and enzymes. Biotechnology allows these useful products to be produced in large quantities and at a much lower cost.

Biotechnology in Industry

- 'Biological' washing powders contain **enzymes** manufactured by bacteria. These enzymes are able to breakdown stains that are protein-based, e.g. blood, grass etc.
- Yeast is used in the **baking and brewing** industry to make bread and alcohol.

Fig. 19.6 Products of biotechnology.

Biotechnology in Medicine

- Bacteria and fungi are used to produce **antibiotics** and other drugs such as insulin.
- Viruses and bacteria are used to produce **vaccines**. An example of a vaccine is the MMR vaccine which protects against measles, mumps and rubella (German measles). Vaccines help to prevent infections.

Biotechnology is the use of living things or parts of living things to produce useful products.

TEST YOURSELF:
Now attempt questions *19.9–19.10* and *W19.5–W19.7.*

What I should know

- Micro-organisms are tiny organisms. They include fungi, bacteria and viruses.
- Fungi cannot make their own food.
- Mushrooms, yeast and bread mould are examples of fungi.
- Fungi produce antibiotics and are used in brewing and baking.
- Some fungi cause diseases, e.g. athlete's foot, and others spoil food.
- Bacteria are made of only one cell.
- Bacteria act as decomposers, produce antibiotics and make yoghurt.
- Three common illnesses caused by bacteria are: tetanus, food poisoning and sore throats.
- Fungi and bacteria can be grown in the laboratory on nutrient agar.
- Viruses are very tiny. They can only reproduce inside other cells.
- Three common illnesses caused by viruses are: the common cold, 'flu' and chicken pox.
- Biotechnology is the use of organisms or parts of organisms to make useful products.
- Two uses of biotechnology in industry are: (i) using yeast in making alcohol and (ii) using enzymes from bacteria in washing powders.
- Two uses of biotechnology in medicine are: (i) using fungi and bacteria to make antibiotics and (ii) using viruses to make vaccines.

QUESTIONS

Write the answers to the following questions into your copybook.

19.1 (a) What is a micro-organism?

(b) Name three types of micro-organism.

(c) Which of these types of micro-organism is the smallest?

19.2 (a) Name a fungus.

(b) Name a place where fungi grow.

(c) Name two useful fungi and what they do.

(d) Name an illness caused by a fungus.

19.3 (a) How many cells are bacteria made of?

(b) Name two things that bacteria need for growth.

(c) Name three human illnesses caused by bacteria.

(d) Name one way in which bacteria are useful to humans.

19.4 Complete the following statements using the words from the following list:

athlete's foot, bacteria, fungus, decomposers, micro-organism, virus

(a) Yeast is a type of ..... and is used in baking and brewing.

(b) ..... foot is a human disease caused by a ..... .

(c) A ..... causes the common cold.

(d) Fungi and ..... act as ..... when they rot down dead plants and animals.

(e) Two kinds of decomposer found in the soil are a ..... and ..... .

19.5 The bacteria which produce sore throats in humans were grown in a medical microbiology laboratory. The number of bacterial colonies were counted and the results obtained shown in the *Table 19.3* below.

TIME (HOURS)	0	6	12	18	24	30	36	42	48
NUMBER OF COLONIES	0	2	6	14	23	30	33	34	35

Table 19.3

(a) Draw a trend graph to illustrate the results. Put time on the horizontal axis.

(b) Use the graph to find how many bacterial colonies were present after (i) 15 hours and (ii) 32 hours.

(c) At which of the following temperatures would you expect these bacteria to grow best: (i) 25°C, (ii) 98°C, (iii) 37°C, (iv) 40°C? Explain your answer.

19.6 Describe how the presence of micro-organisms in the air might be investigated. Include a diagram of any equipment that might be used.

19.7 Describe how you could investigate if there are more micro-organisms in a sample of garden soil than in a sample of sand from the beach.

QUESTIONS

19.8 Using words from the list below, complete the following sentences:

antibiotic, agar, micro-organisms, yeast, fungus, decay

(a) Penicillin is an example of an ..... produced by a ..... .

(b) The baking and brewing industries depend on ..... .

(c) Bacteria, fungi and viruses are examples of ..... .

(d) Bacteria of ..... live in the soil and help to break down dead organisms.

(e) Nutrient ..... provides food for bacteria and fungi to grow in the laboratory.

19.9 (a) What is a virus?

(b) Viruses can be seen with the naked eye. True or false?

(c) Which of the following illnesses are caused by a virus?

(i) measles, (ii) common cold, (iii) potato blight, (iv) SARS, (v) tetanus

19.10 (a) What is meant by biotechnology?

(b) Name two types of organism that are used in biotechnology.

(c) Outline two uses of biotechnology in industry.

(d) Outline two uses of biotechnology in medicine.

Chapter 20 Materials

20.1 Chemistry – What's it all about?

Chemistry may be described as the study of materials. Materials are the substances which make up the world around us. Examples of materials are water, air, earth, rocks, clothes, metals, plastics, etc.

- Some materials are **naturally** occurring, e.g. water, wool, fur, cotton, silk and wood.
- Others materials are **made in the laboratory**, e.g. polythene, glass, nylon, polyester and terylene. Chemistry is often called the **creative science** because chemists are continually making new substances in the laboratory.

Uses of Chemistry

Chemistry has made many improvements to our lives. For example, chemists manufacture items like medicines, plastics, paints, fertilisers, fuels, synthetic clothes, etc. Many of the chemical plants in Ireland manufacture pharmaceuticals, *Fig. 20.1*.

Fig. 20.1 The Novartis pharmaceutical plant in Ringaskiddy, Co. Cork manufactures medicine to help patients suffering from leukaemia.

20.2 States of Matter

The name that scientists use for materials is **matter**.

Matter is anything which occupies space and which has mass.

In other words, any substance which takes up a volume in space and which has mass is referred to as matter. Our common experience tells us that solids and liquids take up space and have a mass. Let's now see if gases do the same, i.e. can gases be classed as matter?

Experiment 20.1

To show that a gas occupies space

Apparatus required: gas jar, beehive shelf, dish, rubber or plastic tubing

Method

1. Set up the apparatus as shown in *Fig. 20.2(a)*.
2. Connect a length of clean rubber or plastic tubing as shown in *Fig. 20.2(b)*. Take a deep breath and blow gently into the water.

Result

As you blow into the gas jar full of water, the gas takes up space. This gas pushes some water out of the gas jar. Therefore, the level of water in the gas jar falls.

Conclusion

A gas occupies space.

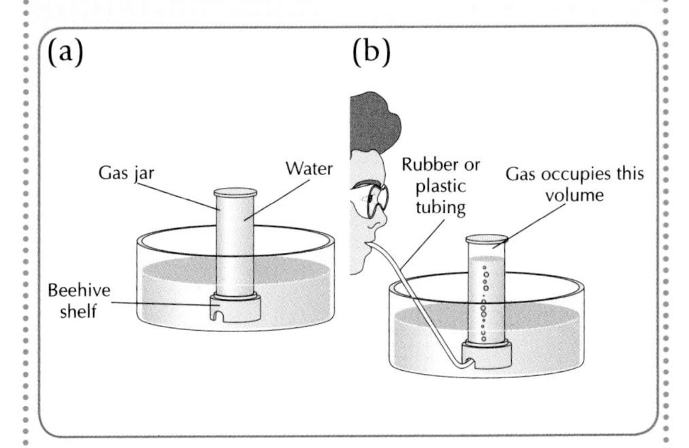

Fig. 20.2 An experiment to show that a gas occupies space.

Experiment 20.2

To show that a gas has mass

Apparatus required: electronic balance (reading to two decimal places); balloon

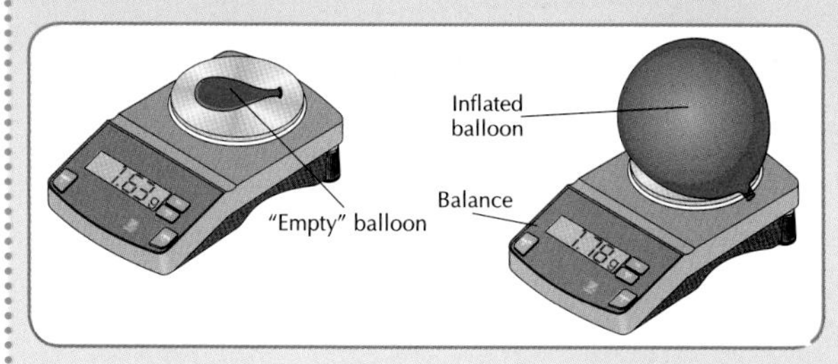

Fig. 20.3(a) Experiment to show that a gas has mass.

Method

1. Place a deflated balloon on a balance, *Fig. 20.3(a)*. Write down the mass of the balloon.
2. Inflate the balloon by blowing into it. Tie a knot to prevent the gas escaping.
3. Write down the new mass of the inflated balloon.

Result

The balloon filled with air has a greater mass than the 'empty' balloon.

Conclusion

Air has mass.

Another way to demonstrate that air has mass is shown in *Fig. 20.3(b)*. The balloon filled with air has a greater mass than the 'empty' balloon.

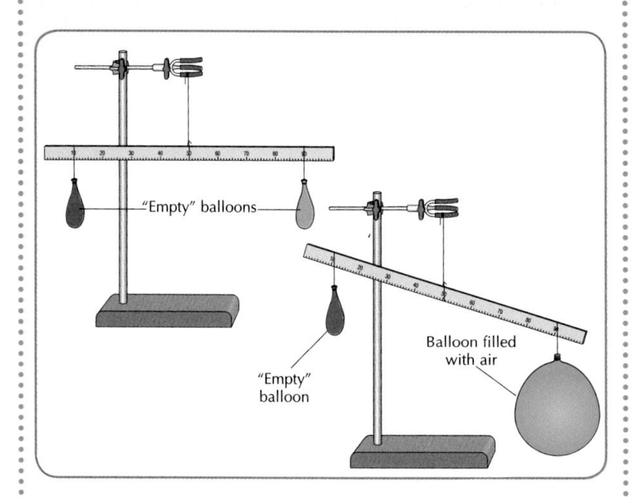

Fig. 20.3(b) Experiment to show that a gas has mass.

Properties of the States of Matter

Every substance in the world around us exists either as a solid, liquid or gas. Wood is a solid, water is a liquid and air is a gas. These three forms of matter are known as the **states of matter**.

The main properties or characteristics of each of the states can be summarised as follows:

(The words 'property' or 'characteristic' are used to describe how a substance behaves.)

Solids

Solids (e.g. iron and wood) have a number of common properties.

- They have a **definite shape**. This means that each solid has its own shape. It will retain this shape unless it is heated or a force causes it to change its shape.
- They have a **definite volume**. This means that it is very difficult to squeeze them into a smaller size. Solids are said to be almost incompressible.
- They do **not flow**. This means that they do not spread over a surface.

Liquids

Liquids (e.g. water and oil) have a number of common properties.

- They have **no definite shape**. This means that a liquid always adopts the shape of the container into which it is placed.
- They have a **definite volume**. This means that liquids, like solids, are almost impossible to squash into a smaller volume.
- They **flow and evaporate**. This means that liquids when spilled on a surface usually spread over that surface. Most liquids evaporate readily from open containers. Evaporation means that the liquid changes to a gas (also called a vapour) at the surface.

Gases

Gases (e.g. air and carbon dioxide) have a number of common properties.

- They have **no definite shape**. This means that gases always take up the shape of the container into which they are placed. Think of the air inside a balloon. Its shape can very easily be changed.
- They have **no definite volume**. This means that gases always spread out in all directions to fill the container into which they are placed. This spreading out of gases to fill all the available space is called **diffusion**. The smell from food being cooked is a gas. We smell food cooking in the kitchen due to diffusion.

Note: Liquids also diffuse but not as quickly as gases. This is demonstrated by using a pipette to place a drop of coloured liquid at the bottom of a beaker of water. The colour slowly spreads throughout the rest of the liquid.

- They **can be compressed easily**. This means that a given volume of gas can be squeezed into a smaller volume. The cylinders carried by the divers shown in *Fig. 20.4* contain compressed air. This means that a lot of air can be squashed into a small volume.

Fig. 20.4 The cylinders carried by the divers contain a lot of air that has been squashed (compressed) into a smaller volume.

Summary of the Properties of the States of Matter

The properties of solids, liquids and gases are summarised in *Table 20.1*.

SOLIDS	LIQUIDS	GASES
Definite shape (Do not flow)	No definite shape (Flow easily)	No definite shape (Diffuse to fill all available space)
Definite volume	Definite volume	No definite volume
Hard to compress	Hard to compress	Easy to compress

Table 20.1 Summary of the properties of solids, liquids and gases.

TEST YOURSELF:
Now attempt questions *20.1* and *W20.1–W20.3*.

Particle Theory and the States of Matter

Scientists have developed a theory to explain the properties of solids, liquids and gases. The theory is called the **particle theory**. It is called the particle theory because we imagine solids, liquids and gases being made up of tiny particles. The particles are so tiny that they can only be seen with special types of microscopes. These microscopes are much more powerful than the ones in your school laboratory.

Solids

- In a solid, the particles are **arranged in a regular way**. They cannot move out of position. Therefore, the solid has a definite shape.
- The particles are packed **closely together**.
- There are **strong forces** holding the particles together.

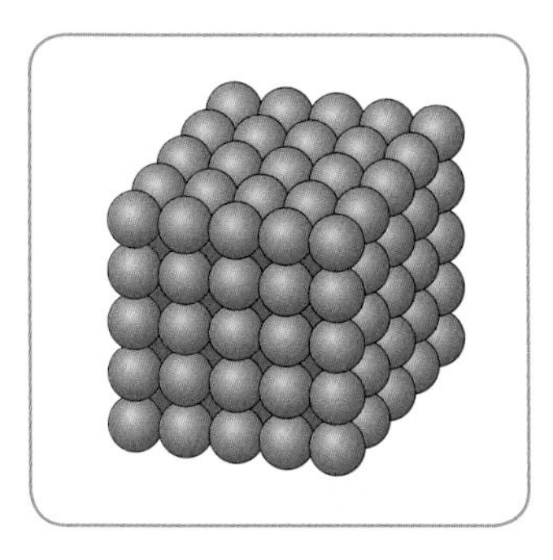

Fig. 20.5(a) The arrangement of particles in a solid. The particles are in fixed positions.

Liquids

- Since there is **no regular arrangement** of particles, the liquid has no shape of its own. A liquid always takes up the shape of its container.
- The particles in the liquid are in contact with each other and can **slide** over each other. This is why a liquid can flow.
- The forces between the particles are **weaker than in a solid**. The particles in a liquid have more freedom of movement than the particles in a solid.

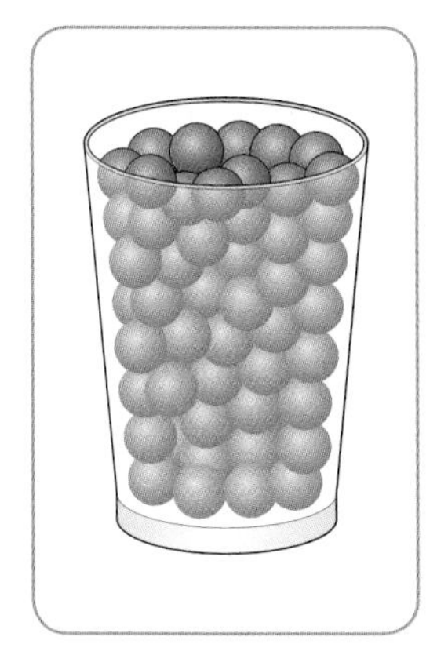

Fig. 20.5(b) The arrangement of particles in a liquid. The particles can slide over one another.

Gases

- In a gas, the particles are **much farther apart than in a liquid or a solid**.
- Particles in a gas can **move quickly into all the space available**.
- There are only very **weak forces** between the particles. When you blow up a balloon, the gas particles collide with each other and with the walls of the balloon and this makes it inflate.

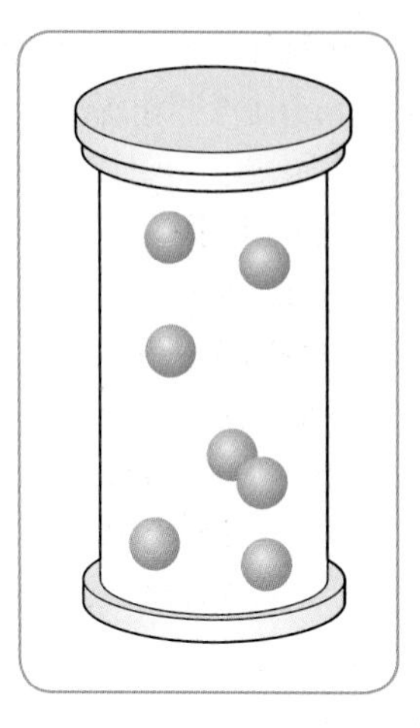

Fig. 20.5(c) The arrangement of particles in a gas. The particles are farther apart and move around in all the space available.

Compression

The particle theory helps us to explain the properties of solids, liquids and gases. For example, the particles in a solid are very close together. This means that it is difficult to squash solids into a smaller volume. However, since the particles in a gas are much farther apart, it is easier to squeeze gases into a smaller volume, *Fig. 20.6*.

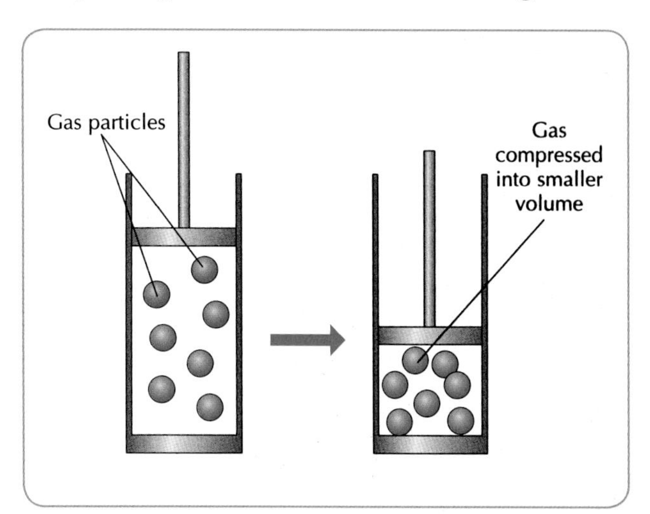

Fig. 20.6 Gases can be squashed or compressed because there is space between the particles of the gas.

Diffusion

The particle theory may also be used to explain diffusion. **Diffusion** is the movement of particles to spread out and fill all of the available space.

- Diffusion takes place **quickly with gases** because the particles of the gas are moving fast and there are only weak forces between the particles.

- Diffusion takes place **more slowly** in **liquids and solids** because there is less movement of particles. The forces holding the particles together in solids and liquids are stronger than those in gases.

Since liquids and gases have the ability to flow, they are called **fluids**.

TEST YOURSELF:
Now attempt questions *20.2* and *W20.4–W20.7*.

20.3 Changes of State

When a solid melts to form a liquid or a liquid evaporates to form a gas we say that there is a **change of state**.

Melting

When a solid is heated, the heat energy causes the particles to vibrate. As heating is continued, the particles vibrate faster and faster. Eventually, they break free from their fixed positions and begin to slide over each other. We say that the solid is melting to form a liquid, *Fig. 20.7*. The temperature at which the solid melts is called the melting point.

The melting point is the temperature at which a solid changes to a liquid.

Evaporation

You may have noticed that puddles on the road dry up quickly on a sunny day. This is because heat from the sun gives some particles near the surface of the liquid extra energy. These particles now have enough energy to escape from the liquid and go into the air. We say that some of the liquid has **evaporated** to form a gas. (The word *evaporate* comes from *vapour* which means gas.)

Evaporation is the changing of a liquid to a gas.

Evaporation is also helped by the wind. If the day is windy, the wind blows away the particles that have already evaporated. This makes room for more to evaporate. If you were hanging clothes out to dry, what would be the ideal conditions to help drying?

Place a few drops of propanone or ether on the back of your hand and notice how quickly it evaporates! Note also how cold your hand feels as heat is taken from you hand to evaporate the liquid.

Boiling

When a liquid is heated, the particles get more energy and move faster and faster. Eventually, all the particles get enough energy to overcome the forces holding them together. The particles break away from the liquid state and form a gas. The liquid is now boiling. The stage is reached where the particles are escaping from the liquid very quickly. This causes bubbles of gas (water vapour in the case of water) to start forming inside the liquid. The boiling point of the liquid has now been reached, *Fig. 20.8*.

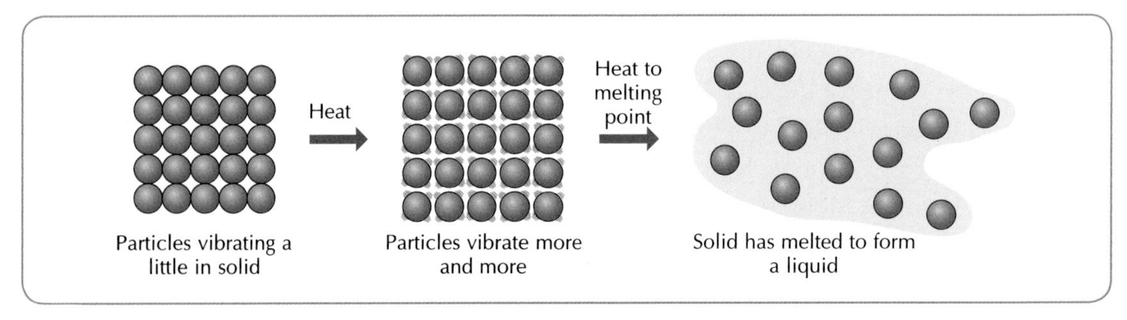

Fig. 20.7 When a solid melts, the particles vibrate so much that they break away from their fixed positions. The solid now starts to become a liquid.

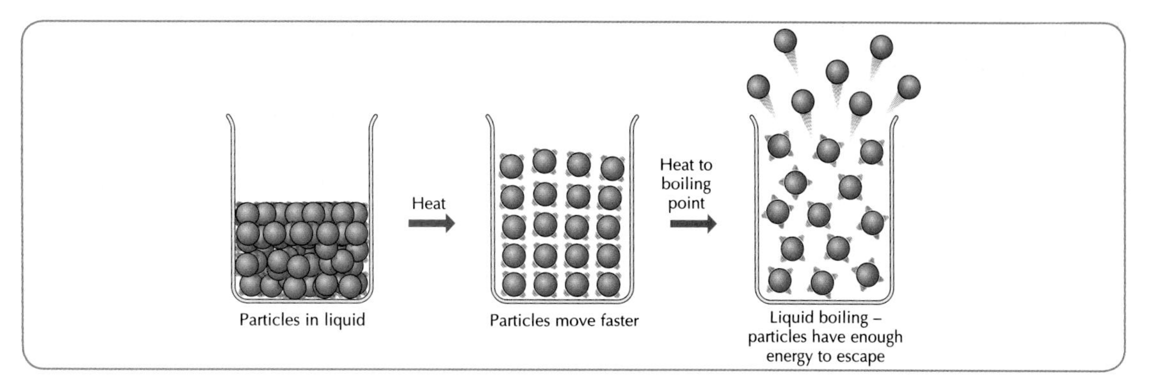

Fig. 20.8 At the boiling point of a liquid, the particles break away from the liquid and form a gas.

The boiling point of a liquid is the temperature at which a liquid changes to a gas throughout the liquid.

Different liquids have different boiling points. For example, the normal boiling point of water is 100 °C and the boiling point of ethanol (found in alcoholic drinks) is 78 °C. The only liquid whose boiling point you need to remember for this course is that of water.

Boiling or Evaporation?

Note the difference between evaporation and boiling. The liquid need not be heated to its boiling point for evaporation to occur. Evaporation from the surface of a liquid is always taking place. It is only when this happens **throughout the liquid** that the liquid is boiling.

Condensation

On cooling a gas, the particles slow down and become closer to each other. Eventually a liquid is formed. The changing of a gas to a liquid is called **condensation**.

Examples of Condensation

- You may notice this happening when you have a **hot bath** or a shower. The water vapour or steam is turning to liquid water on the cold surface of the mirror or window.
- It is also observed in a kitchen when the steam from a **kettle** meets the much colder glass in the window. The hot gas changes to droplets of water on the glass.

The various changes of state are summarised in *Fig. 20.9*.

TEST YOURSELF:
Now attempt questions *20.3–20.7* and *W20.8–W20.10*.

CHEMISTRY

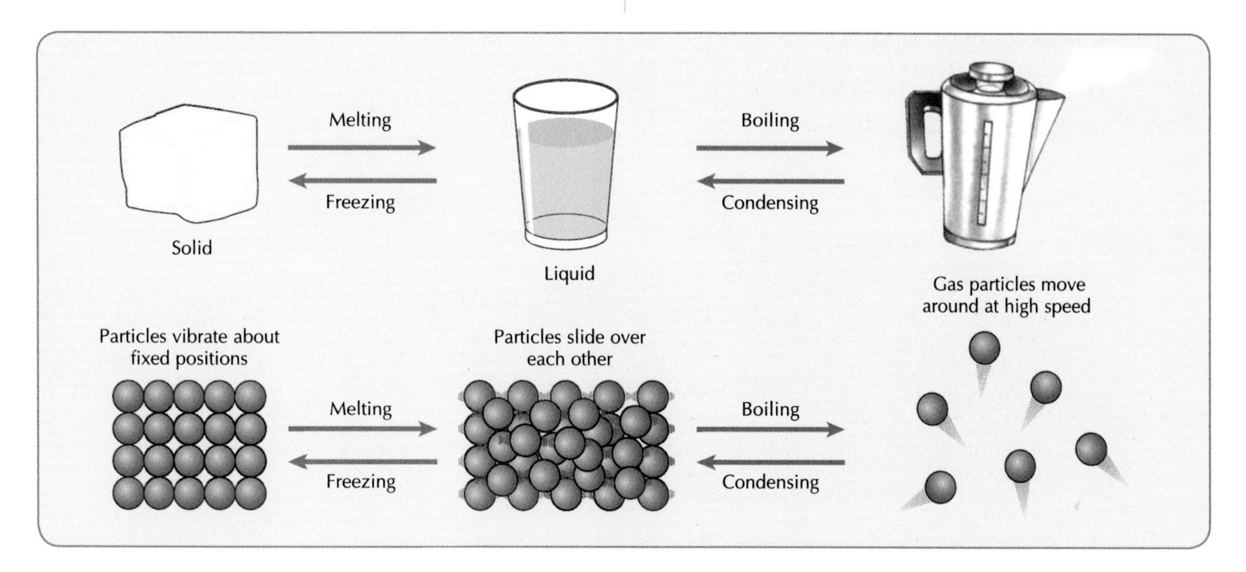

Fig. 20.9 The state of a substance can be changed by heating or cooling.

What I should know

- Matter is anything which occupies space and has mass. There are three states of matter – solid, liquid and gas.
- Solids have a definite shape, definite volume, are hard to compress and do not flow.
- Liquids have no definite shape, a definite volume, are hard to compress and flow easily.
- Gases have no definite shape, no definite volume, are easy to compress and diffuse to fill all the available space.
- The melting point is the temperature at which a solid changes to a liquid.
- Evaporation is the changing of a liquid to a gas.
- The boiling point of a liquid is the temperature at which a liquid changes to a gas throughout the liquid.
- The changing of a gas to a liquid is called condensation.

Questions

Write the answers to the following questions into your copybook.

20.1 (a) The three states of matter are ….., ….. and ….. .

(b) Complete the following statement: 'Gases have no definite volume or shape because …..'.

(c) When a car driver puts a foot on the brake pedal, the brake fluid is pushed down a narrow tube connected to the brake. Since brake fluid is hard to squash or ….. the brakes are put on immediately.

(d) A given volume of gas can be squeezed into a smaller volume i.e. it can be ….. .

(e) When a bottle of perfume is opened, the smell spreads around the room. This spreading is called ….. .

(f) How would you show that a gas occupies space?

20.2 (a) Draw a diagram to show the arrangement of particles in a solid, liquid and gas. Use this model to explain (i) why it is easy to pour a liquid, (ii) why solids have a fixed shape and (iii) why gases fill all the available space.

(b) A student placed some copper sulfate crystals at the bottom of a large graduated cylinder. Then, he gently poured in some water on top of the crystals. After a few days he noticed that the blue colour of the crystals had spread throughout the water. Using the particle theory, explain this observation.

20.3 In *Fig. 20.10* write down the names of the processes represented by the letters A, B, C and D.

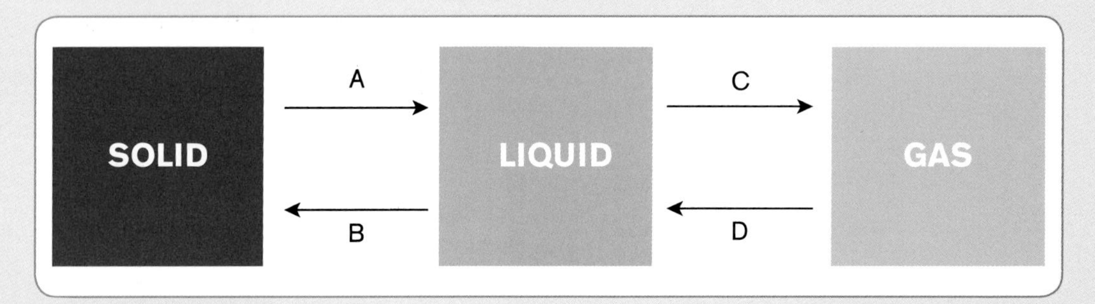

Fig. 20.10

QUESTIONS

20.4 (a) When iron is heated to 1540 °C, it ..... i.e. it changes into a liquid. This means that it can be poured into moulds.

(b) The melting point of a solid is defined as ..... .

(c) The changing of a liquid to a vapour is called ..... .

(d) When evaporation begins to occur throughout the liquid the ..... of the liquid has been reached.

20.5 In each case choose one state of matter from the list on the right which matches the characteristics in the table below.

CHARACTERISTICS	STATE OF MATTER
Has definite shape. Has definite volume. Is not easily compressed.	
Has no definite shape. Has no definite volume. Is easily compressed.	

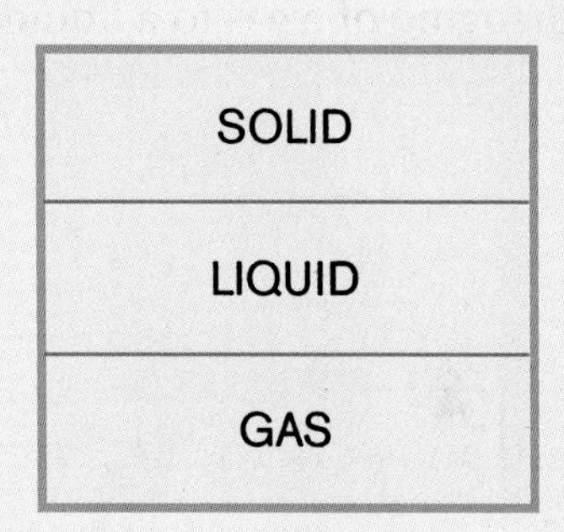

(Junior Cert Science Ordinary Level Examination Question)

20.6 (a) What is meant by the phrase 'gases can be compressed'?

(b) Although it is difficult to change the shape of most solids, give an example of a solid whose shape is easily changed.

(c) Why is the particle theory important in science?

(d) A young pupil in primary school is confused about the difference between evaporation and boiling. How would you explain the difference between these two terms?

(e) A small quantity of acetone (nail varnish remover) was poured on the back of the hand of a student. The student observed that the back of her hand dried quickly . Also, she felt her hand getting cold. Explain what has happened.

20.7 (a) On a cold morning, it was observed that the inside of a car window was covered in moisture. Explain this observation.

(b) Explain why a balloon filled with air has a greater mass than an *empty* balloon?

(c) What is a fluid?

(d) Explain why gases may be compressed, but solids and liquids are difficult to compress.

(e) Give one example where we use a compressed gas in our everyday lives.

Chapter 21 Elements, Compounds and Mixtures

21.1 Elements – Simple Substances

In 1661, a famous Irish scientist called Robert Boyle introduced the word **element** into the language of chemistry. He used the name element to describe a substance that cannot be broken down into a simpler substance.

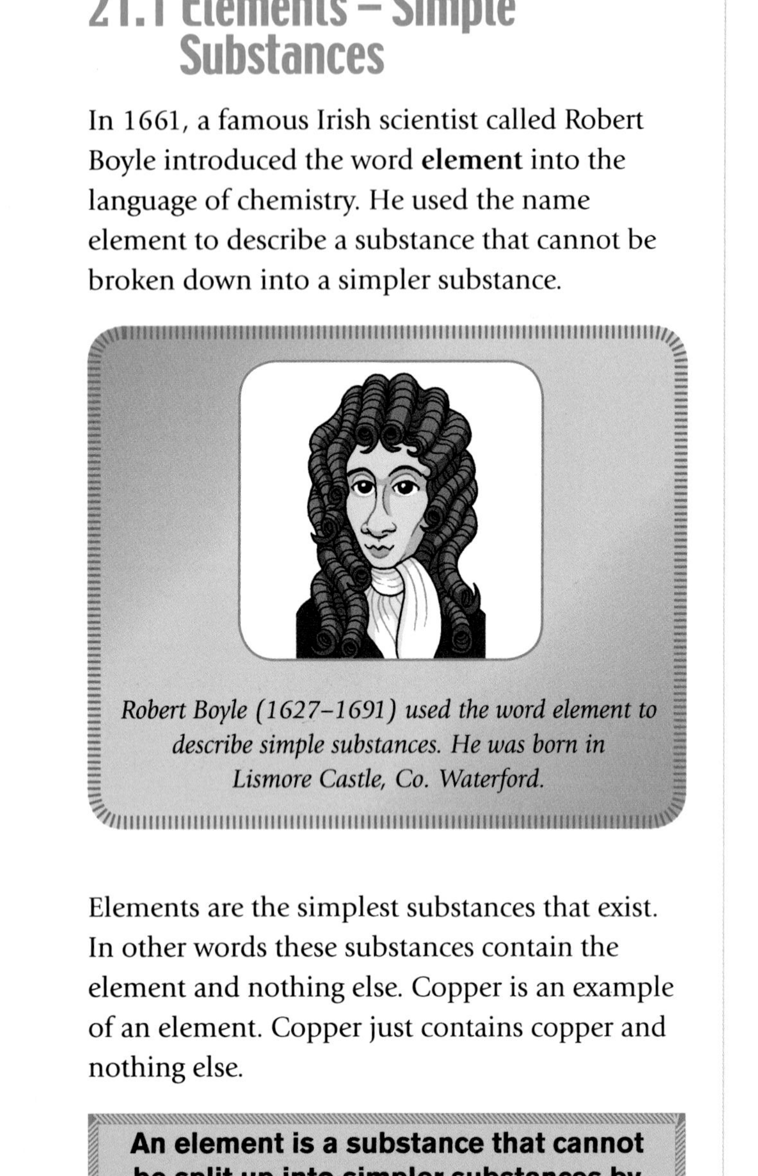

Robert Boyle (1627–1691) used the word element to describe simple substances. He was born in Lismore Castle, Co. Waterford.

Elements are the simplest substances that exist. In other words these substances contain the element and nothing else. Copper is an example of an element. Copper just contains copper and nothing else.

An element is a substance that cannot be split up into simpler substances by chemical means.

The word 'elementary' means 'simple'. When the famous detective Sherlock Holmes found it simple to work out the answer to a problem, he often described it as 'elementary', *Fig. 21.1*.

Fig. 21.1 When Sherlock Holmes says 'Elementary, my dear Watson' he means 'Simple, my dear Watson'. The word 'elementary' means simple. In chemistry, the word 'element' means 'simple substance'. For example, the black substance in burnt toast is the element carbon.

Examples and Uses of Elements

There are many other examples of elements. Some of these are solids, like aluminium and gold; some are liquids, like mercury and bromine; and others are gases, like oxygen and nitrogen. Some elements which you often meet are illustrated in *Fig. 21.2*.

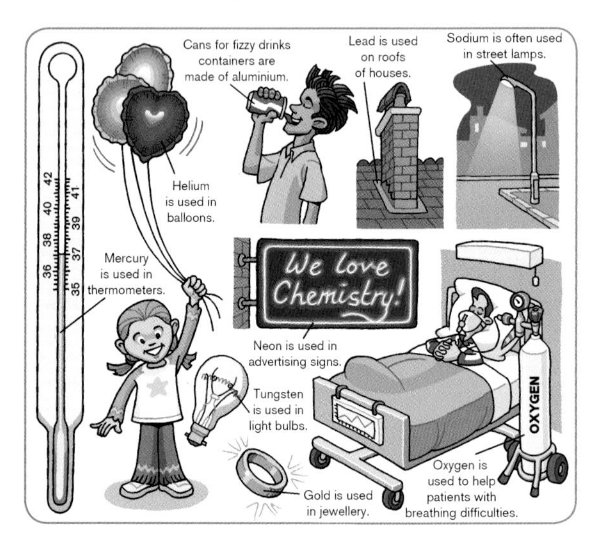

Fig. 21.2 Elements are the simplest substances. They have many uses in our everyday lives.

Number of Elements

If we were to examine all the materials in the world, we would find that there are about 100 elements. For example, Pierre and Marie Currie discovered the elements polonium and radium.

In 1898 Pierre (1859–1906) and Marie Curie (1867–1934) discovered two new elements. One of these elements they called polonium after Madame Curie's native land. The second element they called radium meaning 'giver of rays'.

Everything in the world is made up from these elements. That is why elements are called the **building blocks** from which all types of materials have been built.

Names and Symbols of the Elements

The Periodic Table

All of the elements discovered to the present time are listed in a table called the **Periodic Table** of the elements, *Fig. 21.3,* page 138. The Periodic Table is also printed on the back inside cover of this textbook. So, if you ever want to know if a particular substance is an element, just look at the Periodic Table. If the substance is listed there, it is an element. If it is not listed there, it is not an element!

Note the 'steps of stairs' going from element number 5 to element number 85. This separates the metals (left hand side) from the non-metals (right hand side).

International Symbols

If you examine the Periodic Table, you will see that each element has its own symbol. When each element is given a symbol, chemists around the world recognise the symbols regardless of the language they speak. In other words, the symbols for the elements form an international language. Therefore, a chemist or chemistry student anywhere in the world will know that Mg stands for magnesium. The symbol is a shorthand way of representing the element.

Symbols with One Letter

The symbol for some elements is simply the first letter of the name of the element in the English language, *Table 21.1*.

ELEMENT	SYMBOL
Hydrogen	H
Boron	B
Carbon	C
Nitrogen	N
Oxygen	O
Fluorine	F
Phosphorus	P
Sulfur	S

Table 21.1 The symbol for some elements is simply the first letter of the name of the element.

Symbols with Two Letters

The symbol for other elements is the first letter plus another letter, *Table 21.2*. Why do you think that two letters are necessary for some elements?

ELEMENT	SYMBOL
Helium	He
Lithium	Li
Beryllium	Be
Neon	Ne
Magnesium	Mg
Aluminium	Al
Silicon	Si
Chlorine	Cl
Argon	Ar
Calcium	Ca
Manganese	Mn
Zinc	Zn

Table 21.2 The symbol for some elements is the first letter of the name plus another letter. The first letter is always a capital letter. The second letter is always a small letter.

PERIODIC TABLE OF THE ELEMENTS

					1 H Hydrogen												2 He Helium
3 Li Lithium	4 Be Beryllium											5 B Boron	6 C Carbon	7 N Nitrogen	8 O Oxygen	9 F Fluorine	10 Ne Neon
11 Na Sodium	12 Mg Magnesium											13 Al Aluminium	14 Si Silicon	15 P Phosphorus	16 S Sulfur	17 Cl Chlorine	18 Ar Argon
19 K Potassium	20 Ca Calcium	21 Sc Scandium	22 Ti Titanium	23 V Vanadium	24 Cr Chromium	25 Mn Manganese	26 Fe Iron	27 Co Cobalt	28 Ni Nickel	29 Cu Copper	30 Zn Zinc	31 Ga Gallium	32 Ge Germanium	33 As Arsenic	34 Se Selenium	35 Br Bromine	36 Kr Kypton
37 Rb Rubidium	38 Sr Strontium	39 Y Yttrium	40 Zr Zirconium	41 Nb Niobium	42 Mo Molybdenum	43 Tc Technetium	44 Ru Ruthenium	45 Rh Rhodium	46 Pd Palladium	47 Ag Silver	48 Cd Cadmium	49 In Indium	50 Sn Tin	51 Sb Antimony	52 Te Tellurium	53 I Iodine	54 Xe Xenon
55 Cs Caesium	56 Ba Barium	57 La Lanthanum	72 Hf Hafnium	73 Ta Tantalum	74 W Tungsten	75 Re Rhenium	76 Os Osmium	77 Ir Iridium	78 Pt Platinum	79 Au Gold	80 Hg Mercury	81 Tl Thallium	82 Pb Lead	83 Bi Bismuth	84 Po Polonium	85 At Astatine	86 Rn Radon
87 Fr Francium	88 Ra Radium	89 Ac Actinium															

Fig. 21.3 All of the known elements are listed in the Periodic Table of the elements. This is a simplified form of the Periodic Table. There are 109 elements in the more detailed form of the Periodic Table (see back inside cover).

Symbols from Latin Names

Sometimes the symbol for the element is derived from the Latin name for the element.

- For example, the symbol for **copper** is **Cu** from the Latin word *cuprum*.
- The symbol for **iron** is **Fe** from the Latin word *ferrum*.
- The symbol for **gold** is **Au** from the Latin word *aurum*.
- The symbol for **silver** is **Ag** from the Latin work *argentum*.
- The symbol for **lead** is **Pb** from the Latin word *plumbum*.

Names and Symbols I Should Know

You may find some of these difficult to remember. However, as you go through this chemistry course you will meet them a number of times and this will help you to remember them. The syllabus lists a number of metals and non-metals whose symbols you should know. Study *Table 21.3(a)* and *(b)* carefully.

(a)

METAL	SYMBOL
Copper	Cu
Zinc	Zn
Aluminium	Al
Iron	Fe
Silver	Ag
Gold	Au

(b)

NON-METAL	SYMBOL
Carbon	C
Oxygen	O
Sulfur	S
Hydrogen	H
Nitrogen	N

Table 21.3 Symbols of some metals and non-metals.

TEST YOURSELF:
Now attempt questions *21.1* and *W21.1–W21.2*.

21.2 Compounds

You will not find substances like salt or sugar or water in the Periodic Table. The reason for this is these substances are not elements. Remember that the Periodic Table only lists elements. In fact, substances like salt, sugar and water are composed of more than one element. Substances made by joining together two or more elements are called **COMPOUNDS**.

A compound is a substance that is made up of two or more different elements combined together chemically.

Substances like water and table salt are not elements because they can be broken down into simpler substances.

Examples of Compounds

Some examples of common substances and the elements which they contain are shown in *Fig. 21.4*.

(a) Water contains the elements hydrogen and oxygen.

(b) Salt contains the elements sodium and chlorine.

(c) Sand contains the elements silicon and oxygen.

(d) Sugar contains the elements carbon, hydrogen and oxygen.

(e) Lime contains the elements calcium and oxygen.

Fig. 21.4 Some common compounds and the elements they contain.

CHEMISTRY

Making Compounds

Compounds can be made when elements react or combine together in what is called a **chemical reaction**. For example, to make the compound magnesium oxide, you could burn magnesium in oxygen, *Fig. 21.5*. You will observe a bright light being given off during the reaction. This is why magnesium is used in fireworks. Magnesium oxide contains the elements magnesium and oxygen combined together chemically.

Fig. 21.5 The elements magnesium and oxygen are reacting together. A chemical reaction takes place and a bright light is given off. The white powder formed is called magnesium oxide.

Although there are only about 100 elements, there are millions of compounds. This can be compared to the way that the 26 letters of the alphabet can give us many thousands of words.

Properties of Compounds

The properties of compounds usually differ completely from those of the elements from which they were made. For example, sodium is a soft and very reactive metal. Chlorine is a highly poisonous green gas. Yet, when these elements combine chemically, they form a white, unreactive solid which is essential for life. This white solid is called sodium chloride or common salt, *Fig. 21.6*.

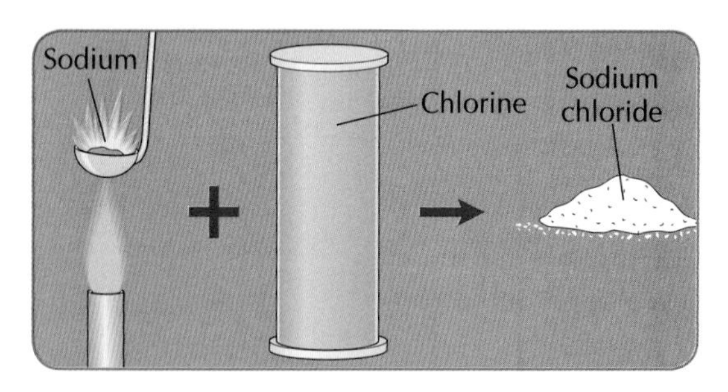

Fig. 21.6 The elements sodium and chlorine react together to form the compound sodium chloride.

There are many examples of compounds with different properties from the elements from which they are formed. The four examples on the syllabus are shown in *Table 21.4*.

COMPOUND	ELEMENT 1	ELEMENT 2
Water (H_2O) ◆ Liquid with refreshing taste. ◆ Essential for life. ◆ Does not burn.	**Hydrogen** ◆ Colourless, odourless, tasteless gas. ◆ Explosive when mixed with air.	**Oxygen** ◆ Colourless, odourless, tasteless gas. ◆ Substances burn very well in it.
Carbon dioxide (CO_2) ◆ Colourless, odourless, tasteless gas. ◆ Does not allow substances to burn in it.	**Carbon** ◆ A black solid.	**Oxygen** ◆ Colourless, odourless, tasteless gas. ◆ Substances burn very well in it.
Magnesium oxide (MgO) ◆ White powder. ◆ Dissolves slightly in water to make *Milk of Magnesia*.	**Magnesium** ◆ A shiny metal. ◆ Burns brightly in air.	**Oxygen** ◆ Colourless, odourless, tasteless gas. ◆ Substances burn very well in it.
Iron sulfide (FeS) ◆ Black solid. ◆ Not attracted to a magnet.	**Iron** ◆ Grey metal. ◆ Attracted to a magnet.	**Sulfur** ◆ Yellow solid.

Table 21.4 The properties of compounds are very different from the elements that react together to form the compound.

Composition of Compounds

In compounds the elements are present in a fixed ratio by number.

- For example, it is found by experiment that the ratio of hydrogen : oxygen in water is always 2 : 1. Hence, chemists represent water by the formula H_2O. This is called the chemical formula for **water**. We will be learning more about chemical formulas later on in this course.
- In **carbon dioxide**, the ratio of carbon : oxygen is always 1 : 2. Therefore, carbon dioxide is represented by the chemical formula CO_2.

TEST YOURSELF:
Now attempt questions *21.2* and *W21.3–W21.6.*

21.3 Mixtures

So far, we have considered just two types of chemical substances – elements and compounds. However, most of the ordinary things we see around us are made up of two or more elements or compounds which are just mingled or jumbled up together. Such a substance is known as a **MIXTURE**.

A mixture consists of two or more different substances mingled with each other but not chemically combined.

Examples of Mixtures

There are many examples of mixtures.

- **Air** is a mixture of a number of gases (*Chapter 25*).
- **Sea water** is a mixture of water, salt and a number of other substances.
- **Crude oil** is one of the most valuable mixtures ever discovered.
- **Earth** is a mixture of stones, sand, clay, water, etc.
- **Petrol** is a mixture of a number of different compounds of hydrogen and carbon.

Some examples of mixtures are shown in *Fig. 21.7*.

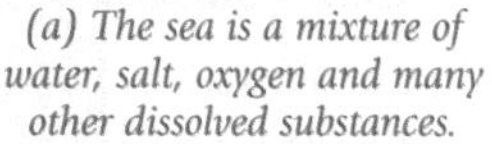

(a) The sea is a mixture of water, salt, oxygen and many other dissolved substances.

(b) Salad dressing is a mixture of oil and vinegar.

(c) Concrete is a mixture of cement, sand, water and various other materials.

(d) Toothpaste is a mixture of various substances like foaming agents, flavourings, abrasive materials, colouring agents, flouride salts etc.

Fig. 21.7 Examples of mixtures.

Iron and Sulfur

In a mixture, the substances which make up the mixture can be present in any amounts. A mixture of iron and sulfur may be made by stirring together some powdered sulfur and iron filings. Any quantities of sulfur and iron filings may be mixed together, e.g. 10 grams of iron with 10 grams of sulfur or 500 grams of iron with 1 gram of sulfur. In the mixture, it is easy to see the yellow sulfur and the silvery iron particles. However, if iron and sulfur are heated together in a test-tube there is a sudden red glow in the test-tube.

This indicates that a chemical reaction is taking place. A new substance called iron sulfide is formed. This reaction is studied in detail in *Experiment 21.1*.

Experiment 21.1

To make a mixture using iron and sulfur and then change the mixture into the compound iron sulfide

Apparatus required: test-tube; retort stand and clamp; pestle and mortar; bar magnet; bunsen burner; cloth; spatula; forceps; hammer (optional)

Chemicals required: iron filings, sulfur

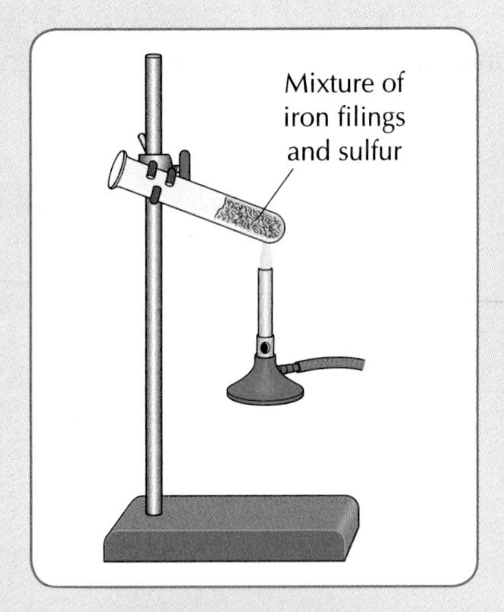

Fig. 21.8 If a mixture of iron and sulfur is heated, the compound iron sulfide is formed.

Method

1. Mix 7 g of iron filings and 4 g of sulfur in a pestle and mortar. Grind the mixture very well.
2. Place some of the mixture on to a piece of paper. Bring a bar magnet up to the mixture and write down what you observe.
3. Place some of the mixture in a test-tube until it is about one third full. Clamp the test-tube as shown in *Fig. 21.8*.
4. Heat the bottom of the test-tube until the mixture begins to glow. Take away the Bunsen burner at this stage and note that the glow spreads through the material. This indicates that a chemical reaction is taking place and that heat is being given off.
5. Continue heating along the test-tube until the material no longer glows.
6. Allow the test-tube to cool. Tip out the contents into a mortar. If the contents do not tip out, you will have to smash the test-tube by placing a cloth over the test-tube in the mortar, and then carefully breaking the test-tube with the pestle or with a hammer. (***Note:*** This should be carried out under the direct supervision of your teacher.)
7. Remove the cloth and use forceps to pick out several pieces of the grey material. This grey material is the compound iron sulfide.
   iron + sulfur = iron sulfide
8. Now compare the properties of the mixture of iron and sulfur with the compound iron sulfide.

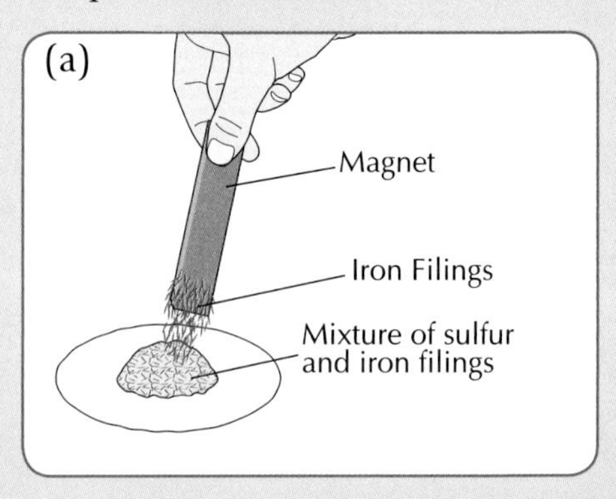

Fig. 21.9(a) When iron filings and sulfur are simply mixed together, no new substance is formed. The individual pieces of iron and sulfur may still be seen in the mixture. The iron may easily be separated from the mixture with a magnet.

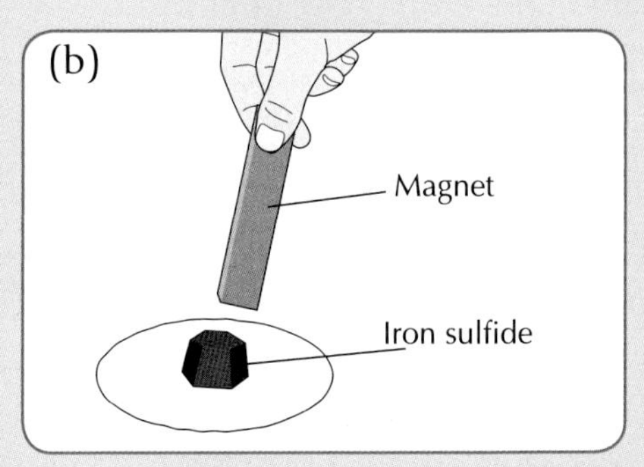

Fig. 21.9(b) The compound iron sulfide is different from the mixture of iron and sulfur. Iron sulfide cannot be separated into iron and sulfur using a magnet. Iron sulfide is not attracted to a magnet.

Results

The substance formed, iron sulfide, looks different from the mixture. It is a dark grey solid with no specks of yellow sulfur in it. It cannot be separated using a magnet into iron and sulfur. Iron sulfide is not attracted to a magnet.

Conclusion

- Iron sulfide is a **compound** of the **two elements** iron and sulfur.
- The iron sulfide **looks different** from the elements iron and sulfur.
- The properties of the compound (iron sulfide) are very different from those of the elements (iron and sulfur) from which it was made. Iron sulfide has **new properties** of its own.

We conclude that when elements combine to form compounds, the elements lose their individual properties.

If we were to examine a large number of mixtures and compounds, we would find that the same differences would show up each time. These differences are summarised in *Table 21.5*.

MIXTURE	COMPOUND
1. The amounts of the substances in the mixture can vary.	The elements in a compound are always present in the same amount.
2. A mixture contains two or more substances.	A compound is a single substance.
3. The properties of a mixture are similar to those of the substances in the mixture.	The properties of a compound are different to those of the elements which reacted to form the compound.
4. It is usually easy to separate the parts of a mixture.	It is usually difficult to separate out the elements in a compound. (Only possible using suitable experiments).
5. There are practically no energy changes when a mixture is made.	Heat is usually given out or taken in when a compound is formed.

Table 21.5 The differences between mixtures and compounds.

TEST YOURSELF:

Now attempt questions *21.3–21.4* and *W21.7–W21.11*.

What I should know

- **Element**. A substance that cannot be split up into simpler substances by chemical means. There are about 100 known elements.
- **Compound**. A substance which is made up of two or more different elements combined together chemically. Millions of compounds exist.
- **Mixture**. Consists of two or more different substances mingled with each other but not chemically combined.
- **Five differences between mixtures and compounds.**
  (1) Amount of substances in the mixture can vary.
  (2) Mixtures contain two or more substances.
  (3) Mixtures have similar properties to the substances in the mixture.
  (4) Substances in the mixture can be easily separated.
  (5) Little energy changes occur when a mixture is made.

QUESTIONS

Write the answers to the following questions into your copybook.

21.1 (a) A substance which cannot be broken down into anything simpler is called an ..... .

(b) Name the Irish scientist who introduced the word *element* into the language of chemistry.

(c) What is the chemical symbol for copper?

(d) Write down the name of the element which has the symbol Au ..... .

(e) Write down the symbols of the following elements oxygen, magnesium, potassium, neon.

(f) In an examination, a student was asked to write down the symbol for the element chlorine. The student wrote down CL. No marks were awarded by the examiner for this answer. Why not?

(g) X is an element. It is a liquid with a silvery colour. It is often used in thermometers. Its name is ..... .

21.2 A number of common compounds and the way that chemists represent them using chemical formulas are shown in *Table 21.6*. Examine the chemical formulas closely and write down the elements in each compound in the third column. The first example has been done for you.

NAME OF COMPOUND	FORMULA	ELEMENTS IN THE COMPOUND
Water	H_2O	Hydrogen and oxygen
Magnesium oxide	MgO	
Iron sulfide	FeS	
Carbon dioxide	CO_2	
Hydrogen chloride (hydrochloric acid)	HCl	
Sodium chloride (common salt)	NaCl	
Calcium oxide (lime)	CaO	
Sugar (common table sugar)	$C_{12}H_{22}O_{11}$	

Table 21.6

21.3 A student was asked to represent an element, a compound and a mixture using the particle theory. He drew three diagrams but forgot to label the diagrams. Put the correct label on each diagram in *Fig. 21.10*. Explain your answer in each case.

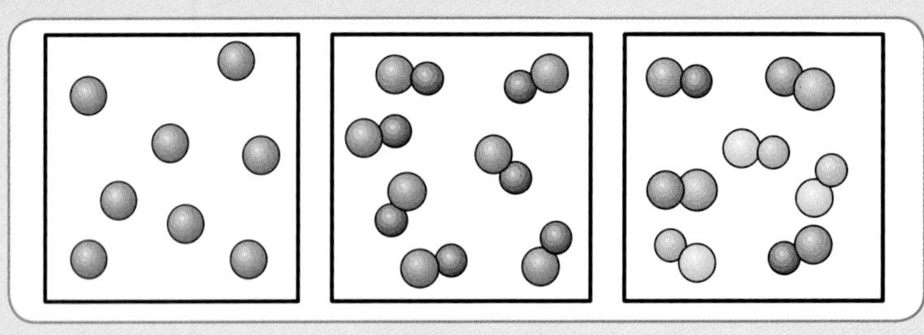

Fig. 21.10

21.4 The following description of an experiment was found in a student's notebook. 'A mixture of iron and sulfur was heated for a few minutes in a boiling tube using a Bunsen burner. The mixture began to glow brightly. The glow spread throughout the boiling tube when the Bunsen burner was removed. The boiling tube was allowed to cool and a dark grey solid was tipped out of it.'

(a) How could the iron and sulfur in the original mixture be separated?

(b) What information in the description tells us that a chemical reaction has taken place?

(c) Write down the name of the compound formed in the reaction.

Chapter 22 Solutions

22.1 Water as a Solvent

One of the most important properties of water is that a very large number of substances dissolve in it. We say that water is an excellent **solvent**.

A solvent is a substance that dissolves other materials to form a solution.

No other compound is as good a solvent as water. Water is often called the universal solvent because so many substances dissolve in it. Since water is such a great solvent, it is very difficult to get pure water. If you examine the label on a bottle of mineral water, you will see that many substances are dissolved in the water, *Fig. 22.1*. It is these substances that give the refreshing taste to water.

STILL MINERAL WATER

Ingredients: 100% Mineral Water

NUTRITIONAL INFORMATION

Average Mineral content (mg/ltr)	
Calcium	82.0
Magnesium	9.7
Sodium	23.9
Potassium	2.2
Bicarbonate	228.0
Chloride	41.0
Sulphate	10.8
Nitrate	5.0

Fig. 22.1 All of the substances listed on this label are dissolved in the spring water.

Solutions

- The substance that dissolves in the water is called the **solute**.
- Water is called the **solvent**.
- The solute and solvent mix together to give a **solution**.

solute + solvent ⟶ solution

Therefore, a solution may be defined as follows:

A solution is a mixture of a solute and a solvent.

Examples of a Solution

- **Cup of coffee**. When you make a cup of coffee you are making a solution.
  coffee + water ⟶ cup of coffee
  (solute) (solvent) (solution)

Substances like coffee or sugar that dissolve in water are said to be **soluble** in water. A substance like sand, which does not dissolve in water, is said to be **insoluble** in water.

- **Copper sulfate solution**. Some of the main terms used when talking about solutions are summarised in *Fig. 22.2*. Copper sulfate is a blue substance that dissolves in water. It is commonly found in science laboratories.

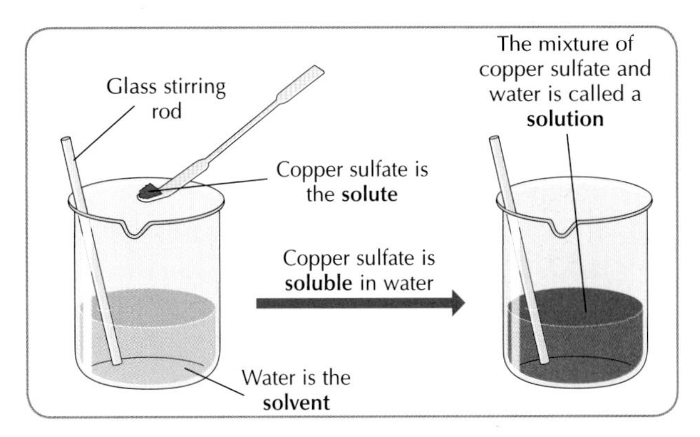

Fig. 22.2 A solute (copper sulfate) is being added to a solvent (water) to form a solution of copper sulfate in water.

- **The sea**. The sea is the biggest solution in the world. Water is the solvent in the sea. If you taste sea water it is obvious there is something (dissolved) in it!

Evidence for a Solute in a Solution

When a substance like salt or sugar is added to water, the substance seems to vanish or disappear in the water.

- We know that the substance has not actually disappeared since the **taste** of salt or sugar is still present in the solution.

- Another piece of evidence to show that the solute is still present is shown in *Fig. 22.3*.

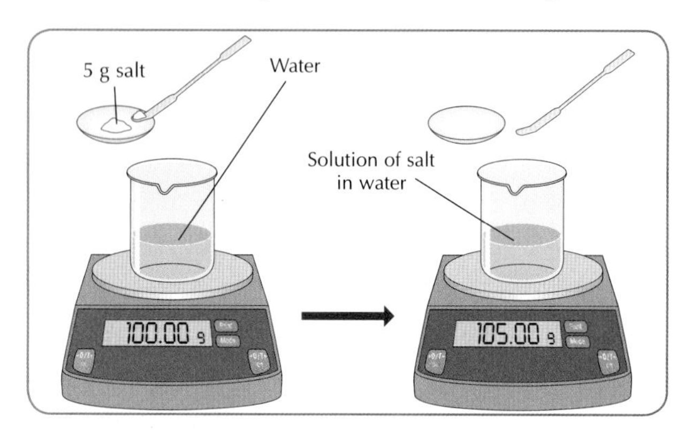

Fig. 22.3 This experiment shows that the salt has not 'disappeared' but is present in the solution. The mass of the solution is the sum of the mass of the solute and solvent.

Note: The **mass** of a solution is the sum of the mass of the solute and solvent. This shows that the solute has not disappeared. The solute still exists in the solution and its mass is shown on the balance.

- Solutions are always **clear**, i.e. if you hold them up to the light, the light passes through them. Examine a solution of copper sulfate in water in this way. (You may not think a cup of coffee looks very clear. However, if you add plenty of water to it, you will see that the light passes through it).

Melting or Dissolving?

Do not get confused between **melting** and **dissolving**. Dissolving involves two substances (a solute and a solvent), e.g. sugar added to water. However, melting needs only one substance, e.g. ice cubes on a warm day.

Particle Theory and Solutions

A solution is a perfect mixture because when you look at a solution, you cannot see the individual solute and solvent. The particle model helps to give us a picture of what happens when a solute dissolves in a solvent, *Fig. 22.4*.

Suspensions

In some cases, a solid added to water becomes scattered throughout the liquid. This often happens if the solid is divided into fine particles, e.g. powdered chalk added to water. Such a mixture is called a **suspension**.

Examples of a Suspension

- **Muddy water** is a suspension of mud particles in water.
- **Milk** is an example of a suspension of tiny droplets of oil (cream) in a watery liquid.
- Another example of a suspension is ***Milk of Magnesia***. This is a suspension of a substance called magnesium hydroxide in water.

Over a period of time, the suspended particles settle to the bottom of the container. Therefore, the instruction 'shake before use' is often printed on suspensions.

It is easy to see the difference between a suspension and a solution. A solution looks clear but a suspension looks cloudy.

TEST YOURSELF:
Now attempt questions *22.1–22.2* and *W22.1–W22.2*.

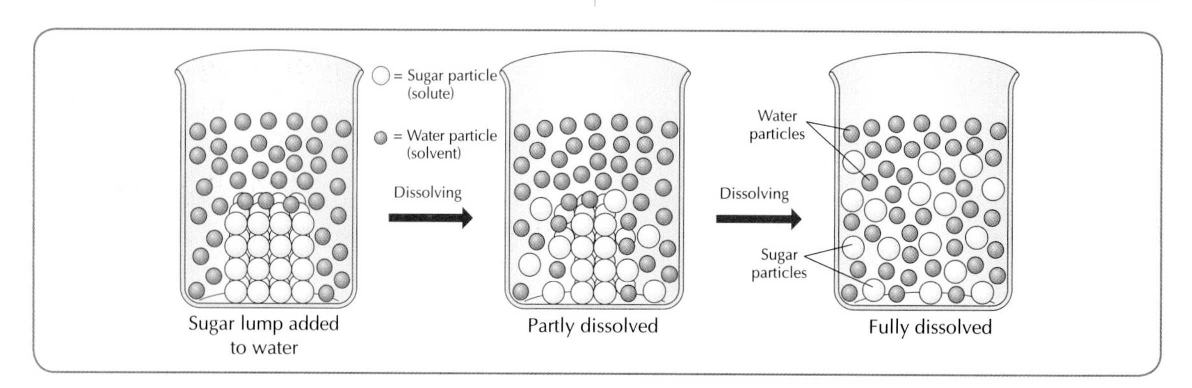

Fig. 22.4 This diagram shows how a solution is formed when a sugar lump is added to water. The water particles move in between the sugar particles and push them apart. The sugar particles begin to break away from each other and mix completely with the water particles.

22.2 Dilute and Concentrated Solutions

You often see the instruction 'dilute to taste' on solutions like orange squash, *Fig. 22.5(a)*. This is because the orange squash as purchased is very **concentrated**, i.e. there is a large amount of solute dissolved in the water. The drink tastes nicer if the amount of orange squash is made smaller relative to the quantity of water, i.e. we need to make the concentrated solution more dilute.

A dilute solution has a small amount of solute in a large amount of solvent.

Two examples of a dilute solution are shown in *Fig. 22.5*.

Fig. 22.5(a) The bottle contains a concentrated solution and the glass contains a dilute solution.

Fig. 22.5(b) The beaker on the left contains a dilute solution of copper sulfate. The beaker on the right contains a concentrated solution of copper sulfate.

On the other hand, if a large amount of solute is dissolved in the solvent, we say that the solution is concentrated.

A concentrated solution has a large amount of solute in a small amount of solvent.

Particle Theory and Concentrations

Using the particle theory, we can see that a dilute solution has only a few dissolved particles in a certain volume, *Fig. 22.6*. A concentrated solution has a lot of dissolved particles in a certain volume. A dilute solution can be made more concentrated by adding more solute.

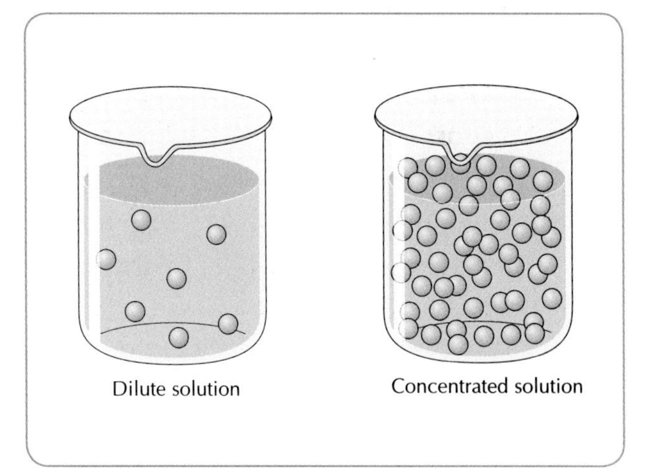

Fig. 22.6 A dilute solute has a small number of dissolved particles in a certain volume. A concentrated solution has many dissolved particles in a certain volume.

Note: In everyday speech we use the word 'weak' to describe a dilute solution, e.g. a cup of weak tea. We also use the work 'strong' to describe a concentrated solution. In science it is best to avoid the use of the terms 'weak' and 'strong' when talking about dilute and concentrated solutions. The reason for this is because the terms 'weak' and 'strong' have different meanings in chemistry when talking about the properties of acids.

22.3 Solubility and Solubility Curves

In this section we will study how well some substances dissolve in water and how temperature affects the amount of the substance that will dissolve.

Mandatory experiment 22.1

To grow crystals of alum or copper sulfate

Apparatus required: pestle and mortar; stirring rod; beaker; Bunsen burner (or hotplate); tripod; wire gauze; evaporating basin; thermometer

Chemicals required: alum or copper sulfate; water

In this experiment we will attempt to grow crystals of alum or copper sulfate. A crystal is a solid substance with a regular shape and smooth faces. Alum is a substance used in the manufacture of leather goods. Copper sulfate is a substance that is used to kill fungi and also for preserving timber.

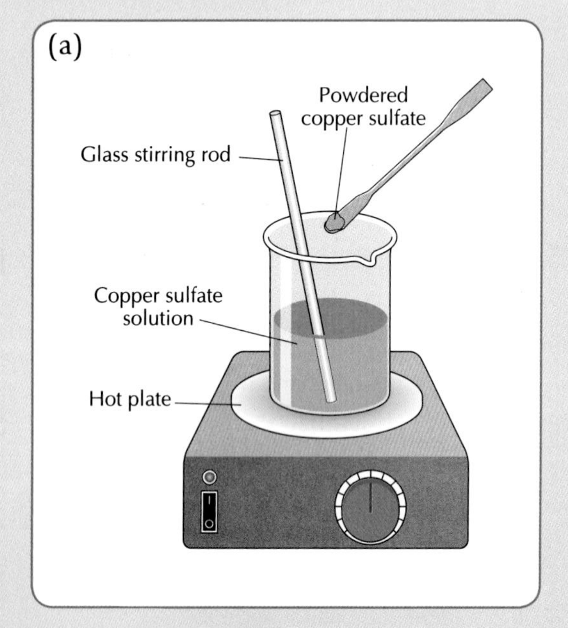

Fig. 22.7(a) Making a saturated solution of copper sulfate using powdered copper sulfate.

Method

1. Grind up a sample of alum or copper sulfate using a pestle and mortar. This will help it to dissolve more quickly.
2. Gradually add the powdered substance to 100 cm^3 of water in a beaker, *Fig. 22.7(a)*. Use a stirring rod to help the powder to dissolve. Continue adding until the solute no longer dissolves but settles at the bottom of the beaker. Using a thermometer, note the temperature of the water.
3. Heat the water to about 60 °C and note what happens to the undissolved copper sulfate. What does this show? Add in more copper sulfate and heat until no more will dissolve.
4. Pour about half of the solution into a warm evaporating basin and put it aside to cool slowly, *Fig. 22.7(b)*. Cool the other half quickly by holding the beaker under running water from the tap. Note what you observe.

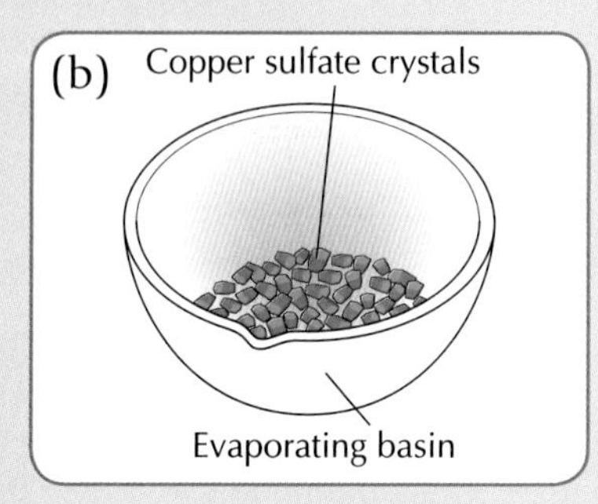

Fig. 22.7(b) Crystals of copper sulfate are formed when the hot saturated solution cools.

5. Compare the size of the crystals obtained by slow cooling of the hot solution with the size of those obtained by fast cooling.

Result

Crystals are formed.

Conclusion

Crystals are formed when a hot concentrated solution is cooled.

Saturation

We can see from the above experiment that, at a certain temperature, we could not get any more copper sulfate to dissolve. In other words, there is a limit to the amount of copper sulfate that can be dissolved at a certain temperature. A solution that contains the maximum amount of solute is called a **saturated solution**. This is similar to a sponge being saturated with water, i.e. it cannot absorb any more water, *Fig. 22.8*.

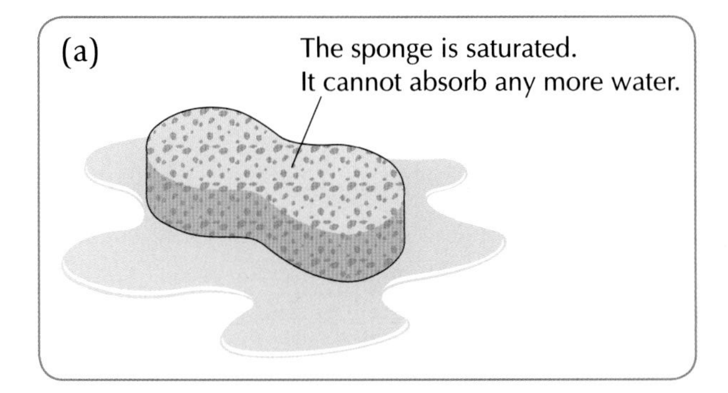

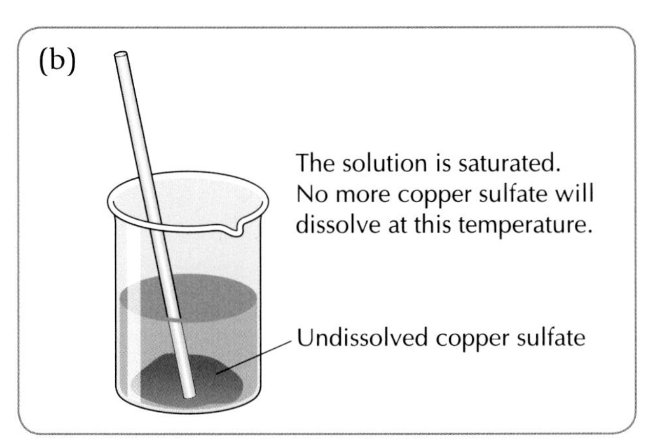

Fig. 22.8 When a sponge is saturated, it cannot absorb any more water. Similarly, when a solution is saturated, it cannot dissolve any more solute at that temperature. If more solute is added, it will simply settle at the bottom of the beaker.

A saturated solution is one that contains as much dissolved solute as possible at a given temperature.

Solubility and Temperature

It is also clear from the above experiment that the **amount of solute that will dissolve in a solvent depends on the temperature**. In general, substances are more soluble in hot solvents than in cold ones.

Examples of solubility and Temperature

- It is better to **wash dishes** in hot water than in cold water. This is because the hot water helps to dissolve the grease more easily than cold water.
- When making a cup of coffee, the **coffee grains** dissolve more easily in hot water than in cold water.
- If you have ever made **jelly**, you will have noticed that the jelly dissolves more easily in hot water than in cold water.

Note: A solution that is not saturated is said to be **unsaturated**. In other words, an unsaturated solution has not dissolved the maximum amount of solute at a given temperature.

How is Solubility Measured?

It is clear from the above experiment that crystals are formed when a hot saturated solution is cooled. Chemists use the term **solubility** to give us an idea of the amount of solute that will dissolve in a certain amount of solvent.

The solubility of a substance is its mass (in grams) that will dissolve in 100 grams of solvent at a fixed temperature.

For example, the solubility of copper sulfate is 40 g at 60 °C. This means that 40 g of copper sulfate will dissolve in 100 g of water at 60 °C. Solubility varies from solute to solute. For example, the solubility of a substance called potassium nitrate is 110 g at 60 °C. This tells us that potassium nitrate is more soluble in water than copper sulfate.

Experiment 22.2

To investigate the effect of temperature on solubility

Apparatus required: beaker; thermometer; stirring rod; Bunsen burner or hotplate; spatula

Chemicals required: copper sulfate (or any substance that dissolves in water); water

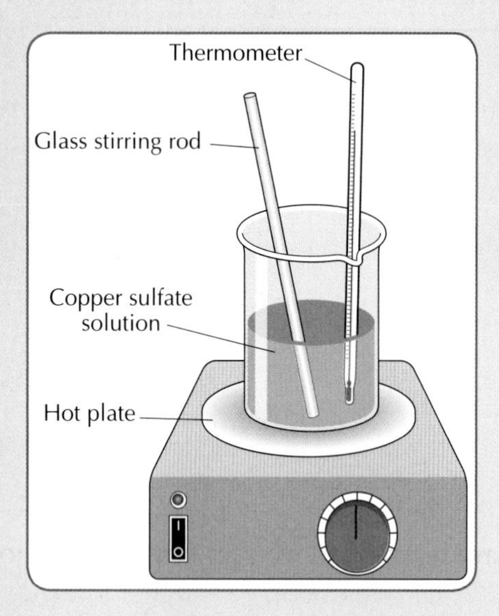

Fig. 22.9 Apparatus to investigate the effect of temperature on solubility.

Method

1. Grind up a sample of the substance (e.g. copper sulfate) with a pestle and mortar.
2. Weigh out 100 g of the copper sulfate.
3. Place 100 cm^3 of water in a beaker and heat the water to about 30 °C.
4. Measure the exact temperature of the water using a thermometer.
5. Add some of the copper sulfate and stir after each addition.
6. Continue adding the copper sulfate until no more will dissolve, i.e. until the salt simply settles at the bottom of the beaker.
7. Weigh the amount of copper sulfate that you have **not** added to the water.
8. By subtraction, calculate the amount of the copper sulfate that has dissolved in the water.
9. Repeat the experiment at various other temperatures.

Result

Summarise your results in the form of a table like that shown in *Table 22.1*. (Some sample results have been added).

TEMPERATURE (°C)	SOLUBILITY i.e. mass of copper sulfate (g) that dissolves in 100 g of water
0	14
10	17
20	20
30	25
40	29
50	34
60	40
70	47
80	56
90	68

Table 22.1 This table shows how the solubility of copper sulfate changes with temperature.

Show your results on a graph similar to that shown in *Fig. 22.10*.

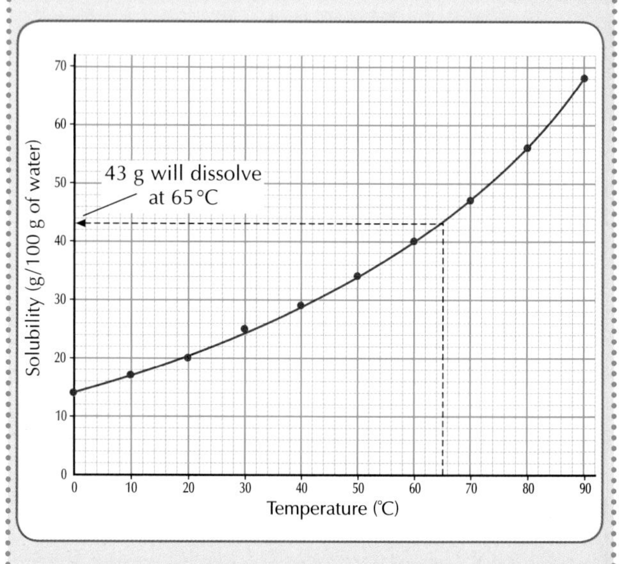

Fig. 22.10 The solubility curve for copper sulfate. A solubility curve shows how solubility changes with temperature. In this case we see that more copper sulfate dissolves at a higher temperature.

Conclusion

The solubility of copper sulfate increases as the temperature is increased.

Solubility Curve

The graph shown in *Fig. 22.10* is called a solubility curve.

A solubility curve is a graph showing how the solubility of a substance varies with temperature

Solubility curves are very useful to chemists. For example, suppose you wanted to know how much copper sulfate will dissolve in 100 g of water at 65 °C. The mass of copper sulfate (43 g) can be read off from the graph as shown in *Fig. 22.10*.

Crystallisation

A solubility curve also helps us to explain why crystals are formed. At 80 °C we see that 56 g of copper sulfate can dissolve in 100 g of water. However, if the solution is cooled to 20 °C, we see that only 20 g of copper sulfate can now dissolve. Therefore, 56 – 20 = 36 g will come out of solution. When this happens, crystals are usually formed. This process is called **crystallisation**. *Fig. 22.11*.

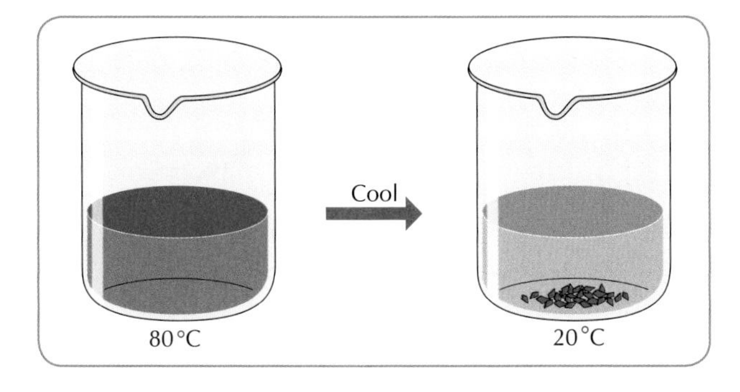

Fig. 22.11 Crystals are formed when the hot saturated solution is cooled. This process is called crystallisation.

Crystals could also be formed by simply allowing the water to evaporate from a solution. Since there is now less water to dissolve the copper sulfate, crystals of copper sulfate will be formed.

Crystallisation is the formation of crystals when a hot saturated solution is cooled or when the solvent is allowed to evaporate from the solution.

Solutes and Solvents as Solids, Liquids and Gases

Solutions are not always made by dissolving a solid solute in a liquid solvent.

- For example, fizzy drinks contain carbon dioxide (a **gas**) dissolved in **water**.
- Whiskey and other alcoholic drinks contain alcohol (a **liquid**) dissolved in **water**.
- Marshmallows contain air (a **gas**) dissolved in sugar paste (a **solid**).

Other Solvents

We have learned above that water is the most common and the most important solvent. Water, however, is not the only type of solvent. Some substances like oil and grease do not dissolve in water. To remove oil or grease stains from clothing one would have to use a solvent such as white spirit or chloroform. In the dry cleaning of clothes, special solvents must be used to remove certain stains. Dry cleaning is so called because it is cleaning without water.

Dangers of Other Solvents

There are, however, problems using solvents other than water.

- Some solvents used in **paint** (e.g. white spirit and alcohol) are flammable and can cause dangerous fires.
- Solvents used in **glue** are extremely poisonous. Breathing their vapour can cause damage to the lungs, brain, liver and other parts of the body.

As paint dries, the solvent evaporates and the solid particles of paint are left on the object being painted. Therefore, when using strongly smelling paints, always work in a well-ventilated area.

TEST YOURSELF:

Now attempt questions *22.3–22.5* and *W22.3–W22.11*.

What I should know

- **Solute**. The substance that is dissolved.
- **Solvent**. The substance that dissolves other materials to form a solution.
- **Solution**. A mixture of a solute and a solvent.
- **Dilute solution**. A solution in which there is a small amount of solute in a large amount of solvent.
- **Concentrated solution.** A solution in which there is a large amount of solute in a small amount of solvent.
- **Saturated solution**. A solution that contains as much dissolved solute as possible at a given temperature.
- **Unsaturated solution.** A solution that has not dissolved the maximum amount of solute at a given temperature.
- **Solubility**. The mass of a substance (in grams) that will dissolve in 100 grams of solvent at a fixed temperature.
- **Solubility curve.** A graph showing how the solubility of a substance varies with temperature.
- **Crystallisation**. The formation of crystals when a hot saturated solution is cooled or when the solvent is allowed to evaporate from the solution.

Questions

Write the answers to the following questions into your copybook.

22.1 (a) The solid which dissolves in a liquid is called the ..... .

(b) Water is a good ..... for salt but not for sand.

(c) When sugar dissolves in water a ..... is formed.

(d) Explain why the can of regular Cola has a greater mass than a can of Diet Cola, *Fig. 22.12*.

Fig. 22.12

(e) Sodium chloride will not dissolve in paraffin oil, i.e. it is ..... in paraffin oil.

(f) The most common solvent is ..... .

(g) A solution looks clear but a suspension looks ..... .

(h) A bottle of calamine lotion (used for treating itchy spots) has an instruction to 'shake the bottle before use' printed on its label. Do you think that this lotion is a solution or a suspension? Give a reason for your answer.

(i) Milk is a suspension of tiny droplets of ..... in water.

22.2 (a) A beaker of water is placed on a laboratory balance, *Fig. 22.13*. The total mass of the beaker and water is 100 g. A student adds 4 g sugar to the water. What is the mass of the solution? Explain your answer.

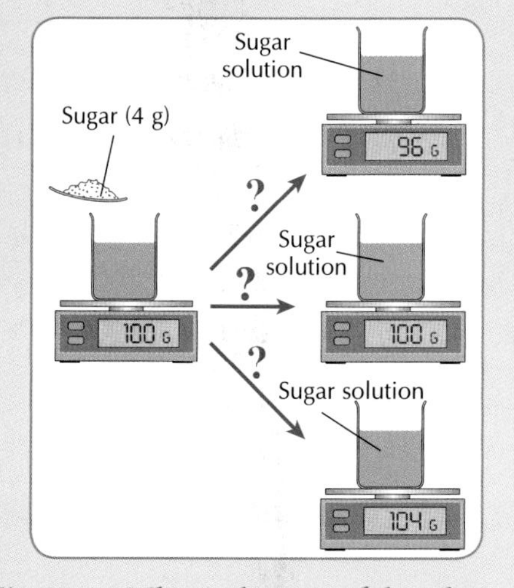

Fig. 22.13 What is the mass of the solution?

QUESTIONS

(b) If you were given two beakers, one containing a solution and the other a suspension, how would you decide which is the solution and which is the suspension?

22.3 (a) Some sugar is dissolved in water. The general name used to describe the sugar is the ....., the water is called the ..... and the mixture of sugar and water is called the ..... .

(b) The most common and the most important solvent in our lives is ..... .

(c) White spirit is a good ..... for paint.

(d) A suntan lotion has instructions printed on it to 'shake before use'. What does this tell us about the type of material in the container?

(e) 'Tincture' of iodine is a disinfectant. It is a ..... of iodine in a solvent called ethanol.

(f) Whiskey contains a solution of ..... in water.

(g) If the cap is left off a bottle of fizzy lemonade it goes 'flat'. This is because there is very little of the gas called ..... dissolved in the lemonade.

(h) Nail varnish is ..... in water but ..... in acetone.

22.4 *Table 22.2* shows how the solubility of a substance called potassium chloride changes with temperature.

TEMPERATURE(°C)	0	10	20	30	40	50	60	70	80	90	100
SOLUBILITY (g/100 g WATER)	28.0	31.0	34.5	37.5	40	43	45.5	48.5	51.0	54.0	56.5

Table 22.2

(a) Putting temperature on the x-axis and solubility on the y-axis, plot the solubility curve for potassium chloride.

(b) What is the solubility of potassium chloride at 55 °C?

(c) At what temperature will 53 g of potassium chloride dissolve in 100 g water?

(d) In this experiment, the solubility of the potassium chloride was measured at several temperatures. Describe, using a labelled diagram, how one of these measurements could have been made.

(e) What would you expect to happen if you cooled a saturated solution of potassium chloride from 60 °C to 20 °C?

22.5 (a) When manufacturing nail varnish, it is important to ensure that the nail varnish is insoluble in water. Explain.

(b) Paint stripper contains a solvent which ..... paint.

(c) Give two examples of substances whose solvents are dangerous to inhale.

(d) Correcting fluids consists of a white solid dissolved in a solvent that evaporates easily. Why is it important that the solvent used in correcting fluid evaporates easily?

(e) Some graffiti was written on a wall using an ink marker. The graffiti could not be removed by water. Suggest one other solvent that may remove the graffiti.

Chapter 23 Separating Mixtures

23.1 Introduction

Many of the ordinary things we see around us are mixtures of a number of different substances.

- **Sea water** not only contains common salt, but also has many other substances dissolved in it.
- Clean **air** is a mixture of a number of different gases.

Making Pure Substances

Chemists often find it necessary to prepare a **pure substance**.

- A pure substance is one that contains only **a single substance** and no impurities.
- A pure substance is either **an element or a compound** – it cannot be a mixture. For example, oxygen is a pure substance but air is not. Air is a mixture of substances.

In this chapter we will examine how pure substances can be obtained from mixtures. When substances are separated from mixtures they are said to be **purified** or **refined**. The method of separation depends on the type of mixture.

23.2 Separation Using Filtration

Examples of Filtration

- **Coffee filters**. You may have used a coffee filter. The filter is simply a piece of paper that separates the coffee solution from the coffee grounds, *Fig. 23.1(a)*.
- **Oil and air filters**. If you are interested in cars, you may know something about oil filters and air filters. In a car engine, these filters separate out small particles of dirt from the oil and air. These small particles of dirt would damage the engine of a car if they were not removed.
- **Surgical masks**. Another type of filter you may have seen are the surgical masks used in operating theatres, *Fig. 23.1(b)*.
- **Kitchen Utensils**. In the kitchen, filtration is used in a tea strainer or tea bags, *Fig. 23.1(c)*. It is also used in a colander when separating vegetables cooked in water from the water, *Fig. 23.2*. Even a chip pan and a fishing net make use of filtration!

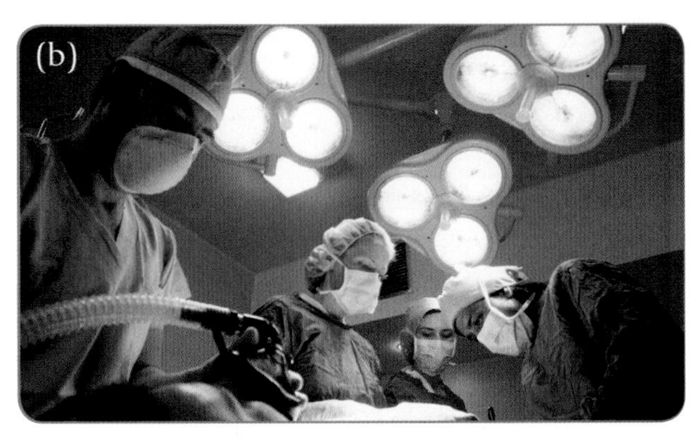

Fig. 23.1(a) Filtration is used in coffee filters to separate the coffee solution from the ground coffee. (b) Filtration is used in surgical masks to let the air through but trapping the germs. (c) Filtration is also used in a tea strainer to stop the tea leaves entering the cup, but not the tea.

The method of separation used in the above examples is known as **filtration**. It is very useful in the science laboratory for separating a solid from a liquid, e.g. particles of chalk from water, soil from water, etc.

Filtration is a method of separating an insoluble solid from a liquid using a material that allows the liquid to pass through but not the solid.

Fig. 23.2 An example of filtration in the kitchen!

In the next experiment, we will use filtration to separate a mixture of water and soil.

Mandatory experiment 23.1

To separate a mixture of water and soil using filtration

Apparatus required: filter paper; filter funnel; glass rod; retort stand and clamp; wash bottle; beaker or conical flask

Chemicals required: mixture of water and soil

Method

1. Take a filter paper and fold it into the shape of a cone as directed by your teacher.
2. Rinse a filter funnel with water and place the cone of filter paper in it as shown in *Fig. 23.3*. Rinsing the filter funnel not only cleans it but also helps the filter paper to stick to it.

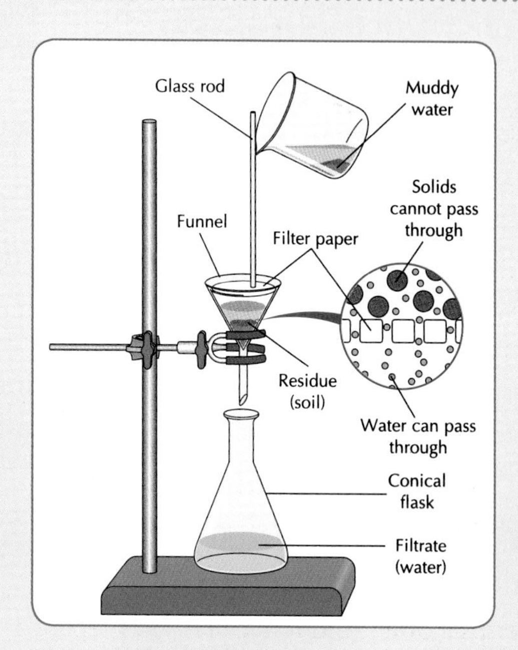

Fig. 23.3 Apparatus to separate water and soil. The soil remains behind on the filter paper. The water passes through the filter paper.

3. Set up the apparatus shown in *Fig. 23.3*. Carefully pour some of the mixture down a glass rod and into the filter funnel, *Fig. 23.3*.
4. Keep adding the mixture until it has all been filtered.
5. Open the filter paper and allow the soil to dry.

Note: There are millions of tiny holes in the filter paper. These holes allow liquids to pass through the filter paper. However, solid particles are too large to pass through. In other words, the filter paper traps any solid particles like those of soil or sand.

Result

The solid remaining on the filter paper is called the **residue**. (The word residue means 'what is left behind'). In this case, soil is the residue. The clear liquid that passes through the filter paper is called the **filtrate**. In this case, water is the filtrate.

TEST YOURSELF:
Now attempt questions *23.1* and *W23.1–W23.3*.

CHEMISTRY

23.3 Separation Using Evaporation

In *Experiment 23.1* we studied how to separate soil (insoluble) from water. We now look at various methods of separating substances dissolved in water from the water itself. We first examine how to separate salt from water.

The simplest way of separating salt from water is to use the process of **evaporation**. Evaporation involves changing a liquid into a gas (or vapour – hence the name *evaporation*).

Evaporation is the changing of a liquid to a gas.

Evaporation to Obtain Sea Salt

In hot countries, evaporation is used to obtain salt from the sea. The chemical name for table salt is **sodium chloride**.

Fig. 23.4 Evaporation is used to obtain salt from the sea. The heat from the sun evaporates the water and crystals of salt are formed. The salt pans shown in the photograph are in Brittany, France.

Evaporation in the Laboratory

In the laboratory, we can bring about evaporation by boiling the solution and driving the water off as steam. We use this technique in the next experiment.

Note: Evaporation is a very useful method of separation provided that only the solid is required. (If the liquid is also required, another method of separation called distillation must be used, *Section 23.5*).

Mandatory experiment 23.2

To separate sodium chloride from a solution of sodium chloride in water

Apparatus required: evaporating basin; tripod and wire gauze; beaker; bunsen burner

Chemicals required: salt solution

Method

1. Pour the salt solution into an evaporating basin. Place the evaporating basin on a tripod as shown in *Fig. 23.5(a)*.

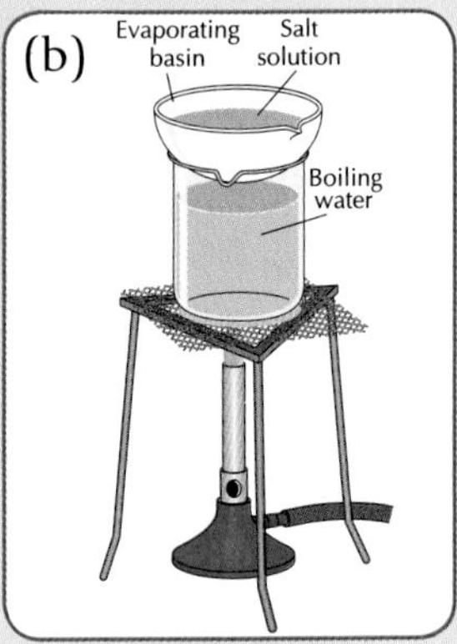

Fig. 23.5(a) Evaporation is a very useful method of separating salt from water provided that only the salt is required. (b) Heating the salt solution using steam.

2. Heat the evaporating basin gently until most of the water is driven off.

Note: It is very important that you wear your safety glasses when evaporating the solution. (The reason for this is that the solution can 'spit'. If this occurs, turn down the gas to make the flame smaller.)

3. Then place the evaporating basin over the beaker of boiling water as shown in *Fig. 23.5(b)*. ***Note:*** Use tongs to transfer the evaporating basin as it may be quite hot. (Heating with the boiling water prevents the salt from crackling and spurting out of the basin during the final stages of evaporation.)

4. Carefully remove the salt from the evaporating basin and show it to your teacher.

23.4 Separation Using both Filtration and Evaporation

Sodium chloride is found in the earth in the form of rock salt. Rock salt is mined in places like Cheshire in England. Rock salt consists of sodium chloride mixed with sand and earth.

Mandatory experiment 23.3

To purify rock salt (sand and salt) using filtration and evaporation

Apparatus required: pestle and mortar; beaker; stirring rod; Bunsen burner; tripod; wire gauze; conical flask; evaporating basin; wash bottle; retort stand and clamp

Chemicals required: rock salt

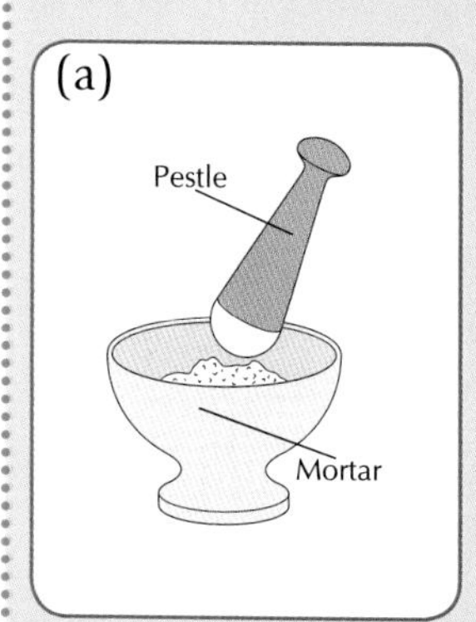

(a) In this experiment we will separate sand and salt from rock salt.

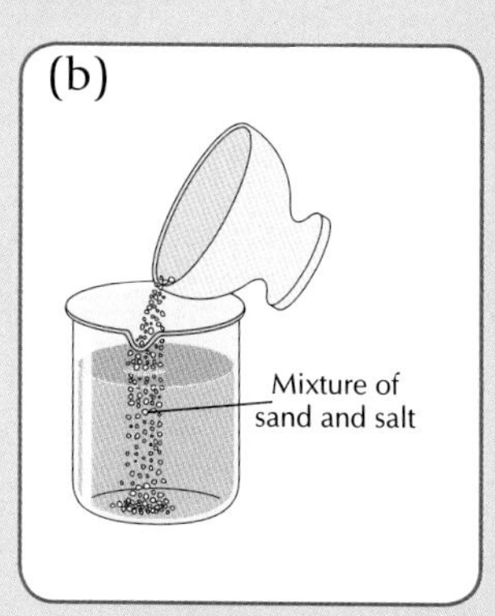

(b) We make use of the fact that salt dissolves in water but sand does not.

(c) Therefore, when we add water to the mixture, the salt dissolves but the sand does not.

(d) If this mixture is filtered, the salt solution passes through the filter paper but the sand remains behind on the filter paper.

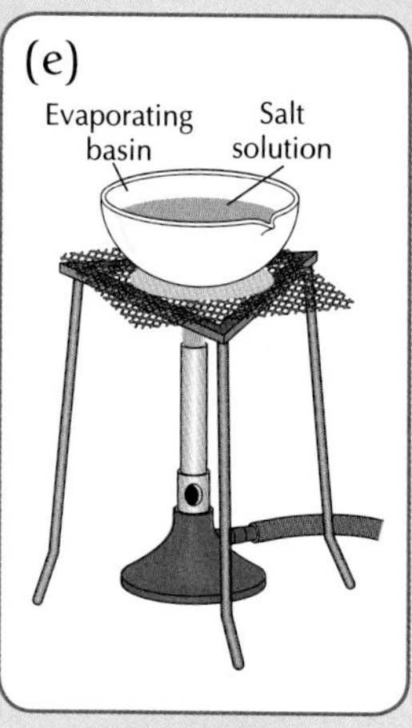

(e) We will then obtain pure salt from the salt solution by evaporation.

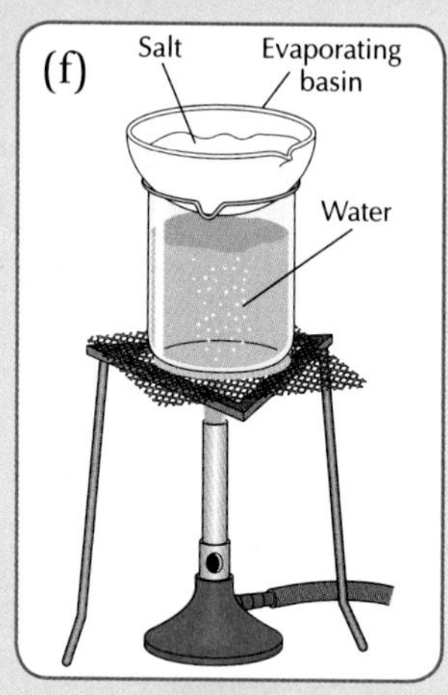

(f) We heat the solution more gently in the last stage to avoid the solution 'spitting'.

Fig. 23.6 Steps in separating sand and salt.

Method

1. Crush some small lumps of rock salt using a pestle and mortar, *Fig. 23.6*. Do not forget to wear your safety glasses.
2. Place the rock salt in a beaker of warm water and stir to dissolve the salt. Add more water, if necessary, to dissolve all the salt and continue heating over the Bunsen burner for about fifteen minutes.
3. Turn off the Bunsen burner and leave the beaker stand until it has cooled down. Filter the warm mixture into a conical flask or an evaporating basin.
4. Evaporate the water from the sodium chloride solution using the method detailed in *Experiment 23.2*. Examine what is left in the evaporating basin. You have produced crystals of pure salt!

TEST YOURSELF:

Now attempt questions *23.2* and *W23.4–W23.5*.

23.5 Separation Using Distillation

We have already seen that we can get salt from sea water by evaporating off the water. When the sea water is evaporated, the water vapour escapes into the atmosphere. Suppose, however, we want to get pure water from sea water. We would need to have some method of changing the vapour back to a liquid again. The process of changing a vapour back to a liquid is called **condensation** (*Chapter 20*). The apparatus which cools the vapour to a liquid is called a **condenser**. A chemist called Justus von Liebig invented a condenser for use in the laboratory. This type of condenser is called a Liebig condenser.

The process of evaporating a liquid and condensing the vapour is called **distillation**. Distillation involves vaporising a liquid by boiling and then condensing the vapour by cooling.

Some countries which lack fresh water use distillation to get pure water from sea water.

Mandatory experiment 23.4

To obtain a sample of pure water from sea water

Apparatus required: beakers; two retort stands and clamps; tripod; wire gauze; Bunsen burner; conical flask; rubber tubing; Quickfit apparatus (27BU); Liebig condenser; thermometer; pear-shaped flask; receiver adaptor; still head

Chemicals required: sea water or salt water; anti-bumping granules

Method

1. Pour some sea water into the distillation flask. Add a few anti-bumping chips to help the liquid to boil more smoothly. Set up the apparatus shown in *Fig. 23.7*.
2. Turn on the water to the Liebig condenser. The purpose of the water is to keep the inner tube cool.
3. Heat the distillation flask gently. Note that the temperature remains at 100°C while the water distils over. Note also that as the steam passes through the Liebig condenser it is being cooled and converted to a liquid again. The pure liquid that is being collected in the receiver is called 'distilled water'.
4. Remove the Bunsen burner when most of the liquid has been boiled off.

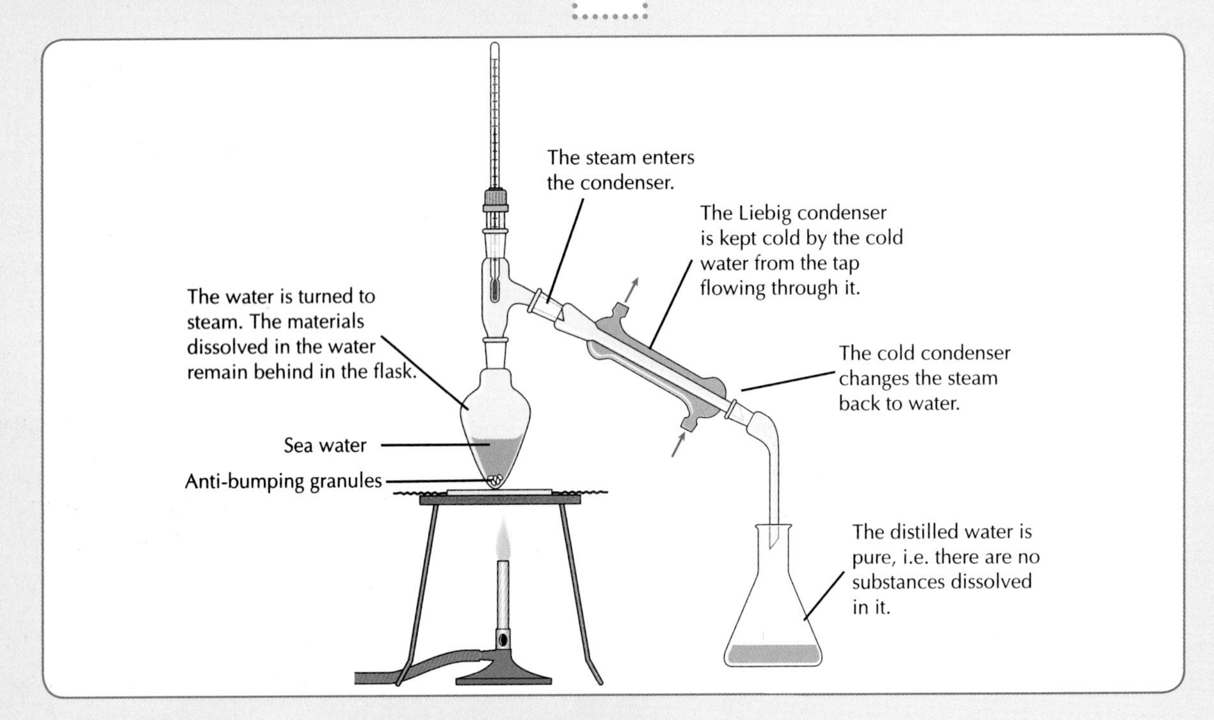

Fig. 23.7 Distillation apparatus to obtain a sample of pure water from sea water.

23.6 Separation Using Paper Chromatography

The word chromatography means 'colour writing'. It was first used to separate coloured materials from plants. Coloured inks, food colourings and dyes in paints are usually made by dissolving various coloured substances in solvents. Chromatography helps us to separate out the colours and find out which ones are present in the ink or dye. This may be demonstrated with the help of a length of paper and suitable solvent.

Paper chromatography is used to separate mixtures of substances in solution.

- **The solvent in which the substances are dissolved is passed along a length of paper.**
- **The dissolved substances separate out on the paper.**

Mandatory experiment 23.5

To separate the dyes in a sample of ink using chromatography

Apparatus required: gas jar; glass rod; dropper or capillary tube; strip of filter paper or chromatography paper; scissors; paper clip

Chemicals required: markers of various colours (black, brown, etc), water (if the dye is water soluble), propanone, alcohol, hexane, etc. (if the dye is not water soluble).

Method

1. Cut a strip of absorbent paper (filter paper or chromatography paper or paper from the side of a newspaper) long enough for a beaker or a gas jar as shown in *Fig. 23.8*.

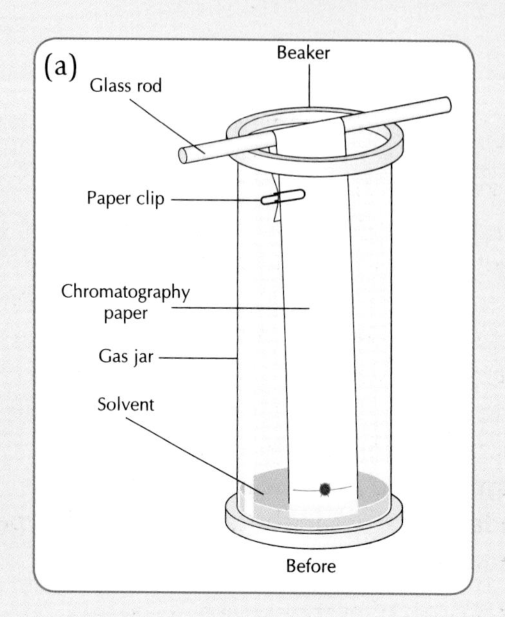

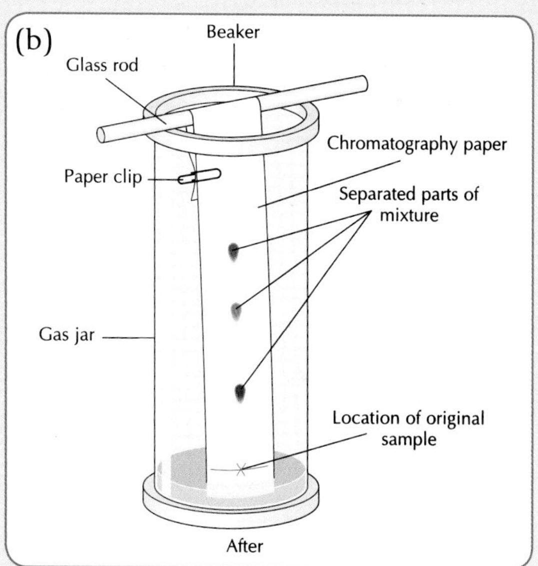

Fig. 23.8 An experiment to separate the various colours in a dye using paper chromatography. Note that the sample is placed just above the level of the solvent.

2. Using a pencil, draw a line about 3 cm from the end of the paper to indicate where the spot will be placed. Using a marker, place a spot of the ink on the line. If the dye is in liquid form, use a dropper or a capillary tube to put a spot of ink on the paper. Using the pencil, write the colour of the ink and solvent on the top of the paper. (Your teacher will tell you what solvent to use.)
3. Pour some solvent into the container to a depth of 2 cm maximum and hang the

paper so that the solvent level is below the ink mark as shown in *Fig. 23.8*.

4. When the solvent reaches the top of the paper, take out the paper and allow it to dry. Examine the chromatogram and note the colours of the dyes that were in the ink.
5. Repeat the experiment using a different colour of ink.

How it Works

1. The solvent carries the various dyes in the ink up the paper.
2. Some dyes are more soluble in the solvent than others.

- Dyes that are **not very soluble** in the water come out of solution **early on** and appear as a colour near the bottom of the filter paper.
- Dyes that are **very soluble** in the water stay in solution and get carried **further up** the paper.

The fact that the dye is separated into various colours proves that the dye is a mixture. The paper showing the separated colours is called a **chromatogram**.

Uses of Paper Chromatography

In the above experiment we saw that chromatography can be used to find the number of dyes present in ink. Chromatography may also be used to **identify the dyes** present.

- **Art**. Scientists working in art galleries often need to know the composition of paints that an artist used. This information is necessary when works of art are being restored. Suppose that the brown pigment *X* was thought to consist of a number of other dyes, *A*, *B*, *C* and *D*, *Fig. 23.9*. Chromatograms of all of the dyes are obtained. From comparing the various chromatograms, we see that the brown pigment *X* was made by mixing dyes *A*, *B* and *D*. Note that these other dyes are all pure dyes since they do not separate into a number of colours.

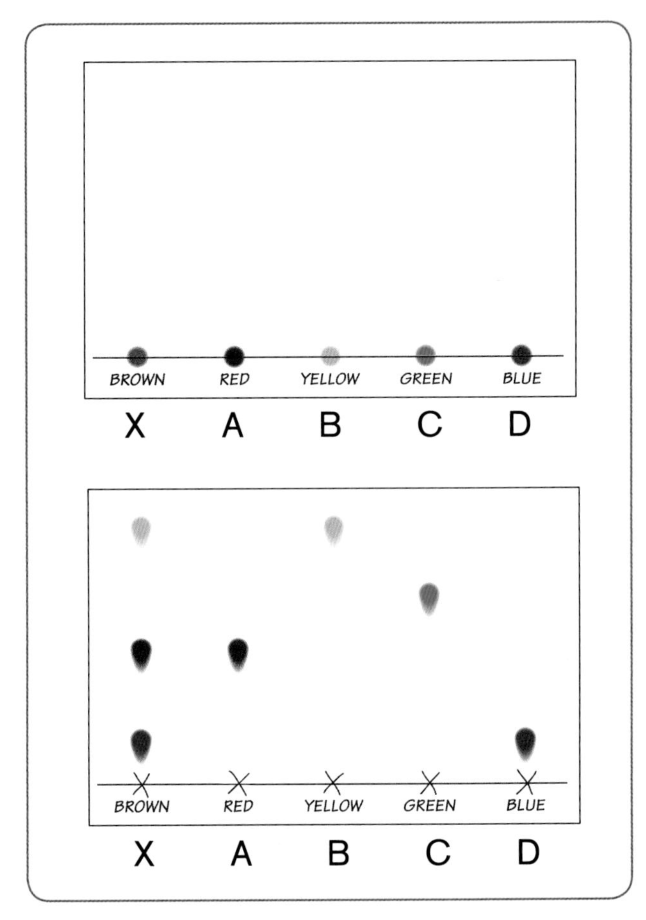

Fig. 23.9 Paper chromatography may be used by scientists who work in art galleries to identify the various pigments used by artists.

- **Forensic science**. Chromatography is also commonly used by forensic scientists who study evidence given to them by the Gardaí. For example, particles of paint found on the clothes of the victim of a hit-and-run accident can be analysed to identify the type of car involved in the crime.
- **Food colourings**. Chromatography is also used by food scientists to examine the food colourings added to food. By studying the chromatograms obtained from various foods, the scientists can check that only permitted food colourings are being used.

TEST YOURSELF:

Now attempt questions *23.3–23.8* and *W23.6–W23.21*.

What I should know

- **Filtration** is used to separate an insoluble solid from a liquid using a material that allows the liquid to pass through but not the solid, e.g. soil and water.
- **Evaporation** is used to separate a soluble solid from a liquid when the liquid is not required e.g. salt and water.
- **Simple distillation** is used to separate a liquid and a soluble solid when the liquid is required e.g. pure water from sea water.
- **Paper chromatography** is used to separate mixtures of substances in solution. The solvent in which the substances are dissolved is passed through a length of paper. The dissolved substances separate out on the paper.

QUESTIONS

Write the answers to the following questions into your copybook.

23.1 (a) A mixture of water and soil may be separated by ..... .

(b) Give two examples of the use of filtration in our everyday lives.

(c) Why is it not possible to separate a solution of salt in water by filtration?

(d) Explain how a filter paper works to separate a mixture of soil and water.

(e) The solid material which remains behind on filter paper after filtration is called the ..... .

(f) The clear liquid which passes through the filter paper is called the ..... .

(g) Masks that are worn by spray painters prevent tiny specks of paint entering the lungs but allow the air to get through, *Fig. 23.10*. What separation process does this demonstrate?

Fig. 23.10

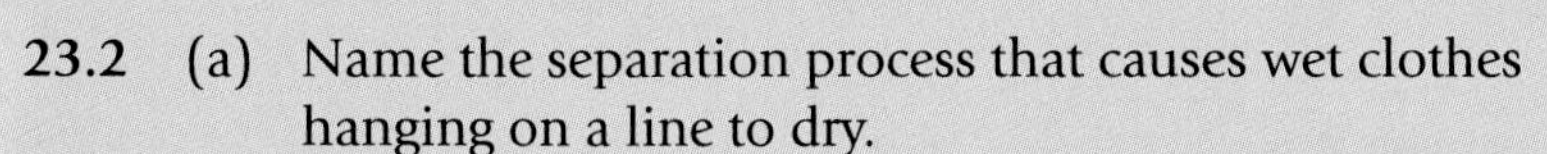

23.2 (a) Name the separation process that causes wet clothes hanging on a line to dry.

(b) When water is evaporated from salt water, ... of salt are left.

(c) In some countries, salt is obtained by evaporating sea water. Why are large shallow pans used rather than small deep pans? Why is salt not produced by evaporation of sea water in Ireland?

(d) When carrying out a filtration experiment, a student suggested that filtration would be speeded up if a hole were put in the filter paper. Is this a good suggestion?

(e) Describe how you would carry out an experiment in the laboratory to obtain a sample of pure salt from seawater.

23.3 (a) The pure liquid which is collected after distilling sea water is called ..... .

(b) Distillation is defined as ..... .

(c) The condenser used in the laboratory is commonly called a ..... condenser.

(d) The condenser changes ..... back to water.

QUESTIONS

(e) In distillation, how is the condenser kept cold?.

(f) In distillation, what is used to help the liquid to boil more smoothly?

23.4 A student carried out a distillation experiment using sea water. The sea water was ..... and the ..... that was formed was passed through a ..... . This piece of equipment is named after a famous scientist called ..... . The purpose of this piece of equipment is to change the ..... back to ..... . The liquid which is collected is called ..... water. It has no substances ..... in it.

23.5 Connect up lines in the following table to make the correct links.

SEPARATION METHOD	EXAMPLE OF USE
Chromatography	Obtain salt from sea water
Filtration	Obtain drinking water from sea water
Evaporation	Separate chalk suspension from water
Distillation	Separate different coloured inks from a mixture

23.6 A lot of coloured substances in inks, paints, dyes, etc. are not ..... colours but are made up of a ..... of colours. To separate the colours, some dye is placed on a piece of ..... paper dipping into a suitable ..... . Some dyes are ..... soluble than others so they are carried further up the paper. Other dyes are ..... soluble and do not travel very far up the paper. This method of separating colours is called ..... .

23.7 (a) The process used to separate dyes or colours in chemistry is called ..... .

(b) To carry out a chromatography experiment, we need a liquid which ..... the colours.

(c) How could you prove that black ink is a mixture of a number of different coloured inks?

(d) When carrying out and experiment on chromatography, a student was told to write on the paper using a pencil and not a pen. Why is it better to use a pencil rather than a pen?

23.8 A forensic scientist is trying to discover the type of car involved in a hit-and-run accident. He is comparing the paint found on the victim's clothing with paint taken from a number of cars, *Fig. 23.11*. Which paint matches the paint found on the victim's clothing? Explain your answer.

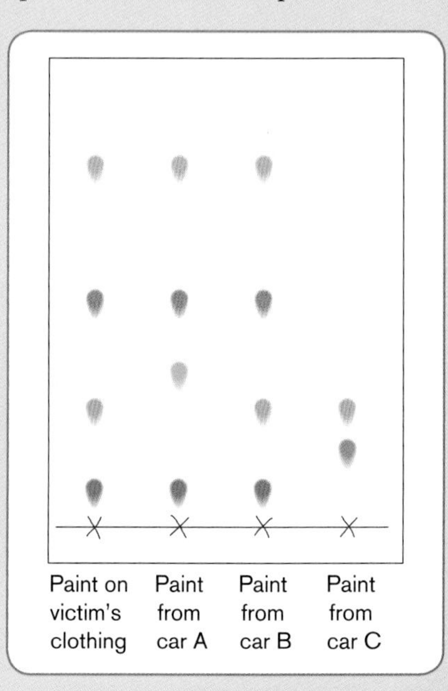

Fig. 23.11 Which paint matches the paint found on the victim's clothing?

Chapter 24 Acids and Bases I

24.1 What are Acids?

Many people think that acids are dangerous substances. Some acids are dangerous, but others are harmless.

- For example, the acid found in car batteries is dangerous. This acid is called sulfuric acid. It is a corrosive substance. This means that it 'eats away' other materials.
- Other substances, however, contain acids that are used in our food. For example, vinegar contains an acid called acetic acid, *Fig. 24.1*. Fizzy drinks contain an acid called carbonic acid. Acetic acid and carbonic acids are not dangerous acids.

Fig. 24.1(a) The acid in a car battery is dangerous. It can burn through materials. (b) The acids in vinegar and fizzy drinks are not dangerous.

Other Examples of Acids

The word **acid** comes from the Latin word **acidus** meaning sour. If you look carefully at the examples in *Fig. 24.2*, you will recognise many substances with a sour or sharp taste.

Fig. 24.2 Some examples of everyday substances that contain acids.

24.2 Acid-Base Indicators

Acids

It is very dangerous to taste certain acids. Instead, we use a substance called an **indicator** to test for the presence of an acid. These substances are called indicators because they indicate (tell us) if a substance is an acid. An example of such a substance is **litmus**. This is a purple dye obtained from certain types of plants called lichens that grow on the bark of trees. Litmus turns red when mixed with an acid, *Fig. 24.3*.

Fig. 24.3 The acid in the lemon turns litmus from blue to red.

There are other substances that also work well as indicators e.g. the juices from red cabbage and blueberries.

Acids turn litmus from blue to red.

Note: To help you remember that litmus is red in an acid, think of red for danger. Some acids are dangerous and must be handled with care. Another way to help you remember this is that the words 're**d**' and 'aci**d**' both contain the letter '***d***'.

Bases

There is another group of chemicals that affect litmus in the opposite way. For example, sodium hydroxide turns litmus from red to blue. A chemical like sodium hydroxide is called a **base**. A base is the 'chemical opposite' to an acid.

Examples of Bases

Examples of common bases are shown in *Fig. 24.4*.

Fig. 24.4 Some examples of everyday substances containing bases.

Alkalis

Bases that dissolve in water are often called **alkalis**. For example, sodium hydroxide (caustic soda) is an alkali. It dissolves in water and is very corrosive – especially to skin!

Always wear safety glasses and gloves when handling sodium hydroxide. It is used to clear blocked drains and works by breaking down the material blocking the drain.

Bases turn litmus from red to blue.

To help you remember that litmus is blue in a base, the words '**b**lue' and '**b**ase' both contain the letter '***b***'.

SAFETY TIP! If you ever spill acids or bases on your skin, wash them off with plenty of water immediately. This helps to dilute them and makes them less harmful to your skin.

Thus, litmus can be used to show whether a substance is acidic or basic. It shows this by means of a colour change.

An indicator shows by means of a colour change whether a substance is acidic or basic.

Neutral Substances

If there is no colour change, we say that the substance is **neutral**. This means that the substance is neither an acid nor a base. Examples of neutral substances are water, sugar and salt. In neutral substances, red litmus stays red and blue litmus stays blue.

Other Indicators

Litmus is found in laboratories as a solution or as litmus paper. Litmus paper is absorbent paper that has been soaked in litmus solution and dried. Litmus is the best known indicator. Other indicators that you may have in your school laboratory are methyl orange and phenolphthalein.

In the next experiment we will study an indicator that is present in red cabbage.

- We will **extract** (take out) this indicator from red cabbage.
- We will then **add** the indicator to an acidic, basic and neutral solution.
- We will then **note** the colour of the indicator in each solution.

Experiment 24.1

To extract an indicator from red cabbage and test it

Apparatus required: beaker; hotplate (or Bunsen burner and tripod); glass rod; rack of test-tubes; labels; dropper

Chemicals required: red cabbage leaves; water; dilute sulfuric acid; dilute hydrochloric acid

Fig. 24.5 The indicator in the cabbage is being extracted by the water.

Method

1. Put some pieces of red cabbage into 50 cm^3 approx. of water in a Pyrex beaker.
2. Set up the apparatus as shown in *Fig. 24.5* and bring the water to the boil. While the water is being heated stir it with a glass rod. You will observe the water turning purple. This is because the indicator in the cabbage is being extracted by the water.
3. Allow the solution to cool.
4. Label the test-tubes: water, dilute sulfuric acid and dilute sodium hydroxide. Fill the test-tubes to about one third of their capacity with these substances.
5. Using a dropper, place about five drops of the indicator in each test-tube.
6. Note the colour in each test-tube. Summarise your results in the form of a table.

SUBSTANCE	COLOUR OF RED CABBAGE INDICATOR
Water (Neutral)	
Sulfuric acid (Acidic)	
Sodium hydroxide (Basic)	

TEST YOURSELF:
Now attempt questions *24.1* and *W24.1–W24.6.*

24.3 The pH Scale

In *Section 24.2* we used indicators to tell us whether a substance is acidic or basic. There are, however, different levels of acidity and basicity. For example, the acid in a car battery is highly acidic, but the acid in orange juice is only mildly acidic. In order to indicate the level of acidity or basicity, chemists use a special scale called the **pH scale**. The pH scale goes from 0 to 14.

- A solution whose pH is **7** is **neutral.**
- A solution whose pH is **less than 7** is **acidic.**
- A solution whose pH is **greater than 7** is **alkaline.**

The lower the pH value below 7, the more highly acidic is the substance. The higher the pH value above 7, the more alkaline is the substance.*

* You will learn where the number 7 comes from when you study chemistry for your Leaving Certificate.

pH Sensors

Litmus tells us if a substance is acidic or basic. It does not, however, tell us the pH of the substance. One way of finding the pH of a solution is to use a pH sensor attached to a datalogger and a computer, *Fig. 24.6*.

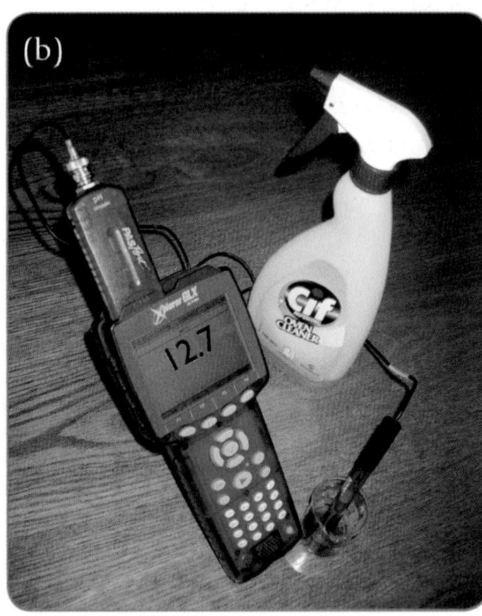

Fig. 24.6 Measuring the pH of (a) vinegar and (b) window cleaner using a pH sensor and a datalogger.

Universal Indicator

If a pH sensor is not available, one could use either pH paper or universal indicator solution. pH paper is simply universal indicator solution which has been soaked on to paper.

How it Works

Universal indicator is a mixture of a number of different indicators.

1. A piece of pH paper or a few drops of universal indicator are placed in the solution whose pH is to be measured.
2. The colour shown by the paper or indicator is then matched against a standard set of colours, *Fig. 24.7*. (If the solution is already coloured, then the pH must be measured using a pH sensor and datalogger or a pH meter).

Fig. 24.7 Universal indicator tells us the pH of a solution.

The range of colours of universal indicator solution are shown in *Fig. 24.8*.

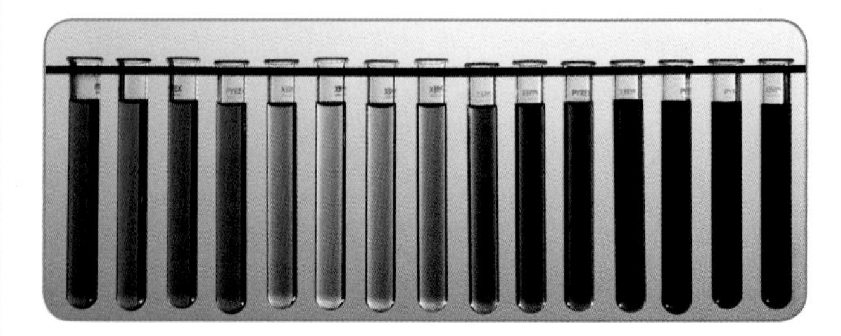

Fig. 24.8 The colour of universal indicator changes in different pH solutions.

Some examples of common substances and their corresponding pH values are shown in *Fig. 24.9*.

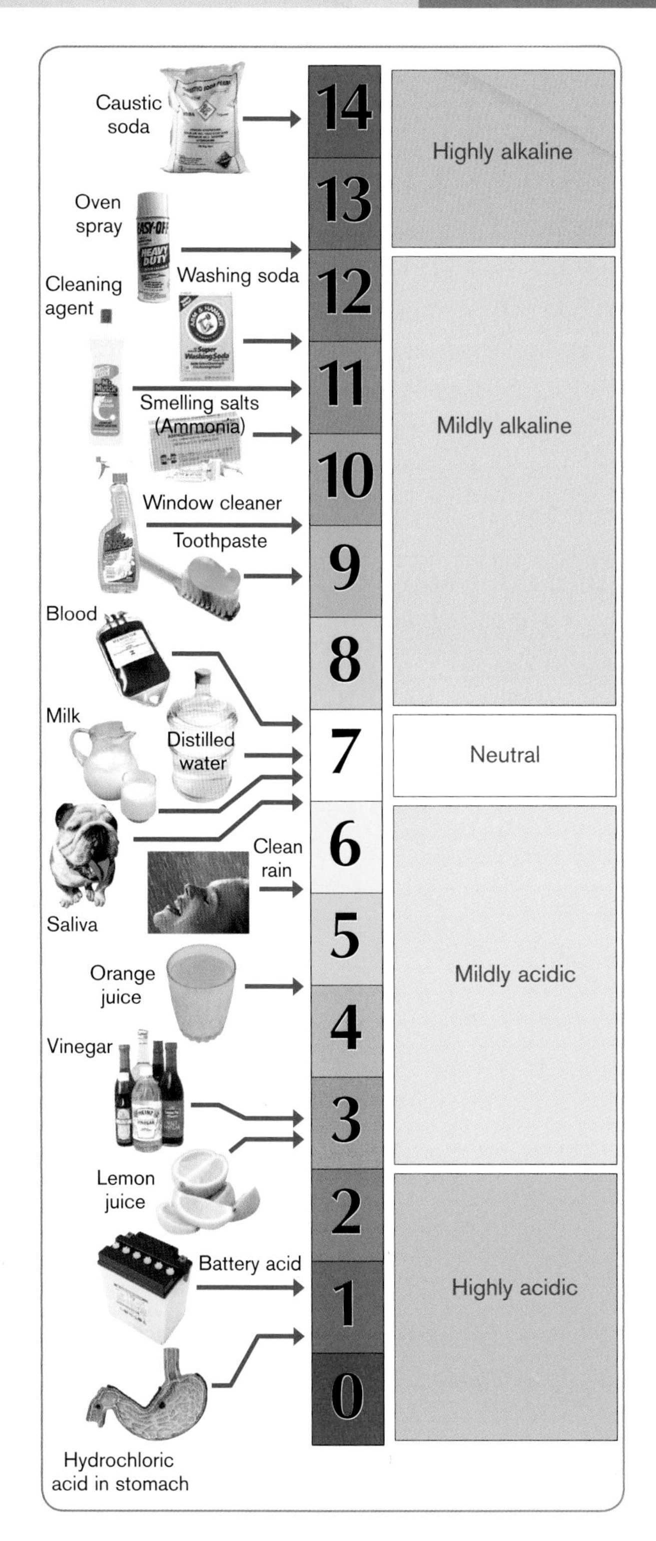

Fig. 24.9 The pH scale of acidity.

Mandatory experiment 24.2

To investigate the pH of a variety of materials

Apparatus required: rack of test-tubes; universal indicator paper 1–14 and various narrow-range indicator papers

Chemicals required: dilute hydrochloric acid; dilute sodium hydroxide solution; vinegar; sodium carbonate; limewater; washing soda; tea; milk; detergent; baking powder; indigestion powder; orange juice; lemon juice; oven cleaner; toothpaste; lemonade; cola; acid drop sweets; ammonia solution; rain water; etc.

Method

1. Use a wide-range universal indicator paper (1–14) to find the approximate pH value of each solution. The substances tested may be taken from the above list of chemicals. If the substance is a solid, dissolve it in water before you test it.
2. Choose a suitable narrow-range paper and measure each pH value more accurately. This is done by matching the colour obtained against the colour chart that comes with the paper.
3. Summarise your results in the form of a number line, *Fig. 24.10*. Show the pH of the substance you have tested on the number line.

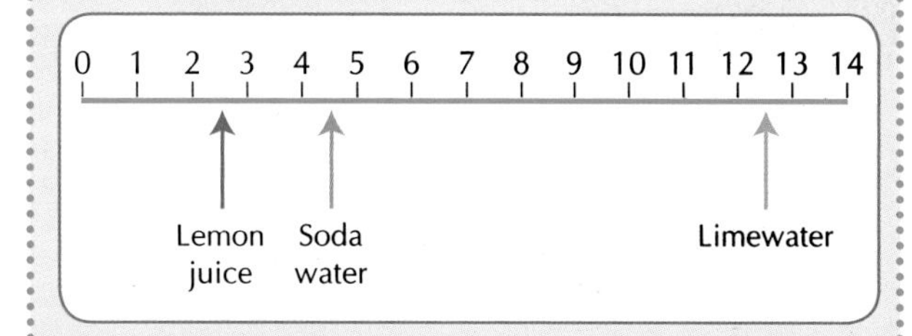

Fig. 24.10 Displaying the results of your investigation.

TEST YOURSELF:
Now attempt questions *24.2 and W24.7–W24.8.*

What I should know

- An indicator shows, by means of a colour change, whether a substance is acidic or basic.
- Acids turn litmus from blue to red.
- Bases turn litmus from red to blue.
- A solution whose pH = 7 is neutral.
- A solution whose pH is less than 7 is acidic.
- A solution whose pH is greater than 7 is basic.
- Universal indicator is a mixture of indicators.

Questions

Write the answers to the following questions into your copybook.

24.1 (a) A bottle of concentrated sulfuric acid was marked with the symbol shown in *Fig. 24.11*. What does this symbol mean?

(b) Soap is a mild alkali. How would you describe what happens when soap gets in your eyes?

(c) What is an indicator?

(d) Indicators change ..... when added to acids or bases.

(e) Litmus is an indicator which is coloured ..... in an acid and ..... in a base.

Fig. 24.11

(f) The chef has accidentally spilled some vinegar over some berries, *Fig. 24.12*. The berries have turned red. Can you explain what has happened?

24.2 (a) The pH scale ranges from ..... to ..... .

(b) The pH of distilled water is ..... . Therefore, water is said to be ..... .

(c) A solution whose pH is less than 7 is said to be ..... and one whose pH is greater than 7 is said to be ..... .

(d) Which is the more acidic solution, one with a pH of 2 or one with a pH of 6?

(e) Name a solution, other than sodium hydroxide, which has a pH greater than 7.

(f) How would you determine the pH of a solution?

Fig. 24.12 Why have the berries changed colour?

Chapter 25 The Air and Oxygen

25.1 The Air – A Mixture of Gases

The Atmosphere

The air is a mixture of gases which forms a blanket around the Earth. Another name for the air is the *atmosphere*. The atmosphere is held near the surface of the Earth by gravity. Part of the atmosphere is shown in *Fig. 25.1*. (There is no definite upper limit to the atmosphere). However, above 6 km the air is so thin that mountain climbers have to carry breathing apparatus. Only astronauts usually pass out of the atmosphere!

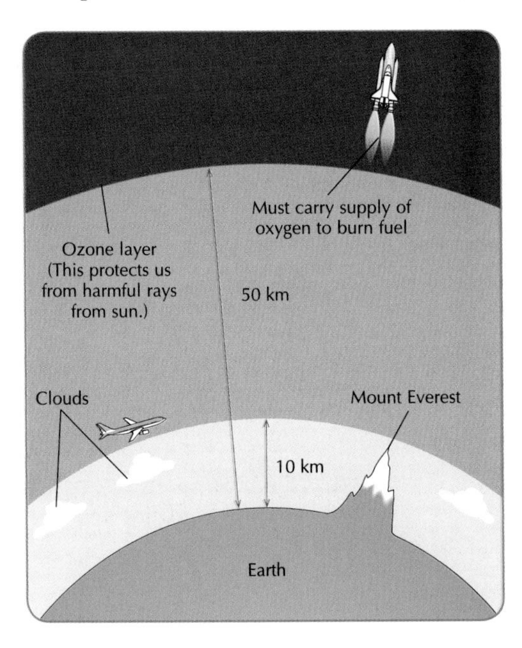

Fig. 25.1 The air (atmosphere) is a mixture of gases. It forms a blanket around the earth.

Oxygen

Not only do we need the air to live, but we also need the air to burn fuels and keep warm. The substance in the air that we need for both burning and breathing is oxygen.

In *Experiment 25.1*, we will investigate what percentage of air is oxygen. We will do this by burning a night light in a fixed volume of air.

Mandatory experiment 25.1

To investigate the percentage of oxygen in the air (using a night light)

Apparatus required: basin; night light; graduated cylinder (100 cm^3); matches

Chemicals required: water

Method

1. Light a night light. Float the night light on the surface of the water in a basin.
2. Place a graduated cylinder over the burning night light, *Fig. 25.2(a)*.

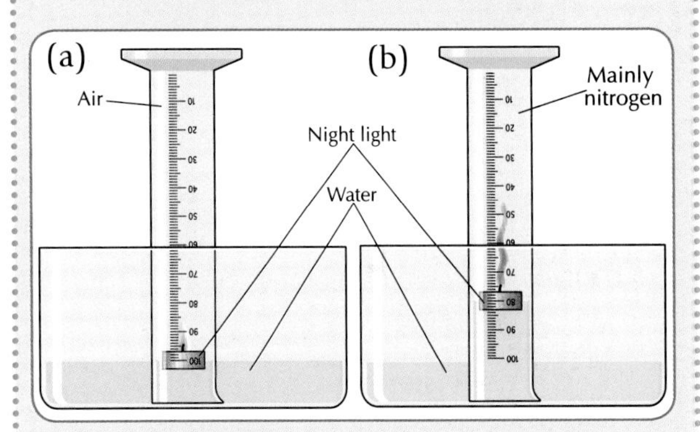

Fig. 25.2 The burning night light uses up the oxygen inside the graduated cylinder. The water rises up inside the graduated cylinder. The water takes up the space previously occupied by the oxygen.

3. Note what happens as the night light burns. The water rises up inside the graduated cylinder. The water is pushed up by the pressure of the atmosphere.
4. Allow the apparatus to cool. Measure the volume of gas left in the graduated cylinder, *Fig. 25.2(b)*.
5. By subtraction, calculate the percentage of oxygen in the air.

Conclusion

Oxygen occupies approximately one fifth of the air.

Although the previous method has been described on exam papers, it does not give very accurate results. A more accurate method of carrying out the experiment is shown in *Experiment 25.2*.

Mandatory experiment 25.2

To measure the percentage of oxygen in air (using gas syringes)

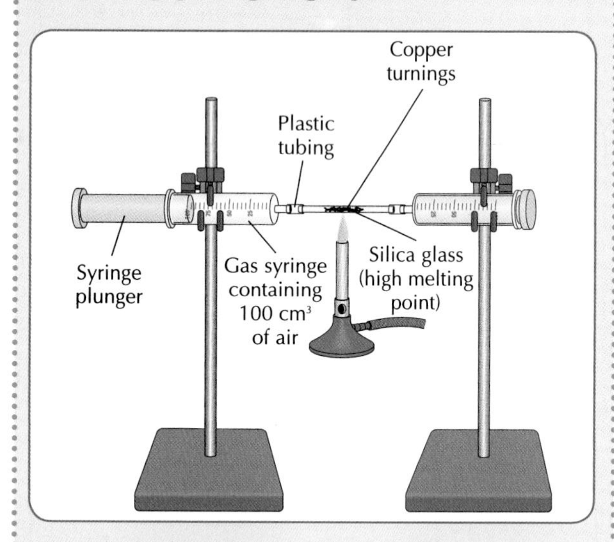

Fig. 25.3 This apparatus is used to find the volume of oxygen in the air. A volume of 100 cm^3 of air is passed over the heated copper a number of times. The copper reacts with the oxygen in the air. This removes the oxygen from the air. From the remaining volume of air left, we can work out the percentage of oxygen in the air.

Apparatus required: two gas syringes; two syringe holders; two retort stands; silica glass tube and connectors; Bunsen burner

Chemicals required: copper turnings

Method

1. Set up the apparatus shown in *Fig. 25.3*.
2. Fill one syringe with 100 cm^3 of air. The other syringe should read 0 cm^3.
3. Gently pass the 100 cm^3 of air from the left-hand syringe over the heated copper into the right-hand syringe. It will be observed that the brown copper begins to turn grey-black. This is because the copper is reacting with the oxygen in the air to form a substance called copper oxide.

   copper + oxygen ⟶ copper oxide

4. Push the air to and fro several times over the hot copper so that all the oxygen in the air reacts with the copper.
5. Allow the apparatus to cool. Push all the gas into the left-hand syringe and measure the volume of air in it.
6. It is found that the total volume of air in the syringe has decreased. This decrease must be equal to the amount of oxygen in the air.

Specimen Results

Volume of air in syringe before heating = 100 cm^3

Volume of gas left after first heating and cooling = 85 cm^3

Volume of gas left after second heating and cooling = 79 cm^3

Volume of gas left after third heating and cooling = 79 cm^3

∴ Decrease in volume of air = 21cm^3

Calculations

$$\text{percentage of oxygen in air} = \frac{\text{decrease in volume of air}}{\text{initial volume of air}} \times 100$$

$$= \frac{21\text{ cm}^3}{100\text{ cm}^3} \times 100$$

$$= 21\%$$

Conclusion

Percentage of oxygen in air = 21%.

Composition of the Air

The gas left in the syringe in *Experiment 25.2* is mainly nitrogen. It is found that this gas occupies 78% of the volume of the air.

The actual percentages of the gases in clean dry air are listed in *Table 25.1*. Clean dry air is air from which moisture and any pollutants have been removed.

GAS	PERCENTAGE CONTENT IN DRY AIR
Nitrogen	78%
Oxygen	21%
Inert gases* + carbon dioxide	1%

Table 25.1 The percentage composition by volume of clean, dry air.

*The word inert means unreactive. It is very difficult to get these gases to take part in any reaction. Therefore, they are called inert gases or noble gases.

The composition of air is also represented in *Fig. 25.4*.

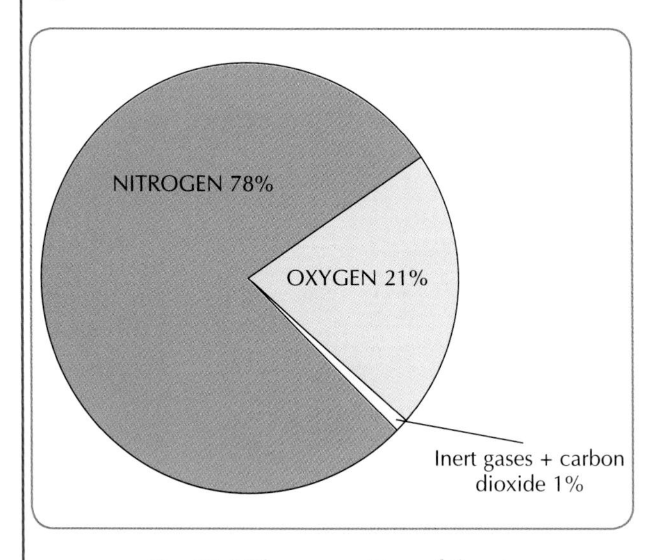

Fig. 25.4 The composition of the air.

Some liquid nitrogen being poured from a thermos flask is shown in *Fig. 25.5*.

Fig. 25.5 This liquid nitrogen has been obtained from the air.

Exam Question

Junior Cert Science Higher Level Examination Question

The composition of air can be investigated in different ways. Two experiments are shown in the diagram.

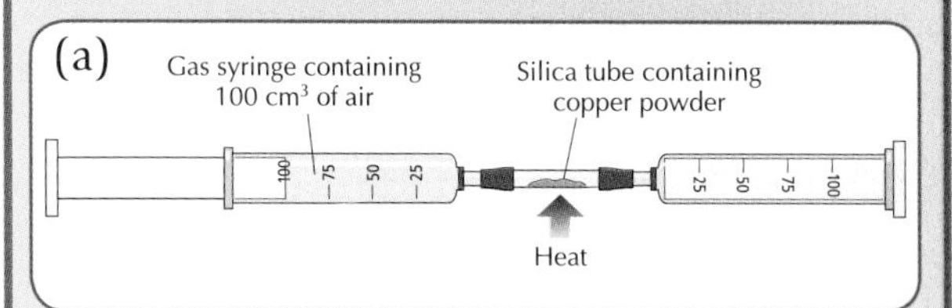

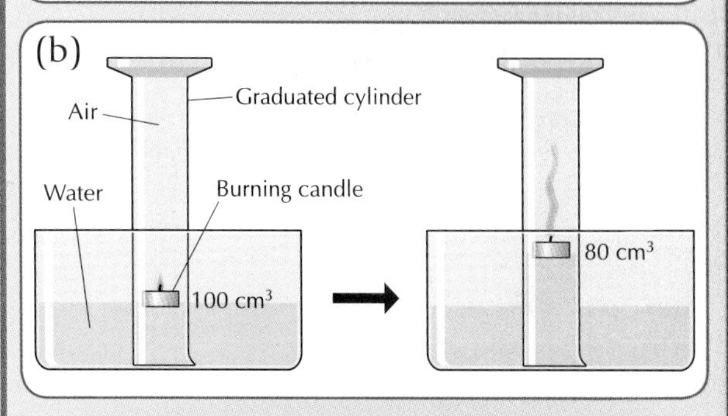

Fig. 25.6

In Experiment A, the air was pushed repeatedly over the heated copper powder and only 79 cm³ of gas remained at the end of the experiment.

(i) Why is it necessary to let the apparatus cool down before measuring the volume of the remaining gas?

A: The apparatus is allowed to cool because gases expand when heated. Therefore, volumes of gases must be compared at the same temperature.

(ii) Why did the volume of gas decrease and then remain steady?

A: The volume of gas decreases and remains steady because the oxygen has been used up by the burning night light.

(iii) What is the remaining gas mainly composed of?

A: Nitrogen.

(iv) Experiment B is less accurate than Experiment A. Give a reason why this is so.

A: In *Experiment B* the candle often goes out before all the oxygen is used up. Since more oxygen is removed in *Experiment A*, a more accurate result is obtained. Also, in *Experiment B* the graduated cylinder does not measure volume as accurately as the gas syringes used in *Experiment A*.

Mandatory experiment 25.3

To show that carbon dioxide is present in air

In this experiment we shall use a special chemical to test for the presence of carbon dioxide. This chemical is called **limewater**. If carbon dioxide is passed through limewater, the limewater turns milky white. (The reason why the limewater turns milky is covered in the next chapter).

Apparatus required: Gas-washing bottle; two-holed rubber stopper; glass tubing; vacuum pump (filter pump)

Chemicals required: limewater

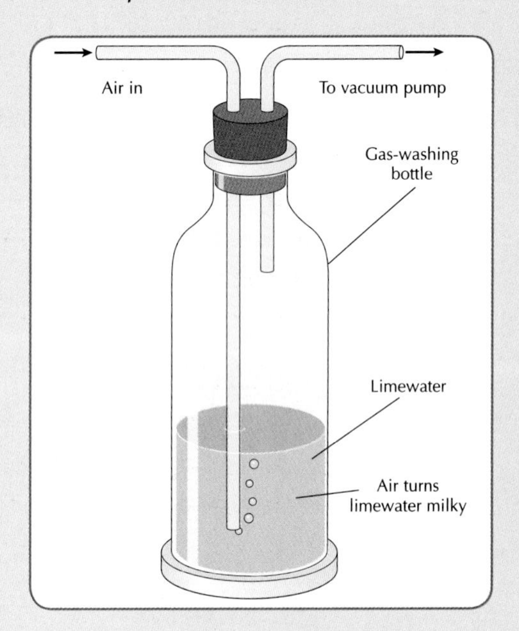

Fig. 25.7 The carbon dioxide in the air turns the limewater milky.

Method

1. Set up the apparatus as shown in *Fig. 25.7*. Take care to connect the vacuum pump to the shorter piece of glass tubing.
2. Turn on the tap to operate the vacuum pump. Allow the air to be drawn through the limewater for about half an hour.

Result

The limewater turns milky.

Conclusion

This shows us that carbon dioxide is present in the air.

Mandatory experiment 25.4

To show that water vapour is present in air

Apparatus required: boiling tube; retort stand; rubber stopper; beaker

Chemicals required: salt; ice; anhydrous copper sulphate or cobalt chloride paper

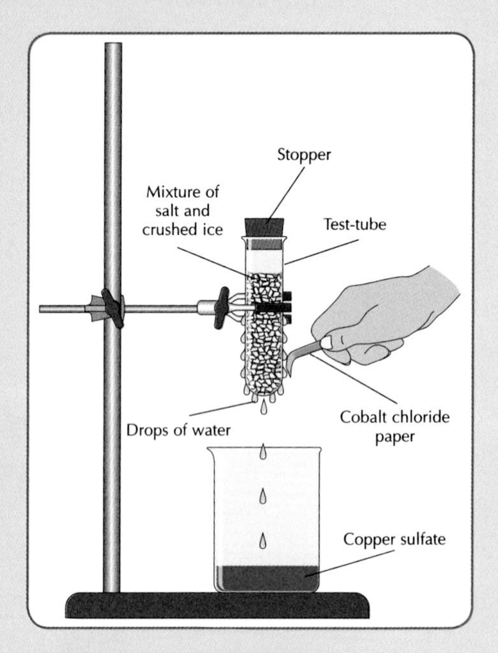

Fig. 25.8 This experiment shows that water vapour is present in the air. The condensed liquid turns the anhydrous copper sulfate blue. It also turns blue cobalt paper pink. Therefore, it must be water.

Method

1. Set up the apparatus as shown in *Fig. 25.8*. Make sure that the outside of the tube is perfectly dry before you begin the experiment.
2. Observe that fine droplets of a colourless liquid condense on the outside of the tube. Collect this liquid and test with anhydrous copper sulfate. Anhydrous

copper sulfate is a white powder. When water is added to anhydrous copper sulfate it turns blue. Another substance used to test for the presence of water is a chemical called cobalt chloride. Cobalt chloride paper is blue when dry, but turns pink in the presence of water.

Results

The liquid turns the white anhydrous copper sulfate blue.

Also, the liquid turns the blue cobalt chloride paper pink.

Conclusion

The liquid that condensed on the outside of the tube is water. Therefore, water vapour is present in air.

TEST YOURSELF:

Now attempt questions *25.1–25.2* and *W25.1–W25.6*.

25.2 Oxygen – Preparation and Properties

Oxygen was discovered in 1774 by Joseph Priestley, an English scientist, and Antoine Lavoisier, a French scientist.

Joseph Priestly (1733–1804) and Antoine Lavoisier (1743–1794) were the scientists who discovered oxygen.

Almost all of the oxygen used in industry is obtained by cooling the air to a liquid and then distilling it. We shall, however, study a more convenient method of preparing oxygen in the laboratory in *Experiment 25.5*.

Laboratory Preparation of Oxygen

One of the simplest methods of preparing oxygen in the laboratory is by decomposing (breaking down) **hydrogen peroxide**. Hydrogen peroxide is a colourless liquid that slowly decomposes to give off oxygen and form water. We can represent this using the following word equation:

Word equation:
hydrogen peroxide ⟶ water + oxygen

Chemical Symbols

Chemists like to use symbols and formulas to express reactions like the above.

- The chemical formula for **hydrogen peroxide** is H_2O_2.
- The chemical formula for **water** is H_2O.
- The chemical formula for **oxygen** gas is O_2.

These formulas are understood by chemists all over the world. For the moment, just accept that these are the correct formulas for the above substances. Later on, we will be studying chemical formulas and chemical equations (*Chapter 32*) in more detail.

Chemical equation: $2H_2O_2 \longrightarrow 2H_2O + O_2$

In words, the above chemical equation tells us that two amounts of hydrogen peroxide break down to give two amounts of water and one amount of oxygen gas.

Catalysts

A compound called manganese dioxide is used to make the hydrogen peroxide decompose more quickly.

- The chemical formula for manganese dioxide is MnO_2.
- The manganese dioxide is **not used up** in the reaction and does not appear in the chemical equation.
- The manganese dioxide is called a **catalyst**.

A catalyst is a substance that changes the speed of a chemical reaction but is not used up in the reaction itself.

Very often the catalyst is written above the arrow in a chemical equation.

Chemical equation:

$$2H_2O_2 \xrightarrow{MnO_2} 2H_2O + O_2$$

Note: Don't get confused between the words manganese and magnesium! These are two different metals. The symbol for manganese is Mn. The symbol for magnesium is Mg. Try to find them in the Periodic Table!

Mandatory experiment 25.5

(a) To prepare a sample of oxygen

(b) To examine the properties of oxygen

Apparatus required: dropping funnel; Buchner flask; delivery tubing; trough; gas jars; gas jar covers (or test-tubes or boiling tubes and stoppers); beehive shelf; wooden splints; deflagrating spoon

Chemicals required: hydrogen peroxide (dilute – '20 volume'); manganese dioxide; water; red and blue litmus paper; charcoal; limewater; magnesium ribbon

(a) To prepare a sample of oxygen

Method

1. Set up the apparatus as shown in *Fig. 25.9*.

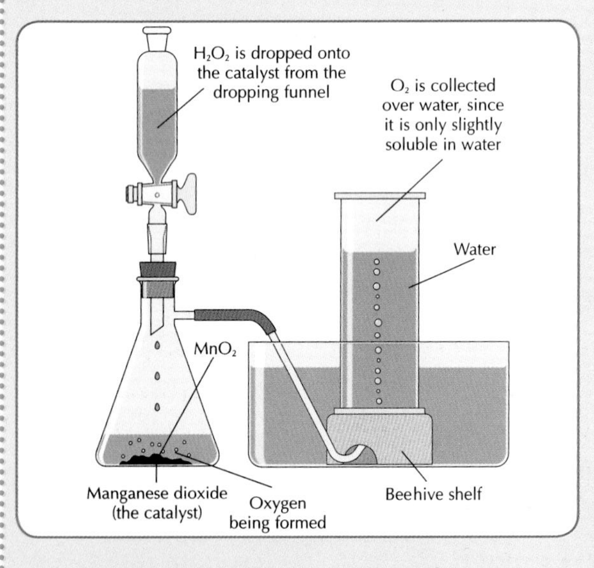

Fig. 25.9 Laboratory preparation of oxygen.

2. Allow the hydrogen peroxide to fall on the manganese dioxide so that oxygen is produced at a fairly fast rate. Wait for about half a minute before collecting the oxygen to allow for the air in the flask to escape. The oxygen is collected in a gas jar or a test-tube or a boiling tube. These are first filled with water and placed on top of the beehive shelf.
3. Collect five gas jars or boiling tubes or test-tubes of the gas.

(b) To examine the properties of oxygen

Method

1. **Litmus and oxygen**. Note that the gas is colourless and odourless. Place pieces of moist red litmus paper and blue litmus paper into a jar of the gas. Note that there is no change in colour of the paper indicating that oxygen is a neutral gas.
2. **Wooden splint and oxygen**. Light a wooden splint. Observe how it burns. Now place the wooden splint in a gas jar of oxygen. Note how it burns more vigorously. Take out the burning splint and shake it so that it is now just glowing. Place the glowing splint in the oxygen again, *Fig. 25.10*.

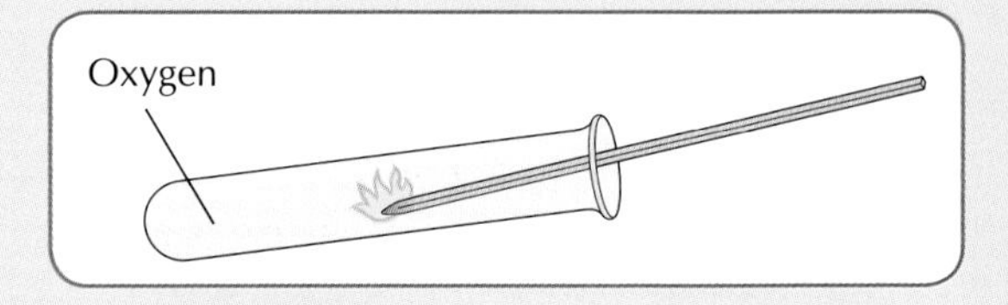

Fig. 25.10 Oxygen relights a glowing splint. This is the characteristic test for oxygen.

The glowing splint is re-kindled and bursts into flames. This is a characteristic test for oxygen. **Oxygen rekindles a glowing splint**. If the experiment is repeated with a candle, a similar observation is made. The candle burns more vigorously in oxygen.

3. **Carbon and oxygen**. Heat a small amount of carbon (charcoal) on a deflagrating spoon. When the charcoal is

glowing, place it in a gas jar of oxygen as shown in *Fig. 25.11(a)*.

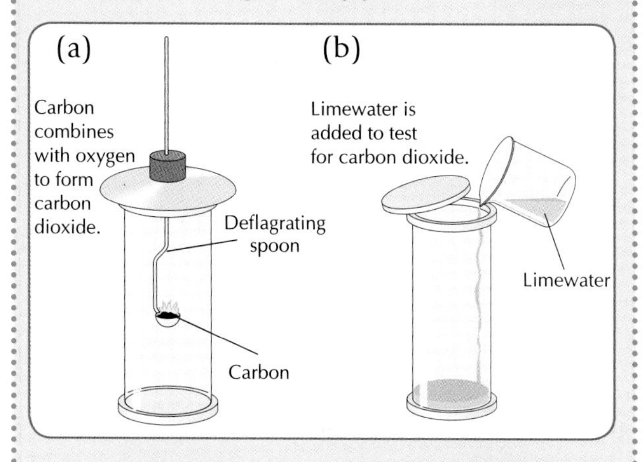

Fig. 25.11(a)(b) Experiment to verify that carbon dioxide is formed when carbon burns in oxygen.

The carbon continues to burn brightly. This is because the following reaction is occurring:

carbon + oxygen ⟶ carbon dioxide

Add limewater to the gas jar. Place a gas jar cover on the gas jar and shake the gas jar. Note that the limewater turns milky. This proves that carbon dioxide has been formed when carbon is burned in the gas jar of oxygen.

4. **Carbon dioxide and water**. Burn another sample of carbon in a fresh gas jar of oxygen. Add some water and some blue litmus paper to the gas jar. (Instead of adding water, you could just add moist blue litmus paper). Replace the gas jar cover and shake the gas jar, *Fig. 25.11(c)*.

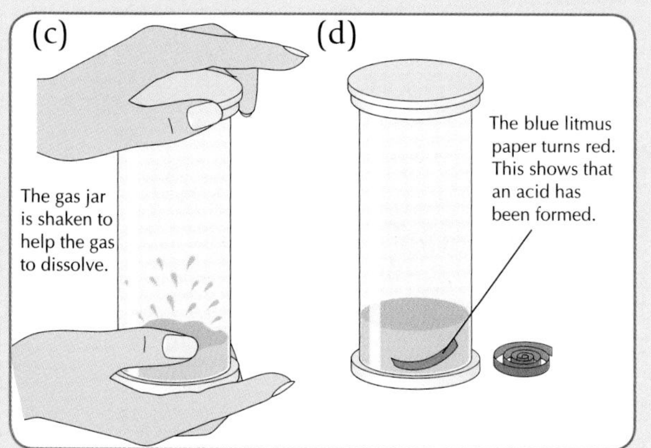

Fig. 25.11(c)(d) Carbon dioxide dissolves in water to form an acidic solution.

Note: The blue litmus paper turns red. This implies that carbon dioxide dissolves in water to form an acidic substance. This acidic substance is called carbonic acid. This is the acid found in fizzy drinks.

carbon dioxide + water ⟶ carbonic acid

5 a. **Magnesium and oxygen**. Using metal tongs, burn some magnesium metal in air. Note that the magnesium burns easily with a bright flame. Do not stare directly at the flame as the light is very intense. You can see why magnesium is used in fireworks! Now investigate what happens when magnesium is burned in a gas jar of oxygen. Wrap a small piece of magnesium around a deflagrating spoon. Ignite the magnesium using the Bunsen burner and quickly place it in a gas jar of oxygen, *Fig. 25.12(a)*. Once again, do not stare directly at the flame as it is very dazzling.

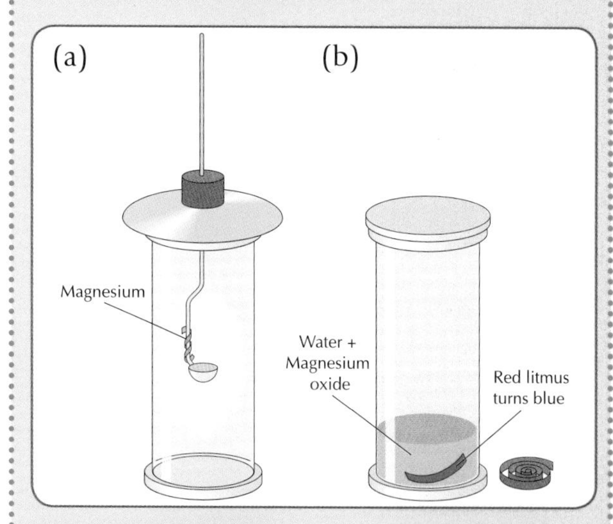

Fig. 25.12 Experiment to verify that a base is formed when magnesium oxide dissolves in water.

Note that the magnesium burns a lot more vigorously in oxygen than it does in air. It burns with a brilliant white flame in oxygen. When the magnesium has stopped burning, note that a white powder has been formed in the gas jar. This white powder is called magnesium oxide.

magnesium + oxygen ⟶ magnesium oxide

5 b. **Magnesium oxide and water**. When the gas jar has cooled down, add water and litmus paper as shown in *Fig. 25.12(b)*. Note that the red colour of the litmus paper slowly turned blue. This shows that a base has been formed. This base is called magnesium hydroxide. Magnesium hydroxide is the base found in *Milk of Magnesia*. It is formed when magnesium oxide reacts with water.

Word equation:
magnesium oxide + water ⟶ magnesium hydroxide

The reactions of carbon and magnesium with oxygen are summarised in *Table 25.2*.

ELEMENT	REACTION WITH OXYGEN	COMPOUND FORMED	EFFECT OF COMPOUND ON MOIST LITMUS PAPER
Carbon	Glows brightly	Carbon dioxide (CO_2)	Blue litmus turns red (Carbonic acid formed)
Magnesium	Burns vigorously – bright light given off	Magnesium oxide (MgO)	Red litmus turns blue (Magnesium hydroxide formed)

Table 25.2 Summary of reactions of carbon and magnesium with oxygen.

The main properties of oxygen are summarised in *Table 25.3*.

PHYSICAL PROPERTIES	CHEMICAL PROPERTIES
1. Colourless, odourless, tasteless gas.	1. Supports combustion – substances which burn in air, burn more vigorously in oxygen
2. Slightly soluble in water	2. Reacts with most elements to form oxides
3. Slightly heavier than air.	3. No effect on litmus

Table 25.3 The properties of oxygen.

25.3 Uses of Oxygen

(1) **Breathing**. Oxygen stored in cylinders is used to support breathing.

- For example, in hospitals it is used for **patients** who have difficulty in breathing, *Fig. 25.13*.
- Oxygen is often used to **revive people rescued** from drowning or who have been dragged out of smoke-filled rooms.
- Cylinders of oxygen are also used by **mountain climbers** and in underwater work.

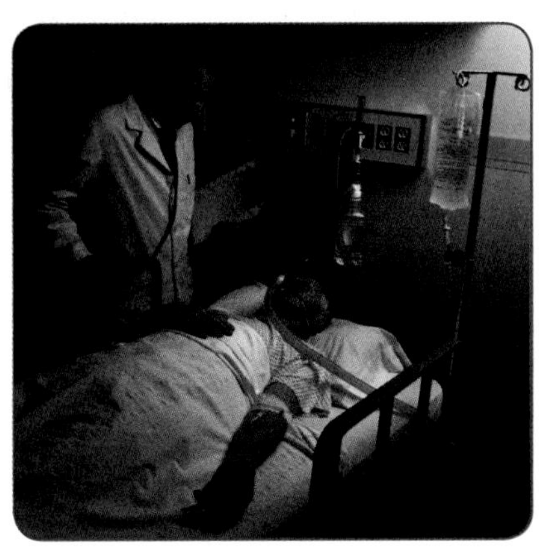

Fig. 25.13 Oxygen is used for patients who have difficulty in breathing.

(2) **Burning and welding**. Acetylene (also called ethyne) is a gas that burns in air. However, it burns much more fiercely in oxygen to produce a very hot flame of about 3000 °C. This flame is hot enough to cut through metals or to melt metals and join them together (welding).

(3) **Space rockets**. Since there is no oxygen in outer space, rockets carry their own oxygen and hydrogen (the fuel). Most rockets carry two tanks. One tank contains the hydrogen and the other tank contains the liquid oxygen.

TEST YOURSELF:
Now attempt questions *25.3–25.7* and *W25.7–W25.11*.

What I should know

- Air is a mixture of gases. It consists of 21% oxygen and 78% nitrogen.
- Small amounts of carbon dioxide and inert gases are also present in dry air.
- When burning occurs, the oxygen in the air is used up.
- To test for water use either anhydrous copper sulfate or cobalt chloride paper.
- Anhydrous copper sulfate changes from white to blue in the presence of water.
- Cobalt chloride paper changes from blue to pink in the presence of water.
- Oxygen is made in the laboratory by decomposing hydrogen peroxide using manganese dioxide as a catalyst.
- A catalyst is a substance that changes the speed of a chemical reaction but which is not used up in the reaction itself.
- Oxygen is a colourless, odourless, tasteless gas, is slightly soluble in water and supports combustion very well. Oxygen rekindles a glowing splint.
- Carbon burns in oxygen to form carbon dioxide. Carbon dioxide dissolves in water to form carbonic acid.
- Magnesium burns in oxygen to form magnesium oxide. Magnesium oxide dissolves in water to form a base called magnesium hydroxide.
- Oxygen is used in medicine, burning and welding and in space rockets.

QUESTIONS

Write the answers to the following questions into your copybook.

25.1 (a) Air consists of a ..... of gases.

(b) Which of the following represents the approximate percentage of oxygen in the air: 0.03%; 1%; 21%; 78%.

(c) Underline which of the following gases are normally present in the atmosphere: nitrogen, ammonia, carbon dioxide, hydrogen.

(d) Cloud, rain and dew are formed when the ..... in air condenses.

(e) Name the gas which is most plentiful in air.

(f) What chemical test would you use to detect the presence of water?

25.2 (a) The two main constituents of air are elements. Name these elements. Give two proofs that air is a mixture and not a compound.

(b) A stoppered bell jar is placed over a lighted candle floating on a dish of water, *Fig. 25.14*. With regard to this experiment, answer each of the following:

(i) What happens to the water as the candle burns?
(ii) Why does the candle stop burning after a while?
(iii) How would you measure the volume of air in the bell jar?
(iv) What does the experiment prove about the composition of the air?

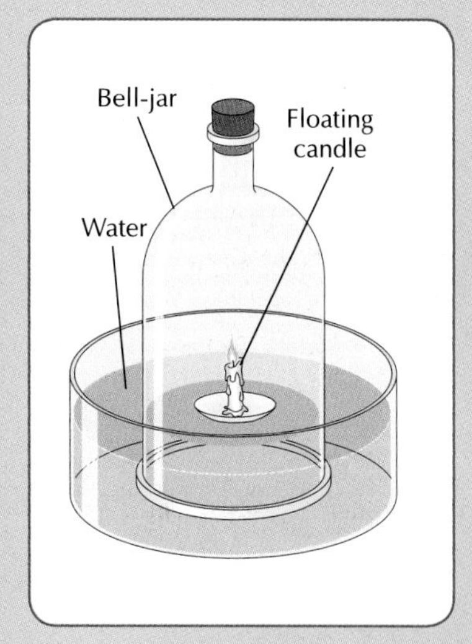

Fig. 25.14

QUESTIONS

25.3 (a) Two scientists are credited with the discovery of oxygen. Name one of them.

(b) Chemists represent oxygen gas using the chemical formula ..... .

(c) The chemicals used to prepare oxygen are ..... and ..... .

(d) One of the two chemicals used to prepare oxygen is only present to speed up the reaction. This chemical is..... .

(e) A chemical which helps to speed up a reaction but is not used up in the reaction is called a ..... .

(f) What colour is manganese dioxide?

25.4 (a) Describe how you would prepare and collect a sample of oxygen gas in the laboratory. Give three physical properties and three chemical properties of oxygen.

(b) Give a characteristic test for the presence of oxygen.

(c) Explain what is meant by the term *catalyst* and give an example of one.

25.5 (a) Why is oxygen rather than air used to burn fuel when a rocket is being launched ?

(b) Describe what you would observe when a piece of burning magnesium is lowered into a gas jar of oxygen?

(c) Give one use for magnesium metal.

(d) Why are cylinders of oxygen normally carried in ambulances?

(e) Some magnesium is burned in oxygen and the white substance formed is dissolved in water to form compound *X*. (i) Name *X*. (ii) Where is *X* commonly found?

25.6 (a) Some carbon is burned in a gas jar of oxygen. Water is added and the gas jar is shaken for a short time. A piece of blue litmus paper is then added. What would you observe? Explain this observation.

(b) Write a word equation for the burning of carbon in oxygen gas.

25.7 Oxygen may be prepared in the laboratory using hydrogen peroxide and another substance *X*.

(a) Name *X*.

(b) Explain the purpose of *X*.

(c) Describe the appearance of *X*.

(d) Write (i) a word equation and (ii) a chemical equation to describe the preparation of oxygen from hydrogen peroxide.

(e) Describe how you would test for the presence of oxygen in a container.

(f) State two uses of oxygen.

Chapter 26 Carbon Dioxide

26.1 Introduction

In the previous chapter we learned that less than 1% of the air consists of carbon dioxide. Although carbon dioxide occupies only a small percentage of the air, its presence in the air is of vital importance. Plants take in carbon dioxide through their leaves and use it to make food. (*Chapter 14*). Without carbon dioxide, plants could not grow. Without plants, animals would starve. Without crops and meat to eat, the human race could not survive.

26.2 Carbon Dioxide – Preparation and Properties

In the previous chapter we saw that carbon dioxide is produced when carbon is burned in oxygen gas. This would be an expensive way to produce large quantities of carbon dioxide since fuel is used to heat the carbon. In the next experiment we use a much easier and cheaper way to prepare carbon dioxide in the laboratory.

Laboratory Preparation of Carbon Dioxide

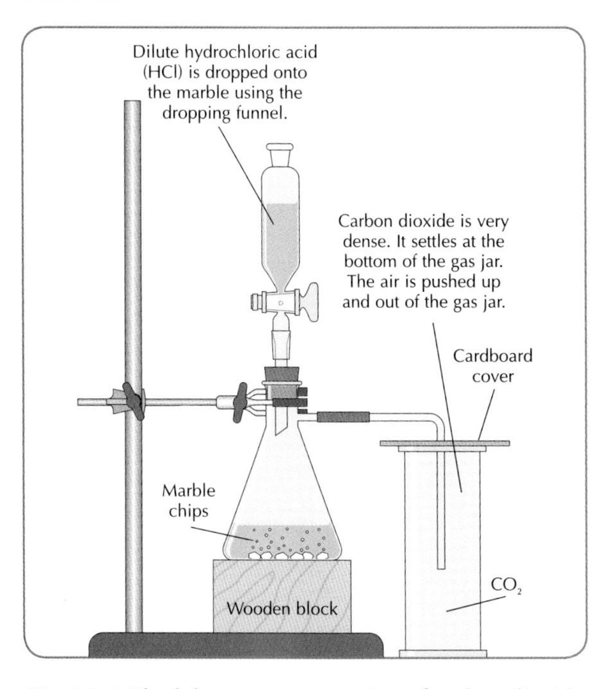

Fig. 26.1 The laboratory preparation of carbon dioxide.

Carbon dioxide is usually prepared in the laboratory by the reaction between dilute hydrochloric acid and marble chips using the apparatus shown in *Fig. 26.1*. The chemical name for marble is calcium carbonate. Chemists use the formula $CaCO_3$ to represent marble. (This is also the chemical found in chalk).

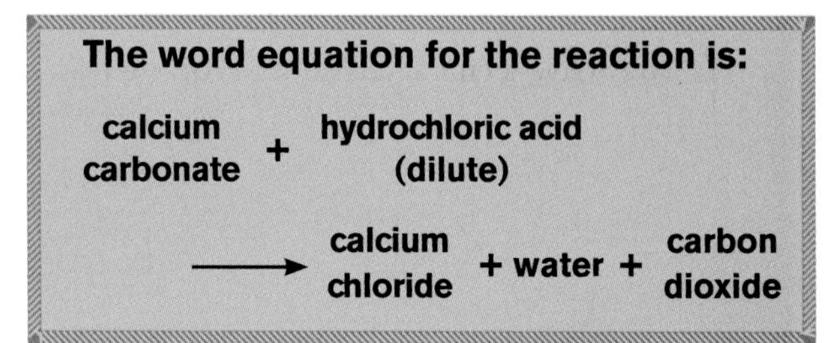

The balanced chemical equation for the reaction is:

$$CaCO_3 + 2HCl \longrightarrow CaCl_2 + H_2O + CO_2$$

Since carbon dioxide is slightly soluble in water, it is not normally collected over water. Instead, it is collected by the upward displacement of air. This means that the gas flows into the gas jar and pushes (displaces) the air out of the gas jar.

Mandatory experiment 26.1

(a) To prepare a sample of carbon dioxide

(b) To examine the properties of carbon dioxide

Apparatus required: dropping funnel; Buchner flask; delivery tubing; gas jars; gas jar covers; (or boiling tubes and stoppers); cardboard, test-tube rack; test-tubes; tapers

Chemicals required: hydrochloric acid (dilute); limewater, marble chips; litmus paper or litmus solution

(a) To prepare a sample of carbon dioxide

Method

1. Set up the apparatus as shown in *Fig. 26.1*.
2. Allow the dilute hydrochloric acid to fall on the marble chips. Note that a rapid 'fizzing' begins as soon as the acid and the marble come in contact.
3. Add a little hydrochloric acid from time to time to keep the reaction going, i.e. to keep the marble fizzing.
4. Collect three jars of the gas. Check that each gas jar is full by slowly lowering a lighted taper into it. If the jar is full, the taper will be extinguished at the mouth, *Fig. 26.2*.

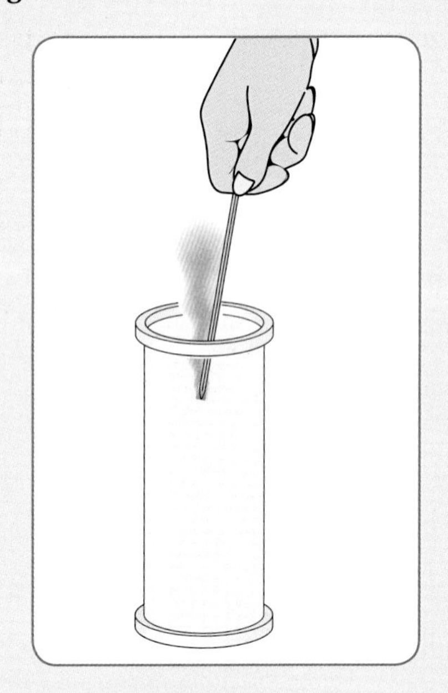

Fig. 26.2 A lighted taper is used to check if the gas jar is full of carbon dioxide. The lighted taper goes out (is extinguished) if carbon dioxide is present.

(b) To examine the properties of carbon dioxide

Method

1. Note that carbon dioxide is a **colourless**, **odourless** and **tasteless** gas.
2. Place a lighted taper in a jar of the gas. Note that it goes out.

   We conclude that **carbon dioxide does not support combustion**, i.e. carbon dioxide does not allow substances to burn in it.
3. Testing for carbon dioxide. The characteristic test for the presence of **carbon dioxide is that it turns limewater milky**. Bubble carbon dioxide through a test-tube containing limewater. Limewater is a solution of a substance called calcium hydroxide. It is observed that the limewater turns milky, *Fig. 26.3*.

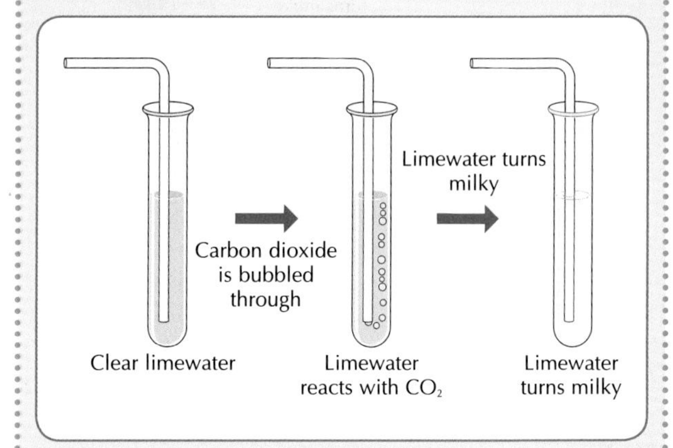

Fig. 26.3 Carbon dioxide turns limewater milky. The milky colour is due to the fact that chalk is formed. This is the characteristic test for the presence of carbon dioxide.

The milkiness is caused by the formation of chalk. The chemical name for chalk or limestone is calcium carbonate. This does not dissolve in water. The following is the reaction that occurs:

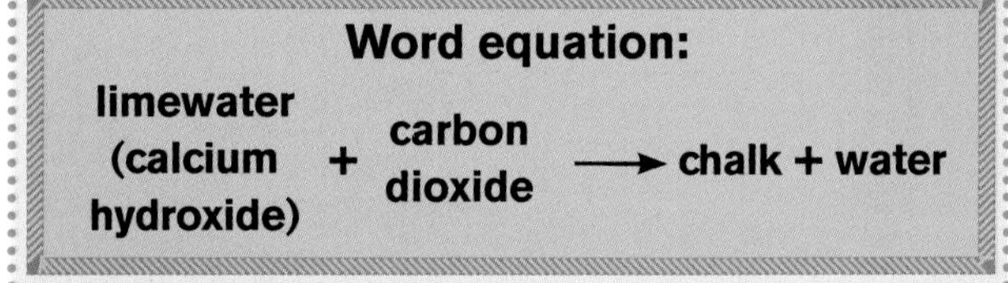

Chemical equation:

$$Ca(OH)_2 + CO_2 \longrightarrow CaCO_3 + H_2O$$

4. **Carbon dioxide is more dense than air**. Pour carbon dioxide from one gas jar into another. Test each gas jar for the presence of carbon dioxide using a lighted taper or limewater, *Fig. 26.4*.

Fig. 26.4 Carbon dioxide is more dense than air and may be poured from one container to another. We test for the presence of the gas in the lower gas jar using a lighted taper or limewater.

Another way of demonstrating that carbon dioxide is more dense than air is shown in *Fig. 26.5*.

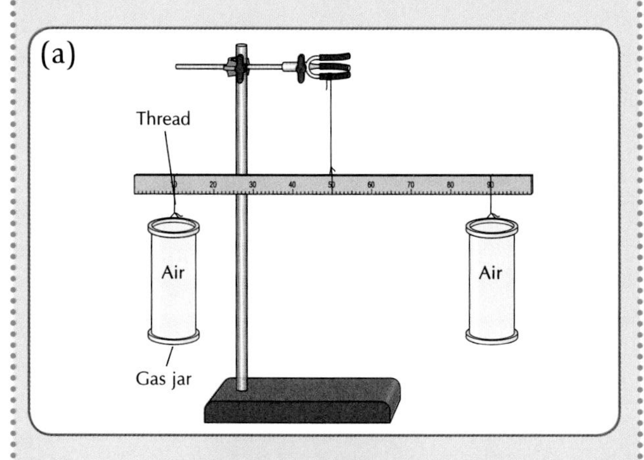

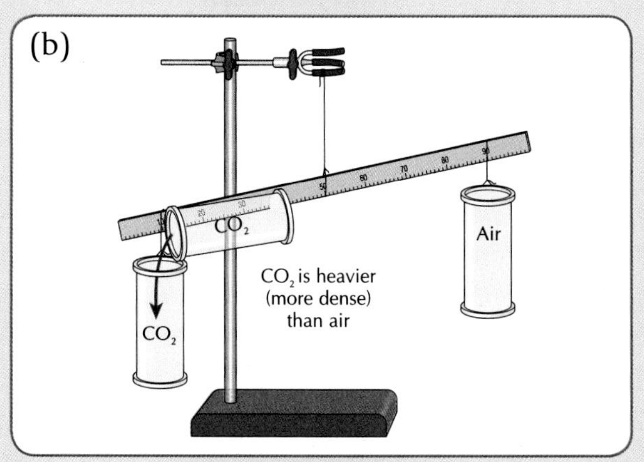

Fig. 26.5 This experiment shows that carbon dioxide is more dense than air.

Also, since carbon dioxide is more dense than air, it may be poured down on a lighted night light. Since carbon dioxide does not support combustion, the candle is extinguished, *Fig. 26.6*.

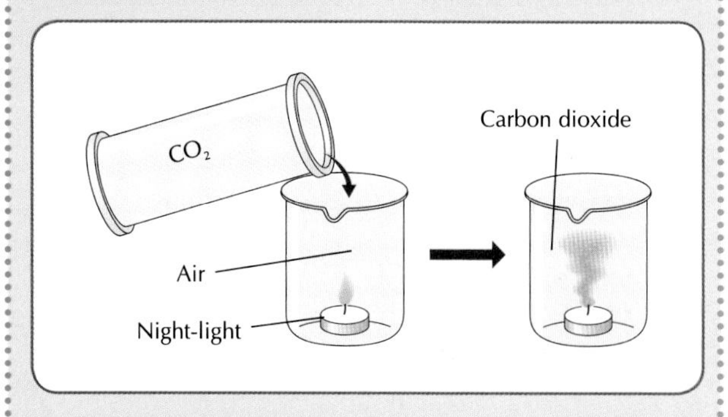

Fig. 26.6 This experiment shows that carbon dioxide is more dense than air. It also shows that carbon dioxide does not allow substances to burn in it. The carbon dioxide flows down and puts out the flame of the night light.

5. **Carbon dioxide dissolves slightly in water to form an acidic solution.** Bubble carbon dioxide through water to which a few drops of litmus solution have been added, *Fig. 26.7*. The litmus solution turns red. Alternatively, a piece of blue litmus paper could be placed in the test-tube of water. The litmus paper changes from blue to red. As stated in the last chapter, the acid formed is called carbonic acid.

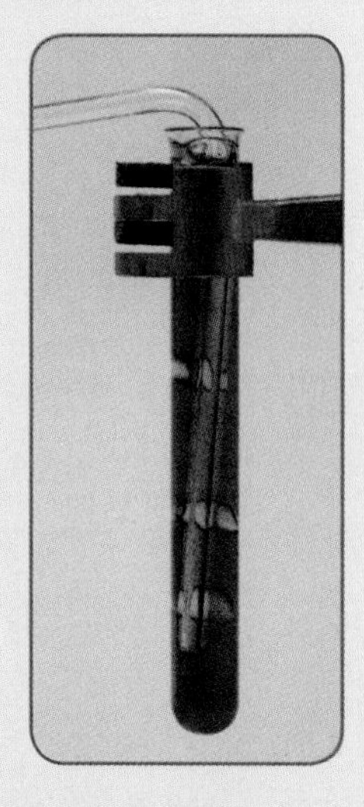

Fig. 26.7 Carbon dioxide dissolves in water to form an acid (carbonic acid). Therefore, the blue litmus indicator turns red when carbon dioxide is bubbled into the water.

The physical and chemical properties of carbon dioxide are summarised in *Table 26.1*.

PHYSICAL PROPERTIES	CHEMICAL PROPERTIES
1. Colourless, odourless, tasteless gas	1. Does not support combustion
2. Slightly soluble in water	2. Turns limewater milky
3. More dense than air	3. Dissolves in water to form carbonic acid

Table 26.1

26.3 Uses of Carbon Dioxide

(1) **Fizzy Drinks**. Carbon dioxide is present in all fizzy drinks. These contain dissolved carbon dioxide under pressure.

(2) **Fire extinguishers**. Carbon dioxide is ideal for use in fire extinguishers. Since it is more dense than air, it flows down on top of the fire. Since carbon dioxide does not support combustion, the layer of carbon dioxide gas puts out the fire.

(3) **Refrigeration**. Carbon dioxide gas, under pressure, is converted to a white solid. This solid is commonly called dry ice. It is called 'dry ice' because it looks like ordinary ice. It is, however, much colder (-78 °C) and sublimes (changes directly from a solid to a gas). This means that it does not go through the messy liquid state – hence the term 'dry'. Therefore, dry ice is ideal for cooling foods when they are being transported.

(4) **Special effects on stage**. Dry ice is also used for special effects on stage. If lumps of dry ice are put into warm water, wet carbon dioxide gas is released. This carries with it clouds of condensed water vapour. This causes the mist effect often used on stage, *Fig. 26.8*. This white cloud is used for special effects.

Fig. 26.8 When dry ice is put into water, it causes a mist effect.

Note: For your examination, you must know two uses of carbon dioxide!

TEST YOURSELF:
Now attempt questions *26.1–26.3* and *W26.1–W26.5*.

What I should know

- Carbon dioxide is prepared in the laboratory by the reaction between dilute hydrochloric acid and marble chips.
- Carbon dioxide is a colourless, odourless, tasteless gas.
- Carbon dioxide does not support combustion.
- Carbon dioxide is more dense than air.
- The characteristic test for carbon dioxide is that it turns limewater milky.
- Carbon dioxide is used in fizzy drinks, fire extinguishers, in refrigeration and for special effects on stage.

QUESTIONS

Write the answers to the following questions into your copybook.

26.1 (a) Name the two substances used to prepare carbon dioxide in the laboratory.

(b) In the preparation of carbon dioxide, what do you observe when the hydrochloric acid falls on the marble chips?

(c) 'Carbon dioxide may be collected by upward displacement of air'. Explain this statement.

(d) What chemical is used to test for the presence of carbon dioxide?

(e) Why is it important to have a tight fitting cap on a bottle of fizzy drink?

(f) State two uses for carbon dioxide.

26.2 (a) Describe with the aid of a diagram, a laboratory method of preparing carbon dioxide. Write down the balanced chemical equation for the preparation.

(b) List three properties of carbon dioxide and describe how you would demonstrate these properties in the laboratory.

(c) Fizzy drinks contain carbon dioxide gas. Why do these drinks only fizz after the bottle or the can is opened?

26.3 (a) You are given three closed gas jars which are not labelled. One contains oxygen, one contains carbon dioxide and the other contains nitrogen. Describe tests you would carry out to label each gas jar correctly.

(b) Some hydrochloric acid is added to a boiling tube containing an indigestion tablet, *Fig. 26.9*. The indigestion tablet contains $CaCO_3$. Bubbles of gas were given off. This gas was passed into limewater and the limewater turned milky.

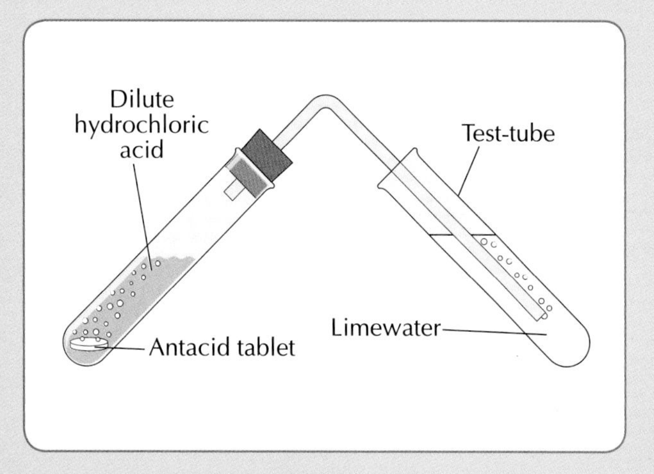

Fig. 26.9

(i) Name the gas given off.
(ii) Give two physical and two chemical properties of this gas.
(iii) Write a word equation and a balanced chemical equation for the reaction that takes place assuming that the tablet contains $CaCO_3$.
(iv) Give two uses for this gas.

Chapter 27 Water

27.1 Introduction

Water plays a very important part in our lives. Without water we could not live. The phrase 'water is life' is a very true one. Each day we use water for drinking, washing, carrying away the sewage from our homes and many other purposes.

Water is something we tend to take for granted. In fact, each of us uses a considerable amount of water each day. You may be surprised to know that, on average, each person in Ireland uses about 170 litres of water every day.

The Water Cycle

About 80 per cent of the earth's surface is covered with water. The water we use at home comes from rivers, lakes and underground wells. Have you ever wondered where all the rain comes from? Water is continuously moving from the earth's surface to clouds and then back again to the earth as rain. This continuous re-circulation of water is called the **water cycle**, *Fig. 27.1*. Study this diagram carefully and see if you can understand why the atmosphere does not run out of water.

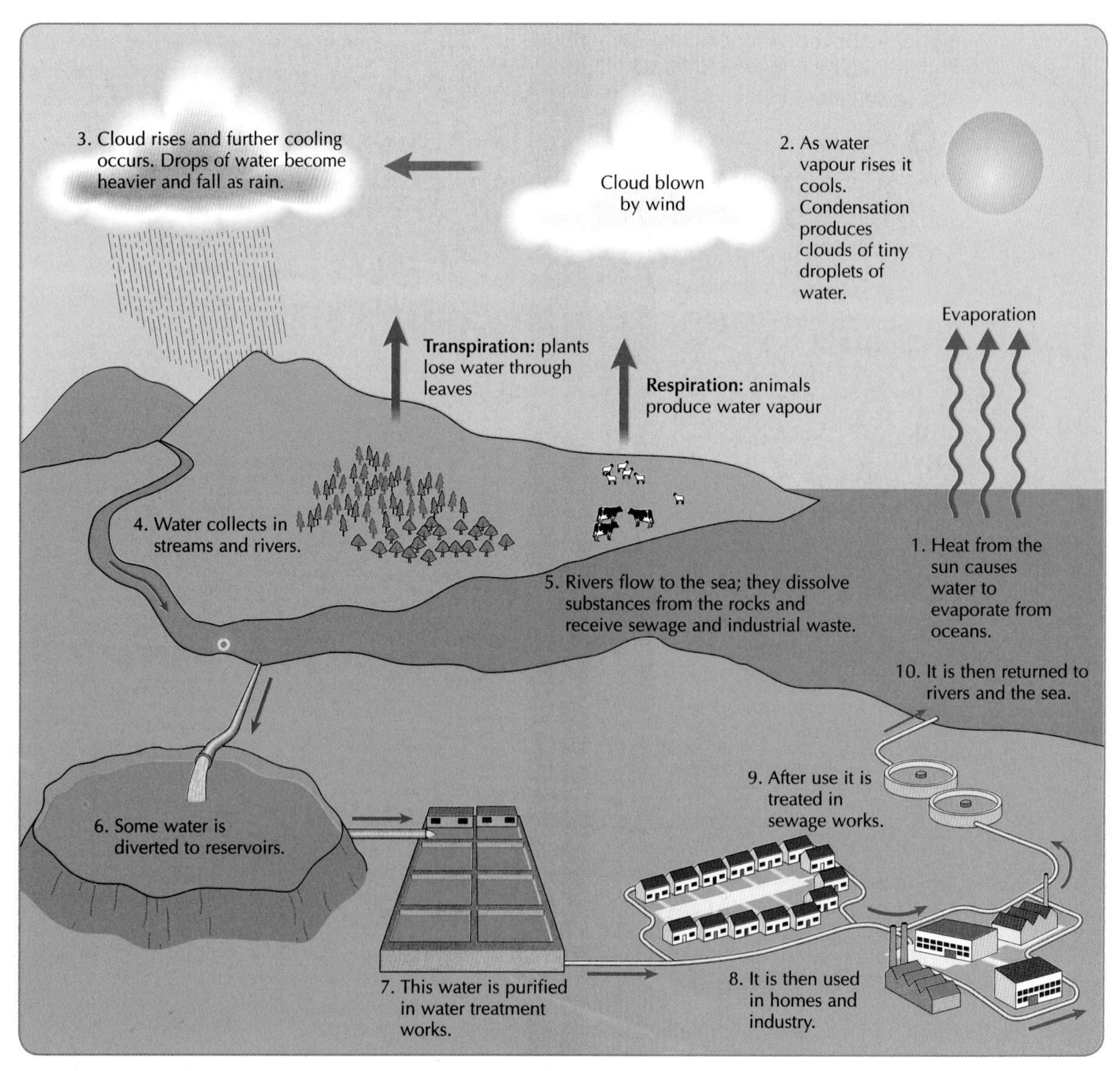

Fig. 27.1 The water cycle.

27.2 Properties of Water

Water is a very unusual liquid. Many of its physical properties are unique, e.g. it is the only substance that we commonly experience in each of its three physical states. We can summarise the properties of water as follows:

- **The freezing point of pure water is 0 °C and its boiling point is 100 °C.** This means that water is a liquid at room temperature. This property is used to identify a substance as pure water.
- **Water expands as it freezes.** All other compounds contract as they freeze. This is the reason why water pipes often burst when the water inside them freezes.
- **The density of water is 1 g/cm^3.** The density of ice (0.92 g/cm^3) is less than that of water. Therefore, ice floats on water.
- **Water is an excellent solvent.** It dissolves a wide variety of substances. No other compound is as good a solvent as water. Since water is such a good solvent, it is very difficult to get pure water. Pure water is water that contains nothing else in it but water particles.
- **Tests for the presence of water.** We have already seen in *Chapter 25* that water may be tested for in two ways: (a) Water turns white anhydrous copper sulfate blue. (b) Water turns cobalt chloride paper from blue to pink.

  Note: These two tests show that the liquid being tested **contains** water. It need not however be pure water. For example, milk would show a positive test for each of the above tests. To prove that a liquid is pure water, you would have to find its freezing point, boiling point and density.

- **Water tends to cling to glass.** The reason for this is because the water particles are more strongly attracted to glass than they are to each other. The attraction of water to glass explains the shape of the **meniscus** of water. The meniscus is the dark line on the surface of the liquid. In a measuring cylinder, the surface of the water is not flat because the water is attracted to the sides of the glass. This results in the shape of the meniscus for water. *Fig. 27.2.*

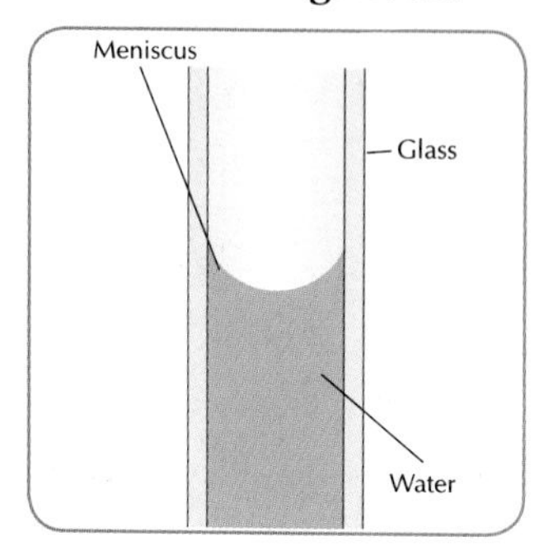

Fig. 27.2 The meniscus is the dark line on the surface of water. The surface is not flat because water clings to glass.

27.3 Water Treatment

Untreated Water

At one time, people took their drinking water from the nearest river or stream. However, as towns and cities grew larger, this led to outbreaks of disease. Domestic and industrial waste were dumped into the same rivers from which people took water to drink. Diseases like cholera, typhoid and river blindness are all carried by dirty water.

The Water Treatment Process

In order to ensure supplies of clean and safe drinking water, county councils and corporations operate water treatment plants throughout the country. Water supplied to our homes must be attractive looking (i.e. colourless), odourless and safe to drink. In order to achieve this, there are a number of steps involved in the treatment of water. Study *Fig. 27.3* carefully as you are required to understand each stage for your examination.

1. **Screening.** This stage involves passing the water through a wire mesh to remove any floating debris like twigs, plastic bags, etc.
2. **Settling.** The water then flows into large tanks called settling tanks or sedimentation tanks. As the water flows into these tanks, aluminium sulfate (commonly called 'alum') is added to the water. The aluminium sulfate takes the cloudiness out of the water and leaves the water clear. The cloudiness in the water is caused by small particles suspended in the water.

3. **Filtration**. The water is then passed through filter beds. These consist of layers of sand and gravel. The purpose of the filter beds is to ensure that even the smallest particles of dirt are removed from the water. A model of a filtration system similar to that used in water treatment plants is shown in *Fig. 27.4*.

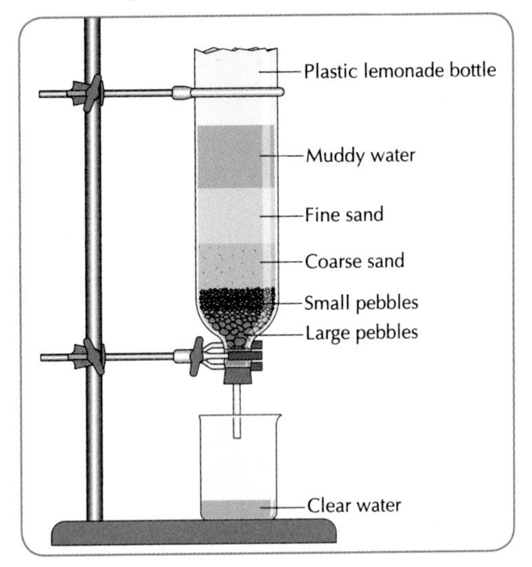

Fig. 27.4 Model of filtration bed.

4. **Chlorination**. The next stage in the treatment of water is to kill any germs that are in the water. Small quantities of chlorine are added to the water to kill bacteria in the water. The chlorine is not present in large enough quantities to smell or taste it. This process is often called **disinfecting** or **sterilising** the water.
5. **Fluoridation**. You may have heard the word 'fluoride' before. Many advertisements for toothpastes refer to the fact that the toothpaste 'contains fluoride'. The presence of fluoride compounds in toothpaste helps to reduce tooth decay. For this reason, in many countries (including Ireland) water authorities are obliged by law to add fluorine to water. This is done by adding small amounts of compounds like sodium fluoride to the water.

Finally, the quality of the water is tested by the laboratory staff of the water treatment plant, *Fig. 27.5*. The water must be certified as being fit to drink.

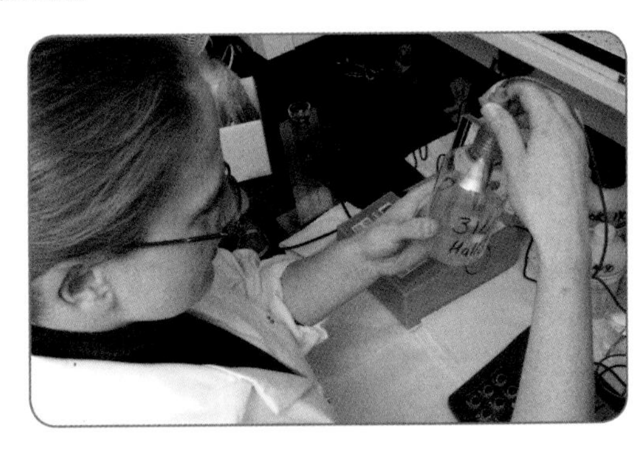

Fig. 27.5 This laboratory technician is carrying out tests on the water to ensure that it is fit to drink.

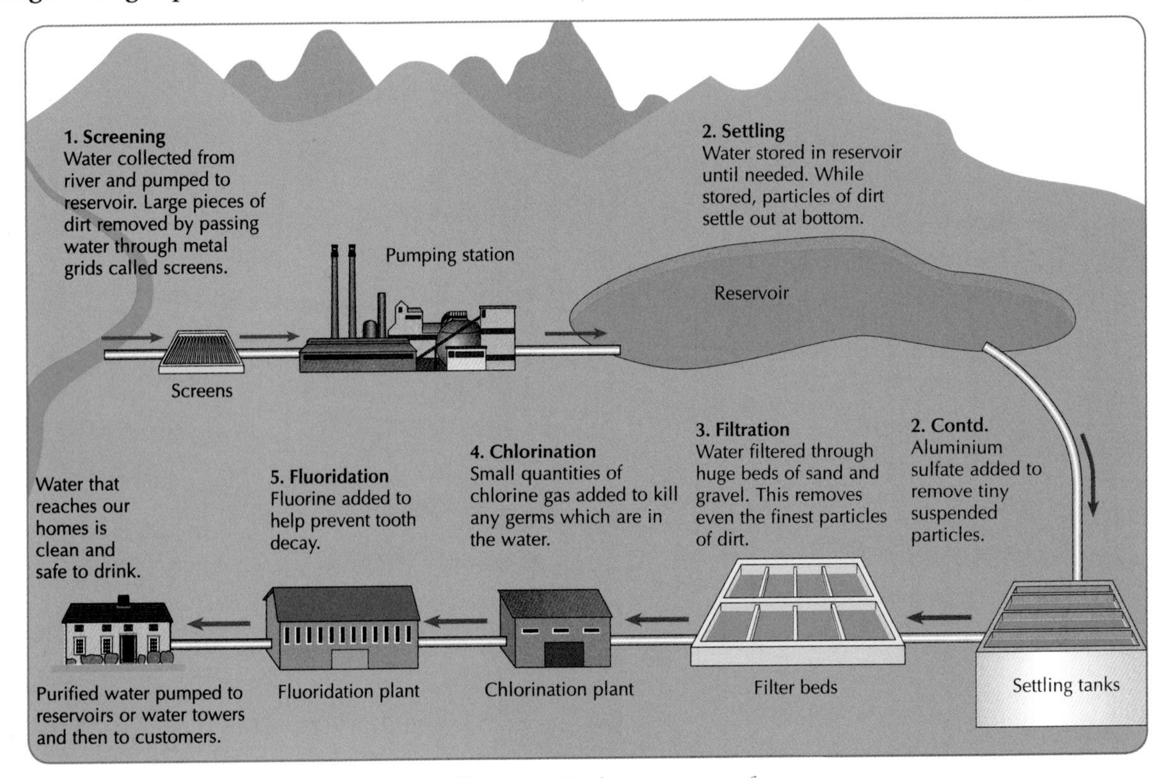

Fig. 27.3 The stages in the treatment of water.

TEST YOURSELF:
Now attempt questions *27.1–27.2* and *W27.1–W27.4.*

27.4 Water – A Pure Substance?

By pure water, a chemist means water that contains no dissolved material. Since water is such a good solvent, it is rarely found in the pure state. The purest form of water is distilled water.

Although water from the tap looks pure, it is in fact a solution of various substances.

- It contains dissolved **gases** like oxygen and carbon dioxide.
- It also contains dissolved **solids** that are picked up by the water as it flows over rocks in rivers and streams.

The presence of dissolved substances is studied in *Mandatory experiment 27.1.*

Mandatory experiment 27.1

To show the presence of dissolved solids in a sample of water

Apparatus required: clock glass; Pyrex beaker; hotplate or bunsen burner and wire gauze; tongs

Chemicals required: water (tap water or sea water or water from a stream or lake)

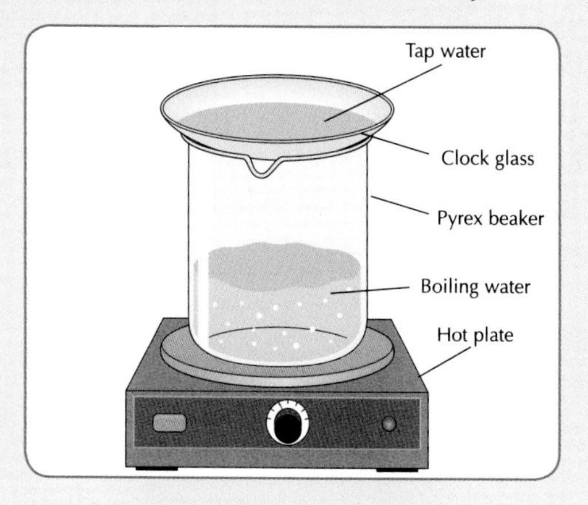

Fig. 27.6 An experiment to investigate if tap water is pure water.

Method

1. Set up the apparatus shown in *Fig. 27.6.* Water from the sea or a stream or a lake may be used instead of tap water.
2. Bring the water in the beaker to its boiling point. Note that the steam produced, evaporates the tap water from the clock glass.
3. Turn off the heat when all the water in the clock glass has evaporated.
4. Using tongs, remove the clock glass from the beaker.

Result

Solid material is left on the clock glass.

Conclusion

This material must have been dissolved in the tap water.

In many cases, these dissolved substances add a refreshing taste to the water. Spring water usually contains many substances dissolved in it, *Fig. 27.7.*

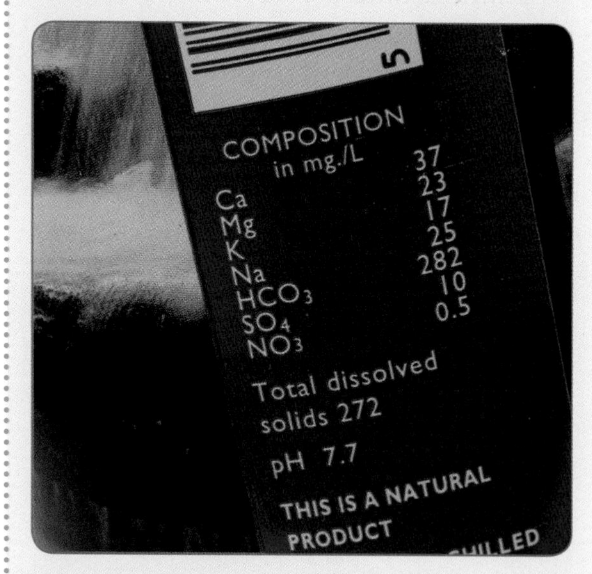

Fig. 27.7 The dissolved substances in spring water give it a refreshing taste.

Sea water contains many substances dissolved in the water. The Dead Sea, because it is an enclosed sea, contains about 275 grams of salts in every kilogram of water, *Fig. 27.8.* This is why the density of water in the Dead Sea is so great.

CHEMISTRY

Fig. 27.8 The water in the Dead Sea contains a lot of dissolved substances. This makes the water more dense than normal and it is easy to float in it.

In the home, water is used mainly for drinking and washing. The fact that substances are dissolved in the water does not matter if the water is being used for drinking. However, some substances can affect the use of water for washing and cause damage to boilers and hot water pipes. We consider this problem in the next section.

27.5 Hard and Soft Water

One of the important uses of water is for washing. When soap dissolves in water it usually forms a lather. In some cases, however, it is found that it is difficult to form a lather; rather, a dirty scum is produced before the lather is formed.

Fig. 27.9 Hard water is water that does not easily form a lather with soap.

Hard water is water that does not easily form a lather with soap.

Water in which soap lathers easily is called **soft** water.

What Causes Hard Water?

When chemists investigated the problem of hardness in water, they found that only water which flowed over rocks containing calcium or magnesium compounds is hard. When hard water was analysed, it was found that the hardness was caused by the presence of charged particles of calcium or magnesium. These charged particles are called **ions**. (We will learn more about ions in *Chapter 31*). In Ireland, hardness is nearly always caused by calcium ions. These ions combine with soap ions to form insoluble calcium compounds. These compounds are the scum that is formed in the water.

calcium ions + soap ions ⟶ scum

If you keep adding soap to the water, eventually all the calcium and magnesium ions will be removed as scum. The soap will then be able to work as a cleaning agent. This, however, wastes a lot of soap.

Limestone

Hard water is generally found in limestone areas. This gives us a clue to understanding how the calcium ions get into the water supply. The chemical name for limestone or chalk is **calcium carbonate**. Limestone does not dissolve in water. How do the calcium ions get into the water? The answer lies in the fact that rain water is slightly acidic. Rainwater is acidic because the **carbon dioxide** in the air dissolves in water to give an acidic solution. The acidic rainwater reacts with the insoluble calcium carbonate and changes it into soluble **calcium hydrogencarbonate**.

calcium carbonate (insoluble) + carbon dioxide + water ⟶ calcium hydrogencarbonate (soluble)

The calcium hydrogencarbonate then dissolves in water and gives rise to the calcium ions that cause hardness.

Mandatory experiment 27.2

To test samples of water for hardness

Apparatus required: rack of test-tubes

Chemicals required: soap flakes; water samples (rain water; tap water; sea water; distilled water; hard water)

Note: Soap flakes may be made using a bar of soap and a grater. Soap solution may be made by dissolving five soap flakes in 100 cm^3 of deionised water.

In this experiment the hardness of different samples of water is compared, *Fig. 27.10*. This is done by adding soap flakes (or soap solution) to the water, a small amount at a time until the water gives a permanent lather. **The harder the water, the more soap flakes are required to give a permanent lather.**

Fig. 27.10 Comparing the hardness of different samples of water.

Method

1. Add the same amount of water from each sample to the test-tubes as shown in *Fig. 27.10*.
2. Add a soap flake to each sample of water. Alternatively, add 2 cm^3 of soap solution from a burette.
4. Put your thumb over the mouth of each test-tube and shake vigorously for a few seconds. Put each test-tube back in the rack and wait for about 20 seconds. If a lather remains after this time, we can say that one soap flake or 2 cm^3 of soap solution was required.
5. If a lather does not remain, add another soap flake and shake again. Wait a further 20 seconds.
6. Continue on in this way and record the number of soap flakes (or volume of soap solution) required to produce a permanent lather. Summarise your results in the form of a table.

WATER SAMPLE	NO. OF SOAP FLAKES (VOLUME OF SOAP SOLUTION)
Rain water	
Tap water	
Sea water	
Distilled water	
Hard water	

Result

Each water sample requires a different number of soap flakes to form a permanent lather.

Conclusion

The water sample that requires the largest number of soap flakes (or largest volume of soap solution) is the hardest. (We expect sea water to be very hard).

The water sample that requires the smallest number of soap flakes (or smallest volume of soap solution) is the softest. (We expect rain water and distilled water to be very soft).

27.6 Removal of Hardness

In Ireland, it is common for hardness to be caused by a substance called calcium hydrogencarbonate dissolved in the water. When the water is boiled, the dissolved calcium hydrogencarbonate is changed to solid limestone (calcium carbonate). This limestone is commonly called 'fur' or 'scale'. It builds up in kettles, boilers, hot water pipes, etc, *Fig. 27.11*. This 'fur' causes problems as it can eventually block pipes.

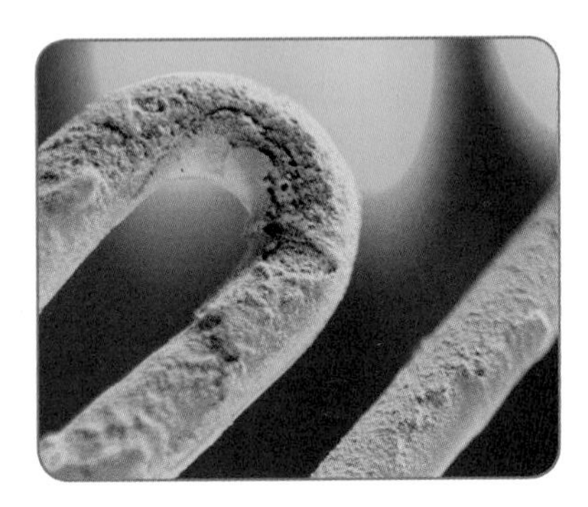

Fig. 27.11 The hardness of the water has caused this heating element of a washing machine to become coated with 'fur'. This 'fur' is limestone (calcium carbonate).

Since hard water causes problems in the home, it is necessary to soften the water. A number of methods may be used to remove the hardness from water. All methods involve the removal from solution of the calcium or magnesium ions, i.e. the ions that cause hardness. In this course, we shall consider two of these methods.

(1) Boiling

This removes hardness due to **calcium hydrogencarbonate only**. As already stated, boiling the water converts the soluble calcium hydrogencarbonate into insoluble calcium carbonate. Thus, the calcium ions are removed from the solution.

(2) Ion Exchange

The easiest way of removing the hardness from water is to use an **ion exchange resin**. Ion exchange resins are man-made materials that exchange ions. They have the ability to swap the ions that cause hardness for ions that do not cause hardness. If you live in a hard water area, you may have a water softener in your home. This water softener contains an ion exchange resin, *Fig. 27.12*.

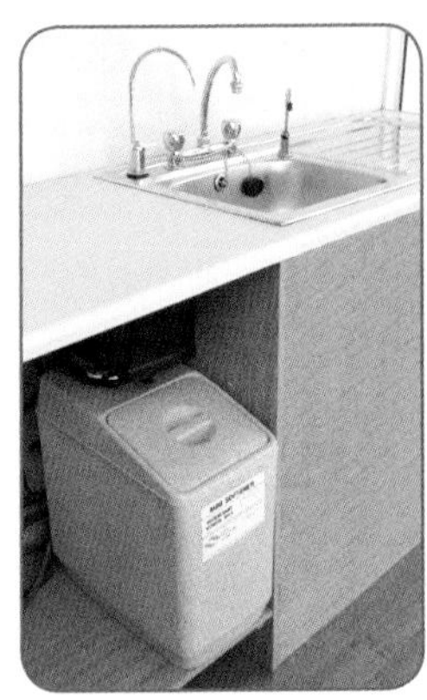

Fig. 27.12 These photographs show a water softener used in homes in hard water areas. It contains ion exchange resin. This resin replaces calcium and magnesium ions in the water with sodium ions.

An ion exchange resin swaps the calcium and magnesium ions in the water for **sodium ions**. Therefore, the water has been softened because the ions that cause hardness have been removed, *Fig. 27.13*.

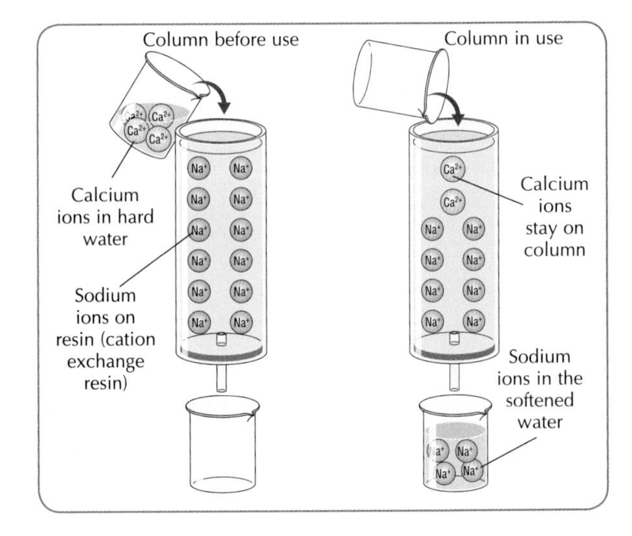

Fig. 27.13 The water is softened because the calcium ions have been replaced by sodium ions.

Deionised Water

In some cases, it is important that **all** ions are removed from the water. For example, water without any ions dissolved in it is often needed for solutions used in medicine or in chemistry experiments. This is done by using other types of ion exchange resin. Water which has **all** its ions removed is called **deionised water**. There is probably a water deioniser in your school laboratory.

Advantages and Disadvantages of Hard Water

The advantages and disadvantages of hard water are summarised in *Table 27.1*.

ADVANTAGES	DISADVANTAGES
1. Provides calcium for teeth and bones	1. Blocks pipes, leaves scale on kettles and boilers
2. Nicer taste	2. Wastes soap
3. Good for brewing and tanning	3. Produces scum

Table 27.1 The advantages and disadvantages of hard water.

TEST YOURSELF:
Now attempt questions *27.3–27.4* and *W27.5–W27.8*.

27.7 Splitting Water into its Elements

When electricity is passed through a liquid like water, it is found that a chemical reaction takes place. For example, consider the following experiment in which electricity is passed through water, *Fig. 27.14*. A little sulfuric acid is added to the water to help it conduct electricity.

The Set-up

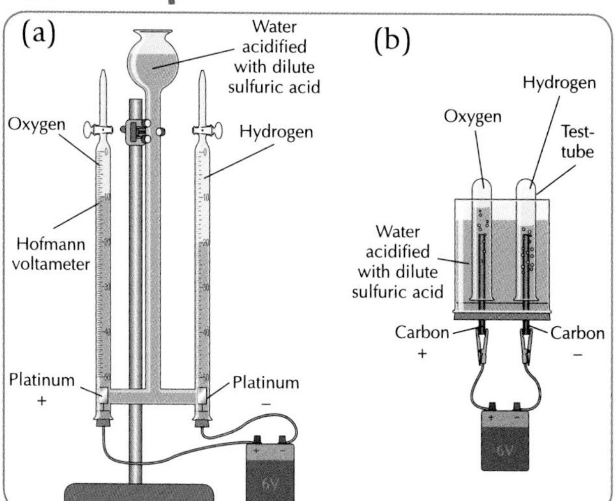

Fig. 27.14 An electric current may be passed through water using (a) a Hofmann voltameter or (b) using test-tubes. This splits the water into hydrogen gas and oxygen gas. The volume of hydrogen gas is twice that of oxygen gas. Therefore, we represent water by the formula H_2O.

This special type of apparatus is called a **Hofmann voltameter**. (Do not confuse the name **voltameter** with the name **voltmeter**. A voltameter is an apparatus used to pass electricity through substances. A voltmeter is an instrument used to measure voltage).

The Hofmann voltameter contains two pieces of platinum metal. Platinum must be used, as this metal does not react with the acid added to the water. (If platinum is not available, carbon could also be used). These two pieces of platinum are connected to the battery.

- The piece of platinum connected to the **positive terminal** of the battery is called the **positive electrode.**
- The piece of platinum connected to the **negative terminal** of the battery is called the **negative electrode**.

The Results

An electric current is allowed to flow for a few minutes through the voltameter. It is found that two gases are formed.

- The gas that is formed at the positive electrode is **oxygen**. We test for the oxygen using a glowing splint.
- If a sample of the gas that collects at the negative electrode is collected, it is found that it burns with a 'pop'. This shows that it is **hydrogen** gas (*Chapter 33*).

If you measure the volumes of gases formed, you will notice that **twice as much hydrogen as oxygen** is formed. This is why water is given the formula H_2O.

Electrolysis

What is actually happening is that the electric current is causing a chemical reaction to take place. Water is being broken down into the elements hydrogen and oxygen.

water ⟶ hydrogen + oxygen

The production of a chemical reaction using electricity is called **electrolysis**. (The word 'electrolysis' comes from the Greek language and means 'splitting up using electricity').

Electrolysis is the production of a chemical change by electricity.

The word electrolysis was introduced into chemistry by Michael Faraday. Electrolysis is very useful in chemistry as it allows chemists to bring about certain chemical reactions using electricity.

TEST YOURSELF:
Now attempt questions *27.5* and *W27.9*.

Michael Faraday (1791–1867) was one of the greatest scientists of all time. It was he who introduced the word electrolysis into the language of chemistry.

What I should know

- Pure water freezes at 0 °C, boils at 100 °C and has a density of 1 g/cm^3.
- Water is an excellent solvent. It expands as it freezes. It is tested for using anhydrous copper sulfate or cobalt chloride paper.
- There are five stages involved in water treatment: screening, settling, filtration, chlorination and fluoridation.
- Hard water is water that does not easily form a lather with soap.
- Hardness in water is caused by the presence of calcium or magnesium ions dissolved in the water.
- Hard water causes scale to build up in kettles, boilers, etc. and it wastes soap.
- Ion exchange resins remove hardness.
- Hard water is useful as it provides a good supply of calcium in our diets and it has a nicer taste than soft water.
- Electrolysis is the production of a chemical change by electricity.
- Water is split up into hydrogen and oxygen by electricity.

Questions

Fill in the following answers in your copybook.

27.1 (a) Give five uses of water in your everyday life.

(b) A piece of blue cobalt paper left exposed to the air slowly turned pink. What substance in the air caused this to happen?

(c) Write down (i) the melting point of water and (ii) the boiling point of water.

(d) Why is the surface of water in glass tubes curved?

(e) What is the dark line on the surface of water called?

27.2 (a) When water is being taken from a river or lake, what is the purpose of 'screening'?

(b) After being taken from a river, water is often stored in reservoirs. How does storing water in reservoirs help to purify the water?

(c) What is the purpose of fluoridating water supplied to our homes?

QUESTIONS

(d) Name the substance added in water treatment plants to help remove small suspended particles from water?

(e) In some countries which do not have water treatment plants it is recommended that all water is boiled before being drunk. What is the reason for this?

(f) What substance is added to water in a water treatment plant to make sure the water is safe to drink?

27.3 (a) Why is it difficult to form a lather with soap if the water is hard?

(b) Would you expect rain water to be hard or soft? Explain your answer.

(c) Limestone does not dissolve in pure water. Why does it dissolve in rain water?

(d) What ions cause hardness in water?

(e) Name the substance formed in a kettle when hard water is boiled.

(f) Give two disadvantages of hard water.

(g) Why is it healthier to drink hard water than soft water?

(h) Why is it preferable for power stations to use soft water in their boilers?

(i) How is hardness removed from water?

27.4 (a) Two unlabelled bottles are known to contain soft water and hard water. Describe how you would distinguish between these samples of water.

(b) A house in the Burren has its water supply passing through a container labelled 'Ion Exchanger'.

(i) What is the purpose of the ion exchanger?
(ii) What does it contain?
(iii) Name one type of substance removed from the water.
(iv) With the aid of a diagram, explain how an ion exchanger works.
(v) What is the difference between the ion exchanger used in a house and the ion exchanger used in a laboratory?

27.5 (a) Explain the meaning of the term *electrolysis*.

(b) Name the famous scientist who introduced the term electrolysis into science.

(c) Look at the apparatus shown in *Fig. 27.15* used to carry out the electrolysis of water. (i) Name the gas *X* shown in the diagram. (ii) What test would you perform on this gas to identify it?

(d) What substance is added to water to help it carry the electric current?

(e) The electrodes of the apparatus are usually made of a certain element. Name the element. Why would iron not be a suitable material to use for the electrodes?

(f) Write a word equation and a balanced equation to describe what happens when the electrolysis of water is carried out.

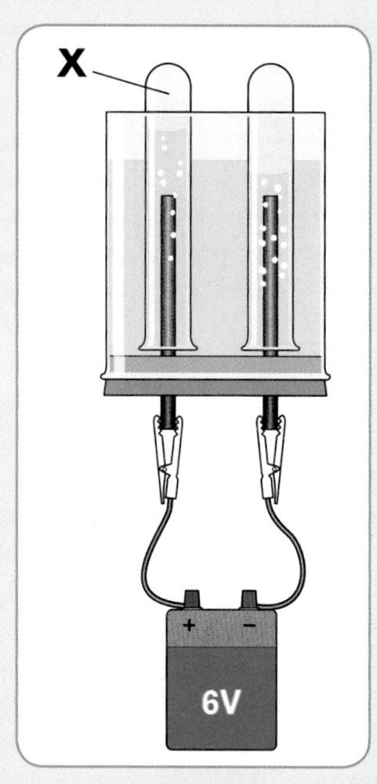

Fig. 27.15

CHEMISTRY

Chapter 28 Acids and Bases II

28.1 Introduction

In *Chapter 24* we studied some of the properties of acids and bases. We learned the following:

- Acids and bases affect the **colour of indicators**.
- Bases which dissolve in water are called **alkalis**.
- The **pH scale** gives us an indication of the level of acidity or basicity in a solution.

In **this chapter** we will take a closer look at the properties of acids and bases.

- We will learn what happens when **acids and bases react** with each other.
- We will study the problem of **acid rain** on the environment.

28.2 Common Laboratory Acids and Bases

Acids

The two most common laboratory acids are hydrochloric acid (HCl) and sulfuric acid (H_2SO_4), *Fig. 28.1*. These acids are represented by the chemical formulas as shown in the diagram.

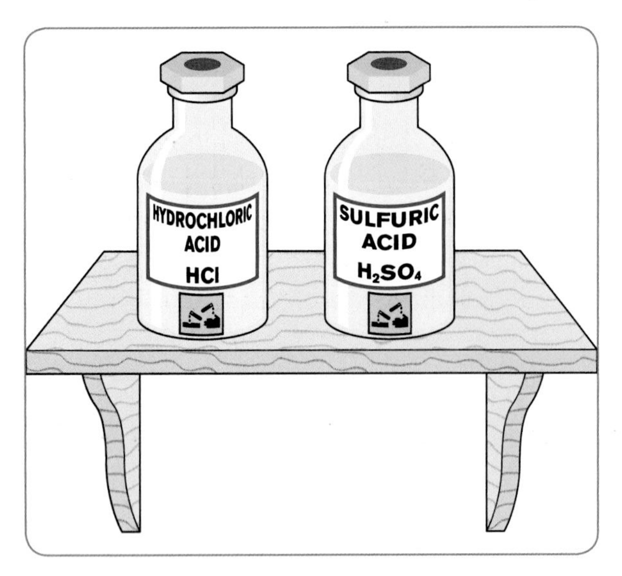

Fig. 28.1 Two common laboratory acids.

Warning! Always take great care when handling acids in the laboratory. Keep them well away from your eyes, mouth and any cuts. You should never handle concentrated solutions of acids or bases in the laboratory without the permission of your teacher.

Sulfuric acid and hydrochloric acid are **strong** acids. Strong acids release a lot of hydrogen ions (H^+) in solution. Acids that release a small number of hydrogen ions are **weak** acids. Examples of weak acids are acetic acid and citric acid.

Bases

The most common bases that are used in the school laboratory are sodium hydroxide and calcium hydroxide.

Sodium Hydroxide

- The chemical formula for sodium hydroxide is **NaOH**.
- Sodium hydroxide is commonly called **caustic soda**.
- It is often used to **clear blocked drains**, *Fig. 28.2*.

It is a very corrosive material, i.e. it 'eats away' other substances. Always make sure you wear gloves and safety glasses when handling sodium hydroxide.

Fig. 28.2 Sodium hydroxide is the most common base. It is used to clear blocked drains. It corrodes or 'eats away' the material that is blocking the drain.

Calcium Hydroxide

You may remember that we used a base called calcium hydroxide in *Chapter 25*.

- The formula for calcium hydroxide is **$Ca(OH)_2$**.
- A solution of calcium hydroxide in water is called **limewater**.
- Limewater is used to **test for** the presence of **carbon dioxide** in air.

28.3 Reaction of an Acid with a Base – Neutralisation

Fig. 28.3 Antacid tablets are used to react with excess acid in the stomach.

You may have seen the word 'antacid' on a medicine bottle or box of tablets, *Fig. 28.3*. 'Antacid' means 'against acid'. Antacids are bases that cure stomach upsets by reacting with the excess acid in the stomach. Excess acid means that there is too much acid in your stomach. It is found that the properties of an acid are counteracted or neutralised by a base. Therefore, this type of reaction is called a **neutralisation reaction**. This is because the substance formed is often neutral (i.e. neither acidic nor basic).

Examples of Neutralisation

There are many examples of neutralisation reactions in everyday life.

- **Toothpaste** is a basic substance. It neutralises the **acids that damage our teeth**.

Fig. 28.4 Toothpaste is a basic substance. It neutralises the acids that damage our teeth.

- When a **bee** stings you, it injects an acidic liquid into the skin. The sting is neutralised using **baking soda**. Baking soda is a base.
- **Vinegar** is used to neutralise the alkaline sting of **wasps** (remember 'VINEGAR FOR VASP STINGS!').
- **Dock leaves** are used to neutralise the acidic sting from **nettles**.

Formation of Salts

When an acid and a base neutralise each other, a salt is formed; for example, sodium chloride. This may be demonstrated in the laboratory as shown in *Fig. 28.5*.

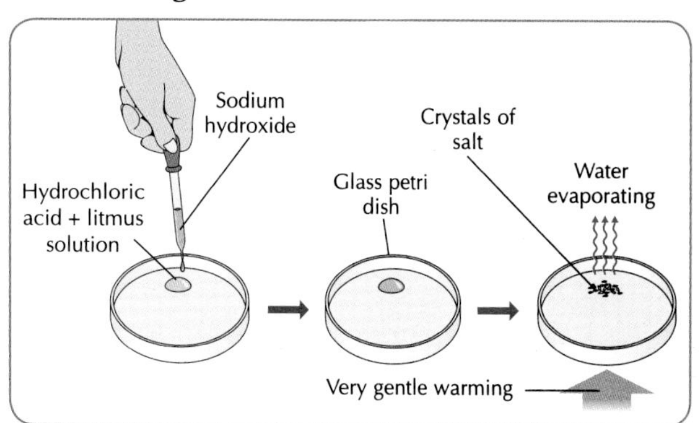

Fig. 28.5 The acid reacts with the base to form a salt. In this case, the salt is sodium chloride, NaCl.

Place five drops of dilute hydrochloric acid on a petri dish or clock glass. Add a drop of litmus solution. From another dropper, keep adding dilute sodium hydroxide solution until the indicator just changes from red to blue. Allow the water to evaporate and you will observe the salt that is formed.

In general, a neutralisation reaction may be represented as:

acid + base ⟶ salt + water

What is a Salt?

When you hear the word salt, you probably think of something you put on food. This is just one example of a salt. The salt you put on food is called sodium chloride. However, in science the word salt applies to many other substances as well. In fact, there are lots of different salts!

A salt is formed when the hydrogen in an acid is replaced by a metal.

For example, if the H in HCl is replaced by a metal like sodium, we form the salt sodium chloride. This is shown in *Fig. 28.6*.

Acids and Limestone

Recall: We have already met one other property of acids. We saw that hydrochloric acid reacts with limestone (marble chips) to form carbon dioxide (*Chapter 26*). We used this fact to prepare carbon dioxide. The salt calcium chloride and water were also formed.

calcium carbonate (marble) + hydrochloric acid ⟶ calcium chloride + water + carbon dioxide

$$CaCO_3 + 2HCl \longrightarrow CaCl_2 + H_2O + CO_2$$

Laboratory Preparation of Sodium Chloride

As discussed above, salts may be prepared by reacting together an acid and a base. To make the salt, you must react exactly the right amounts of acid and base. If we add too much acid, some acid will be left over. If we add too little acid, some base will be left over. We use an indicator to find out when the acid has been exactly neutralised.

Titration

The general procedure is that the acid (in a burette) is slowly added to the base (in a conical flask). The acid is added until the indicator starts to change colour. A burette is used because it is a very accurate way of measuring volumes of liquids. The method of adding one solution from a burette to another solution in order to find out how much of the two solutions will just react with each other is called **titration,** *Fig.28.7*.

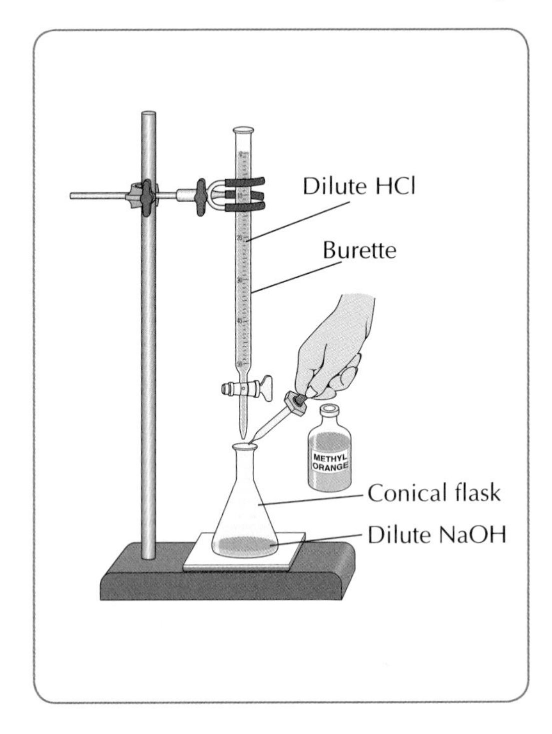

Fig. 28.7 Titration apparatus.

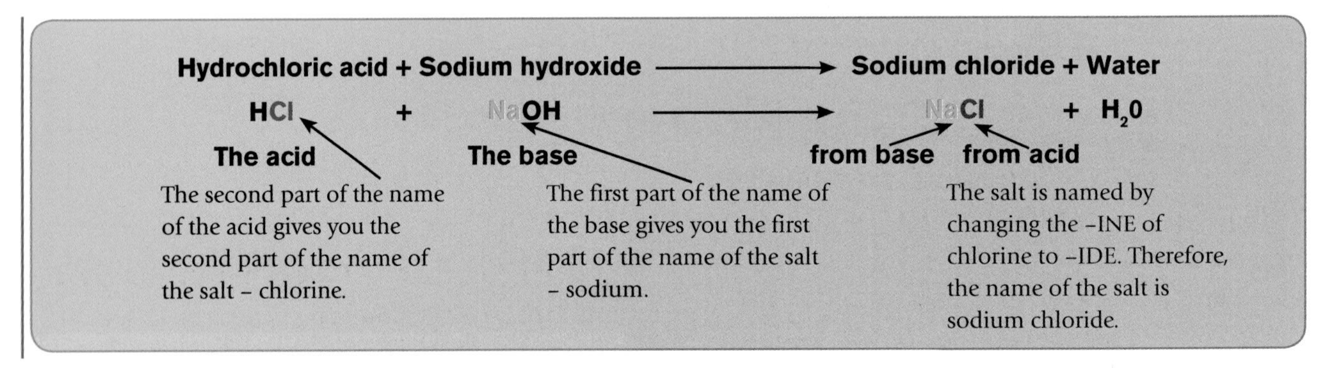

Fig. 28.6 Sodium chloride is formed when hydrochloric acid reacts with sodium hydroxide.

Mandatory experiment 28.1

To titrate hydrochloric acid (HCl) against sodium hydroxide (NaOH) and prepare a sample of sodium chloride (NaCl)

Apparatus required: burette; conical flask; retort stand and clamp; pipette; pipette filler; white tile; funnel; wash bottle; dropper; evaporating dish; hotplate (or bunsen and tripod); beakers

Chemicals required: dilute hydrochloric acid (1 M); dilute sodium hydroxide (1 M); methyl orange indicator or litmus indicator (methyl orange is preferable as the colour change at the end point is clearer)

hydrochloric acid + sodium hydroxide → sodium chloride + water

$$HCl + NaOH \rightarrow NaCl + H_2O$$

Method

1. Set up the apparatus as shown in *Fig. 28.10*. ***Note:*** the acid is placed in the burette and the base is placed in the conical flask. (ACID ABOVE, BASE BELOW). Always use a pipette filler when using the pipette.
2. Carry out each step (a) to (j) as shown in *Fig. 28.10*.
3. When you have finished the titration, pour the contents of the conical flask into an evaporating dish. Evaporate the water almost to dryness over a water bath.

Note 1: The tap of the burette should be operated with the thumb and first two fingers of your left hand, *Fig. 28.8*. Using a wash bottle, wash down the sides of the conical flask from time to time. This ensures that any drops of acid stuck to the side of the flask are washed into the flask.

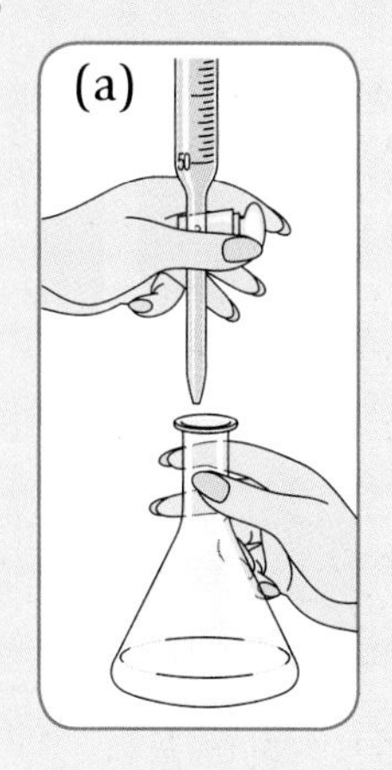

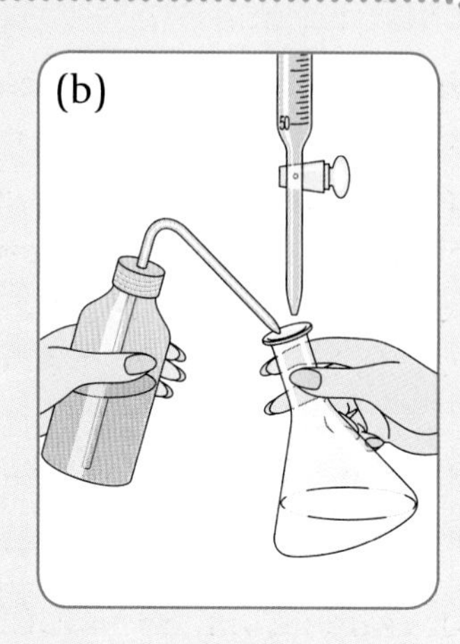

Fig. 28.8(a) The left hand is used to operate the tap of the burette. The right hand is used to shake the flask. (b) During the titration, the sides of the flask are washed down from time to time with deionised water.

Note 2: When reading the burette, always read from the bottom of the meniscus. Make sure your eye is level with the meniscus, *Fig. 28.9*.

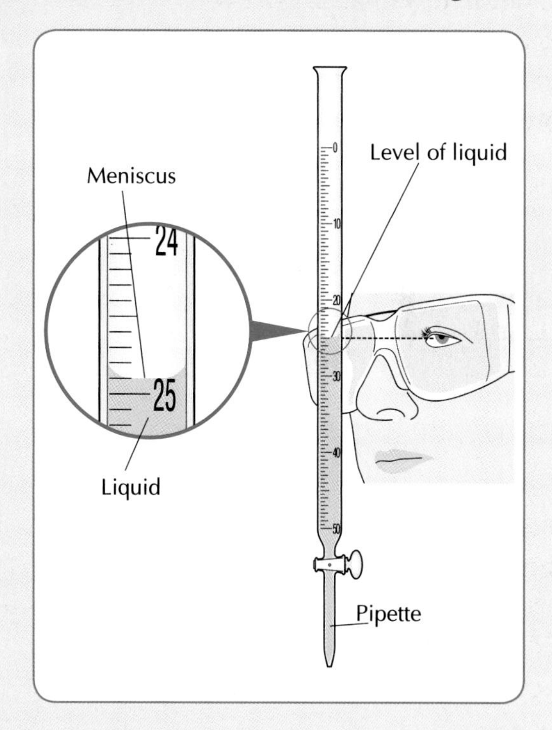

Fig. 28.9 When reading the volume of liquid, make sure your eye is level with the bottom of the meniscus

Result

White crystals of sodium chloride are formed.

Conclusion

Hydrochloric acid and sodium hydroxide react to form sodium chloride.

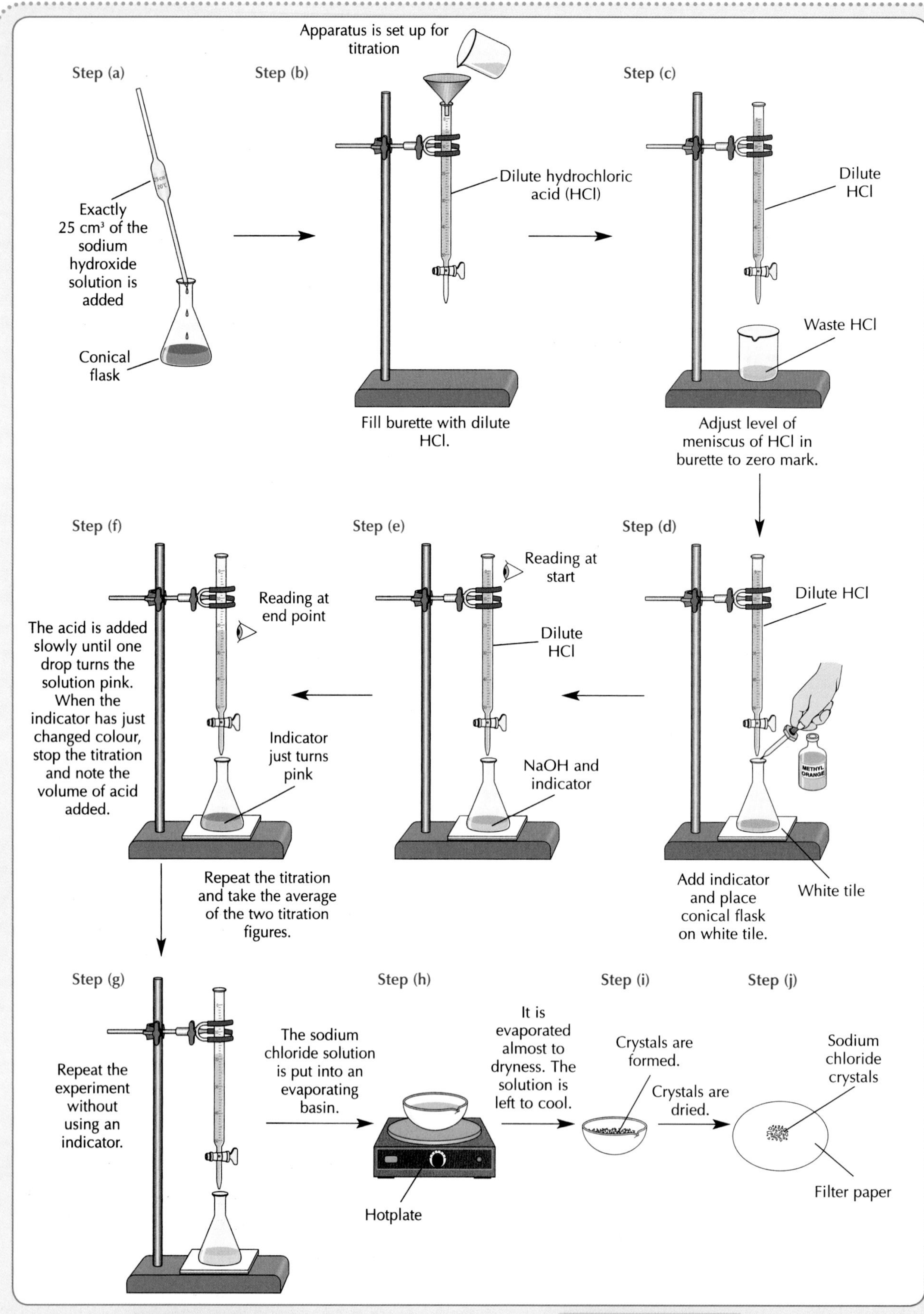

Fig. 28.10 Dilute hydrochloric acid is added from the burette into the conical flask containing the sodium hydroxide solution. When the indicator just changes colour, the solution is neutral.

TEST YOURSELF:

Now attempt questions *28.1–28.2* and *W28.1–W28.3*.

28.4 Fossil Fuels

Burning was one of the first chemical reactions used by humans. Any substance that burns in oxygen to produce heat is called a **fuel**.

A fuel is any substance that burns in oxygen to produce heat.

We now understand that burning is a chemical reaction in which a fuel is reacting with oxygen to give out heat. Burning is still a very important process in today's world. We burn fuels to keep us warm, to cook food, to generate electricity, to run our cars, etc.

Examples of Fuels

Each of us is familiar with many examples of fuels.

- Examples of **solid** fuels are **coal**, **turf**, **wood**, etc.
- Examples of **liquid** fuels are **petrol**, **diesel**, **paraffin oil**, etc.
- A common example of a fuel that is a **gas** is natural gas. Natural gas is mainly **methane**. The chemical formula for methane is **CH_4**.

Formation of Fossil Fuels

The most commonly used fuels in the world are coal, oil and natural gas. The energy stored in these fuels originally came from the sun.

- **Coal.** Two hundred million years ago, huge trees and ferns grew on swampy land trapping the sun's energy by the process of photosynthesis. When the plants died, layers of earth and rock pressed them down to form coal, *Fig. 28.11*.
- **Oil and natural gas.** Also at this time, the sea was inhabited by huge numbers of tiny shell creatures. The pressure of the earth and sea on these decaying remains turned them into our supply of oil and natural gas. This is the reason why oil and natural gas are often found together.

Coal, oil and natural gas are the three main **fossil fuels**.

Fossil fuels are fuels that were formed from the remains of plants and animals that lived millions of years ago.

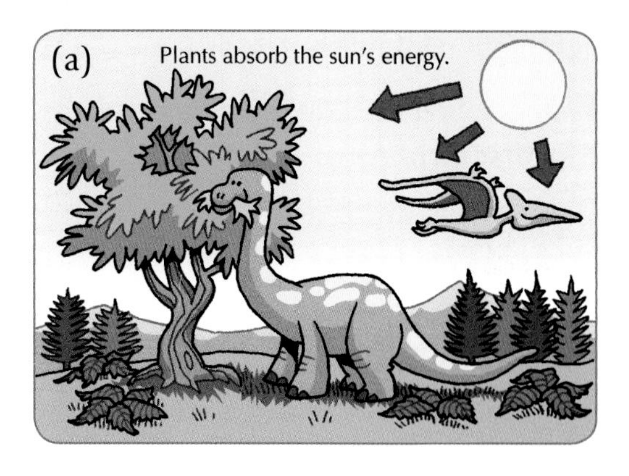

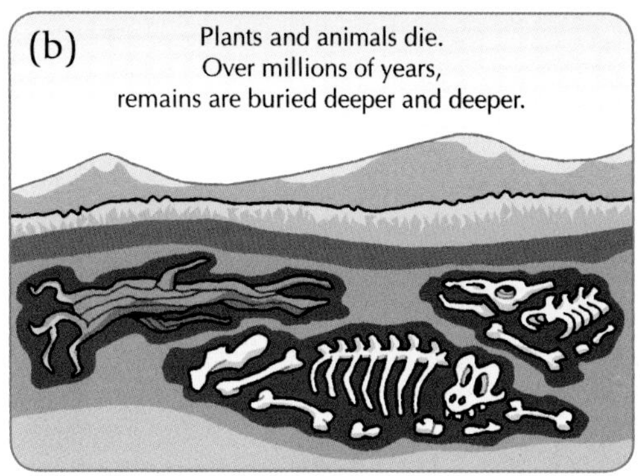

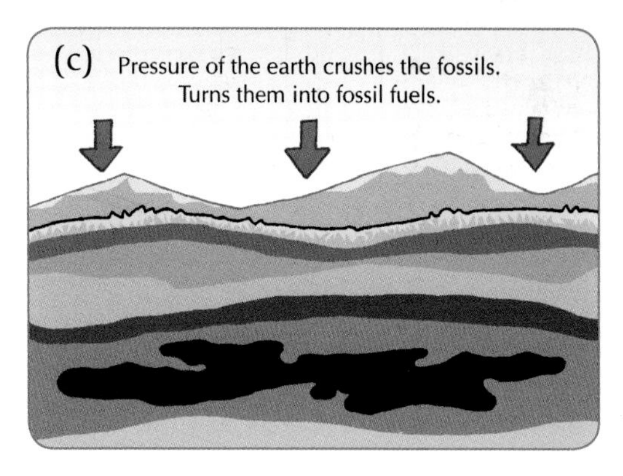

Fig. 28.11 Fossil fuels were formed over millions of years from the remains of plants and animals. The three most common fossil fuels are coal, oil and natural gas.

Hydrocarbons

Although the three types of fossil fuel look completely different, they all contain the same sort of chemicals. These chemicals are called **hydrocarbons**.

Fossil fuels contain hydrocarbons. Hydrocarbons are compounds consisting of hydrogen and carbon only.

Burning Hydrocarbons

When hydrocarbons are burned, carbon dioxide and water are always formed. This is shown in *Fig. 28.12*. The candle wax that is used to make the night light contains hydrocarbons.

- **Water.** When the liquid in the U-tube is tested, it turns cobalt chloride paper from blue to pink. It also turns anhydrous copper sulfate from white to blue. Therefore water has been formed.
- **Carbon dioxide.** The limewater turns milky showing that carbon dioxide has also been formed.

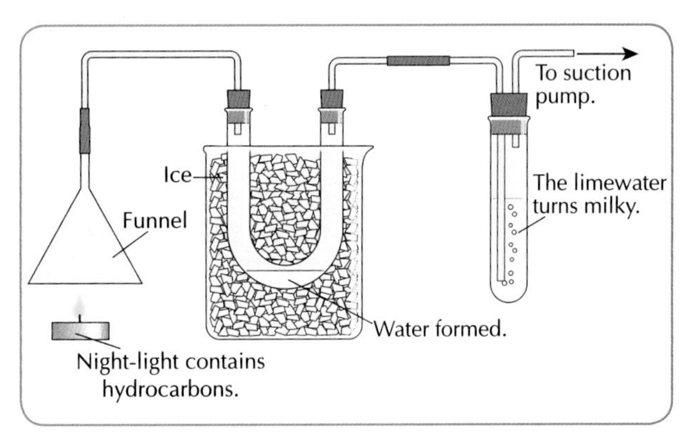

Fig. 28.12 This experiment shows that carbon dioxide and water are formed when hydrocarbons are burned. The liquid in the U-tube turns cobalt chloride paper from blue to pink: this shows that water has been formed. The limewater turns milky: this shows that carbon dioxide has been formed.

We can summarise our results as follows:

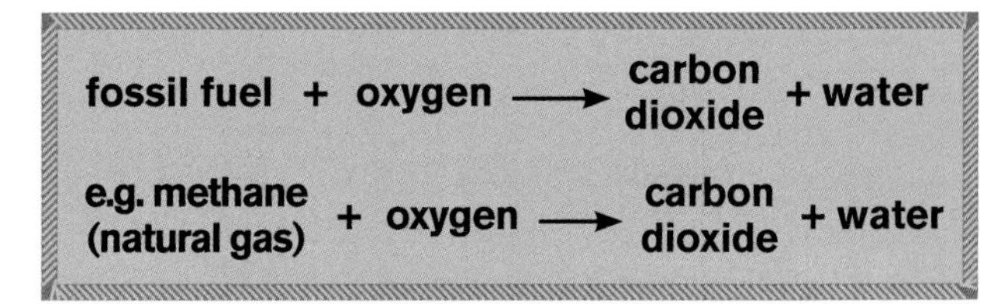

TEST YOURSELF:
Now attempt questions 28.3 and W28.4–W28.5.

28.5 Acid Rain

Rain is naturally slightly acidic. The very small amount of carbon dioxide in the air reacts with rain to form a dilute solution of carbonic acid with a pH of about 5.5. In central Europe, however, rain is far more acidic with an average pH of around 4.1. Rainwater with a pH of less than 5.5 is described as **acid rain**. This rain contains dilute sulfuric acid and nitric acid. We now discuss how these two acids get into the air.

Sulfuric Acid

Sulfuric acid comes from sulfur dioxide (SO_2) in the air. How does this sulfur dioxide get into the air? Most of the sulfur dioxide in the air comes from burning fossil fuels, *Fig. 28.13*.

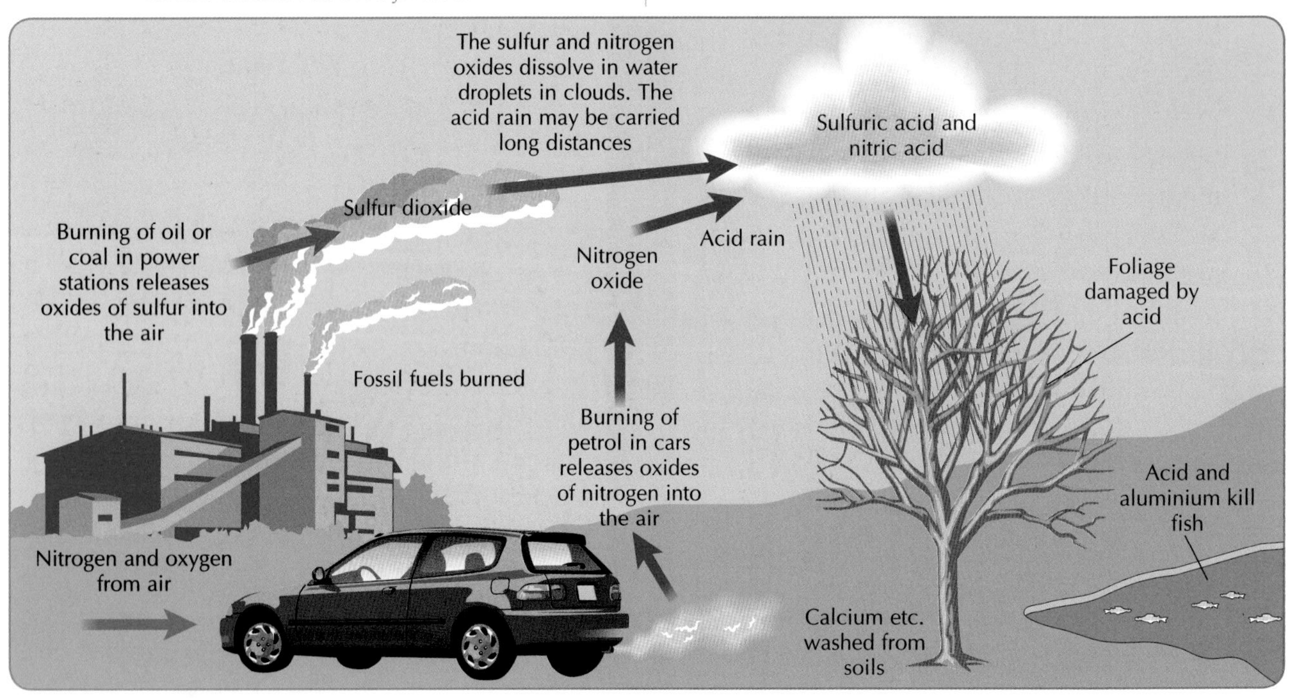

Fig. 28.13 Acid rain is formed when sulfur dioxide and oxides of nitrogen dissolve in water. Acid rain contains sulfuric acid and nitric acid.

Formation of Sulfuric Acid

Although natural gas contains little sulfur, some oil and coal can contain between 1–3% sulfur.

1. When **coal and oil are burned** (in power stations and factories), **sulfur dioxide** is produced. Sulfur dioxide is a colourless, choking gas that irritates our eyes and lungs.
2. Sulfur dioxide is very soluble in water. It dissolves in **water** to eventually form **sulfuric acid**.

Nitric Acid

Where does the nitric acid in acid rain come from? The nitric acid comes from the emission from exhausts of motor cars.

Formation of Nitric Acid

1. Car engines take in **nitrogen and oxygen** from the air. Car exhausts emit **oxides of nitrogen**.
2. These oxides of nitrogen dissolve in **water** vapour in clouds. This forms **nitric acid**.

Problems of Acid Rain

Acid rain gives rise to many problems:

- **Acid rain destroys lakes (killing fish)**. It washes large quantities of aluminium salts into streams and lakes. The aluminium interferes with the operation of the gills of the fish and they die as a result. Clouds may be carried hundreds of kilometres before they drop their water. Scientists have made a link between the pollution blown across the sea from Britain and other parts of Europe with acidified **lakes and streams** in the Scandinavian countries. Fish life is disappearing from these waters, *Fig. 28.14*.

Fig. 28.14 Lime is being sprayed on this lake to reduce the acidity in the lake.

- **Acid rain harms trees**. It damages foliage and removes many minerals from the soil. These minerals should be absorbed by the roots of trees. Therefore, many trees die.
- **Acid rain attacks stone and iron**. Many **buildings** suffer from acid rain because limestone reacts with acid rain. The acid rain causes the limestone to wear away. A similar problem occurs with statues made of limestone, *Fig. 28.15*.

Fig. 28.15 This statue is made from limestone. Acid rain has damaged it as acids react with limestone. This wears away the statue.

Solving the Problems of Acid Rain

What is the solution to the problem of acid rain?

(1) Many countries are installing chemical plants to **remove the sulfur dioxide** from emissions of power stations.
(2) A second way is to **reduce the sulfur** content of fuels like oil and gas. This, however, makes these fuels more expensive.
(3) A third way is to install **catalysts** in the exhaust systems of cars. These devices are called catalytic converters. They contain catalysts that remove the harmful oxides of nitrogen from the exhaust fumes.
(4) A fourth way is to **burn less fossil fuels** and use other forms of energy, e.g. wind energy, hydroelectric energy, etc.

TEST YOURSELF:

Now attempt questions *28.4* and *W28.6*.

What I should know

- The two common acids in school laboratories are hydrochloric acid and sulfuric acid.
- The most common bases are sodium hydroxide and calcium hydroxide.
- acid + base ⟶ salt + water
- The reaction of an acid with a base is called a neutralisation reaction.
- A salt may be prepared in the laboratory using the process of titration.
- A fuel is any substance that burns in oxygen to produce heat.
- Fossil fuels are fuels that were formed from the remains of plants and animals that lived millions of years ago.
- The three main **fossil fuels** are coal, oil and natural gas.
- Fossil fuels contain hydrocarbons. Hydrocarbons are compounds that contain hydrogen and carbon only.
- Hydrocarbons burn to form carbon dioxide and water.
- Acid rain is a mixture of sulfuric acid and nitric acid.
- Acid rain destroys lakes (killing fish), harms trees and attacks stone.

Questions

Write the answers to the following questions into your copybook.

28.1 (a) Name two acids commonly found in the school laboratory.

(b) Give an example of (i) a strong acid (ii) a weak acid.

(c) What sorts of substances are used to cure indigestion?

(d) The treatment for a wasp sting is to dab it with some vinegar. What can you deduce about a wasp sting from this information?

(e) Why is *Milk of Magnesia* used to treat indigestion?

(f) Some acid from a car battery is spilled on the garage floor. Name one common household substance which could be used to neutralise the acid.

(g) Adding an acid to marble chips gives off a gas. Name the gas.

(h) Give the chemical formula for hydrochloric acid.

(i) What is the common name given to sodium hydroxide?

(j) Calcium hydroxide is used to test for the presence of carbon dioxide. What name is commonly used to describe a solution of calcium hydroxide in water?

28.2 (a) Ant stings contain methanoic acid. How would you treat these stings?

(b) Bacteria on our teeth produce acidic substances. What do these substances do to our teeth? How does brushing our teeth help to get rid of the problem?

(c) The old saying 'vinegar for vasps (wasps), bicarbonate for bees' is based on scientific fact. Explain the science behind the old saying.

(d) Explain why rubbing dock leaves on a nettle sting helps to stop the pain.

(e) Complete the following: acid + base ⟶ ..... + ...

QUESTIONS

(f) A road tanker carrying sulfuric acid was involved in a crash. This caused some of the acid to spill on the road, *Fig. 28.16*. Explain why the firemen are shovelling lime on to the spill.

Fig. 28.16 Why are the firemen putting lime on the spilled sulfuric acid?

28.3 (a) What is a fuel?

(b) Give an example of a solid fuel.

(c) Name the gaseous fuel commonly used for cooking in the home.

(d) Oil and natural gas are usually found together. Why is this?

(e) What are hydrocarbons?

(f) Name the hydrocarbon found in natural gas.

(g) What two substances are formed when hydrocarbons are burned?

(h) Why is coal called a 'fossil fuel'?

28.4 (a) Why is unpolluted rain water normally acidic.

(b) How is acid rain formed?

(c) Where does sulfur dioxide in the air come from?

(d) How do cars give rise to the problem of acid rain?

(e) How are countries like Norway and Sweden affected by the acid rain problem in England?

(f) Name two ways in which the amounts of acid rain can be reduced.

(g) How does acid rain affect fish in lakes?

(h) What effect does acid rain have on limestone in buildings?

(i) Where would you normally find a catalytic converter?

Chapter 29 Atomic Structure

29.1 Where Did the Idea of Atoms Come from?

Scientists have discovered that the materials around us are made up of millions of tiny particles. These particles are called **atoms**. You may have heard the word atom before. A lot is heard nowadays about atomic power and atomic bombs.

Definition of the Atom

The word atom comes from the Greek for 'cannot be split up' or 'cannot be divided'. For example, a piece of copper could be cut up into smaller and smaller pieces until eventually, a stage would be reached when the smallest possible piece of the element copper is produced. This smallest piece of copper is called an **atom** of copper.

An atom is the smallest particle of an element which still retains the properties of that element.

For example, an atom of copper is the smallest piece of copper which shows all the properties of copper.

Each element is made up of atoms of that element only. We learned in *Chapter 21* that there are about 100 elements in the Periodic Table. Therefore, there are about 100 different kinds of atoms.

Size of the Atom

Atoms are very small. The full stop at the end of this sentence contains about 10^{16} atoms of carbon. This number is greater than the entire population of the world!

Seeing is Believing

Because atoms are so small, it was difficult for scientists to prove they existed. In the past, it was not possible to see atoms even with the aid of a powerful microscope. In recent times, however, it has been possible to see some atoms. To see atoms a special type of microscope called an 'electron microscope' is used.

Dalton's Theory of Atoms

An English scientist called John Dalton put forward his own ideas about atoms in 1808. Dalton performed many experiments. He put forward his theory of atoms to explain the results of his experiments.

John Dalton (1766–1844) put forward a theory about atoms in 1808.

Dalton imagined atoms to be like marbles or billiard balls, only very much smaller, *Fig. 29.1*. This was a very simple model. He imagined an atom as being something tiny, round and hard. He also thought that atoms could not be divided into anything simpler.

Fig. 29.1 Dalton imagined atoms to be like tiny billiard balls. He thought that atoms could not be divided into anything simpler.

29.2 Particles Inside Atoms

In the 1890s, a number of scientists began to explore atoms in more detail. They found that Dalton was incorrect when he said that atoms could not be split up into anything simpler. They found that there were tiny particles inside an atom. These particles are called **sub-atomic particles**. The word 'sub-atomic' means 'inside the atom'.

- One of these particles was discovered in 1897. It had a negative charge on it and was called the **electron**.
- Another particle with a positive charge was later discovered: this particle was called the **proton**.
- In 1932 it was discovered that there was a third particle inside the atom. This particle had no charge on it – it is neutral. For this reason it was called the **neutron**.

Properties of Sub-atomic Particles

A lot of information is now known about protons, neutrons and electrons. The properties of these particles are summarised in *Table 29.1*.

NAME OF PARTICLE	WHERE SITUATED IN ATOM	RELATIVE MASS	RELATIVE CHARGE
Proton	Nucleus	1 unit	+1 unit
Neutron	Nucleus	1 unit	0
Electron	Outside the nucleus	$^1/_{1840}$ unit	−1 unit

Table 29.1 Summary of the properties of the sub-atomic particles.

Mass

The masses of these particles are so small that, if expressed in grams, there would be about 23 zeros after the decimal point before the first digit of the number (mass of proton = 1.6×10^{-24} g). Therefore, we use a new unit of mass called the **atomic mass unit** (a.m.u.). On this scale, the mass of the proton or the neutron is 1 a.m.u. The electron has the smallest mass of the sub-atomic particles. The mass of the electron is only $^1/_{1840}$ of the mass of the proton, *Fig. 29.2*.

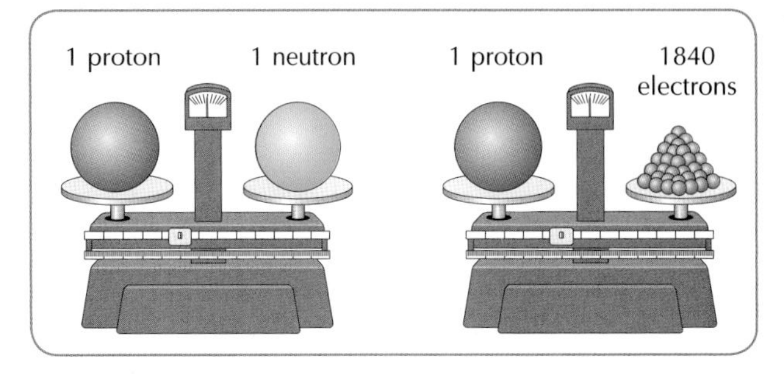

Fig. 29.2 Protons and neutrons have the same mass. The electron has a very small mass compared to that of the proton and neutron.

Charge

The electron carries a small charge of electricity. We say that the electron carries a unit of negative charge or a charge of −1. The proton also carries a unit of electrical charge. In this case the charge is positive, +1. The neutron has no charge.

TEST YOURSELF:
Now attempt questions *29.1–29.3* and *W29.1–W29.3*.

29.3 Structure of the Atom

The Planetary Model

We have already learned that the protons and neutrons are found in the nucleus of the atom. In 1913, a Danish scientist called Niels Bohr proposed that the electrons move around the nucleus in fixed paths called **orbits**. (These orbits are also called shells).

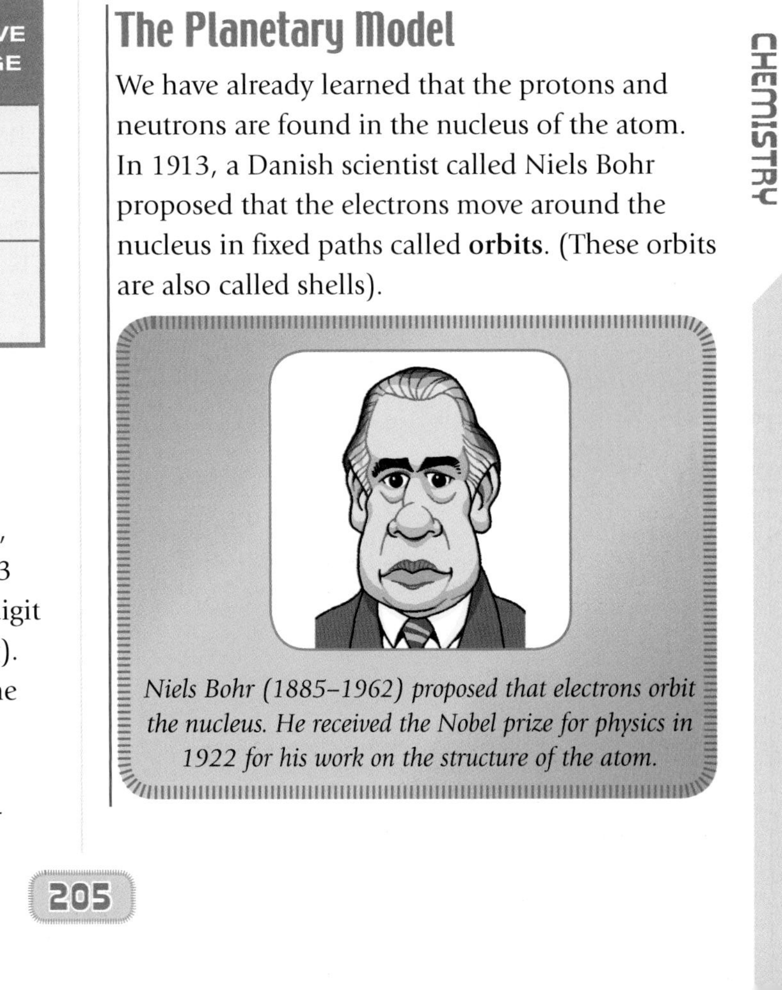

Niels Bohr (1885–1962) proposed that electrons orbit the nucleus. He received the Nobel prize for physics in 1922 for his work on the structure of the atom.

Bohr's model is often called the planetary model. The electrons revolve around the nucleus rather like the planets move around the sun. The movement of the electrons around the nucleus gives rise to an **electron cloud**.

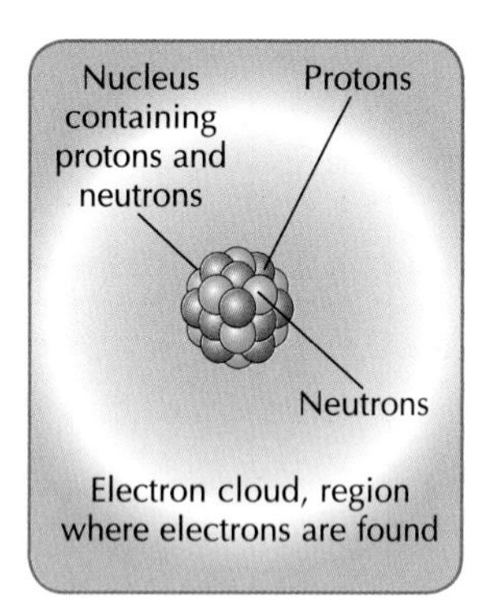

Fig. 29.3 The protons and neutrons are in the nucleus of the atom. The electrons revolve around the nucleus.

Size of the Nucleus

Scientists have calculated how the size of the nucleus compares to the size of the atom itself. They have concluded that most of the atom consists of empty space. To understand this, imagine an atom made bigger until it is the size of Croke Park. The nucleus could be represented by a marble at the centre of the pitch. The electrons would be like specks of dust at various parts of the stands. Everything in between would be just empty space. The electrons are free to move in this space. (For this reason, any diagrams drawn in this textbook to indicate the structure of the atom are purely descriptive and not to scale).

29.4 Atomic Structure of the First 20 Elements

There are a number of important points about atoms. These are summarised as follows:

Number of Electrons

All atoms are neutral. Therefore, the number of protons (positive) must be the equal to the number of electrons (negative).

Atomic Number

The number of protons in the nucleus of an atom is given a special name. This number is called the **atomic number** of the atom.

The atomic number of an atom is the number of protons in the nucleus of that atom.

Each element has its own atomic number. For example, the atomic number of carbon is 6. This means that an atom of carbon has 6 protons in the nucleus. The atomic number of copper is 29. This means that an atom of copper has 29 protons in the nucleus.

Mass Number

The mass of an atom is obtained by adding together the number of protons and neutrons. (The number of electrons is ignored since electrons have such a small mass). The **mass number**, as the name implies, gives us information about the mass of an atom. The mass number of an atom is the sum of the number of protons and neutrons.

mass number = number of protons + number of neutrons

Nuclear Formula

The atomic number and the mass number are often written with the symbol of the element as shown for aluminium in *Fig. 29.4*.

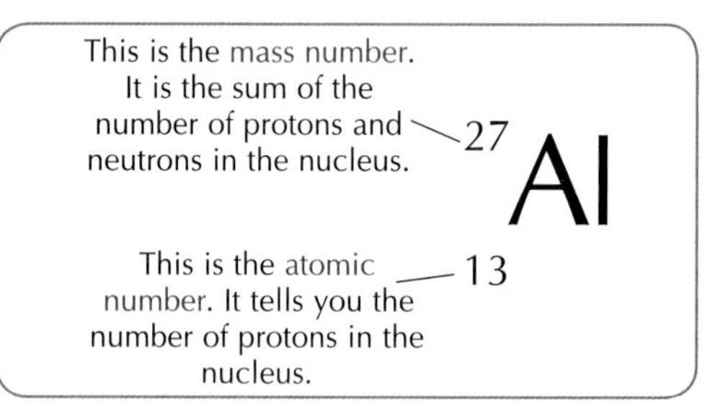

Fig. 29.4 Writing the nuclear formula for an atom gives us information about its structure.

This shorthand method of showing the atomic number and the mass number of an atom is called the nuclear formula. It is so called because it gives us information about the nucleus of the atom. For example, an atom of aluminium may be written as $^{27}_{13}Al$. This gives us the following information:

- **Position in the periodic table**. Since the atomic number is 13, this means that aluminium is the 13th element in the Periodic Table.

- **Number of protons and electrons**. Atomic number of aluminium = 13. This means that an atom of aluminium contains 13 protons and 13 electrons.
- **Number of protons and neutrons**. Mass number of aluminium = 27. This means that the sum of protons plus neutrons = 27.
- **Number of neutrons**. We know that the mass of the atom is 27. We know there are 13 protons in the atom. Therefore, there must be 14 neutrons with the protons so that their sum will be 27. Number of neutrons = 27 – 13 = 14.

In short, the atomic number subtracted from the mass number tells us the number of neutrons.

number of neutrons = mass number - atomic number.

Structure of First 20 Elements

For your examination, you must be able to work out the number of protons, neutrons and electrons in the atoms of the first 20 elements. The structure of the atoms of the first 20 elements are given in *Table 29.2*. Study this table carefully. Note that hydrogen is the only element that has no neutrons in its nucleus.

Do not worry about how to write down the nuclear formula of each element. This can be done using the Periodic Table. This will be explained in the next chapter.

TEST YOURSELF:

Now attempt questions *29.4–29.6* and *W29.4–W29.6*

ELEMENT	ATOMIC NUMBER	NUMBER OF PROTONS	NUMBER OF ELECTRONS	MASS NUMBER	NUMBER OF NEUTRONS (= MASS NO.–ATOMIC NO.)
$^{1}_{1}H$	1	1	1	1	1–1 = 0
$^{4}_{2}He$	2	2	2	4	4–2 = 2
$^{7}_{3}Li$	3	3	3	7	7–3 = 4
$^{9}_{4}Be$	4	4	4	9	9–4 = 5
$^{11}_{5}B$	5	5	5	11	11–5 = 6
$^{12}_{6}C$	6	6	6	12	12–6 = 6
$^{14}_{7}N$	7	7	7	14	14–7 = 7
$^{16}_{8}O$	8	8	8	16	16–8 = 8
$^{19}_{9}F$	9	9	9	19	19–9 = 10
$^{20}_{10}Ne$	10	10	10	20	20–10 = 10
$^{23}_{11}Na$	11	11	11	23	23–11 = 12
$^{24}_{12}Mg$	12	12	12	24	24–12 = 12
$^{27}_{13}Al$	13	13	13	27	27–13 = 14
$^{28}_{14}Si$	14	14	14	28	28–14 = 14
$^{31}_{15}P$	15	15	15	31	31–15 = 16
$^{32}_{16}S$	16	16	16	32	32–16 = 16
$^{35}_{17}Cl$	17	17	17	35	35–17 = 18
$^{40}_{18}Ar$	18	18	18	40	40–18 = 22
$^{39}_{19}K$	19	19	19	39	39–19 = 20
$^{40}_{20}Ca$	20	20	20	40	40–20 = 20

Table 29.2 The structure of atoms of the first 20 elements.

What I should know

- An atom is the smallest particle of an element which still retains the properties of that element.
- There are about 100 different kinds of atoms.
- Atoms contain protons, neutrons and electrons.
- Protons and neutrons are found in the nucleus of the atom.
- Electrons move about the nucleus in paths called orbits or shells.
- Protons and neutrons have a mass of 1 a.m.u. Electrons have a mass of $^{1}/_{1840}$ a.m.u.
- Protons carry a charge of +1, electrons carry a charge of –1, neutrons carry no charge.
- There are equal numbers of protons and electrons in an atom of an element.
- The atomic number of an atom is the number of protons in the nucleus of that atom.
- The mass number of an atom is the sum of the number of protons and neutrons.
- The number of neutrons in an atom is calculated by subtracting the atomic number from the mass number.

Questions

Write the answers to the following questions into your copybook.

29.1 (a) What is meant by the term *atom*?

(b) The particles inside an atom are referred to as ..... .

(c) Protons and neutrons are found in the ..... .

(d) A proton has a ..... charge, but an electron has a ..... charge.

(e) The neutron is so called because it is a ..... particle.

(f) Both the proton and the neutron have a mass of ..... but the mass of the electron is only ..... of the mass of the proton.

29.2 Fill in the blanks in *Table 29.3*.

NAME OF PARTICLE	WHERE SITUATED IN ATOM	RELATIVE MASS	RELATIVE CHARGE
Proton			
	Nucleus		0
	Outside the nucleus		-1 unit

Table 29.3

29.3 (a) ..... was a famous scientist who put forward an atomic theory in 1808.

(b) There are about ..... kinds of atoms.

(c) Give two words to describe Dalton's picture of the atom.

(d) The central part of the atom is called the ..... .

(e) Dalton said that atoms were indivisible. Was he correct? Explain your answer.

(f) An atom must contain equal numbers of positively charged ..... and negatively charged ..... .

QUESTIONS

(g) Which sub-atomic particle has a negative charge?

(h) Of the three sub-atomic particles, the lightest one is the ..... .

(i) Give the name of the sub-atomic particle that has no charge.

29.4 (a) Name the Danish scientist who proposed that electrons move around the nucleus.

(b) The paths in which electrons move around the nucleus are called ..... .

(c) What is meant by the term *atomic number*?

(d) The shorthand method used to show the atomic number and the mass number of an atom is called the ..... .

(e) If an atom has 6 protons, how many electrons does it have?

(f) What information does the mass number give about the atoms of an element?

29.5 (a) To calculate the number of neutrons in an atom, you subtract the atomic number from the ..... number.

(b) The atomic number of carbon is 6. Give two pieces of information that this gives you about the carbon atom.

(c) In a neutral atom the number of ..... is always equal to the number of ..... .

(d) There is only one element in the Periodic Table which has no neutrons in its nucleus. Name this element.

(e) Write down the atomic number of (i) oxygen ..... and (ii) magnesium ..... .

(f) There are ..... electrons in an atom of chlorine.

29.6 (a) An atom with 17 protons and 18 neutrons has an atomic number of ..... and a mass number of ..... .

(b) An element has a mass number of 16 and an atomic number of 8. Write down the nuclear formula for the element.

(c) The atoms of a certain element contain 15 protons and 16 neutrons. What is the mass number of the element? Name the element.

(d) List two items of information which the atomic number gives us about the atoms of an element.

(e) The nuclear formula of potassium is $^{39}_{19}K$. This tells us that there are ..... protons and ..... neutrons in the nucleus of the atom.

(f) An atom of element *X* has an atomic number of 18 and a mass number of 40. How many (i) protons; (ii) neutrons; and (iii) electrons are there in the atom. What element is *X*?. Write the nuclear formula of the atom.

Chapter 30 The Periodic Table

30.1 What is the Periodic Table?

In *Chapter 21* we learned that all of the elements that have been discovered are listed in a table called the Periodic Table. A copy of the Periodic Table is listed on the inside back cover of this textbook. A simplified version of the Periodic Table is shown in *Fig. 30.1*.

Repeating Properties

The word 'periodic' means 'at regular intervals'. You may have heard the word 'periodical' before. A magazine that is published at regular intervals, e.g. every month or every week, is called a 'periodical'. The Periodic Table is so called because the properties of the elements repeat at regular intervals. For example, the properties of lithium (element number 3) are repeated in sodium (element number 11) and also in potassium (element number 19).

Order of Atomic Number

The Periodic Table is simply a list of elements in order of the number of protons in the nucleus of each element, i.e. in order of atomic number.

The Periodic Table is an arrangement of elements in order of increasing atomic number.

This type of table was first drawn up in 1869 by a Russian chemist called Dmitri Mendeleev.

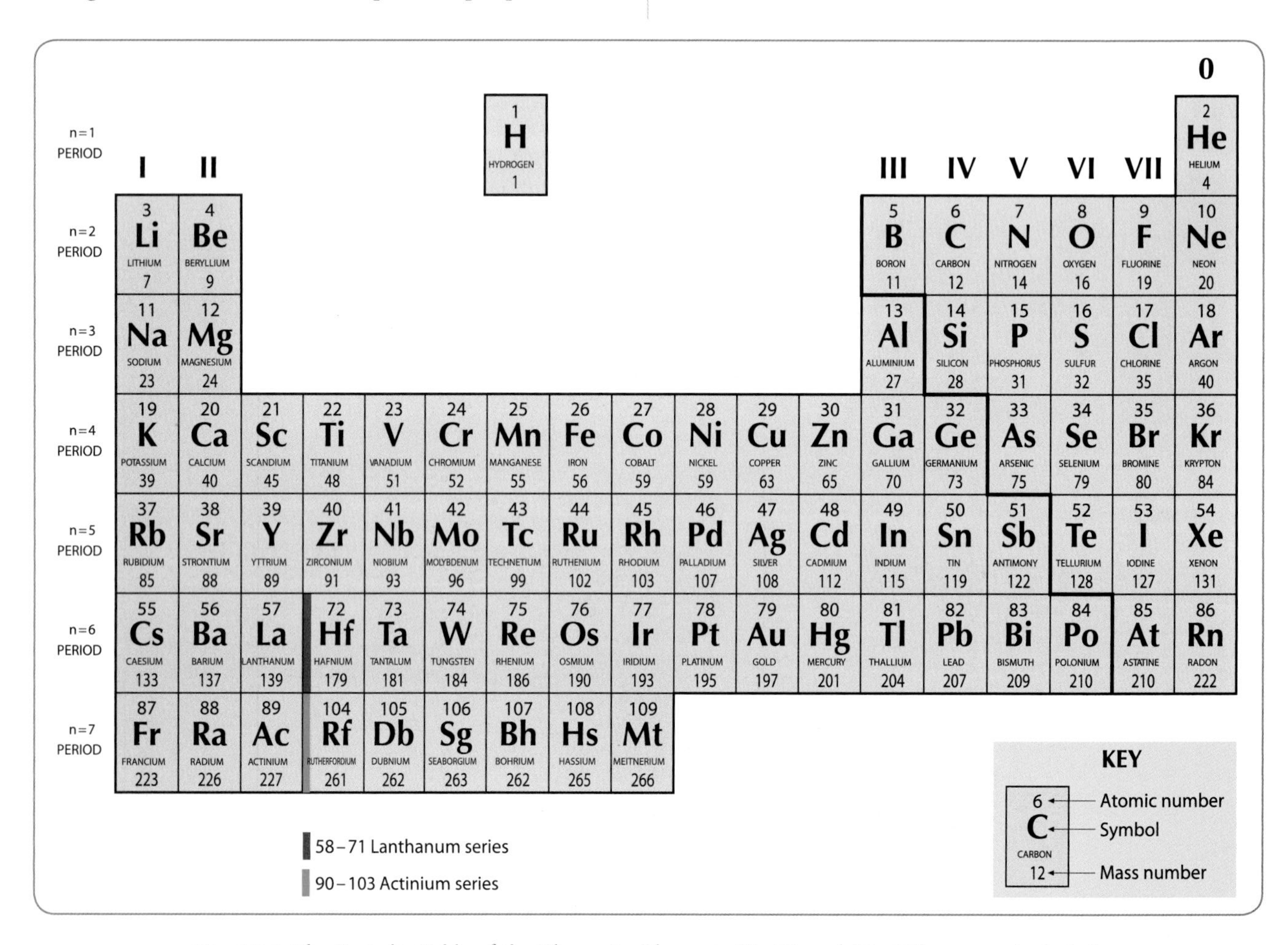

Fig. 30.1 The Periodic Table of the Elements. Elements 58–71 and 90–103 are not included.

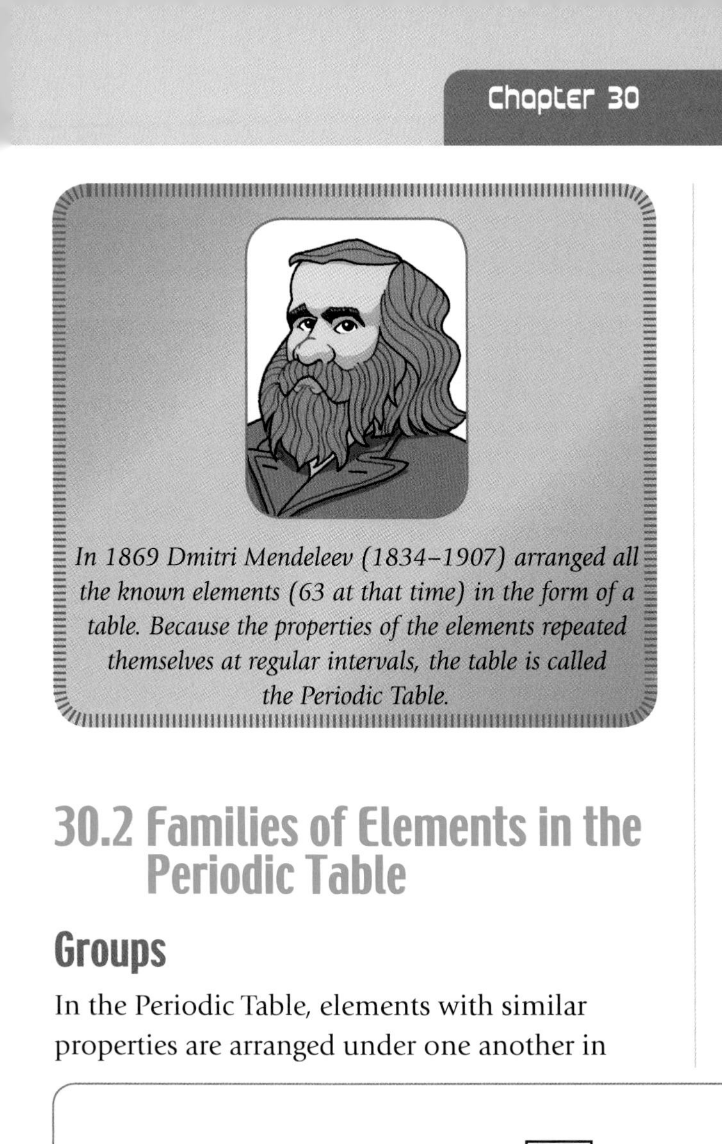

In 1869 Dmitri Mendeleev (1834–1907) arranged all the known elements (63 at that time) in the form of a table. Because the properties of the elements repeated themselves at regular intervals, the table is called the Periodic Table.

30.2 Families of Elements in the Periodic Table

Groups

In the Periodic Table, elements with similar properties are arranged under one another in vertical columns. These vertical columns are called **groups** or **families**, *Fig. 30.2.*

- If you study the Periodic Table, you will see that there are **eight main groups** in the Periodic Table.
- The groups are often numbered using the old **Roman numerals** I, II, III, IV, V, VI, VII.
- The group on the extreme right of the Periodic Table is called either Group **VIII** or Group **0**.
- All elements in the same group have **similar chemical properties**. They are like members of the same family that have similar appearance, e.g. colour of hair, colour of eyes, etc.
- All elements in the same group have the **same number of electrons in their outer orbit or shell**. It is these electrons which determine the chemical properties of the elements. (We will learn more about this in the next chapter.)

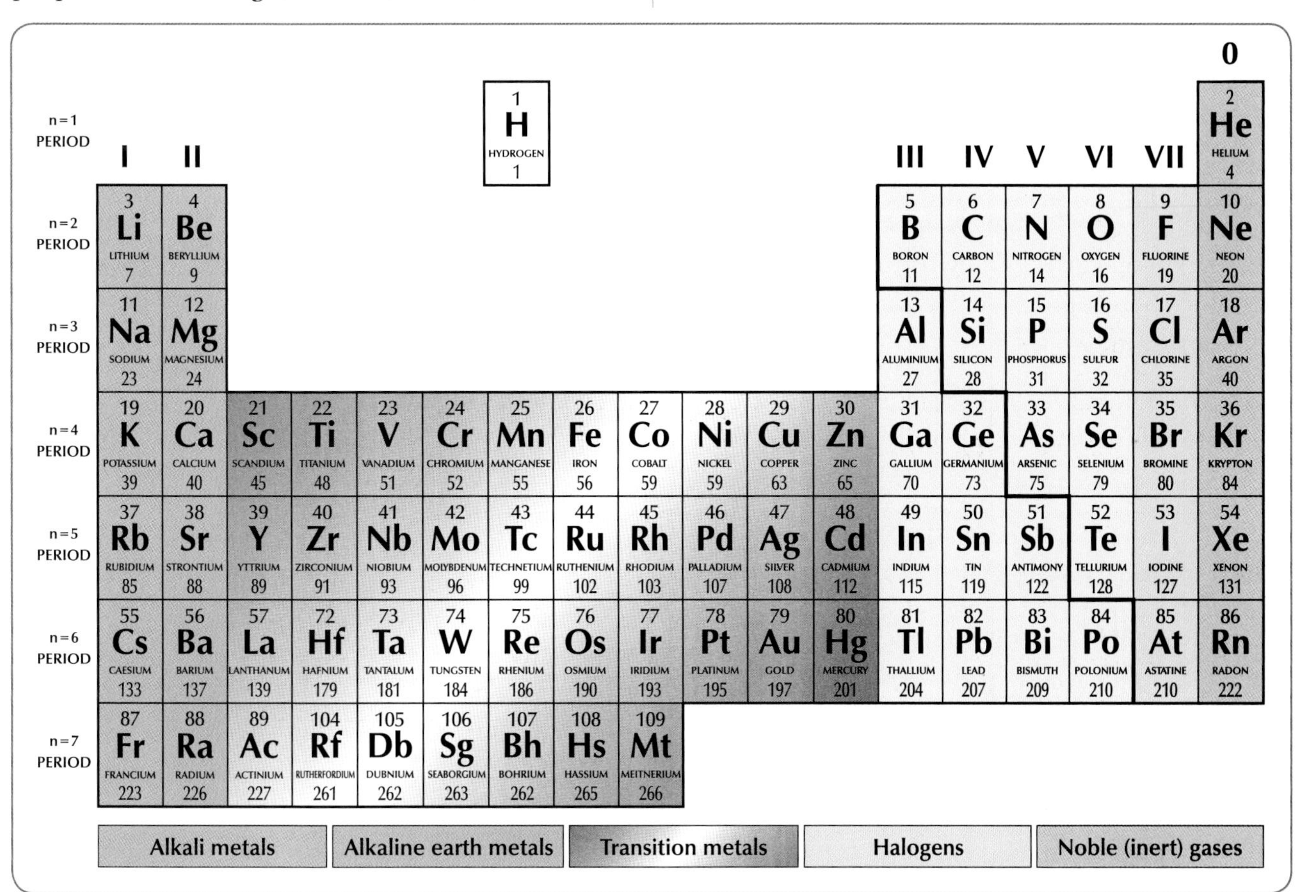

	I	II											III	IV	V	VI	VII	0
n=1 PERIOD							1 H HYDROGEN 1											2 He HELIUM 4
n=2 PERIOD	3 Li LITHIUM 7	4 Be BERYLLIUM 9											5 B BORON 11	6 C CARBON 12	7 N NITROGEN 14	8 O OXYGEN 16	9 F FLUORINE 19	10 Ne NEON 20
n=3 PERIOD	11 Na SODIUM 23	12 Mg MAGNESIUM 24											13 Al ALUMINIUM 27	14 Si SILICON 28	15 P PHOSPHORUS 31	16 S SULFUR 32	17 Cl CHLORINE 35	18 Ar ARGON 40
n=4 PERIOD	19 K POTASSIUM 39	20 Ca CALCIUM 40	21 Sc SCANDIUM 45	22 Ti TITANIUM 48	23 V VANADIUM 51	24 Cr CHROMIUM 52	25 Mn MANGANESE 55	26 Fe IRON 56	27 Co COBALT 59	28 Ni NICKEL 59	29 Cu COPPER 63	30 Zn ZINC 65	31 Ga GALLIUM 70	32 Ge GERMANIUM 73	33 As ARSENIC 75	34 Se SELENIUM 79	35 Br BROMINE 80	36 Kr KRYPTON 84
n=5 PERIOD	37 Rb RUBIDIUM 85	38 Sr STRONTIUM 88	39 Y YTTRIUM 89	40 Zr ZIRCONIUM 91	41 Nb NIOBIUM 93	42 Mo MOLYBDENUM 96	43 Tc TECHNETIUM 99	44 Ru RUTHENIUM 102	45 Rh RHODIUM 103	46 Pd PALLADIUM 107	47 Ag SILVER 108	48 Cd CADMIUM 112	49 In INDIUM 115	50 Sn TIN 119	51 Sb ANTIMONY 122	52 Te TELLURIUM 128	53 I IODINE 127	54 Xe XENON 131
n=6 PERIOD	55 Cs CAESIUM 133	56 Ba BARIUM 137	57 La LANTHANUM 139	72 Hf HAFNIUM 179	73 Ta TANTALUM 181	74 W TUNGSTEN 184	75 Re RHENIUM 186	76 Os OSMIUM 190	77 Ir IRIDIUM 193	78 Pt PLATINUM 195	79 Au GOLD 197	80 Hg MERCURY 201	81 Tl THALLIUM 204	82 Pb LEAD 207	83 Bi BISMUTH 209	84 Po POLONIUM 210	85 At ASTATINE 210	86 Rn RADON 222
n=7 PERIOD	87 Fr FRANCIUM 223	88 Ra RADIUM 226	89 Ac ACTINIUM 227	104 Rf RUTHERFORDIUM 261	105 Db DUBNIUM 262	106 Sg SEABORGIUM 263	107 Bh BOHRIUM 262	108 Hs HASSIUM 265	109 Mt MEITNERIUM 266									

Alkali metals | Alkaline earth metals | Transition metals | Halogens | Noble (inert) gases

Fig. 30.2 Some groups in the Periodic Table are given special names as shown in this diagram. Groups III, IV, V, and VI are often referred to simply as the boron group, carbon group, nitrogen group and oxygen group, respectively.

All elements in the same group of the Periodic Table have similar chemical properties.

Therefore, if you know the chemical properties of one element in a group, you could make a good guess about the properties of the other elements in the group. For example, knowing that sodium reacts vigorously with water implies that potassium does likewise, since both are found in Group 1.

Periods

The horizontal rows of elements are called **periods**.

- There are **seven** periods in the Periodic Table.
- The **first** period contains only **hydrogen** and **helium**.
- The **second** period contains the elements from **lithium to neon**.
- The periods are usually **labelled** n = 1, n = 2, etc.

Transition Metals

The block of elements in the centre of the Periodic Table are known as the **transition metals**. These contain many common metals e.g. iron, copper, gold, etc. We will ignore these elements for the moment.

Note: There are over 100 elements in the Periodic Table. In this course, we will only be dealing with the first 20 elements.

30.3 Drawing Bohr Structures of Atoms

For your examination, you need to be able to draw the Bohr structures of the first 20 elements in the Periodic Table. To draw the Bohr structure of an atom you need to be able to do two things:

(1) Use the Periodic Table to work out the number of protons and neutrons in the nucleus of the atom.
(2) Use the Periodic Table to show the arrangement of electrons in the atom.

(1) Working out the Number of Protons and Neutrons

Atomic Number

The atomic number of an element may easily be worked out from its position in the Periodic Table. For example, carbon is the sixth element in the Periodic Table. Therefore, its atomic number is six. In other words, it has six protons in its nucleus.

Mass Number

The mass number of an element may be deduced from the number underneath the symbol in the Periodic Table. For example, the mass number of carbon is 12 since the number 12 appears in the box underneath the symbol for carbon, *Fig. 30.3(a)*.

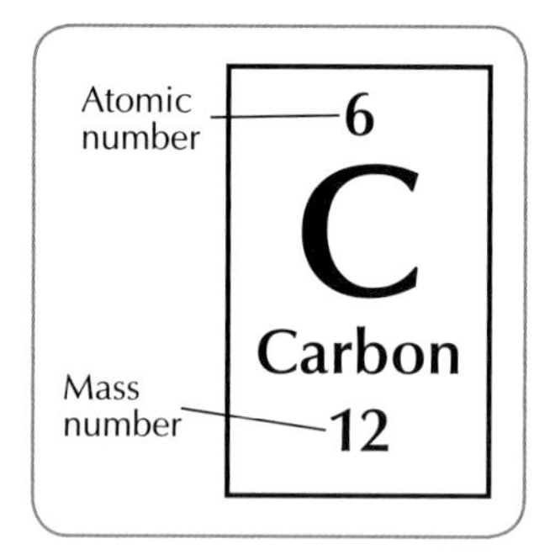

Fig. 30.3(a) The way an atom of carbon is represented in the Periodic Table.

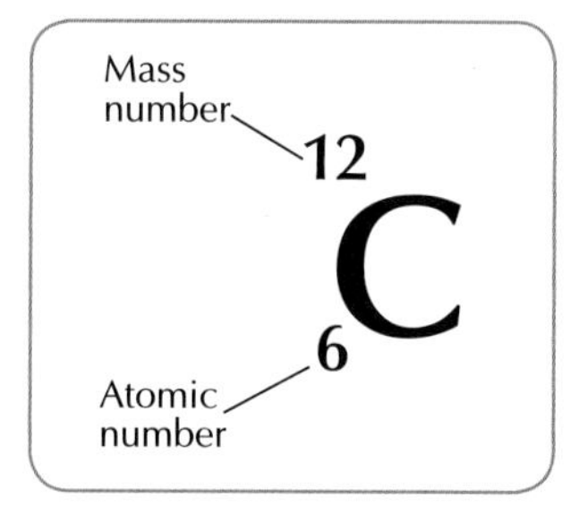

Fig. 30.3(b) The nuclear formula for an atom of carbon.

Note: In the Periodic Table, the atomic number appears on top of the symbol and the mass numbers appears below the symbol. This is different to the way that nuclear formulas are written, *Fig. 30.3(b)*. You will not get confused if you remember that the atomic number is always the smaller number.

Whole Numbers

Look closely at the first 20 elements in the Periodic Table inside the back cover of this textbook. Most of the numbers underneath the symbols are not whole numbers. To find out the mass number of an element, just round off the number under the symbol to the nearest whole number. For example, to find the mass number of helium, change the 4.003 to 4. To find the mass number for oxygen change the 15.999 to 16. The reason why the numbers are not whole numbers is explained in *Section 30.4*.

Number of Neutrons

To work out the number of neutrons in the atom, subtract the atomic number from the mass number as explained in the last chapter (*Table 29.2*, page 207).

(2) Show the Arrangement of Electrons in an Atom

We have already seen in *Chapter 29* that Niels Bohr proposed the idea of electrons orbiting the nucleus.

- The letter ***n*** is used to indicate the orbit. Thus, the first orbit is the n = 1 orbit, the second orbit is the n = 2 orbit, etc.
- The electron arrangement of an element is the number of electrons in each orbit. Another name for electron arrangement is **electron configuration**.
- The electron configuration of an element may easily be worked out from its position in the Periodic Table.

Example 1

Draw the Bohr structure of an atom of carbon. To do this carry out the following steps:

Step 1
Carbon is element number 6, i.e. its atomic number is 6. This means that there are six protons in the nucleus of carbon. Since the atom is neutral, there must also be six electrons in the atom. The mass number of carbon is 12. Therefore, there are 12 – 6 = 6 neutrons in the nucleus.

We must now work out how the six electrons are arranged around the nucleus. This is done by studying the Periodic Table.

Step 2
Carbon is in the second period. Therefore, we must draw two orbits.

1. Note that in the n = 1 period, there are only two elements, *Fig. 30.1*. This means that the n = 1 orbit can only hold two electrons.
2. Start at the second period and count along from lithium. Carbon has four electrons in the n = 2 orbit.

Step 3
We now know that carbon has two electrons in the n = 1 orbit and four electrons in the n = 2. Therefore we can write the electron configuration of carbon as follows:

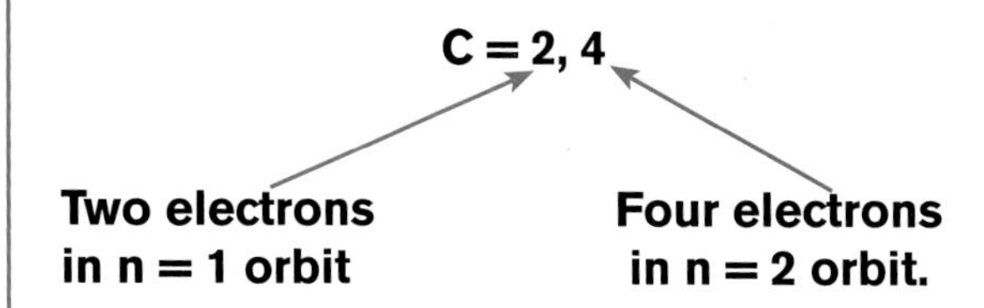

We can now draw the Bohr structure of an atom of carbon as shown in *Fig. 30.4*.

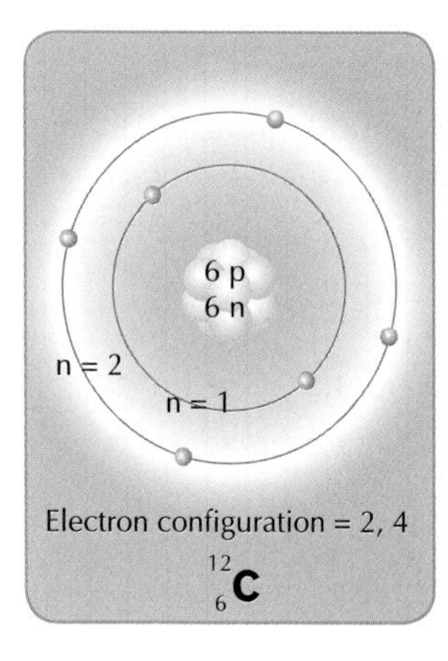

Fig. 30.4 Bohr structure of atom of carbon.

Note: When drawing Bohr structures, always write the number of protons and neutrons next to the nucleus. Also, write the nuclear formula and the electron configuration under the diagram. This makes it easier for the person marking your examination paper! Otherwise, the examiner will have to count the number of electrons you have drawn in each orbit and this will not make the examiner very happy!

Example 2

Draw the Bohr structure of an atom of argon.

Step 1
Find argon in the Periodic Table, *Fig. 30.1*. Argon is element number 18, i.e. its atomic number is 18. This means that there are 18 protons in the nucleus of the atom. Since the atom is neutral, this means that there are also 18 electrons in the atom. The mass number of argon is 40. Therefore, there are 40 – 18 = 22 neutrons in the nucleus.

We must now work out how the 18 electrons are arranged around the nucleus. This is done by studying the Periodic Table.

Step 2
Argon is in the third period. Therefore, we must draw three orbits.

1. Go to the beginning of the Periodic Table. There are **two electrons in the n = 1 orbit**.
2. Now start at the second period and count along from lithium. **There are eight electrons in the n = 2 orbit**.
3. Start at the third period and count along from sodium until you reach argon. Sodium (1), magnesium (2), aluminium (3), silicon (4), phosphorus (5), sulfur (6), chlorine (7) and argon (8). In other words, argon has **eight electrons in the third orbit.**

Step 3
We now know that argon has two electrons in the n = 1 orbit, eight electrons in the n = 2 orbit and eight electrons in the n = 3 orbit. Therefore we can write the electron configuration of argon as follows:

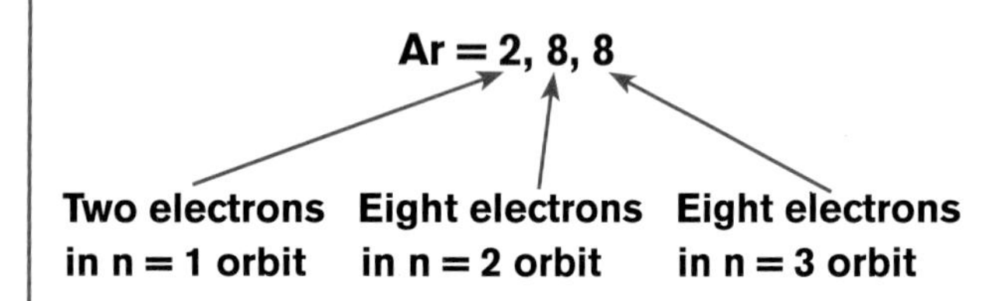

Check that this is correct by adding up these numbers (2 + 8 + 8). This sum should be equal to the atomic number of argon given on the Periodic Table. We can now draw the Bohr structure of an atom of argon as shown in *Fig. 30.5*.

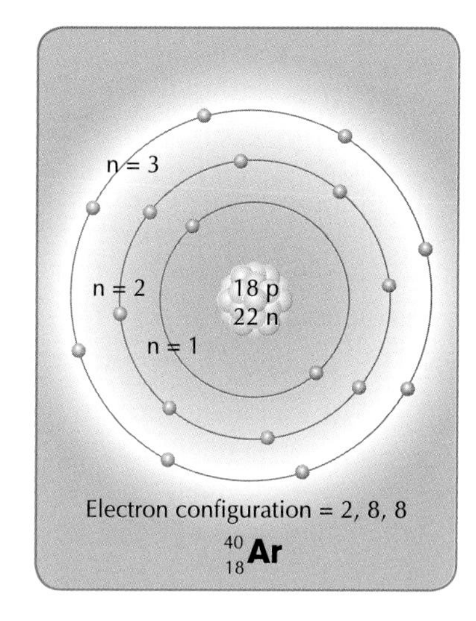

Fig. 30.5 Bohr structure of atom of argon.

Note: Do not forget to write the number of protons and neutrons next to the nucleus. Also, write the nuclear formula and the electron configuration under the diagram.

If you understand the above method, you should be able to draw the Bohr structures of the first 20 elements, *Fig. 30.6*.

Note: From *Fig. 30.6*:

(1) The **period number** (n = 1, n = 2, etc) tells you the number of orbits to draw.
(2) The **group number** tells you the number of electrons in the outer orbit.

A summary of electron configurations of the first 20 elements is given in *Table 30.1*, page 216.

TEST YOURSELF:
Now attempt questions *30.1–30.3* and *W30.1–W30.6*.

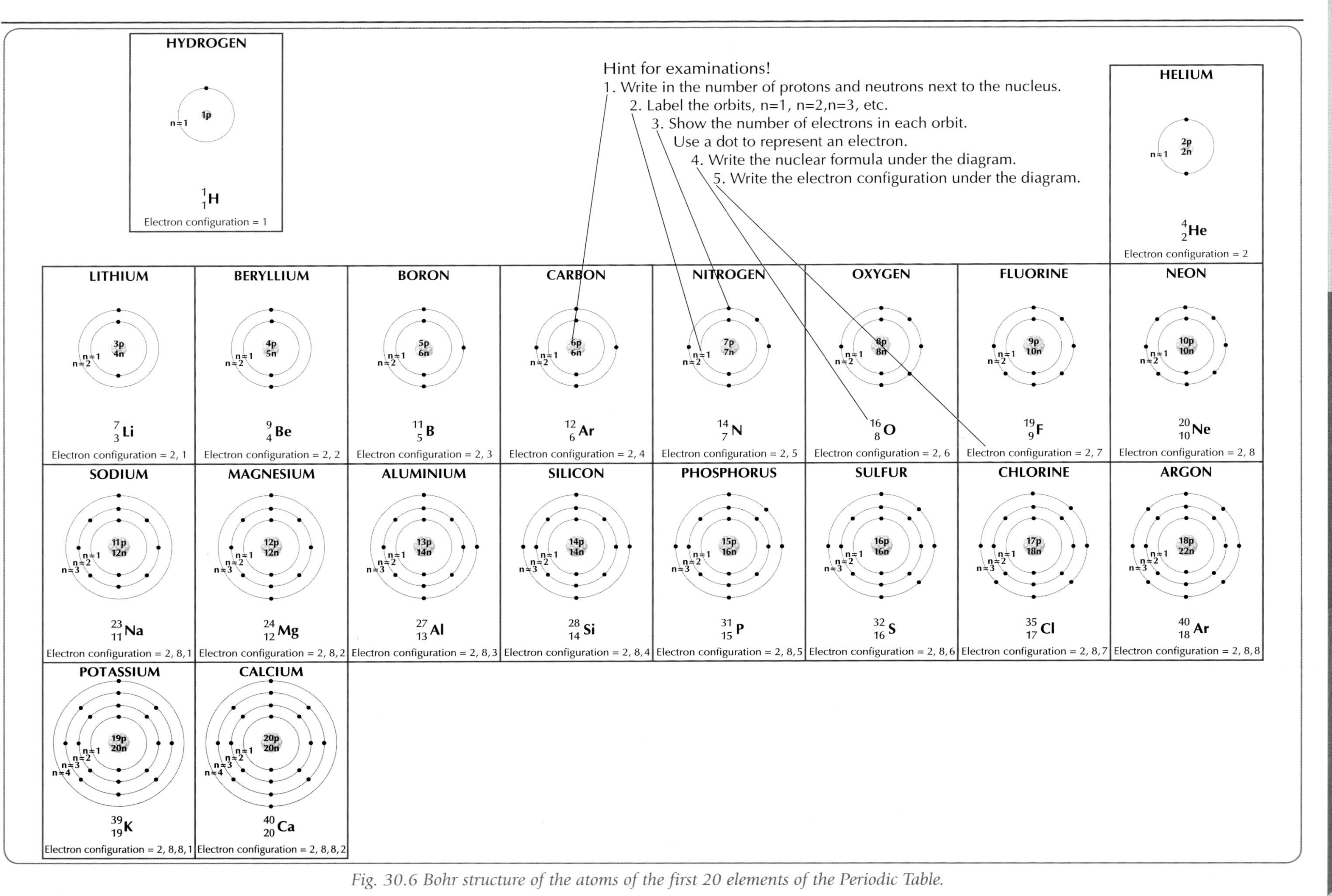

Fig. 30.6 Bohr structure of the atoms of the first 20 elements of the Periodic Table.

ELEMENT	ATOMIC NUMBER	ELECTRONS IN EACH ORBIT				ELECTRON CONFIGURATION
		n = 1	n = 2	n = 3	n = 4	
Hydrogen	1	1				1
Helium	2	2				2
Lithium	3	2	1			2,1
Beryllium	4	2	2			2,2
Boron	5	2	3			2,3
Carbon	6	2	4			2,4
Nitrogen	7	2	5			2,5
Oxygen	8	2	6			2,6
Fluorine	9	2	7			2,7
Neon	10	2	8			2,8
Sodium	11	2	8	1		2,8,1
Magnesium	12	2	8	2		2,8,2
Aluminium	13	2	8	3		2,8,3
Silicon	14	2	8	4		2,8,4
Phosphorus	15	2	8	5		2,8,5
Sulfur	16	2	8	6		2,8,6
Chlorine	17	2	8	7		2,8,7
Argon	18	2	8	8		2,8,8
Potassium	19	2	8	8	1	2,8,8,1
Calcium	20	2	8	8	2	2,8,8,2

Table 30.1 The electron configurations of the first 20 elements.

30.4 Isotopes

You may have noticed that most of the numbers under the symbols in the Periodic Table are not whole numbers. The reason for this is that many elements consist of atoms with different masses. All atoms of the same element have the same number of protons. However, some atoms may have different numbers of neutrons. These atoms are called **isotopes**.

Isotopes are atoms of the same element that have different numbers of neutrons.

In other words, isotopes have the same atomic number but different mass numbers. Isotopes can be compared to Easter Eggs. Imagine an Easter egg made by the same company and containing red and green sweets in the centre. However, there are two varieties of the same Easter Egg. One variety contains 10 red and 10 green sweets. The other variety contains 10 red and 15 green sweets. This makes one Easter egg heavier than the other Easter egg.

Carbon

An example of an element that has isotopes is carbon. Carbon has three isotopes, ${}^{12}_{6}C$, ${}^{13}_{6}C$ and ${}^{14}_{6}C$, *Fig. 30.7.*

You may have heard of the carbon-14 isotope. The amount of this isotope in a substance can be used to tell the age of it. This is known as *carbon dating*. For example, scientists used carbon dating to verify that the Dead Sea Scrolls were written during the period 200 B.C. to 68 A.D.

TEST YOURSELF: Now attempt questions *30.4* and *W30.7.*

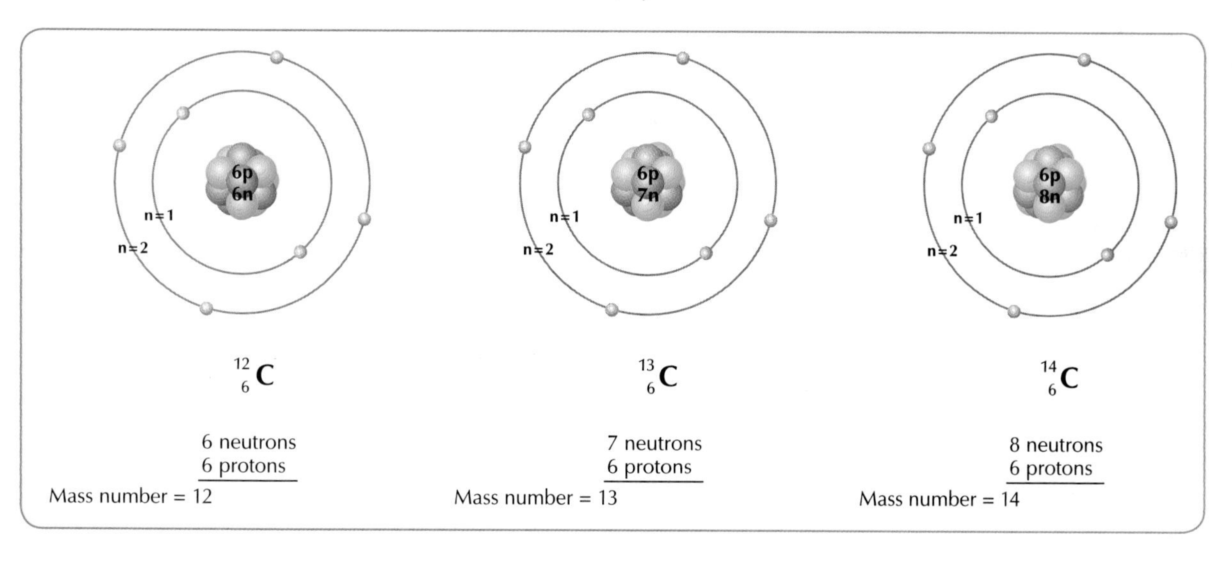

Fig. 30.7 Carbon has three isotopes. These are commonly called carbon-12, carbon-13, and carbon-14. The average mass of an atom of carbon is 12.011. This number is printed in the Periodic Table.

What I should know

- The Periodic Table is an arrangement of elements in order of increasing atomic number.
- The vertical columns are called groups. The horizontal rows are called periods.
- The Periodic Table is very useful for drawing the Bohr structures of the elements.
- Elements of similar chemical properties appear in the same group of the Periodic Table.
- Isotopes are atoms of the same element that have different numbers of neutrons.

QUESTIONS

Write the answers to the following questions into your copybook.

30.1 (a) Name the famous Russian chemist who laid the foundation for the modern Periodic Table.

(b) The Periodic Table is an arrangement of elements in order of increasing ..... .

(c) Elements in the same group of the Periodic Table have ..... chemical properties.

(d) In the Periodic Table what name is given to a column of elements?

(e) Name the scientist who proposed the idea that electrons move around the nucleus in fixed paths called orbits or shells.

(f) The only element in the Periodic Table which has no neutron in its nucleus is ..... .

30.2 (a) Underline which of the following elements are in the same group.

lithium, oxygen, aluminium, sulfur, calcium

(b) Underline which of the following elements are in the same period.

nitrogen, magnesium, helium, chlorine, calcium

(c) The elements of Group I are commonly known as ..... .

(d) To what family of elements do chlorine and bromine belong?

(e) Write down the names of two noble gases.

(f) What name is given to the block of metals in the centre of the Periodic Table?

30.3 (a) How many electrons are in the $n = 1$ orbit?

(b) How many electrons are in the outer orbit of (i) magnesium and (ii) chlorine.

(c) How many (i) protons and (ii) neutrons are in an atom of beryllium?

(d) Write down the electron configurations of (i) sodium and (ii) potassium.

(e) Write down the number of electrons in the outer orbit of (i) sodium and (ii) potassium.

(f) Draw a Bohr diagram to show the structure of an atom of the element whose symbol is Mg.

(g) Name the elements with the following electron configurations (i) 2, 8, 3 and (ii) 2, 8.

30.4 (a) What is meant by the term isotope?

(b) Name any one isotope of carbon.

(c) Two isotopes of chlorine have masses of 35 and 37. How many (i) protons and (ii) neutrons are in the atoms of each isotope?

(d) What is the difference between the isotopes $^{16}_{8}O$ and $^{18}_{8}O$?

(e) Draw the Bohr structures for each of the following isotopes:

(i) Carbon with 6 protons and 8 neutrons.
(ii) Argon with 18 protons and 22 neutrons.

(f) Give one use of isotopes.

Chapter 31 Chemical Bonding

31.1 Introduction

In *Chapter 25* we learned that O_2 is the formula for oxygen gas. This formula is used because chemists have discovered that oxygen gas consists of particles in which two atoms of oxygen are stuck together. Chemists say that the two atoms of oxygen are joined or **bonded** together. This is similar to the way that glue can be used to join or bond various substances together.

But why is the formula for oxygen gas O_2? Why is the formula not O_3, or O_4? Why is it that just two atoms of oxygen join together? Why don't five or six of them join together? In order to answer these sorts of questions, chemists had to develop a theory. This theory is called **Chemical Bonding**. It can be a difficult idea for many students to understand. Therefore, pay close attention to your teacher and fasten your safety belts!

31.2 Molecules

We have learned above that oxygen gas consists of particles in which two oxygen atoms are joined together. A group of atoms that are joined together is called a **molecule**.

A molecule is a group of atoms joined together. It is the smallest particle of an element or compound that can exist independently.

Examples of Molecules

Some examples of molecules on your course are shown in *Fig. 31.1*.

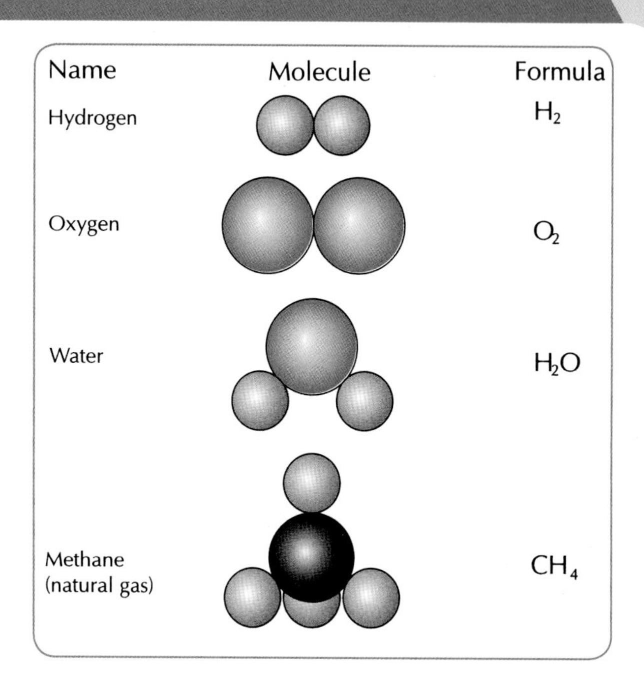

Fig. 31.1 Some examples of molecules.

- The chemical formula for water is H_2O. It is given this formula because a molecule of water consists of **two hydrogen atoms** joined to **an oxygen atom**.
- The formula for methane (natural gas) is CH_4. It has this formula because a molecule of methane consists of **an atom of carbon** joined to **four atoms of hydrogen**.

Chemical Bonds

The atoms in a molecule are held together by forces of attraction called **chemical bonds**. For your examination, you must be able to describe the bonding in H_2, O_2, H_2O and CH_4.

Our understanding of chemical bonding is made a lot easier if we first study a rule called the Octet Rule.

31.3 Stability of the Noble Gases – The Octet Rule

If you look at the Periodic Table, you will see that the noble or inert gases are in group 0 (also called group VIII) of the table. All of the noble gases are very unreactive. They form practically no compounds and consist of single atoms. A sample of helium gas, *Fig. 31.2*, consists entirely of helium atoms. Helium molecules do not exist.

Fig. 31.2 All of these balloons contain helium. Helium is an example of a noble gas. All of the noble gases are very unreactive.

A Stable Octet

Why are the noble gases so unreactive? Chemists tried to find the answer to this question by studying the structure of the atoms of these gases. If you look at the electron configuration of some noble gases, *Table 31.1*, you will observe that they all (except helium) have eight electrons in their outer orbit.

ELEMENT	ELECTRON CONFIGURATION
Helium	2
Neon	2, 8
Argon	2, 8, 8

Table 31.1 The electron configurations of some noble gases.

Since the noble gases are so unreactive, it follows that this arrangement of electrons must be very stable, i.e. difficult to change.

Elements that have eight electrons in their outer orbit are very unreactive.

Therefore, the noble gases are stable because their outer orbits contain eight electrons.

This fact helps us to understand chemical bonding. The eight electrons in the outer orbit are often referred to as a **stable octet**. This is called after the eight electrons (an octet) in the outer orbit of most of the noble gases.

OCTET RULE:
When bonding occurs, atoms try to have eight electrons in the outer orbit.

The octet rule is **not** a strict chemical law but is a useful guide to understanding bonding.

Exceptions to the Octet Rule

- **Helium** does not have eight electrons in its outer orbit. However, its first and only occupied orbit of electrons is full. Its properties are so similar to those of the inert gases that it is always classified with them.
- **Hydrogen** and **lithium** tend to reach the electron configuration of helium when chemical bonding occurs.

Chemical reaction takes place so that the elements can obtain the very stable arrangement of the noble gases. When elements react together to form compounds, it is found that two main types of chemical bonds are formed – the **covalent bond** and the **ionic bond**.

31.4 Covalent Bonding

The word 'covalent' means 'sharing'. In this type of bonding, atoms try to get eight electrons in the outer orbit by **sharing** electrons with one or more other atoms. The shared electrons form a bond between the two atoms. This type of bond is called a **covalent bond**.

A covalent bond is a bond that consists of shared electrons.

Examples of Covalent Compounds

There are four examples of molecules containing covalent bonds on your course. Study the following examples carefully.

Note:

- In the diagrams that follow, different **symbols** (• and ×) are used to represent electrons on different atoms. All electrons are identical but the different symbols help to make clear what is happening.

- Chemical bonding **only involves electrons**. Therefore, when using diagrams to explain bonding, there is no need to draw in the structure of the nucleus.

(1) The Hydrogen Molecule – H_2

A sample of hydrogen gas does not contain any individual hydrogen atoms. It consists of hydrogen molecules i.e. two hydrogen atoms joined together, *Fig. 31.1*, page 219.

To understand the reason for this, consider the electron configuration of an atom of hydrogen:

$$H = 1$$

Two atoms of hydrogen approach close enough so that their outer orbits overlap. Each atom shares the electron of the other atom, *Fig. 31.3*. This gives each atom the stable electron configuration of helium (two electrons in the outer orbit).

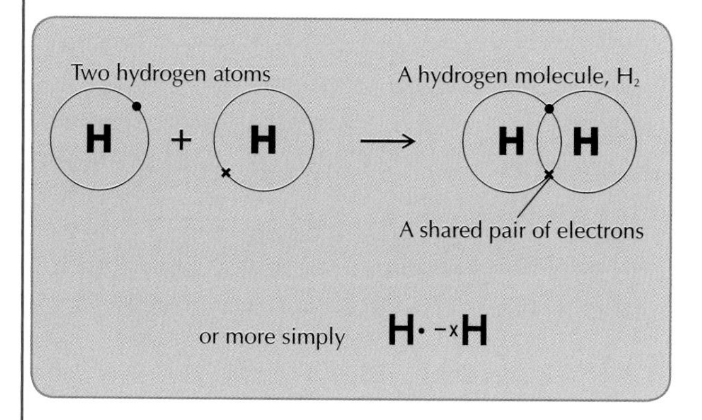

Fig. 31.3 A covalent bond is formed in a molecule of hydrogen.

Thus, each atom of hydrogen now has two electrons in its outer orbit. It has its own electron and the shared electron. The sharing of the pair of electrons is known as a **single covalent bond**. Chemists represent a covalent bond by a line, e.g. H – H.

(2) The Methane Molecule – CH_4

We learned in *Chapter 28* that natural gas consists mainly of methane. A molecule of methane consists of one atom of carbon joined to four atoms of hydrogen. We can understand how the CH_4 molecule is formed if we consider the electron configuration of the atoms involved in bonding.

$$C = 2, 4$$

$$H = 1$$

Carbon has four electrons in its outer orbit. It needs four more electrons in its outer orbit to get noble gas configuration. Therefore, the carbon atom shares each of its four outer electrons with a hydrogen atom. This is illustrated in *Fig. 31.4*.

(3) The Water Molecule – H_2O

A molecule of water consists of one atom of oxygen joined to two atoms of hydrogen. We can understand how the H_2O molecule is formed if we study the electron configuration of the atoms involved in bonding.

$$O = 2, 6$$

$$H = 1$$

Oxygen has six electrons in its outer orbit. It needs two more electrons in its outer orbit to get noble gas configuration. Therefore, the oxygen atom shares two of its six electrons with two

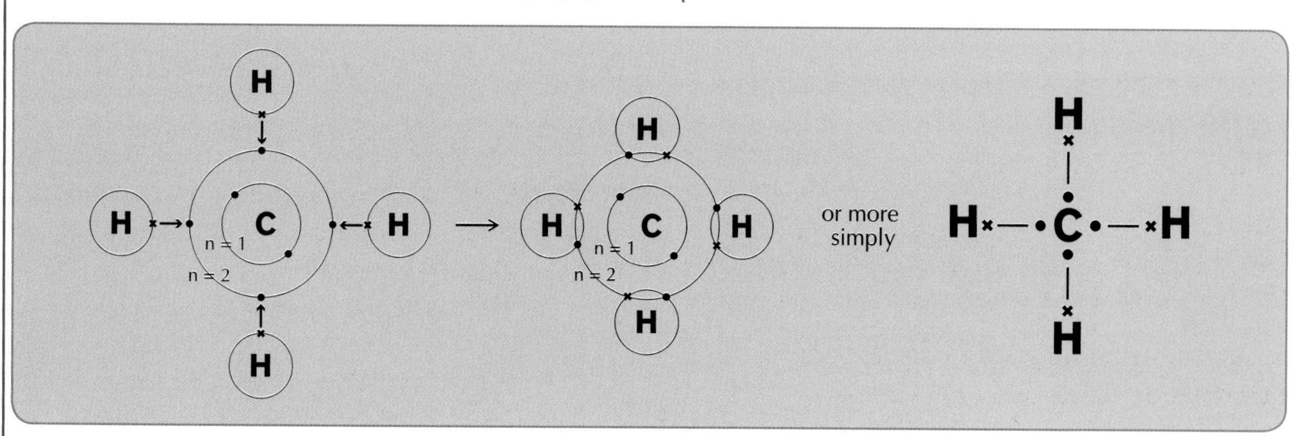

Fig. 31.4 In the methane molecule, a carbon atom is joined by covalent bonds to four hydrogen atoms.

CHEMISTRY

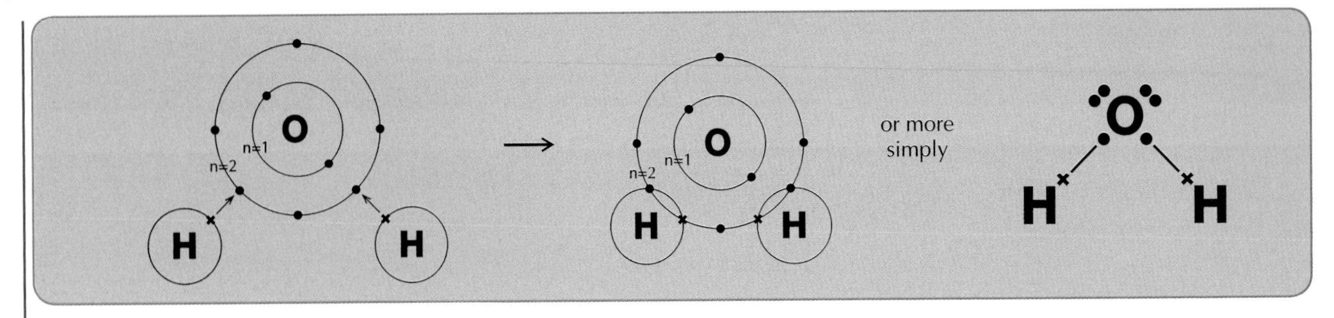

Fig. 31.5 In the water molecule, an oxygen atom is joined by covalent bonds to two hydrogen atoms.

hydrogen atoms. This is illustrated in *Fig. 31.5*.

Single and Double Bonds

In all chemical bonds we have met so far, just one pair of electrons has been shared. These bonds are called **single bonds**. For example:

- The bond in the H_2 **molecule** is a single bond.
- The bond between the **carbon** atom and the **hydrogen** atoms in methane are all single bonds.
- The bond between the **oxygen** atom and the two **hydrogen** atoms in water are single bonds.

A bond in which **two** pairs of electrons are shared is called a **double bond**.

A single bond is formed when one pair of electrons is shared.

A double bond is formed when two pairs of electrons are shared.

An example of a molecule with a double bond is the O_2 molecule.

(4) The Oxygen Molecule – O_2

The oxygen gas that you prepared in *Chapter 25* consists of millions of molecules of oxygen. A molecule of oxygen consists of two oxygen atoms joined together, *Fig. 31.1*. We can understand how the O_2 molecule is formed if we study the electron configuration of the atoms involved in bonding.

$$O = 2, 6$$

Oxygen has six electrons in its outer orbit. It needs two more electrons in its outer orbit to get noble gas configuration. Study *Fig. 31.6* carefully. Note that each oxygen atom has its own six electrons in the outer orbit. Also, each oxygen atom obtains a share of two electrons from the other oxygen atom.

A double bond is often represented by a double dash, e.g. O = O, *Fig. 31.6*.

TEST YOURSELF:
Now attempt questions *31.1–31.3* and *W31.1–W31.2*.

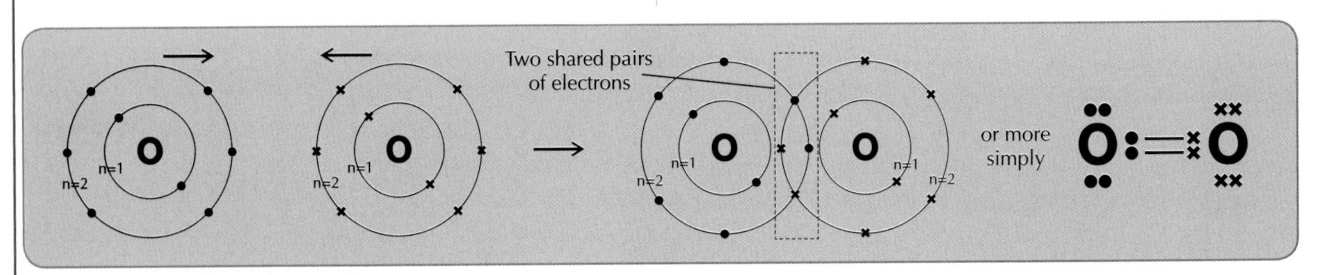

Fig. 31.6 In the O_2 molecule, two oxygen atoms are joined together by a double bond.

31.5 Ionic Bonding

Ions

Not all compounds contain covalent bonds. For example, when X-rays were passed through a crystal of table salt, it was found that the salt consisted of particles called **ions**.

> **An ion is a charged atom or group of atoms.**

The ions in table salt are represented as Na^+ and Cl^-.

- Na^+ is a sodium atom that has lost an electron. Chemists call this a **sodium ion**. Remember that electrons are negatively charged. Therefore, if a neutral atom loses an electron, it is left with a positive charge.
- Cl^- is a chlorine atom that has gained an electron. Chemists call this a **chloride ion**. It has a negative charge because it has one 'extra electron'.

Ionic Bonds

By experiment, it is found that sodium chloride has a very high melting point. This tells us that there must be a strong attraction between the positive ions and negative ions. The force of attraction holding the ions together is called an **ionic bond**.

> **An ionic bond is the force of attraction between positive and negative ions in a compound.**

Examples of Ionic Compounds

There are two examples of ionic compounds on your course.

(1) Sodium Chloride – NaCl

We can understand how the ionic bond is formed if we look at the electron configurations of the atoms involved:

$$Na = 2, 8, \xrightarrow{\text{loses one electron}} Na^+ = 2, 8$$

$$Cl = 2, 8, 7 \xrightarrow{\text{gains one electron}} Cl^- = 2, 8, 8$$

The sodium atom could be left with eight electrons in its outer orbit by losing its one outer electron. Consider what happens if the chlorine atom accept this one electron. Chlorine would now have eight electrons in its outer orbit. Therefore, the reaction to form sodium chloride involves the complete transfer of an electron from an atom of sodium to an atom of chlorine, *Fig. 31.7*.

Fig. 31.7 An ionic bond is formed by the complete transfer of an electron from one atom to another.

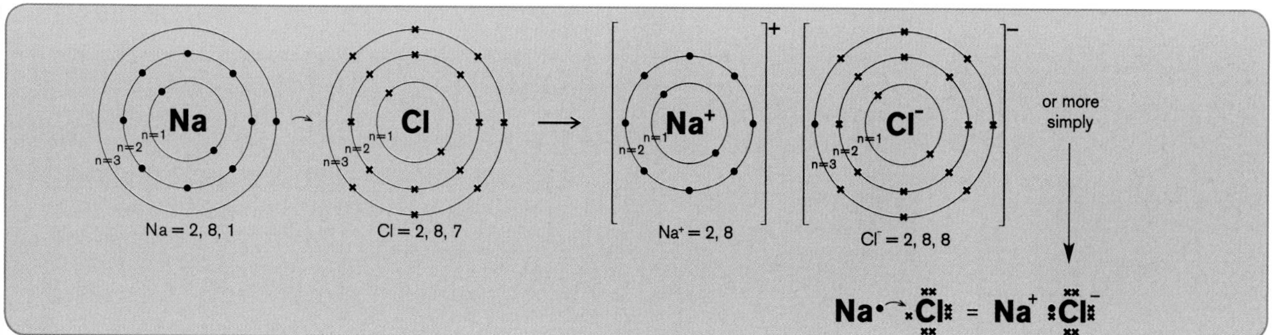

Fig. 31.8 A sodium atom gives an electron to a chlorine atom to form sodium chloride. The attraction between oppositely charged ions is what holds the ions together. This attraction is called an ionic bond.

The formation of the ionic bond in sodium chloride is shown in *Fig. 31.8*, page 223. In an examination, be sure to show the structure of the ions as well as the atoms.

There is only one other ionic compound for which you must be able to describe the bonding. So keep the seat belts fastened, we are nearly finished!

(2) Magnesium Oxide, MgO

The bright light that comes from fireworks is caused by magnesium metal burning in oxygen, *Fig. 31.9*. You may recall the fun you had seeing magnesium burn in *Chapter 25*.

Fig. 31.9 The ionic compound magnesium oxide is formed when magnesium is burned in air. When the ionic bond is formed, a lot of energy is given out in the form of light and heat.

When the magnesium in fireworks is burned, a white powder called **magnesium oxide** is formed. The formula for magnesium oxide is **MgO**. When this powder is analysed, it is found to consist of **ions** of magnesium and oxygen.

Let us now try to understand what is happening. Consider the electron configurations of magnesium and oxygen:

$$Mg = 2, 8, 2 \xrightarrow{\text{loses two electrons}} Mg^{2+} = 2, 8$$

$$O = 2, 6 \xrightarrow{\text{gains two electrons}} O^{2-} = 2, 8$$

- Each Mg atom loses its two outer electrons to an oxygen atom. As the Mg atom has **lost two** electrons, it is written as **Mg^{2+}**. This is called a **magnesium ion**.
- Since the oxygen atom has **gained two** electrons it is written as **O^{2-}**. This is called an **oxide ion**.

The formation of the ionic bond is shown in *Fig. 31.10*.

TEST YOURSELF:
Now attempt questions *31.4–31.5* and *W31.3–W31.4.*

31.6 Properties of Ionic and Covalent Substances

Ionic compounds consist of positive and negative ions. Covalent compounds nearly always consist of separate individual molecules. Covalent bonding **within** a molecule is **very strong**. However, the bonding **between** molecules is **much weaker**. This helps to explain many of the properties of covalent compounds.

Properties of Ionic Compounds

The general properties of ionic compounds may be summarised as follows:

(1) **Ionic compounds are usually solid at room temperature**. This is because of the strong attraction between the millions of oppositely charged ions in the substance. The ionic

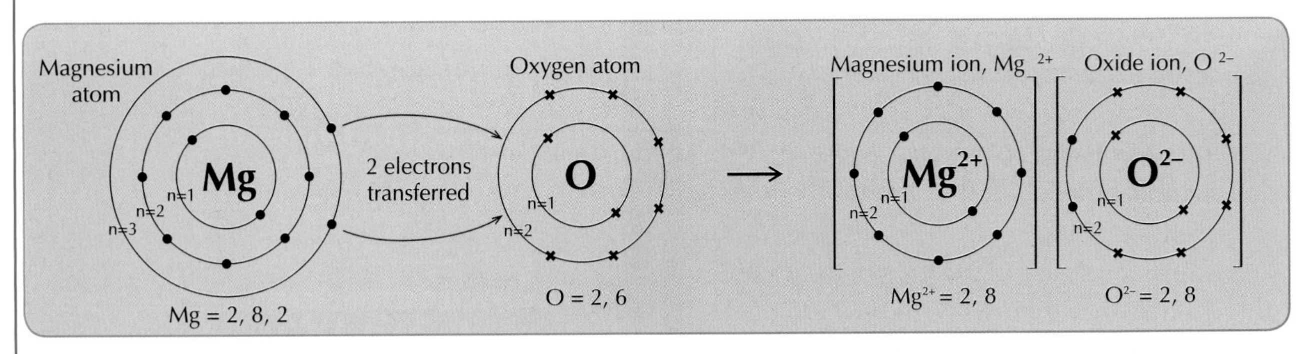

Fig. 31.10 An ionic bond is formed when magnesium burns in oxygen.

bonding in sodium chloride is very strong. In fact, it pulls all the ions together to form crystals with a definite shape.

(2) **Ionic compounds usually have high melting points and boiling points**. This is because there is a strong attraction between the ions. Therefore a lot of energy is needed to separate them from each other e.g. the melting point of sodium chloride is over 800 °C and its boiling point is almost 1500°C.

(3) **Ionic compounds usually dissolve in water**. Water molecules are attracted to ions and cause the ions to be pulled away from the ionic substance and go into solution.

(4) **Ionic compounds usually conduct electricity when molten (melted) or when dissolved in water**. When the ionic material is molten or dissolved in water, the ions are free to move. These ions then carry the electric current. (Ionic compounds cannot conduct electricity when solid because the ions are not free to move).

Properties of Covalent Compounds

The general properties of covalent compounds may be summarised as follows:

(1) **Covalent compounds are usually either a liquid or a gas at room temperature**. The atoms in each individual molecule are held together by covalent bonds. However, there are no strong forces of attraction between the separate molecules. Therefore, unlike ionic compounds, covalent compounds are usually either a liquid or a gas at room temperature, e.g. water is a liquid and methane is a gas.

(2) **Covalent compounds usually have low melting points and boiling points**. This is because the attractive forces between covalent molecules are not very strong. Therefore not much heat energy is needed to separate them from each other, e.g. the melting point of ice is 0 °C and the boiling point is 100 °C.

(3) **Many covalent compounds (but not all) do not dissolve in water**. The reason for this is because covalent compounds do not contain ions. Water molecules are not normally attracted to covalent molecules. There are, however, many exceptions to this generalisation.

(4) **Covalent compounds do not conduct electricity**. This is because there are no ions present. This is studied in more detail in *Experiment 31.1*.

Summary of the Properties of Ionic and Covalent Compounds

The properties of ionic and covalent compounds are summarised in *Table 31.2*.

IONIC COMPONDS	COVALENT COMPOUNDS
1. Consist of positive and negative ions.	1. Usually consist of individual molecules.
2. Usually solid at room temperature.	2. Usually liquid or gas at room temperature.
3. Usually high melting and boiling points.	3. Usually low melting points.
4. Usually soluble in water.	4. Usually insoluble in water.
5. Conduct electricity when melted or when dissolved in water.	5. Do not conduct electricity.

Table 31.2 The properties of ionic and covalent compounds.

Experiment 31.1

To investigate the ability of ionic and covalent substances to conduct electricity

In this experiment we try to distinguish between ionic and covalent substances. To do this, we look at the ability of these substances to conduct electricity. **If ions are present in a solution, the solution will conduct an electric current**. If no ions are present in a solution, the solution will not conduct an electric current.

Apparatus required: power supply or battery; bulb; leads; crocodile clips; carbon electrodes; wash bottle

Chemicals required: water; copper sulfate solution; methylated spirits; sodium chloride solution; table sugar solution; paraffin oil

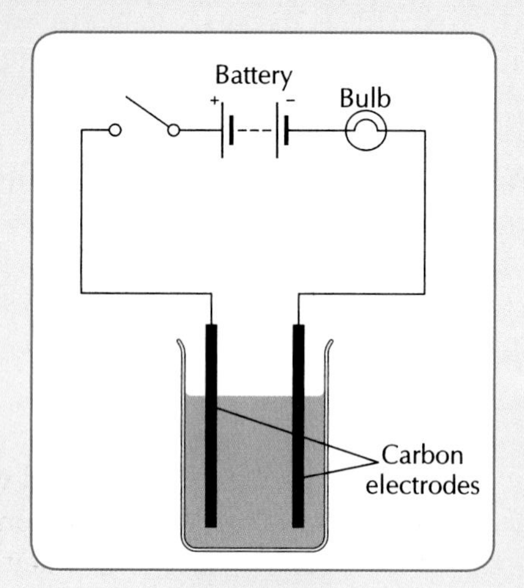

Fig. 31.11 Apparatus to test for presence of ions in solution.

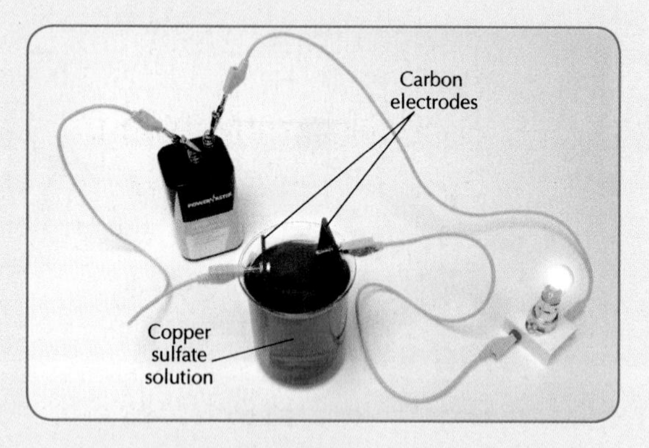

Fig. 31.12 Apparatus to test for presence of ions in solution.

Method

1. Set up the apparatus as shown in *Fig. 31.11*. A photograph of the apparatus is shown in *Fig. 31.12*.
2. Insert the carbon electrodes into a solution of copper sulfate.
3. Write down what you observe.
4. Repeat the experiment with various other solutions given to you by your teacher. Using a wash bottle, wash the carbon rods before using the next solution.

Results

Summarise your results in the form of a table.

NAME OF SOLUTION	CONDUCT ELECTRICITY (YES/NO)
Copper sulfate	
Methylated spirits	
Sodium chloride	
Paraffin oil	
Table sugar	

Conclusion

Solutions that contain ions conduct an electric current.

TEST YOURSELF:

Now attempt questions *31.6* and *W31.5–W31.8*.

What I should know

- A molecule is a group of atoms joined together. It is the smallest particle of an element or compound that can exist independently.
- A covalent bond is a bond that consists of shared electrons.
- A double bond is formed when two pairs of electrons are shared.
- An ionic bond is the force of attraction between positive and negative ions in a compound.
- Ionic compounds consist of ions.
- Covalent compounds consist of individual molecules.
- Ionic compounds are usually solid at room temperature, have high melting points and boiling points, dissolve in water and conduct electricity in the molten state or when dissolved in water.
- Covalent compounds are usually either liquids or gases at room temperature, have low melting and boiling points, most do not dissolve in water and they do not conduct electricity.

QUESTIONS

Write the answers to the following questions into your copybook.

31.1 (a) The forces of attraction which hold atoms together in a compound are called ….. .

(b) The two main types of bond are ionic and ….. bonds.

(c) A useful rule which helps us in our study of chemical bonding is the ….. rule.

(d) Neon is a very unreactive element due to the fact that it has ….. electrons in its outer orbit.

(e) Covalent bonds are formed when electrons are ….. between atoms.

(f) The type of bond formed in the hydrogen molecule is a ….. bond.

(g) Two examples of covalent compounds are ….. and ….. .

31.2 Fill in the blanks in the following statements about the elements of group 0.

The elements in group 0 or group ….. are commonly called the ….. or ….. gases. They are all very ….. and so there are very few compounds formed by these elements. The elements in this group all have ….. properties. All of these elements (except helium) have ….. electrons in their outer orbit.

31.3 (a) What is a single covalent bond?

(b) Give an example of a molecule in which single bonds are formed.

(c) With the aid of a diagram, describe the bonding in the molecule you have named in (b).

31.4 (a) In the formation of an ionic bond electrons are always ….. from one atom to another.

(b) An atom or group of atoms which has lost or gained electrons is called an ….. .

(c) Write down the symbol for (i) a sodium ion (ii) an oxide ion.

(d) An atom which gains an electron becomes ….. charged.

(e) In a negatively charged ion, the number of ….. orbiting the nucleus is always ….. than the number of ….. in the nucleus.

(f) Magnesium and oxygen combine to form magnesium oxide. Which of the following statements are correct? (i) each magnesium atom loses two electrons; (ii) each magnesium atom gains two electrons; or (iii) the magnesium and oxygen atoms share electrons.

31.5 Explain what is meant by a covalent bond and an ionic bond. State the type of bond in each of the following: (i) magnesium oxide, (ii) oxygen, (iii) sodium chloride.

With the aid of diagrams, explain how the bonding is formed in each of the above.

31.6 (a) Covalent compounds usually have ….. melting points and ….. boiling points.

(b) ….. compounds do not normally conduct an electric current.

(c) Both salt and sugar dissolve in water but only the salt solution conducts electricity. Explain.

(d) Ionic substances usually ….. in water.

(e) Compound *X* is a solid. It has a melting point of 700 °C. A solution of *X* in water conducts electricity. The bonding in *X* is likely to be ….. .

(f) Which of the following would not conduct electricity: copper sulfate solution, distilled water and sea water?

Chapter 32 Chemical Formulas and Chemical Equations

32.1 Introduction

In this course we have met some chemical formulas and chemical equations. In this chapter we will do the following:

- Revise the chemical formulas we have met.
- Discuss the reasons why these formulas are written in a certain way.
- Revise the balanced chemical equations that we have met in this course.
- Study the rules for writing balanced chemical equations.

32.2 What is a Chemical Formula?

Chemical Symbols

We have seen that an **element** is represented by a chemical **symbol**, e.g. carbon is represented by C and sodium is represented by Na. A symbol may also be used to indicate one atom of an element.

When a symbol is followed by a small number, this represents a **molecule** of that element, e.g. H_2 represents a molecule of hydrogen. This tells us that a molecule of hydrogen consists of two atoms of hydrogen.

Chemical Formulas

A **chemical formula** is used to represent a **compound** or a molecule of a compound. A chemical formula consists of the symbols of the elements in the compound. For example, in *Chapter 21* we learned that the formula for iron sulfide is FeS. In *Chapter 25* we learned that the formula for the compound formed when magnesium is burned in oxygen is MgO.

Atomic Ratios

A chemical formula also indicates the ratio of the different atoms in a compound. For example, in magnesium oxide, MgO, there are equal numbers of magnesium and oxygen atoms. A molecule of water, however, consists of two hydrogen atoms and one oxygen atom. Therefore, a molecule of water is represented as H_2O. Note that the subscript two multiplies the H in front of it by two, i.e. there are two hydrogen atoms in the molecule. In other words, the number after the symbol tells us the number of atoms present. Note also that the '1' is never written in a chemical formula. When studying acids and bases, we learned that the formula for sulfuric acid is H_2SO_4, *Fig. 32.1*.

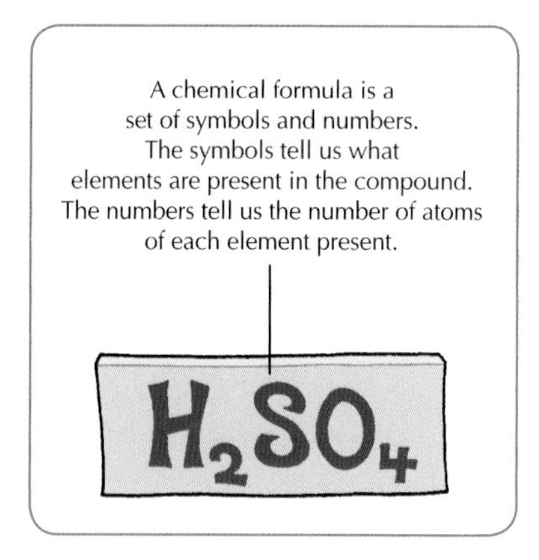

Fig. 32.1 The chemical formula for sulfuric acid tells us that there are two hydrogen atoms, one sulfur atom and four oxygen atoms in the molecule.

32.3 Formulas of Covalent Compounds

The formulas of common covalent compounds that we have met in this course are given in *Fig. 32.2*.

NAME	FORMULA	MODEL OF MOLECULE
Hydrogen Chloride	HCl	
Water	H_2O	

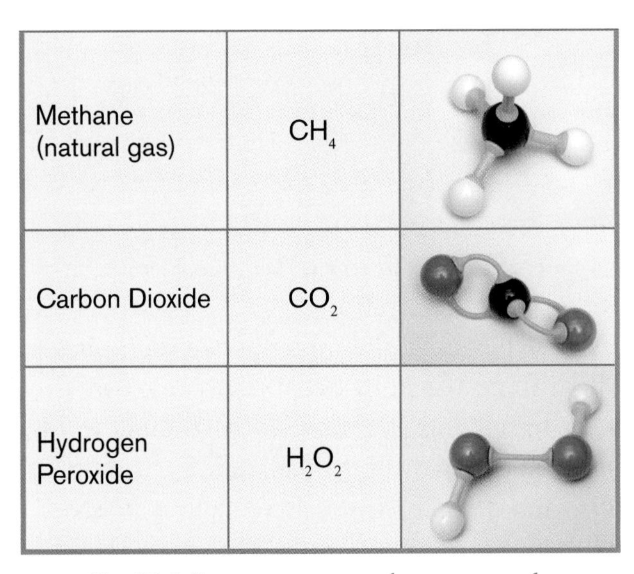

Methane (natural gas)	CH_4	
Carbon Dioxide	CO_2	
Hydrogen Peroxide	H_2O_2	

Fig. 32.2 Some common covalent compounds.

The number of covalent compounds that you meet in this course is quite small. Therefore, study *Fig. 32.2* carefully.

32.4 Formulas of Ionic Compounds

We have met a number of ionic compounds in this course. For example, we have learned that the formula of sodium chloride (common table salt) is NaCl. In the preparation of carbon dioxide, we learned that the formal for marble (chalk) is $CaCO_3$. The formulas for common ionic compounds that we have met in this course are given in *Table 32.1*.

NAME	FORMULA
Sodium chloride	NaCl
Magnesium oxide	MgO
Iron sulfide	FeS
Calcium chloride	$CaCl_2$
Copper sulfate	$CuSO_4$
Calcium carbonate (marble, chalk)	$CaCO_3$
Manganese dioxide	MnO_2
Sodium hydroxide	NaOH
Calcium hydroxide	$Ca(OH)_2$

Table 32.1 Formulas of some ionic compounds.

We will be meeting one other ionic compound in the next chapter. It is called zinc chloride and has the formula $ZnCl_2$.

TEST YOURSELF:
Now attempt questions *32.1–32.2* and *W32.1–W32.2*.

32.5 Equations in Chemistry

When you hear the word 'equation' you probably think of the equations you use in mathematics. As you are aware from your study so far, chemists also use equations. We now look a bit more closely at how chemists use equations to describe chemical reactions. There are two types of equations in chemistry: **word equations** and **chemical equations**.

(1) Word Equations

When hydrogen gas is burned in oxygen (or air) it is found that droplets of water are formed. The word equation for this chemical reaction is:

hydrogen + oxygen ⟶ water

In this word equation the + sign means 'and'. The ⟶ sign means 'changes into'. Sometimes, the arrow may be replaced by the '=' sign.

Reactants

Hydrogen and oxygen are the chemicals that react together in the chemical reaction. These chemicals are called the **reactants**. The reactants are always the starting chemicals. The reactants are always written on the left-hand side of the equation.

Products

In the above example, the reaction between hydrogen and oxygen produces water. Water is said to be the **product** of the reaction. The products are the chemicals that are produced in a chemical reaction.

The products are always written on the right-hand side of the equation.

reactants ⟶ products

(2) Chemical Equations

A chemical equation is similar to a word equation except that it uses symbols (for elements) and formulas (for compounds) in the equation.

Example 1

When preparing oxygen (*Chapter 25*), we burned some carbon in oxygen to form carbon dioxide.

Word equation: carbon + oxygen ⟶ carbon dioxide

Chemical equation: $C + O_2 \longrightarrow CO_2$

A chemical equation is just like a simple sentence written with chemical symbols, i.e. equations are chemical sentences.

Example 2

When preparing sodium chloride (*Chapter 28*) we reacted together hydrochloric acid and sodium hydroxide.

hydrochloric acid + sodium hydroxide ⟶ sodium chloride + water

$HCl + NaOH \longrightarrow NaCl + H_2O$

32.6 Balancing Chemical Equations

When you hear the word 'balance' you probably think of a tight-rope walker or the old style laboratory balance. When chemists talk of balanced equations, they mean something quite different.

Example 3

Consider the equation describing the burning of carbon in oxygen:

carbon + oxygen ⟶ carbon dioxide

$C + O_2 \longrightarrow CO_2$

This is said to be a **balanced** equation. Studying *Fig. 32.3* will help you to understand this.

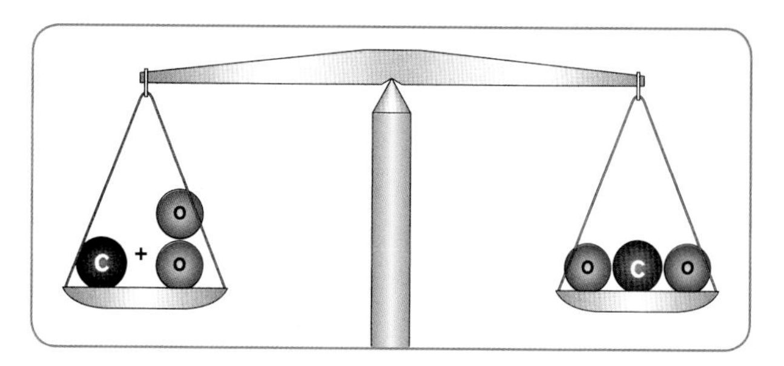

Fig. 32.3 An equation is balanced when there is the same number of each type of atom on both sides of the equation.

Note: In this equation there is one carbon atom on each side of the equation, i.e. carbon is balanced. Also, there are two oxygen atoms on each side of the equation, i.e. oxygen is also balanced. Therefore, we say that the equation is balanced.

A balanced equation is one in which the total number of atoms of each element on the left-hand side of the equation is equal to the total number of atoms on the right-hand side of the equation.

This simply means that in a chemical reaction atoms cannot vanish or appear from nowhere. All the atoms at the start of a reaction must still exist at the end of a reaction.

Example 4

When preparing oxygen (*Chapter 25*) we burned magnesium in oxygen to form magnesium oxide. Let's write a balanced equation for this reaction.

magnesium + oxygen ⟶ magnesium oxide

$Mg + O_2 \longrightarrow MgO$

Is this equation balanced? Referring to *Fig. 32.4* we see that the equation is not balanced. There are two oxygen atoms on the left-hand side, but only one oxygen atom on the right-hand side.

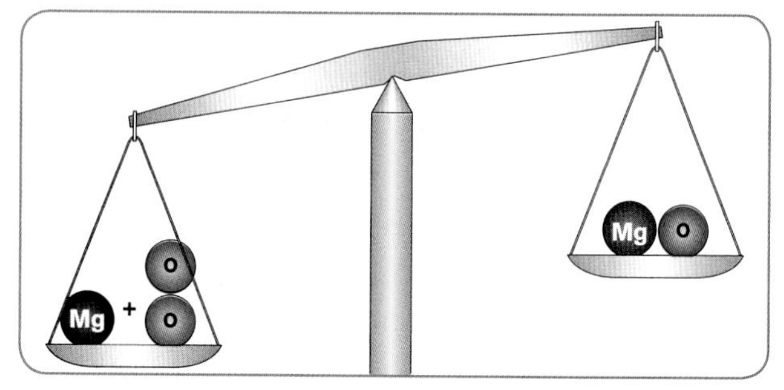

Fig. 32.4 The equation $Mg + O_2 \longrightarrow MgO$ is not balanced.

How can we balance this equation? We must have two O atoms on the right-hand side of the equation. You might be tempted to do this by writing the formula of magnesium oxide as MgO_2. However, **this is wrong** as the formula of Magnesium oxide is MgO and not MgO_2. This is an important point to remember, i.e. *in balancing an equation, formulas cannot be altered in any way.*

In order to have two oxygen atoms on the right-hand side, we must put a '2' before the MgO. This means that two molecules of MgO are formed.

i.e. $Mg + O_2 \longrightarrow 2MgO$

This means that two molecules of magnesium oxide are formed.

This equation, however, is still not balanced. There is one magnesium atom on the left-hand side but two magnesium atoms on the right-hand side. The final step is to put a '2' before the Mg on the left-hand side. This means that two magnesium atoms take part in the reaction, *Fig. 32.5*.

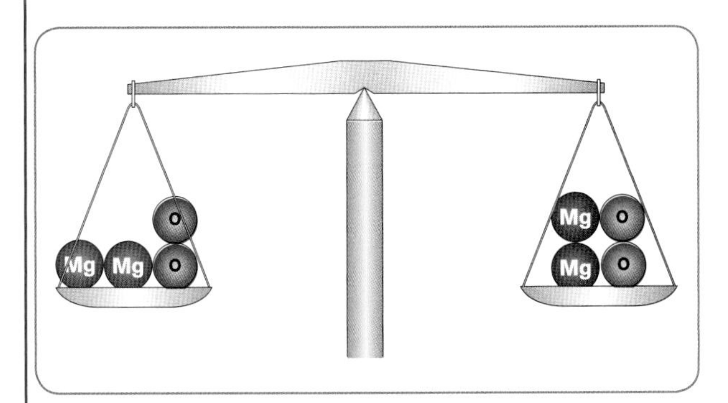

Fig. 32.5 The equation $2Mg + O_2 \longrightarrow 2MgO$ is now balanced.

So the final balanced equation is:

$$2Mg + O_2 \longrightarrow 2MgO$$

Note: When balancing equations it may be helpful to use a pencil to fill in the numbers in front of the symbols or formulas. As you go through the steps in balancing the equation, you may need to change the numbers. When you have the equation fully balanced, write the numbers in ink.

Exam Questions

To show you the type of examination questions you must answer on this topic, two examples from past examination papers are now studied.

Example 5

Junior Cert Science Higher Level Examination Question

A gas jar of oxygen was prepared by decomposing hydrogen peroxide using a suitable catalyst. This preparation may be described as follows:

$H_2O_2 \longrightarrow H_2O + O_2$

Balance the above equation.

A: $2H_2O_2 \longrightarrow 2H_2O + O_2$

Example 6

Junior Cert Science Higher Level Examination Question

Carbon dioxide turns limewater milky. Complete the chemical equation for the reaction of carbon dioxide with limewater.

$Ca(OH)_2 + CO_2 \longrightarrow$

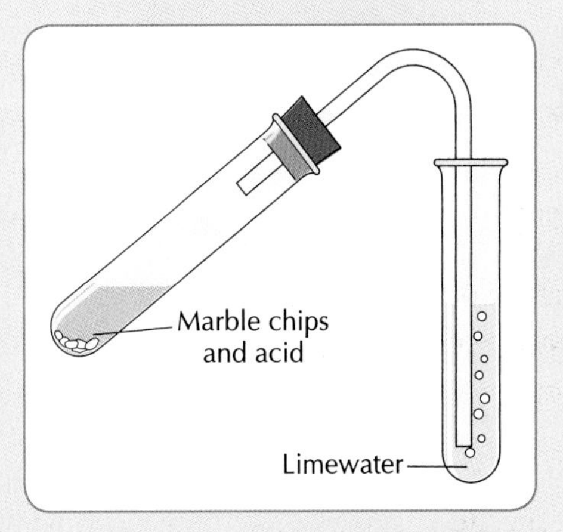

Fig. 32.6

A: $Ca(OH)_2 + CO_2 \longrightarrow CaCO_3 + H_2O$

TEST YOURSELF:
Now attempt questions *32.3–32.4* and *W32.3–W32.6.*

CHEMISTRY

What I should know

- A chemical formula shows which elements are present in a compound. It also shows how many atoms of each element are present in a molecule of the compound.
- The reactants are the chemicals that react together in a chemical reaction.
- The products are the chemicals that are produced in a chemical reaction.
- A chemical equation must always be balanced, i.e. there must be the same number of each particular type of atom on each side of the equation.
- When balancing an equation, numbers are put in front of formulas, *but the formulas themselves must never be changed.*

QUESTIONS

32.1 (a) How many atoms are present in a molecule of CH_4?

(b) The formula for sulfuric acid is ..... .

(c) There are ..... atoms in a molecule of $C_2H_4Cl_2$.

(d) A tiny crystal of calcium chloride, $CaCl_2$, contains three million calcium ions. How many chloride ions does it contain?

(e) Calcium hydroxide is commonly called limewater. The formula for calcium hydroxide is ..... .

(f) The chemical name for chalk or marble is calcium carbonate. Its formula is ..... .

32.2 Write down the formula for each of the following compounds:

(a) Water
(b) Sodium chloride
(c) Carbon dioxide
(d) Magnesium oxide
(e) Methane
(f) Calcium chloride
(g) Hydrogen chloride
(h) Sodium hydroxide
(i) Sulfuric acid
(j) Calcium hydroxide
(k) Manganese dioxide
(l) Hydrogen peroxide

32.3 (a) There are two types of equations in chemistry: ..... equations and ..... equations.

(b) The chemicals which react together in a chemical reaction are called the ..... .

(c) The ..... are the chemicals which are produced in a chemical reaction.

(d) A chemical equation is similar to a word equation except that it uses ..... and formulas instead of words.

(e) When writing a chemical equation, what does the arrow sign represent?

(f) In balancing an equation ..... cannot be altered in any way.

32.4 Balance the following equations:

(a) $H_2 + O_2 \longrightarrow H_2O$
(b) $Na + Cl_2 \longrightarrow NaCl$
(c) $K + O_2 \longrightarrow K_2O$
(d) $Zn + HCl \longrightarrow ZnCl_2 + H_2$
(e) $HCl + NaOH \longrightarrow NaCl + H_2O$
(f) $H_2SO_4 + Zn \longrightarrow ZnSO_4 + H_2$
(g) $H_2 + Cl_2 \longrightarrow HCl$
(h) $Ca(OH)_2 + CO_2 \longrightarrow CaCO_3 + H_2O$

Chapter 33

Metals and Plastics

33.1 Metals and Non-metals

The Periodic Table can be divided by 'steps of stairs' into metals and non-metals, *Fig. 33.1*. The elements on the left of the steps of stairs are called metals. The elements on the right are called non-metals.

Symbols of Metals and Non-metals

In *Chapter 21* we studied the symbols of various metals and non-metals.

- The metals whose symbols you must remember are **copper**, **zinc**, **aluminium**, **iron**, **silver** and **gold**.
- The non-metals whose symbols you must remember are **carbon**, **oxygen**, **sulfur**, **hydrogen** and **nitrogen**.

Metals in our Lives

Metals play a very important part in our lives. Just think of all the uses we have for metals!

Uses of Metals when Getting Ready for School or Work

- In the morning we are awakened by an **alarm clock**, part of which is made of metal.
- We press a **switch** to pass electric current along a copper wire.
- The electric **light bulb** contains a filament made of tungsten.
- We wash in water that has come to us through **copper pipes**.
- We shave using a stainless steel **razor blade**.
- We cook our breakfast on a **stove** made of steel.
- **Cooking utensils** are made of aluminium.
- We eat our food using stainless steel **knives and forks**.
- We travel to work in a **car** 80 per cent of which is made of metals.

	I	II											III	IV	V	VI	VII	0
n=1 PERIOD				1 H HYDROGEN 1														2 He HELIUM 4
n=2 PERIOD	3 Li LITHIUM 7	4 Be BERYLLIUM 9											5 B BORON 11	6 C CARBON 12	7 N NITROGEN 14	8 O OXYGEN 16	9 F FLUORINE 19	10 Ne NEON 20
n=3 PERIOD	11 Na SODIUM 23	12 Mg MAGNESIUM 24											13 Al ALUMINIUM 27	14 Si SILICON 28	15 P PHOSPHORUS 31	16 S SULFUR 32	17 Cl CHLORINE 35	18 Ar ARGON 40
n=4 PERIOD	19 K POTASSIUM 39	20 Ca CALCIUM 40	21 Sc SCANDIUM 45	22 Ti TITANIUM 48	23 V VANADIUM 51	24 Cr CHROMIUM 52	25 Mn MANGANESE 55	26 Fe IRON 56	27 Co COBALT 59	28 Ni NICKEL 59	29 Cu COPPER 63	30 Zn ZINC 65	31 Ga GALLIUM 70	32 Ge GERMANIUM 73	33 As ARSENIC 75	34 Se SELENIUM 79	35 Br BROMINE 80	36 Kr KRYPTON 84
n=5 PERIOD	37 Rb RUBIDIUM 85	38 Sr STRONTIUM 88	39 Y YTTRIUM 89	40 Zr ZIRCONIUM 91	41 Nb NIOBIUM 93	42 Mo MOLYBDENUM 96	43 Tc TECHNETIUM 99	44 Ru RUTHENIUM 102	45 Rh RHODIUM 103	46 Pd PALLADIUM 107	47 Ag SILVER 108	48 Cd CADMIUM 112	49 In INDIUM 115	50 Sn TIN 119	51 Sb ANTIMONY 122	52 Te TELLURIUM 128	53 I IODINE 127	54 Xe XENON 131
n=6 PERIOD	55 Cs CAESIUM 133	56 Ba BARIUM 137	57 La LANTHANUM 139	72 Hf HAFNIUM 179	73 Ta TANTALUM 181	74 W TUNGSTEN 184	75 Re RHENIUM 186	76 Os OSMIUM 190	77 Ir IRIDIUM 193	78 Pt PLATINUM 195	79 Au GOLD 197	80 Hg MERCURY 201	81 Tl THALLIUM 204	82 Pb LEAD 207	83 Bi BISMUTH 209	84 Po POLONIUM 210	85 At ASTATINE 210	86 Rn RADON 222
n=7 PERIOD	87 Fr FRANCIUM 223	88 Ra RADIUM 226	89 Ac ACTINIUM 227	104 Rf RUTHERFORDIUM 261	105 Db DUBNIUM 262	106 Sg SEABORGIUM 263	107 Bh BOHRIUM 262	108 Hs HASSIUM 265	109 Mt MEITNERIUM 266									

Metals | Non-metals

Fig. 33.1 The elements of the Periodic Table may be divided into metals and non-metals by the 'steps of stairs' going from boron to astatine.

Other Uses of Metals

During the day we make many other uses of metals.

- We use **doors** and **windows** made of aluminium.
- We use **coins** made of metals like copper, zinc, nickel, etc.
- We use **jewellery** made of silver and gold.

For your examination, you must remember some uses of metals in your everyday lives. How many of the above examples can you remember? Can you think of any other applications of metals?

33.2 Properties of Metals and Non-metals

(1) Metals

You probably know some properties of metals already.

- Many metals are shiny, e.g. gold and silver. Scientists use the word **lustrous** to describe the fact that metals are shiny. Therefore, they are very useful in jewellery and in making mirrors.
- You may also know that metals can be stretched and made into wires, e.g. copper wire. Since metals can be stretched, we say that they are **ductile**. Copper wires carry the electricity around your home.
- Another property of metals is that they can be beaten into various shapes. For example, panel beaters have to beat the steel in a crashed car back into shape. Scientists use the word **malleable** to describe the fact that metals can be beaten into various shapes.

The three words highlighted above are specifically mentioned on the syllabus so do your best to remember them!

Other Properties of Metals

The other general properties of metals may be summarised as follows:

- Metals are **strong**. The fact that steel is so **strong** means it can support a heavy load. Therefore, it is used for building bridges, as beams (girders) for construction of tall buildings, in scaffolding around buildings, etc.
- Metals usually have **high melting and boiling points**. This is why all of them (except mercury) are solid at room temperature.
- Metals are **fairly heavy** materials. This is because many metals have a high density, e.g. the density of mercury is 14 g/cm^3 approximately and that of gold is 20 g/cm^3 approximately.
- Metals are **good conductors of heat**, i.e. they allow heat to pass through them easily. This means that metals are very useful for making pots and pans.
- Metals are **good conductors of electricity**, i.e. they allow electricity to pass through them easily. Therefore, metals are very useful as wiring in electrical circuits.
- Many metals make a **ringing sound** when you strike them.

The general properties of metals are summarised in *Fig. 33.2*.

Fig. 33.2 Some general properties of metals.

Exceptions

These general properties are not true for **all** metals. For example, sodium and potassium are less dense than water. Also, sodium and potassium are so soft that they may be cut with a knife. Lead is a very dull metal. Not all metals are solid at room temperature, e.g. mercury is a liquid at room temperature.

(2) Non-metals

Non-metals are not as useful in our everyday lives as metals are. Some properties of **non-metals** are:

- The non-metals that are solid at room temperature tend to be **dull**, e.g. carbon and sulfur do not shine.
- They are **inelastic**, not ductile, i.e. they do not stretch.
- The non-metals that are solid tend to be **brittle**, not malleable, i.e. they cannot be hammered into a different shape but shatter instead, e.g. carbon rods.
- They usually have **low melting and boiling points**, making many of them **gases** at room temperature.
- They are **light**. They usually have low density.
- They are **not good conductors of heat or electricity**.
- **None** of them is **magnetic**.

Exceptions

There are exceptions to the above. For example, carbon in the form of graphite is a good conductor of electricity. Carbon in the form of diamond is very hard, has a high melting point and shines.

A Metal or a Non-metal?

One of the easiest ways of testing whether a substance is a metal or a non-metal is to see if it conducts electricity. By experiment, it has been found that the only common solids that conduct electricity well are metals and graphite. This experiment may be done using the apparatus in *Fig. 33.3*.

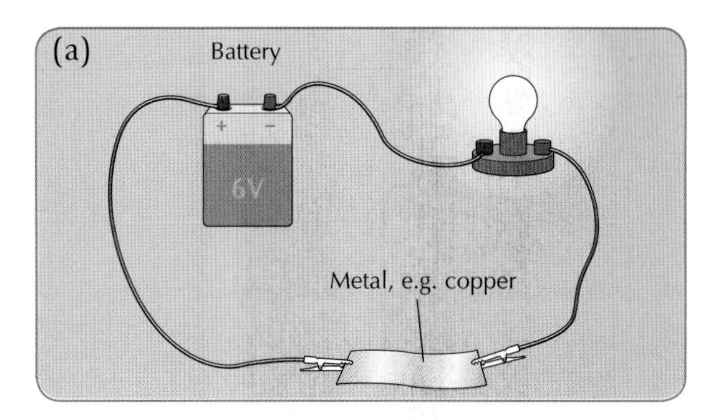

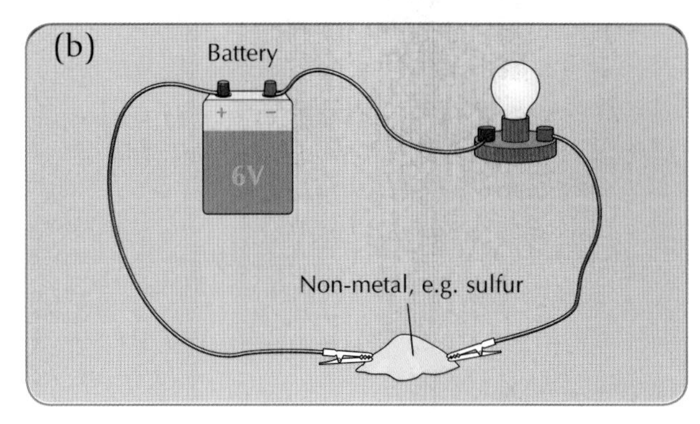

Fig. 33.3(a) Metals are good conductors of electricity. Therefore the bulb lights. (b) Non-metals do not conduct electricity. Therefore the bulb does not light.

The bulb lights when the substance being tested is a metal. When non-metals (except graphite) are tested, the bulb does not light. This is because non-metals do not conduct electricity.

Examples of Metals and Non-metals

For your examination, you must be able to give the state and colour of two examples of metallic elements and two examples of non-metallic elements. Some examples of these are illustrated in *Fig. 33.4*.

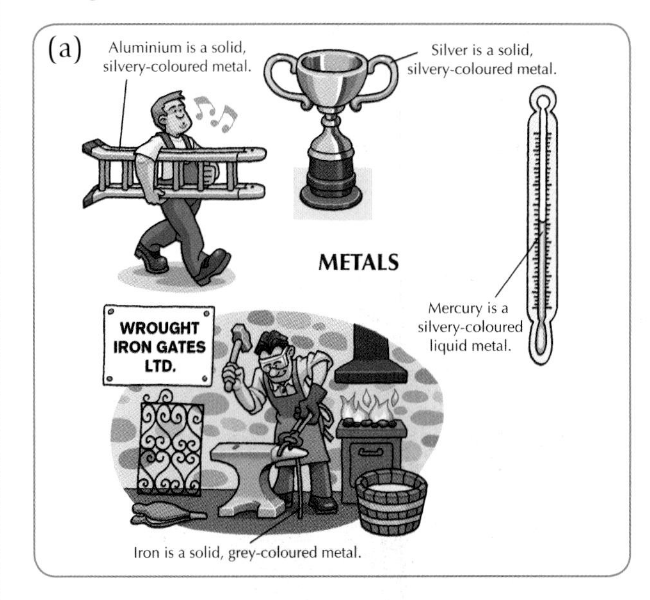

Fig. 33.4(a) Some examples of metals.

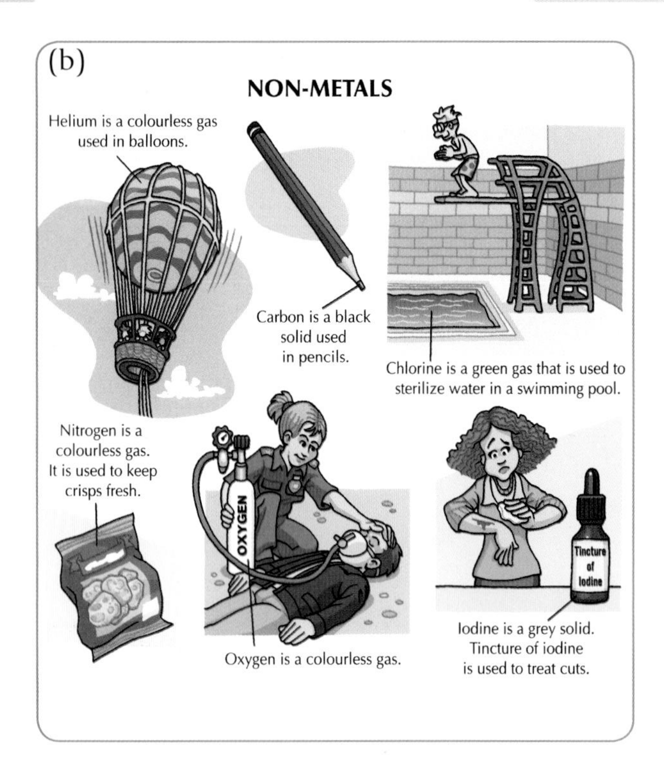

Fig. 33.4(b) Some examples of non-metals.

TEST YOURSELF:
Now attempt questions *33.1* and *W33.1–W33.3*.

33.3 Alloys

Some of the metallic substances that we meet in our everyday lives are not made of pure metals. For example, the coins you use every day are mixtures of metals. The €1 and €2 coins are made of a mixture of copper, nickel and zinc. Such mixtures of metals are called **alloys**.

An alloy is a mixture of metals.

Examples of Alloys

You may have heard of alloy wheels on cars. These wheels are made of aluminium mixed with various other metals. Some alloys contain the non-metal carbon, e.g. steel is a mixture of iron and carbon. There are four alloys on your course. These alloys are **brass**, **bronze**, **steel** and **solder**. **Brass** is a mixture of copper and zinc. **Bronze** is a mixture of copper and tin. **Steel** is a mixture of iron and carbon. **Solder** is a mixture of lead and tin. You do not need to know the composition of these alloys for your examination.

Uses of Alloys

For your examination, you must be able to state one use of each of the above alloys. Study *Fig. 33.5* carefully.

Fig. 33.5 Some uses of the alloys on our course.

Reasons for Making Alloys

Why do we use alloys? Why not simply use pure metals? The reason is because alloys usually have different properties than the metals from which they are made.

Hard Alloys

For example, chemists have found that alloying one metal with another, increases the strength of the metal.

- At one time **coins** were made out of silver and gold. However, these metals were not hard enough and the coins wore away. Therefore, modern coins are made of alloys.
- Pure iron is not hard enough or strong enough for use in **construction**. In addition, pure iron rusts easily. However, when carbon is mixed with iron, it becomes harder and stronger.

TEST YOURSELF:
Now attempt questions *33.2 and W33.4–W33.5.*

33.4 Group I Metals – The Alkali Metals

The alkali metals are the elements of group I of the Periodic Table. As expected, since they are all in the same group, they have similar properties. The name alkali is used because these metals react with water to form alkaline or basic solutions. Our study of the alkali metals will be confined to lithium, sodium and potassium, as these are the alkali metals most commonly found in school laboratories. All of the alkali metals are soft, silvery solids.

Reactivity of the Alkali Metals

- The alkali metals do not occur freely in nature, i.e. they are not found lying in the earth like gold is. They are so reactive that they quickly combine with other elements to form **compounds**.
- They are very reactive because they want to **lose the single outer electron**. Losing this electron will give them a stable electron configuration.
- The alkali metals must always be **stored under oil**. This prevents them reacting with the moisture and oxygen in the air.

Physical Properties of the Alkali Metals

The physical properties of the alkali metals are summarised in *Fig. 33.6*.

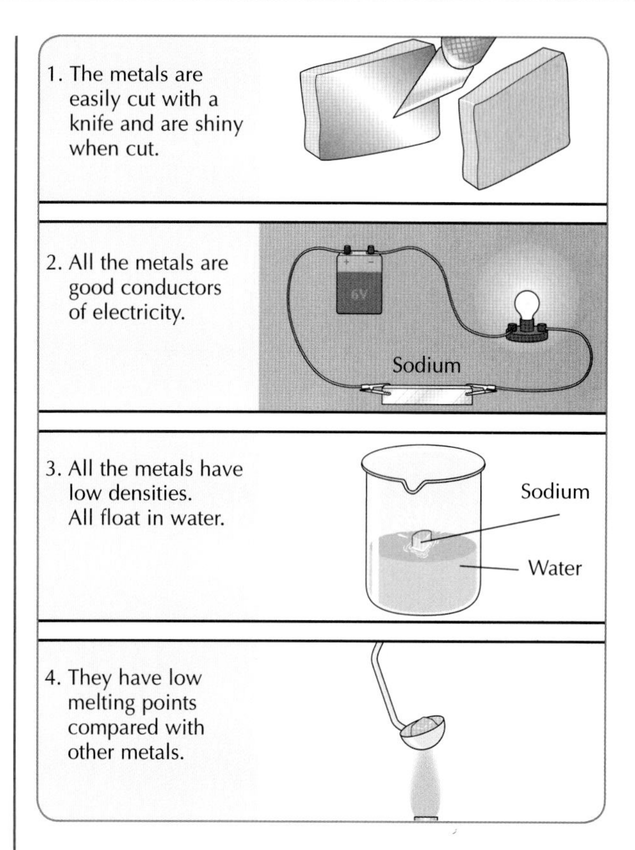

Fig. 33.6 The physical properties of the alkali metals.

Chemical Properties of the Alkali Metals

There are two chemical properties of the alkali metals on your course. These two reactions are (1) reaction with air and (2) reaction with water.

(1) Reaction with Air

All of the alkali metals lose their shiny appearance when exposed to air. We say that they tarnish in air. The alkali metals react with the oxygen in the air as follows:

lithium + oxygen $\longrightarrow$ lithium oxide

sodium + oxygen $\longrightarrow$ sodium oxide

potassium + oxygen $\longrightarrow$ potassium oxide

This is one of the reasons why the alkali metals are stored under oil. If they were allowed to come in contact with air, they would change into a white powder.

(2) Reaction with Water

Sodium

When a piece of sodium is added to water, the metal floats on the surface of the water and melts

CHEMISTRY

into a ball. A gas called hydrogen is formed. This gas pushes the sodium around the surface of the water. A clear solution of sodium hydroxide is formed. The following is the word equation for the reaction that occurs:

sodium + water ⟶ sodium hydroxide + hydrogen

Potassium

When potassium is added to water, a similar reaction occurs. In this case, the hydrogen catches on fire and sends sparks flying, *Fig. 33.7*.

potassium + water ⟶ potassium hydroxide + hydrogen

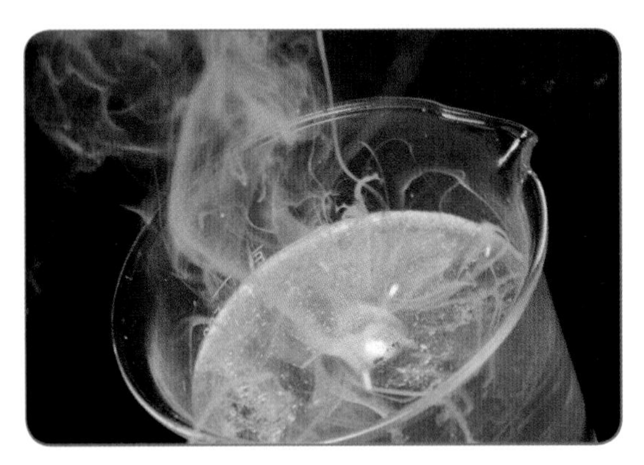

Fig. 33.7 Potassium reacts vigorously with water. The hydrogen formed catches fire!

Lithium

The least reactive of the alkali metals is lithium. This reacts slowly with water to form lithium hydroxide and hydrogen. Like sodium and potassium, it is less dense than water and floats on it.

lithium + water ⟶ lithium hydroxide + hydrogen

We can summarise the above as follows:

The alkali metals react vigorously with water. Hydrogen gas is given off and a hydroxide compound is formed.

Safety glasses and a safety screen should always be used when studying the reactions of the alkali metals with water.

Summary of the Chemical Properties of the Alkali Metals

The chemical properties of the alkali metals are summarised in *Fig. 33.8*.

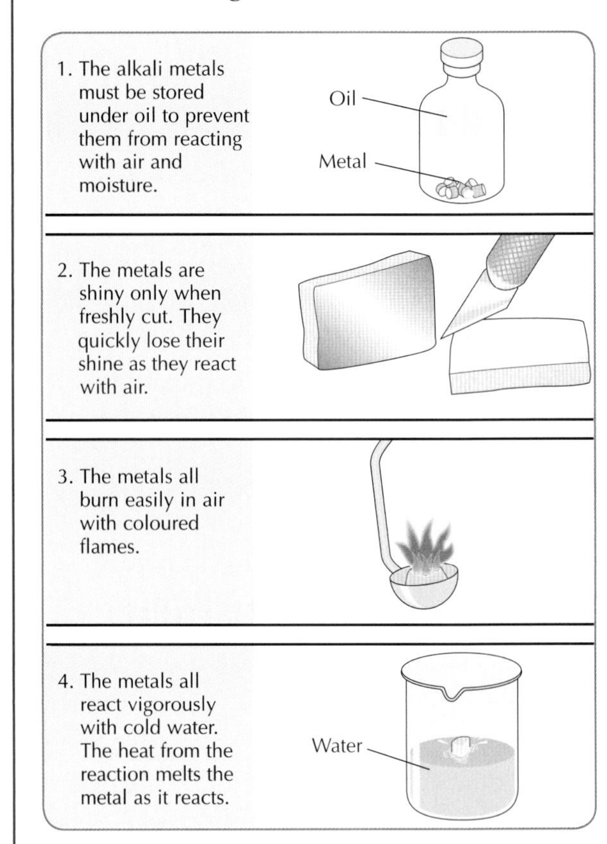

Fig. 33.8 The chemical properties of the alkali metals.

Uses of the Alkali Metals

- **Lithium** is used in the manufacture of **batteries** for watches and cameras.
- **Sodium** is used widely in **street lighting** since the light emitted has a low dazzling effect.
- **Potassium** metal has little use in everyday life but its compounds are widely used in **fertilisers**.

TEST YOURSELF:

Now attempt questions *33.3* and *W33.6–W33.7.*

33.5 Reactions of Metals with Acid and Water

In this section we will investigate the reactions of various metals with water and acid.

(1) Zinc

We first investigate the reaction between zinc and hydrochloric acid and test for hydrogen gas.

Mandatory experiment 33.1

(a) To investigate the reaction between zinc and hydrochloric acid (b) To test for hydrogen

Acids react with many metals to form a salt and hydrogen gas. In this experiment we will study the reaction between hydrochloric acid and zinc metal. The word equations and balanced equations for this reaction are:

Word equation:

zinc + hydrochloric acid ⟶ zinc chloride + hydrogen

Chemical equation:

$$Zn + HCl \longrightarrow ZnCl_2 + H_2$$

Warning! Hydrogen is a highly flammable gas. It forms an explosive mixture with air. Under no circumstances should any flame be brought near the gas preparation apparatus. Tests on the hydrogen should be performed well away from the apparatus.

Apparatus required: dropping funnel; Buchner flask; delivery tubing; trough; test-tubes or boiling tubes; stoppers; beehive shelf; tapers

Chemicals required: hydrochloric acid (dilute); zinc granules; water

(a) To investigate the reaction between zinc and hydrochloric acid

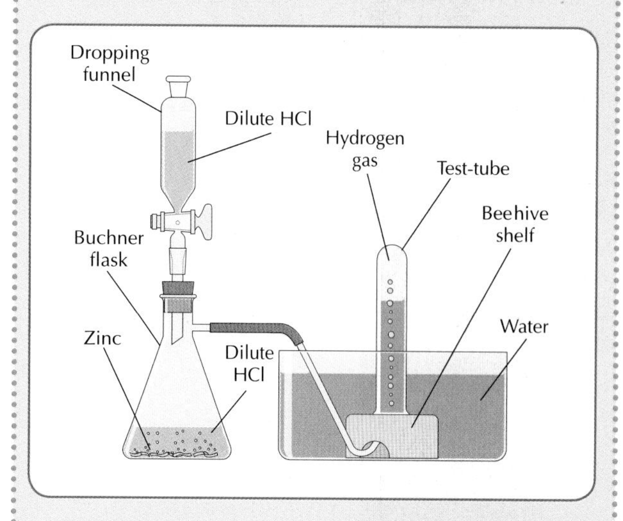

Fig. 33.9 Zinc reacts with hydrochloric acid to form zinc chloride and hydrogen gas.

Method

1. Set up the apparatus as shown in *Fig. 33.9*.
2. Collect about six test-tubes or boiling tubes of the gas.

(b) To test for hydrogen

Method

1. Note that the gas is **colourless** and **odourless**. Place pieces of moist red litmus paper and blue litmus paper into a test-tube of the gas. Note that there is no change in colour of the paper indicating that hydrogen is a **neutral** gas.
2. Light a wax taper. Keep the test-tube upside down for this test. Remove the stopper from the test-tube and bring the lighted taper up to the mouth of the test-tube, *Fig. 33.10*. If the gas is pure, it will burn with a blue flame. If mixed with air, it will burn with a **loud pop**. What is actually happening is that hydrogen gas is combining with oxygen gas to form water.

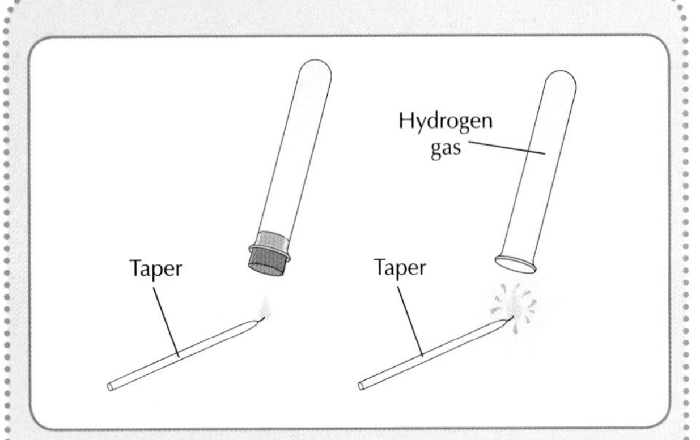

Fig. 33.10 Hydrogen burns with a loud pop when it is mixed with air.

3. Place an empty test-tube over one containing hydrogen. Remove the stopper from the bottom test-tube. Use a lighted taper to test the top test-tube for the presence of hydrogen gas. ***Note:*** Hydrogen gas is **less dense than air**. In fact, hydrogen is the lightest gas known.

(2) Calcium, Magnesium, Zinc and Copper

We now investigate the relative reactivities of a number of metals. Two of these metals are **Alkaline Earth** metals – magnesium and calcium. You may recall from *Chapter 30* that the Alkaline Earth metals are the elements in Group II of the Periodic Table. The Alkaline Earth metals are less reactive than the alkali metals.

Experiment 33.2

To investigate the relative reactivities of four metals with water and acid

Apparatus required: test-tubes; test-tube rack

Chemicals required: calcium; magnesium; zinc; copper; water; dilute HCl

To help us compare the reactivities of these metals, we will add the metals to water and also to dilute acid.

Warning! The addition of calcium to the acid should only be carried out under the direct supervision of your teacher. Safety glasses must be worn for all these experiments.

(a) To investigate the relative reactivities of four metals with water

Method

1. Set up the apparatus as shown in *Fig. 33.11*. To have a fair test, use the same amount of water in each test-tube. Also, choose pieces of metal of about the same size.

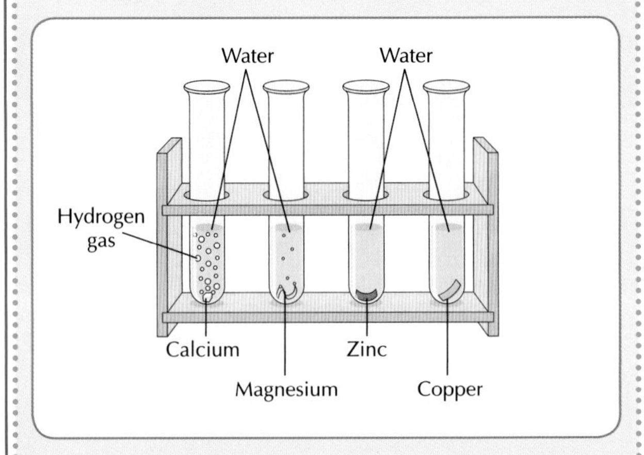

Fig. 33.11 Calcium reacts quickly with water. Magnesium reacts very slowly with water. Zinc and copper do not react.

2. Study how reactive the metals are by noting how quickly bubbles of gas are given off.
3. Using an inverted test-tube, collect a sample of the gas given off, *Fig. 33.12(a)*. Test the gas with a lighted taper. Alternatively, the lighted taper can be held at the mouth of the test-tube, *Fig. 33.12(b)*.

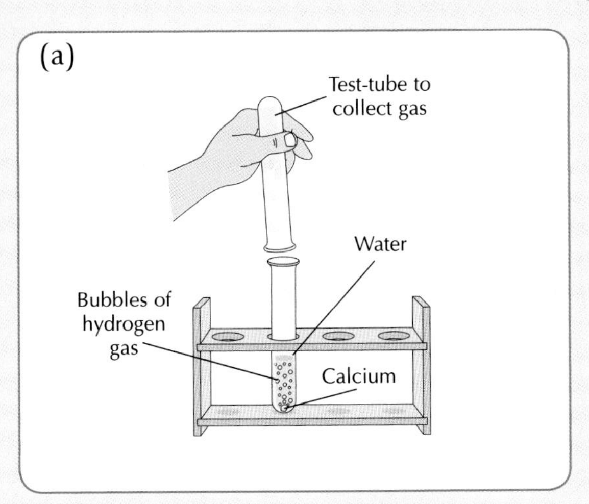

Fig. 33.12(a) The gas given off when calcium reacts with water is collected in an inverted test-tube.

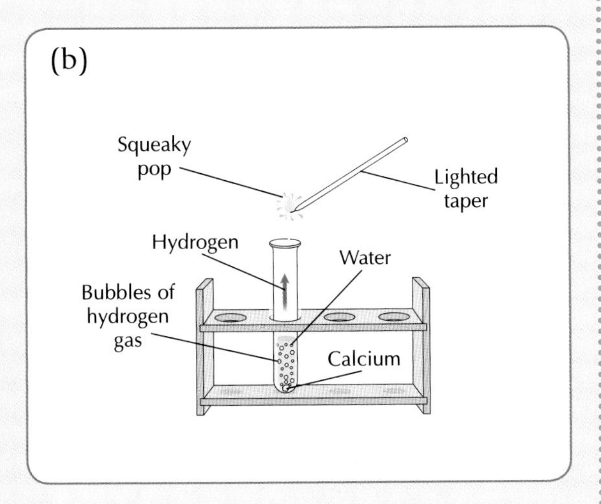

Fig. 33.12(b) The gas can also be tested for as it is being given off.

Result

The reaction between the calcium and water is the fastest. The gas burns with a pop. Therefore, hydrogen gas is given off. The magnesium does not appear to react with water. (In fact, it does react with water, but very slowly. It takes several days for a few cm^3 of hydrogen gas to be collected from the reaction of magnesium with water). The zinc and the copper do not react at all with cold water.

Conclusion

Calcium is the most reactive metal of the four metals studied.

(b) To investigate the relative reactivities of four metals with acid

Method

1. Set up the apparatus as shown in *Fig. 33.13*. To have a fair test, use the same amount of acid and the same concentration of acid in each test-tube. Also, choose pieces of metal of about the same size.

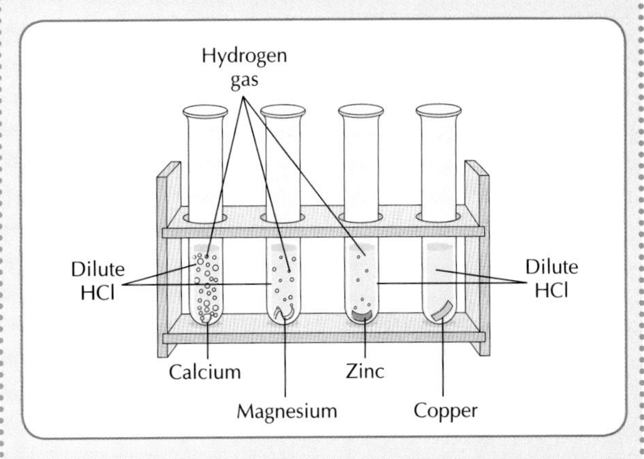

Fig. 33.13 When added to dilute HCl, calcium reacts quickly, magnesium reacts less quickly, zinc reacts slowly and copper does not react at all.

2. Study how reactive the metals are by noting how quickly the bubbles of gas are given off.
3. Test the gas with a lighted taper taking all the precautions already outlined.

Result

Calcium reacts most vigorously, followed by magnesium and then zinc. Copper does not react at all. The fact that the gas burns with a pop, shows that hydrogen gas is given off.

Conclusion

Calcium is the most reactive, magnesium is next most reactive, zinc is next and copper is the least reactive, *Fig. 33.14*.

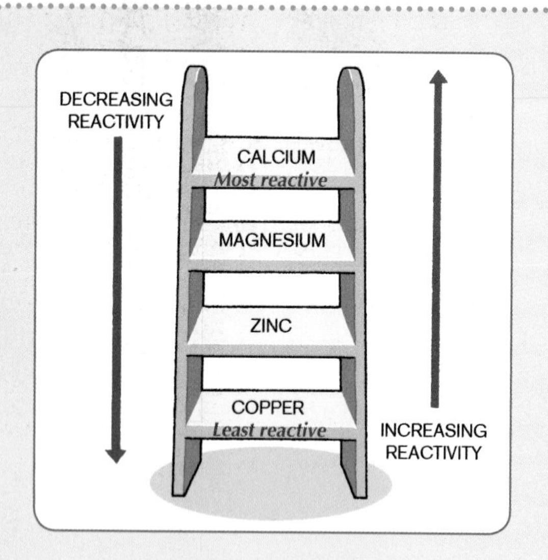

Fig. 33.14 Calcium is the most reactive of the four metals and copper is the least reactive.

Don't get confused in an exam if you are asked for order of decreasing or increasing reactivity.

Decreasing reactivity: calcium, magnesium, zinc, copper.

Increasing reactivity: copper, zinc, magnesium, calcium.

TEST YOURSELF:

Now attempt questions *33.4* and *W33.8–W33.10*.

33.6 Corrosion of Metals

One of the problems of using metals is that they tend to corrode. When a metal corrodes, it is converted to one of its compounds. You are probably familiar with the problem of rusting.

The corrosion of iron and steel is called rusting.

An example of corrosion is shown in *Fig. 33.15*.

Fig. 33.15 The corrosion of iron and steel is called rusting.

The chemical name for rust is iron oxide. Rusting is a chemical reaction in which iron is changed into a new substance (rust). The conditions for rusting to occur are studied in the next experiment.

Note: Not all metals corrode. Some metals (e.g. gold) are very unreactive and do not react with the oxygen or the water in the air.

Mandatory experiment 33.3

To carry out an experiment to demonstrate that oxygen and water are necessary for rusting

In this experiment, three nails are placed in different conditions.

- One nail is in contact with both **air and water**.
- The second nail is in contact with **air only** (no water).
- The third nail is placed in contact with **water only** (no air). From observing which nails rust, we can deduce the conditions necessary for rusting.

Apparatus required: test-tubes (3); test-tube rack; rubber stopper; steel nails (3)

Chemicals required: calcium chloride; oil; water

Method

1. Set up three test-tubes as shown in *Fig. 33.16*.

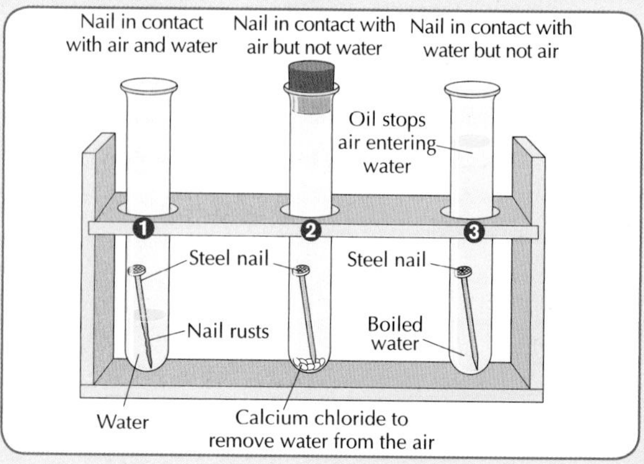

Fig. 33.16 This experiment proves that both air and water are necessary for rusting to occur. The only nail which rusts is that in the first test-tube. Only this nail is in contact with both air and water.

2. In test-tube 1 place a nail. This test-tube contains both air and water.
3. Place another nail in a test-tube that contains some calcium chloride (test-tube 2). Calcium chloride is a substance that absorbs water. Therefore, there is no water in the air in test-tube 2. Stopper the test-tube to prevent moisture from the air entering.
4. Place a third nail in a test-tube containing some boiled water (test-tube 3). While the water is still hot, cover the water with some oil. This prevents any air dissolving in the water when the water cools down.
5. Leave the apparatus to stand for a few days.

Results

It is found that the only nail that rusts is the nail in test-tube 1. This is the only nail that is in contact with both oxygen and water.

Conclusion

Oxygen and water are necessary for rusting.

How is Rusting Prevented?

Replacing rusted steel costs this country many millions of euro each year. Most methods of rust prevention involve coating the metal with some material to prevent air and water coming in contact with the metal. The most important ways of rust prevention are:

- **Painting**. This prevents air and water coming into contact with the metal. The paint forms a layer over the metal. This method is used to prevent corrosion of bridges, ships, gates, cars, bicycles, etc. If the paint gets scratched, the exposed metal may rust.
- **Oiling or Greasing**. The oil or grease forms a layer over the surface of the metal. This prevents air and water coming in contact with the metal. This method of rust prevention is usually used for moving parts of machines. If they were painted, the paint would be scratched off. This is useful for bicycle chains, car engines, etc.
- **Galvanising**. Buckets, dustbins, sheets of corrugated iron, wire for fencing, etc. are often covered with zinc. This is done by dipping the objects into molten zinc. The metal zinc is used because it does not corrode. The coating of zinc prevents the metal from coming in contact with air or water. Iron coated with zinc is often called **galvanised iron**.

The methods of rust prevention are summarised in *Fig. 33.17*.

Fig. 33.17 Painting, oiling (greasing) and galvanising are three important methods of rust prevention.

TEST YOURSELF: Now attempt questions *33.5* and *W33.11–W33.12.*

33.7 Plastics

The word plastic was first used to describe substances that became soft when warm. It means 'able to have its shape changed easily'. Plastics are man-made materials and most of them are made from crude oil. Crude oil is separated into various useful substances in an oil refinery. Approximately 4 per cent of products from oil refineries are used to manufacture plastics.

Examples and Uses of Plastics

Plastics are extremely useful materials. Think of all the materials around you that are made of plastic. Some common plastics that you use are polythene, polystyrene, polypropylene, nylon and PVC. Some examples of items made from these are illustrated in *Fig. 33.18*.

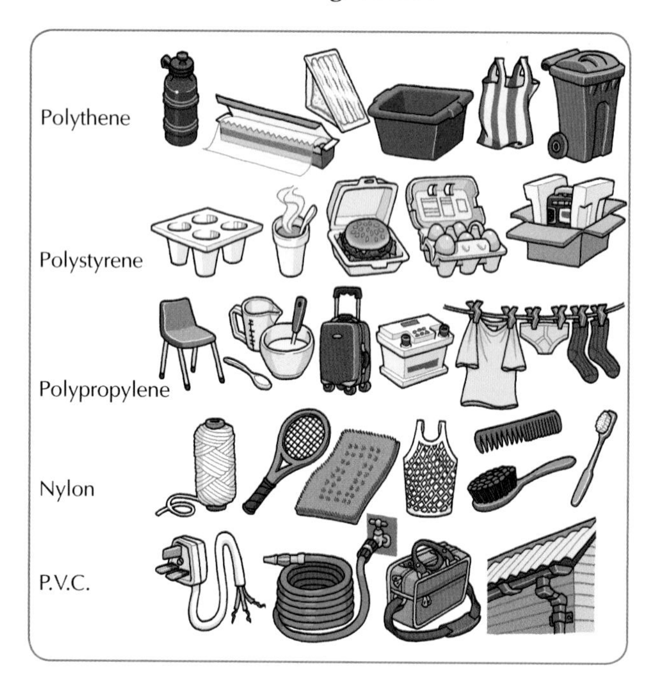

Fig. 33.18 Some plastics and their uses.

Note: Some plastics occur naturally e.g. rubber is obtained from a tree. However, most plastics we use are made from crude oil.

Properties of Plastics

Plastics have replaced traditional materials such as wood and metal for many purposes. For example, PVC is often used instead of wood for making windows. It is also used instead of metal for making gutters on buildings.

Our everyday experience tells us that plastics have many useful properties:

- They can be **moulded** into many different shapes and sizes.
- They are **easy to maintain**. They do not rust or rot.
- They are **not very expensive**.
- They do **not** tend to be **very heavy,** i.e. they have low density.
- They are **good insulators** for both heat and electricity.
- They **can be dyed** and made in many different colours.

Relating Properties of Plastics to their Uses

- Since plastics do **not conduct electricity**, they are used for making **plug tops** and **insulating material** around copper wire.
- Since plastics are good **heat insulators**, they are used for making **drinking cups** for hot liquids, containers for takeaway food, insulation for houses, etc.
- The fact that plastics have **low density** means they are used for **bottles** (instead of glass), packaging, shopping bags, milk crates, etc.
- The ease with which plastics can be **moulded** means that all sorts of shapes can be produced, e.g. **buckets**, dishes, jugs, beakers, etc.
- The fact that plastics do **not rot**, makes them very useful for making **window frames**, doors, ropes, etc.
- Since plastics can be **easily dyed**, they are often used for making **children's toys**. Since plastics do not rust, steel buckets that were used in the past have now been replaced with plastic buckets.

Plastics and the Environment

Even though plastic materials are very useful to us, we must realise that they also have disadvantages.

(1) Non-biodegradable

One of the biggest problems with plastics is that most of them are **non-biodegradable**.

> **Most plastics are non-biodegradable, i.e. they are not broken down in the environment by the action of micro-organisms.**

In other words, they do not rot away like a piece of wood or cardboard does. Items like wood and cardboard rot away when buried in a landfill site. Since a lot of plastic material is used for packaging, this plastic packaging can cause litter problems, *Fig. 33.19*.

Fig. 33.19 Most plastics are non-biodegradable. Therefore, they cause litter problems as they do not rot away.

In general, plastics can cause a serious litter problem unless disposed of carefully. Efforts are being made to produce more plastics with additives that make them biodegradable. In addition, in many countries plastics are recycled and made into other materials.

(2) Poisonous Fumes

Another problem caused by plastics is that many of them soften and melt when heated. When these plastics burn, they give off poisonous fumes. This is very serious in the case of fires as these fumes can cause death if inhaled.

TEST YOURSELF:
Now attempt questions *33.6* and *W33.13–W33.14*.

Exam Question

Junior Cert Science Higher Level Examination Question

Different plastics have different properties. The dust pan and brush set, Fig. 33.20, is made from two different plastics. The bristles are made of type A and the other parts are made of type B plastic. Give one property of type A and one property of type B plastic that make them suitable for their use in this product.

Fig. 33.20

Property of type A: The bristles of the brush must be hard-wearing and flexible (any one property).

Property of type B: The dust pan must not corrode, can be moulded in various shapes, is light to carry, is rigid and can be made in various colours (any one property).

What I should know

- Metals are usually strong, hard and shiny. They have high densities, high melting points and boiling points. They are good conductors of heat and electricity, can be hammered into different shapes and can be pulled into wires.
- Non-metals are not good conductors of electricity, have low melting and boiling points, are brittle and cannot be stretched.
- An alloy is a mixture of metals.
- The alkali metals are soft metals and they are shiny when cut but tarnish in air. They react with air to form oxides. The alkali metals react vigorously with water. Hydrogen gas is given off and a hydroxide compound is formed.
- Zinc reacts with hydrochloric acid to form hydrogen gas.
- Calcium is more reactive than magnesium which is more reactive than zinc which is more reactive than copper.
- The corrosion of iron and steel is called rusting.
- Both oxygen and water must be present for rusting to occur.
- Most plastics do not rot. They are non-biodegradable.

QUESTIONS

33.1 (a) Heat and electricity easily flow through metals. We say that metals are good ..... of heat and electricity.

(b) Give two general properties of metals.

(c) Most metals can support large weights, i.e. we say that metals are ..... .

(d) What is meant by the term malleable?

(e) The tungsten filament in an electric light bulb is an example of a very fine wire. What name is given to the property of metals to be drawn into thin wires?

33.2 (a) Metals can be strengthened by mixing them with one or more other elements. These mixtures are called ..... .

(b) The medals awarded in the Olympic Games are made of the metals ..... and silver and the alloy ..... .

(c) Brass is used for ..... .

(d) Bronze is used for ..... .

(e) The melting point of solder is lower than the melting points of the metals from which it is made. Why is this lower melting point an advantage when soldering?

33.3 The elements in Group I are called the ..... metals. The Group I metals that are found in the school laboratory are ..... , ... and ..... . They are stored under oil to protect them from ..... and ..... . These metals have unusual ..... . For example, they are ..... to cut. When cut they have a ....appearance on the inside. They react ..... with water. They ..... on top of the water. A gas called ..... is given off. The solution formed turns litmus paper from ..... to ..... . This means that the solution is ..... . In the case of ....., the hydrogen gas catches on fire and sparks are produced.

33.4 (a) Complete and balance the following equation:

$Zn + HCl \longrightarrow$

(b) A piece of zinc metal is added to dilute hydrochloric acid. Bubbles are seen to form.

(i) Name the gas given off.
(ii) What test would you perform to identify the gas?
(iii) Draw a labelled diagram to show how the gas is collected.
(iv) Give two properties of this gas.

33.5 (a) What term is used to describe the unwanted conversion of a metal to one of its compounds?

(b) What is rusting?

(c) List the conditions necessary for rusting to take place.

(d) Why does painting a metal help prevent it rusting?

(e) Galvanising means covering a metal with a layer of ..... .

33.6 (a) Plastic materials are usually made from ..... oil.

(b) A label on a plastic dustbin warned about putting hot ashes into it. Why was this warning given?

(c) Name two plastic materials commonly used in everyday life.

(d) When camping, polythene dishes have now replaced enamelled ones. Why do you think that this has happened?

(e) Plastic cups are often used for hot drinks from vending machines. Why is plastic a good material to use for this purpose?

Chapter 34 Measurement and Units

34.1 Physics and the Scientific Method

Physics deals with things that happen in the world around us. It tries to explain **how** and **why** everyday things behave as they do. Scientists are **curious** people. They ask questions (see *Fig. 34.1*). They carry out **investigations**. They take **measurements**. They gather **evidence**. They put forward **theories** or suggestions to try to explain how things work. Later, they gather further evidence. They then often have to amend or improve their theories. They gradually build up an amount of **knowledge**. This often leads to new **discoveries** or **inventions**.

Fig. 34.1 A scientist constantly questions things and tries to come up with answers.

What Skills are Needed to Make a Good Scientist?

- Scientists usually work in teams. They must be able to **communicate** well.
- They must **use information** gathered by other scientists. They must **write clear reports** so that other scientists can understand their work.
- Scientists need good powers of **observation**. They rely on their **senses** for this*. These senses are sight, touch, hearing, smell and taste.
- Scientists must be able to use measuring devices called **instruments.** These help make accurate observations. *Fig. 34.2* shows some everyday **measuring instruments** used in the home.

Fig. 34.2 Common measuring instruments found in the home. Can you identify what each of these measures?

Note:* Scientists cannot always fully rely on their **senses. *Fig. 34.3* shows that sometimes our sense of sight can give us confusing messages.

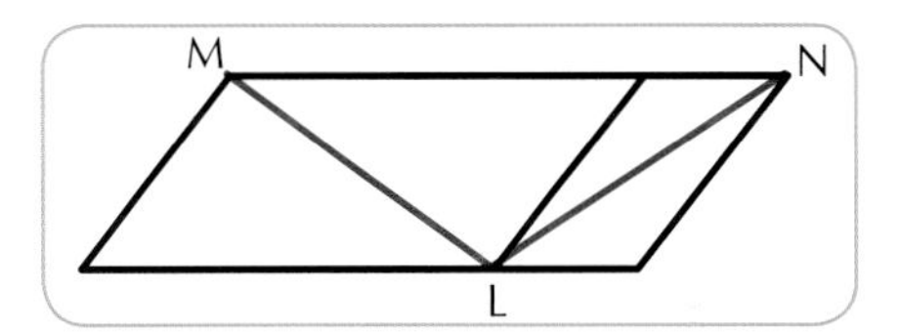

Fig. 34.3(a) Which line do you think is longer, |LM| or |MN|? (Guess first then measure them.)

Fig. 34.3(b) Which person is the tallest? Check the height of each person with a ruler. Can you explain this?

TEST YOURSELF:
Now attempt questions *34.1–34.3* and *W34.1–W34.3.*

PHYSICS

34.2 Measuring Length

Length is measured in units called **metres** (*symbol*: **m**). Other commonly-used units of length are:

kilometre (*km*)	=	**1000 m**		
centimetre (*cm*)	=	**$^{1}/_{100}$ m**	=	**0.01 m**
millimetre (*mm*)	=	**$^{1}/_{1000}$ m**	=	**0.001 m**

Below is a list of the various **instruments** used to measure length.

(1) Metre Stick (or ruler)

(*Accuracy:* To the nearest mm)

When reading a metre stick or ruler, it is important that your head is **directly over** the point being measured. This is true, in fact, when reading any instrument. It avoids what is known as '**the error of *parallax***'. It helps too if the metre stick is on its **edge** when measuring length.

(2) Measuring Tape

(*Accuracy:* To the nearest mm)

Measuring tapes usually come in two varieties: metallic and cloth, *Fig. 34.2*, page 247.

- Builders, carpenters, plumbers, etc., mostly use the **rigid** metallic measuring tape. They use it to measure the lengths of walls, windows, doors, pipes, etc.
- Cloth measuring tapes are more **flexible**. They are used in clothes shops to measure waist size, etc. When using a cloth measuring tape, it is important to keep the tape tight (or taut).

(3) Callipers

(*Accuracy:* To the nearest mm)

Callipers are used along with rulers or metre sticks. They can measure the diameter of objects like pipes, beakers, cans, tennis balls, etc.

Fig. 34.4 shows how callipers are used to find the outside (external) and inside (internal) diameter of a beaker.

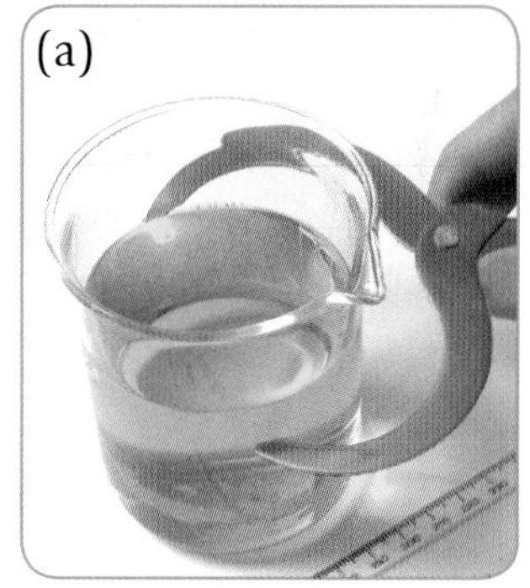

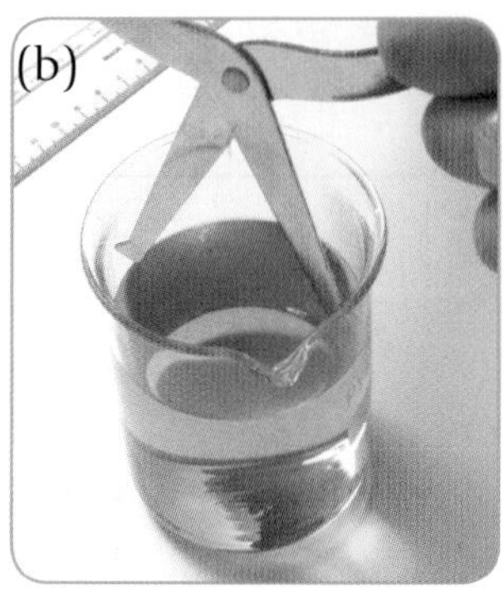

Fig. 34.4 If you measured the outside diameter of the beaker in (a) and its inside diameter in (b), how would you use this information to measure the thickness of the glass? (Think carefully before you answer.)

(4) Vernier Callipers

(*Accuracy:* To the nearest 0.1 mm)

These are a more accurate form of the callipers above. They are used to measure **small lengths** or **thicknesses**. They can also measure inside and outside diameters. They contain two scales.

- The first is a fixed **main scale** measured in centimetres.
- The second is a smaller scale, called a **Vernier scale**. The Vernier scale can slide over the main scale.

Using Vernier Callipers

1. Place the ball to be measured securely between the jaws of the Vernier callipers, *Fig. 34.5(a)*.
2. Take the first reading where the **zero** or first line of the Vernier scale meets the main scale. We can see from the photograph that the diameter of this ball is between 3.2 and 3.3 cm.
3. For the final decimal place, look closely at the Vernier scale, *Fig. 34.5(b)*. You must find the first line on the Vernier scale that is directly underneath a line from the main scale. In this case, the line at '8' best satisfies this condition. Therefore, the diameter of the ball is 3.28 cm.

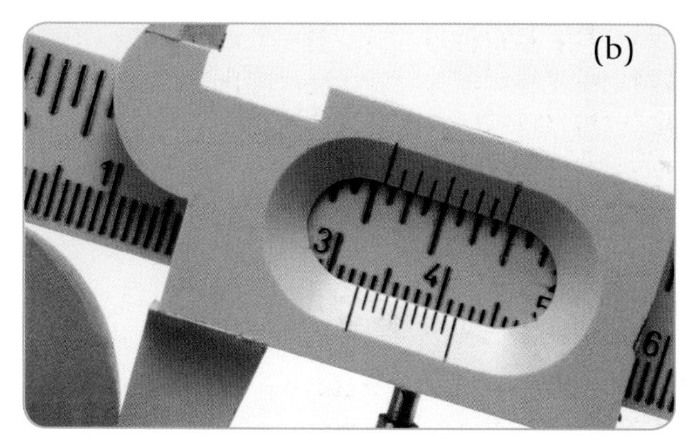

Fig. 34.5 Vernier callipers measuring the diameter of a ball.

(5) Opisometer

(*Accuracy:* To the nearest mm, when used with a ruler or metre stick)

Opisometers are used to measure the length of **curved lines**.

Using an Opisometer

To use an opisometer (*Fig. 34.6*), the following steps should be taken:

1. Turn the wheel of the opisometer until it sticks to the pointer at the side of the axle.
2. Carefully roll the opisometer along the whole length of the line it is measuring.
3. Place the opisometer at the zero of a metre stick.
4. Then roll the opisometer wheel in the opposite direction until it comes back and sticks to the pointer again.
5. Note the reading on the metre stick, where the wheel stops. This is the length of the line.

Using String or Thread

Curved lines can also be measured with a piece of **string** or thread. The string is made to follow the path of the curved line exactly. It is then straightened out along a ruler to measure its length.

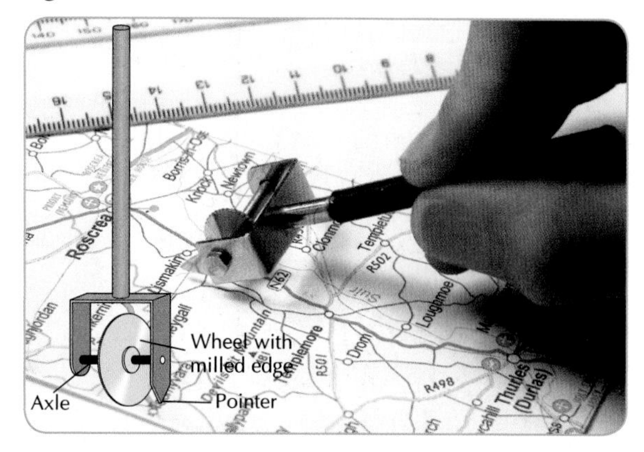

Fig. 34.6 An opisometer being used to measure the distance between two towns on a map. (The actual distance between the towns can later be calculated if the scale of the map is known.)

(6) Trundle Wheel

(*Accuracy*: To the nearest cm)

The trundle wheel is like a large scale model of an opisometer. But, it has its **own scale** marked on it, *Fig. 34.7*. Trundle wheels are used to measure **longer lengths**. They are often used to measure curved or straight lines on football pitches or athletic tracks.

Fig. 34.7 A trundle wheel.

TEST YOURSELF:
Now attempt questions *34.4–34.6* and *W34.4–W34.9*.

34.3 Measuring Mass, Time and Temperature

(1) Mass

The mass of an object is the amount of matter or stuff in it.

Measuring the Mass of a Solid

We measure the mass of an object using an electronic balance or scales, *Fig. 34.8*. It is **important that the balance or scales is clean and dry before switching it on.**

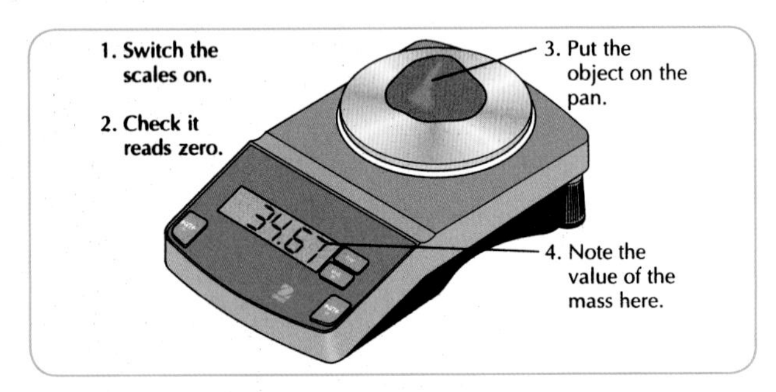

Fig. 34.8 How to measure mass.

Mass is measured in kilograms (kg) or **grams (g)**.
1 kilogram = 1000 grams.

Measuring the Mass of a Liquid

To measure the mass of a liquid, carry out the following steps:

1. **First find the mass of an empty dry beaker (*Fig. 34.9(a)*).**
2. **Add the liquid to the dry beaker.**

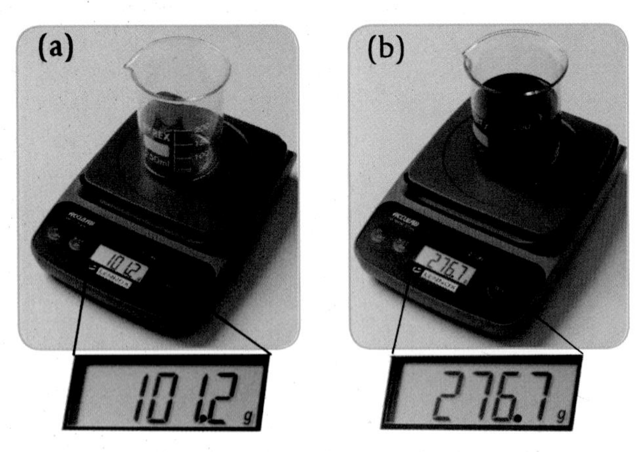

Fig. 34.9 Can you calculate the mass of the liquid in the second photograph? The first photograph gives the mass of the empty beaker.

3. **Read the new mass, i.e. the mass of beaker + liquid (*Fig. 34.9(b)*).**
4. Then subtract the two masses. The result is the mass of the liquid. (On some balances the mass of the added liquid can be got by re-zeroing the balance.)

(2) Time

Time is measured in **seconds (s)**. Minutes, hours, days, years, etc., may also be used.

Instruments to Measure Time

- We use **watches** or **clocks** to measure time. **Stopwatches** or stopclocks are often used to measure time in the laboratory, *Fig. 34.10*.
- **Electronic timers** are used when very accurate results are needed. In the Olympic games, for example, race times can now be measured correct to $^{1}/_{1000}$ of a second.
- Mobile phones, cookers, washing machines, dryers, computers, etc., have their **own timers.**

Fig. 34.10 How many seconds have elapsed on this stopwatch?

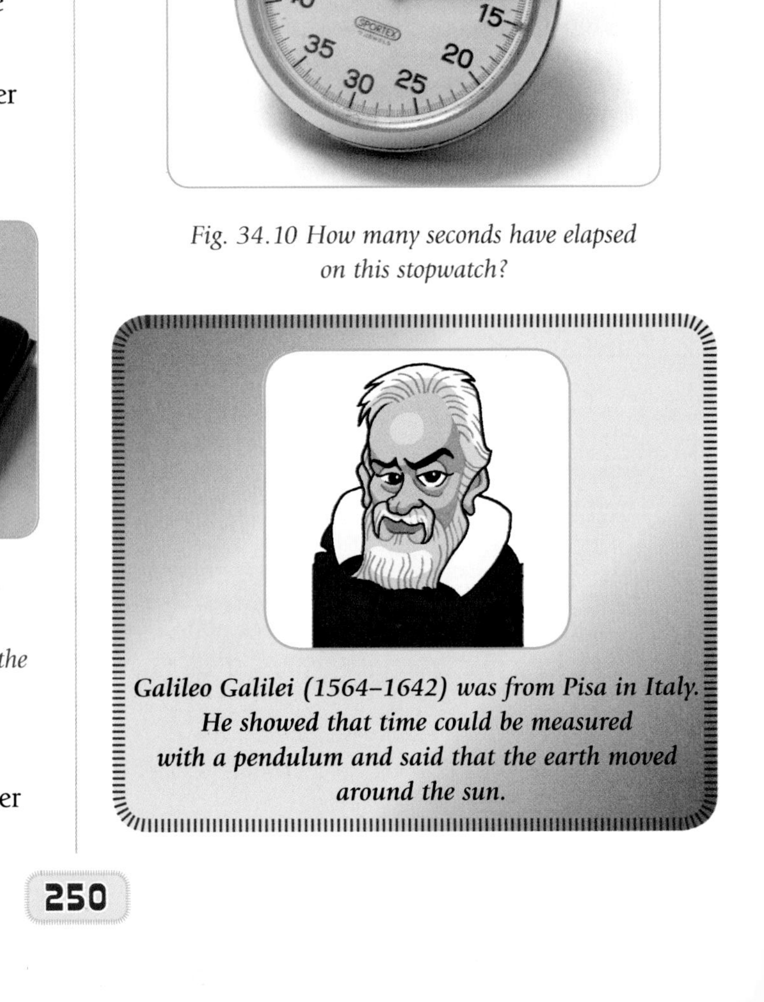

Galileo Galilei (1564–1642) was from Pisa in Italy. He showed that time could be measured with a pendulum and said that the earth moved around the sun.

(3) Temperature

The temperature of an object is a measure of how hot or cold the object is.

Temperature is measured in **degrees Celsius (°C)**. It is measured with a **thermometer**.

Examples of Temperature

- On the Celsius scale, the temperature of the **human body** is 37 °C.
- Pure **water** freezes at **0 °C** and boils at **100 °C**.
- Sometimes temperatures may be below the freezing point of water. They then have a **minus** value (e.g. –4 °C).

Types of thermometers

(1) **Mercury thermometers** (or alcohol thermometers) are used in many school laboratories, *Fig. 34.11 (a)*.

(2) **The thermoscan ear thermometer,** *Fig. 34.11 (b)*, is used in many hospitals. The thermoscan is placed in the patient's ear. The temperature can be read after just 1 second.

(3) **Electronic digital thermometers** are used in many homes to measure body temperature, *Fig. 34.11 (c)*. They take about 60 s to reach the correct temperature. They usually beep when ready.

(4) **Temperature sensors** with data-logging equipment are also commonly used, *Fig. 34.11 (d)*.

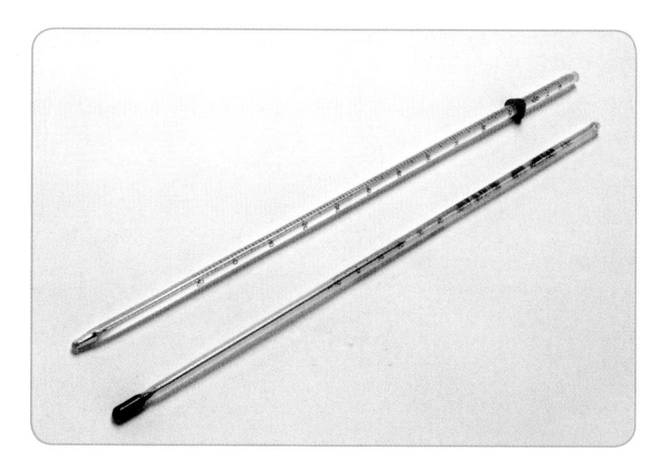

Fig. 34.11 (a) Mercury thermometer (left) and alcohol thermometer (right).

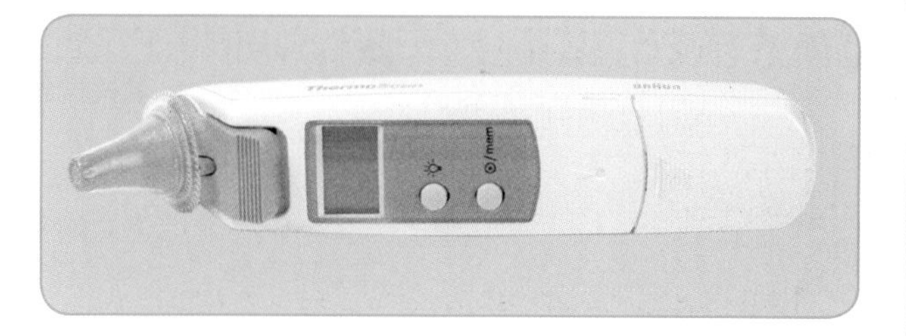

Fig. 34.11 (b) Thermoscan ear thermometer.

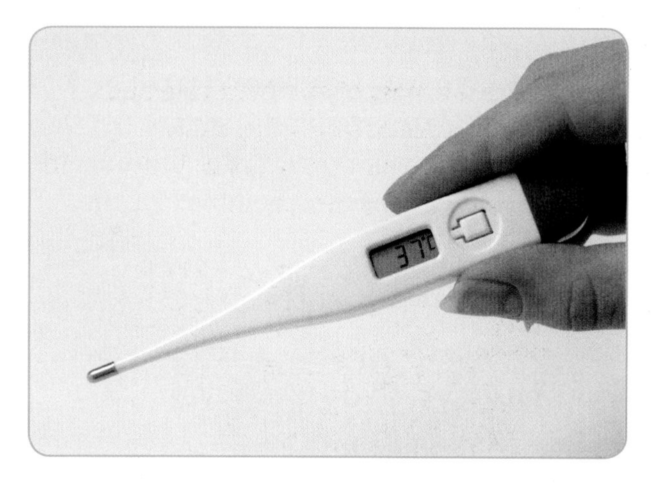

Fig. 34.11 (c) Electronic digital thermometer.

Fig. 34.11 (d) Temperature sensor and data-logging equipment: the investigation here is finding the boiling point of water (with a dye added). The graph shows how the temperature of the water increases with time. When it reaches boiling point, i.e. 100 °C, the temperature remains constant.

34.4 Units and Prefixes

It is important that people use the **same standard units** to measure various quantities. Most of the world now uses the **S.I. system** (Systéme Internationale or International System) of units, *Table 34.1*.

QUANTITY	UNIT	SYMBOL
Length	metre	m
Mass	kilogram	kg
Time	second	s
Temperature	degree Celsius	°C

Table 34.1 S.I. system of basic units.

A **prefix** is often used with units. It is put in front of the unit. For example, **'milli-'** (*symbol*: m) means one thousandth or $^1/_{1000}$. So, **1 millimetre** (or 1 mm) equals one thousandth of a metre.

- What is a **milligram (mg)**?
- What is a **millisecond (ms)**?

Table 34.2 shows some of the common prefixes used with their symbols and meanings.

We must be **careful** when adding numbers with different units or prefixes. For example, **2 m + 3 m** is obviously equal to **5 m**. But, **2 m + 3 kg** simply does not make sense and cannot be added.

2 m and **3 cm** can be added, but only if brought to the **same basic unit** first. You can either add **2 m** and **0.03 m** *or* add **200 cm** and **3 cm**. The correct answer is **2.03 m** or **203 cm**.

PREFIX	SYMBOL	FACTOR BY WHICH UNIT IS MULTIPLIED
milli-	m	$^1/_{1000}$ or 0.001 or 10^{-3}
centi-	c	$^1/_{100}$ or 0.01 or 10^{-2}
deci-	d	$^1/_{10}$ or 0.1 or 10^{-1}
hecto-	h	100 or 10^2
kilo-	k	1000 or 10^3
mega-	M	1 000 000 or 10^6

Table 34.2 Common prefixes, their symbols and their meanings. (A list of prefixes is also given on Pages 4 and 5 of the Mathematical Tables. *Notes on the scientific notation used in this table can be found in Appendix II)*

TEST YOURSELF:

Now attempt questions *34.7–34.10* and *W34.10–W34.16*.

What I should know

- Physics deals with how and why things behave as they do.
- Scientists try to answer questions to explain what is going on in the natural world. They collect evidence by observing and using information collected by other scientists.
- Finding out how and why things work has led to many discoveries and inventions.
- Observation is done using our senses. But, our senses often need the help of measuring instruments.
- There are many instruments to measure length. These include a metre stick, ruler, measuring tape, callipers, Vernier callipers, opisometer and trundle wheel.
- The mass of an object is the amount of matter or stuff in it.
- The mass of an object is found using a balance or scales.
- Time is measured using a stopwatch or stopclock.
- The temperature of an object is a measure of how hot or cold the object is.
- Temperature is measured using a thermometer or temperature sensor and data-logging equipment.
- The melting point of ice is 0 °C and the boiling point of water is 100 °C. The normal temperature of the human body is 37 °C.
- The S.I. units of length, mass, time and temperature are the metre, kilogram, second and degree Celsius, respectively.

What I should know

- A prefix is often combined with a unit to multiply or divide the unit by a power of ten. Common prefixes include *kilo* (k) meaning 1000, *centi* (c) meaning $^1/_{100}$ and *milli* (m) meaning $^1/_{1000}$.

MEASURING TASK	INSTRUMENT(S)	UNITS USED
short straight length	ruler, metre stick, Vernier callipers	m or cm or mm
long straight length	metre stick, measuring tape, trundle wheel	m
diameter (inside or outside)	callipers or Vernier callipers	m or cm
length of short curved line	opisometer or string with ruler	cm or mm
length of long curved line	trundle wheel	m
mass	scales, electronic balance	kg or g
time	stopwatch, stopclock, timers	s or min
temperature	thermometer, temperature sensor	°C

Table 34.3 How various items are measured and the units they are measured in.

QUESTIONS

Write the answers to the following questions into your copybook.

34.1 (a) The branch of science called ..... deals with how and why things behave as they do.

(b) Scientists collect evidence by ..... and by using ..... collected by other scientists.

(c) Finding the answers to questions about how and why things happen in the world around us has led to many new ..... .

(d) To help us make accurate observations, our senses often need the help of measuring devices called ..... .

34.2 Name the five senses with which we observe the world around us.

34.3 Give an example to show how one of your senses might lead you to a wrong conclusion or result.

34.4 (a) When reading a metre stick, it is important to keep your head over the point being measured to avoid the error of ..... . The error is best avoided by using the metre stick on its ..... when measuring.

(b) The two scales on Vernier callipers are called the ..... scale and the ..... scale.

(c) Vernier callipers are used to accurately measure short ..... . They are also used to measure the inside and outside ..... of cylinders or beakers.

(d) Short curved lines can be measured using a ..... or an ..... along with a ruler. Longer curved lines are measured with a ..... .

QUESTIONS

34.5 *Fig. 34.12* shows a poor attempt at measuring the length of a pencil with a ruler. Mention at least *three* errors being made in taking this measurement?

Fig. 34.12 Not the sharpest method of measuring length!

34.6 *Fig. 34.13* shows the main scale and Vernier scale of Vernier callipers. What is the correct reading on this scale?

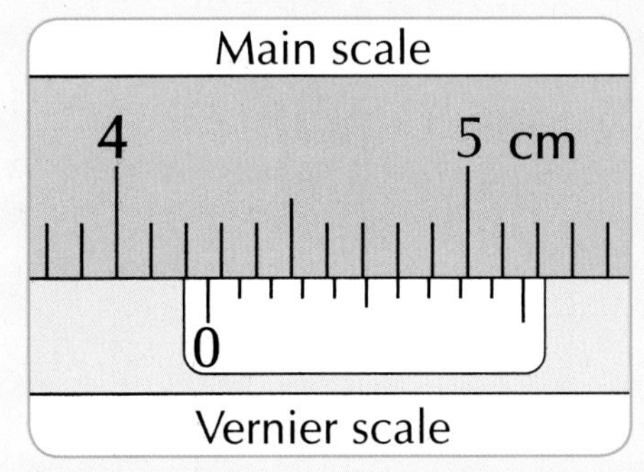

Fig. 34.13 What is the reading on these Vernier callipers?

34.7 (a) The ..... of an object is the quantity of matter in the object.

(b) The instrument used to measure the mass of an object is an ..... or ..... .

(c) Time is measured in the lab using a ..... .

(d) The temperature of an object is a measure of how ..... the object is.

(e) Temperature is measured using a ..... or a temperature ..... with data-logging equipment.

(f) The S.I. unit for length is the ..... and the symbol for this unit is ..... .

(g) The normal unit used for temperature is the ..... . The symbol for this unit is ..... .

34.8 You are given 100 light pins. They are all exactly the same. You are asked to find the mass of a single pin. But, when you place just one pin on the balance, it is too light to give a reading. Describe how you would find the mass of one single pin.

34.9 Add up each of the following:

(a) 13.11 kg + 9090 g

(b) 2.9 m + 7610 mm + 149 cm

34.10 Using *Table 34.2*, page 252, fill in the missing unit symbol in each of the following:

(a) $^{1}/_{100}$ g = 1 .....   (b) $^{1}/_{1000}$ s = 1 .....

(c) 1000 g = 1 .....   (d) 100 g = 1 .....

(e) 1 000 000 m = 1 .....

Chapter 35 Area, Volume, Density, Speed, Velocity and Acceleration

In *Chapter 34*, we dealt with the basic quantities of measurement. These were length, mass, time and temperature. In this chapter, we will look at quantities that are derived (or got) from those basic quantities.

35.1 Area

The area of a shape is the amount of surface enclosed within its boundary lines.

The area of a triangle is the size of the surface that is enclosed within its three lines. If you were painting the triangle, it is the amount of surface you would have to paint. Area is measured in **square metres** or **metres squared**, written as $\mathbf{m^2}$. Area can also be measured in **square centimetres** ($\mathbf{cm^2}$).

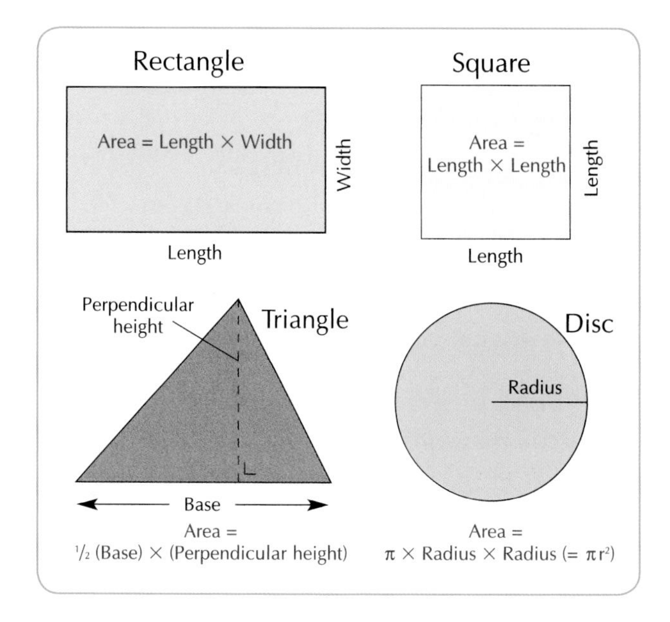

Fig. 35.1 How to find the areas of regular-shaped surfaces.

Note: A square of side 1 m has an area of 1 m × 1 m = $\mathbf{1\ m^2}$. But, the area of this square is also 100 cm × 100 cm or $\mathbf{10\ 000\ cm^2}$. Therefore $1\ m^2 = 10\ 000\ cm^2$.

Example 1

Which of the following has the greatest area:

(i) a rectangle of length 12.5 m and width 4 m, (ii) a square of side 7 m, (iii) a triangle of base 13 m and height 8 m?

(i) Area of rectangle = length × width
= 12.5 × 4
= 50 m²

(ii) Area of square = length × length
= 7 × 7
= 49 m²

(iii) Area of triangle
= ½ (base) × (perpendicular height)
= ½ (13) × (8) = 52 m²

Therefore the triangle has the greatest area.

Example 2

A circular pizza, X, has a radius of 20 cm. Another circular pizza, Y, has a radius of 10 cm. How many times is pizza X bigger than pizza Y?

Area of pizza, X, of radius 20 cm
$= \pi (20)^2$
$= 400\pi\ cm^2$

Area of pizza, Y, of radius 10 cm
$= \pi (10)^2$
$= 100\pi\ cm^2$

∴ area of pizza, X, is FOUR times bigger than the area of pizza, Y.

35.2 Volume

The volume of an object is the amount of space it takes up.

The S.I. unit of volume is the **cubic metre** ($\mathbf{m^3}$). Volume is also very often measured in **cubic centimetres** ($\mathbf{cm^3}$). Another commonly-used unit is the **litre (l)**, which is equal to 1000 cm³. Thus, 1 ml (millilitre) = 1 cm³. *Fig. 35.2* shows examples of volumes being measured in everyday life.

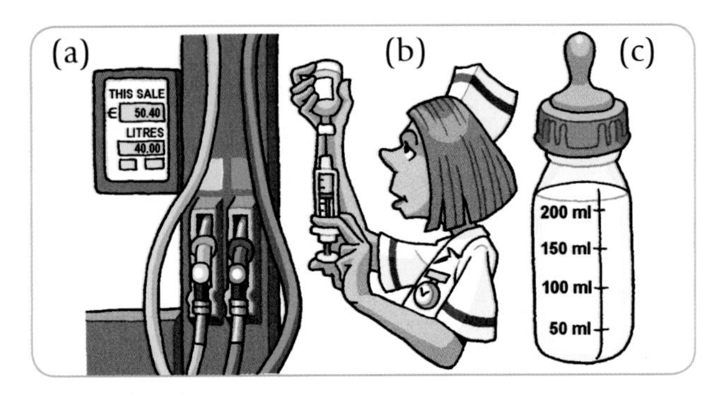

Fig. 35.2(a) A petrol pump measures the volume of petrol it delivers. (b) A nurse must measure the volume of fluid for an injection accurately. (c) A small baby will require different volumes of milk as it grows older.

Measuring the Volume of a Liquid

- At home, we measure the volume of a liquid in a **measuring jug**.
- In the lab, we measure the volume of a liquid using a **measuring cylinder** as follows:
  1. Simply **pour** the liquid into the measuring cylinder, *Fig. 35.3*.
  2. Then, **read its volume**. But, be careful, the surface of the liquid is slightly curved. It looks as if there is a bubble at the top. This is called a **meniscus**. When reading the level of water in a container, read at the **lowest part** of the meniscus.

Fig. 35.3 Which of these three measuring cylinders would be best to accurately measure 5 ml of liquid? Why?

- A **burette** and a **pipette** can also be used to accurately measure the volume of a liquid. They are used to transfer a definite amount of liquid to a container. Practice is needed to use them properly. Pictures of both are included with the laboratory equipment in *Chapter 1*.

Measuring the Volume of a Solid

(1) Measuring the Volume of a Regular Shape

The volume of a regular shape, like a rectangular block of wood, is got by measuring its length and its width and its height. Then the following formula is used:

volume = length × width × height

A cube of side 1 m has a **volume** of (1 m × 1 m × 1 m) = **1 m³**. But, the volume of this cube is also (100 cm × 100 cm × 100 cm) or **1 000 000 cm³**. Therefore 1 m³ = 1 000 000 cm³.

Example 3

What is the volume of a rectangular box of dimensions 4 cm, 2.5 cm and 2 cm?

volume of rectangular box
= length × width × height
= 4 cm × 2.5 cm × 2 cm = 20 cm³

(2) Measuring the Volume of an Irregular Shape

The volume of an irregular shape, like a stone, is got by the method of **displacement**. In *Fig. 35.4(a)*, the dotted space is taken up by water. In *Fig. 35.5(b)*, a stone is put into the water. It pushes the water that was in the dotted space out of the way. The stone now occupies the same amount of space that the pushed out water used to occupy. So, the volume of the stone is equal to the volume of the water pushed out or displaced.

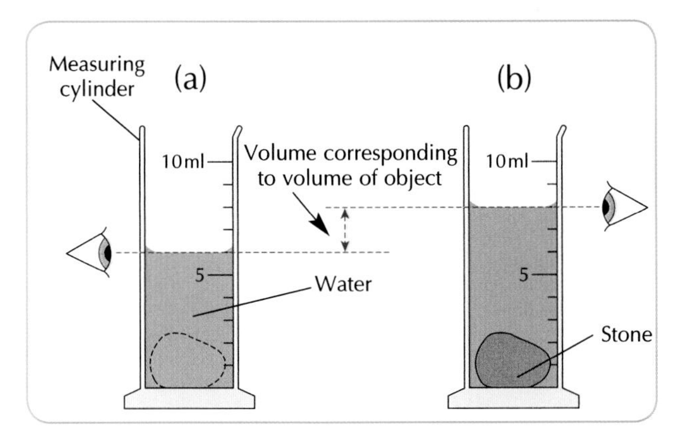

Fig. 35.4 Note the levels of water in both cylinders and calculate the volume of the stone from these readings.

Experiment 35.1

To measure the volume of a large stone

Apparatus required: an overflow (or displacement) can; a piece of thread; a measuring cylinder; a large stone; water

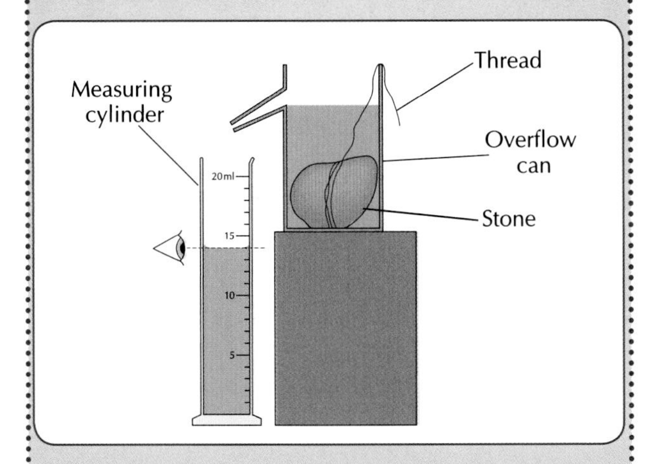

Fig. 35.5

Method

1. Fill the overflow can with water to a point above the spout. Wait until the water stops dripping from the spout.
2. Now place a dry measuring cylinder underneath the spout.
3. Gently lower the stone into the overflow can, using a thread.
4. Note the volume of the displaced water in the measuring cylinder, *Fig. 35.5*. Take this reading at eye level and read from the bottom of the meniscus. This reading is also the volume of the stone.

TEST YOURSELF:
Now attempt questions *35.1–35.3* and *W35.1–W35.5.*

35.3 Density

Which is heavier, lead or water? This is **not** a **fair** question. The answer depends on how much lead and how much water is present. So, what is a **fair** way of comparing materials? The answer is to compare the mass of the **same amount** of each.

Now, the mass of 1 cm^3 of lead is 11.2 g. The mass of 1 cm^3 of water is 1 g. So, the **mass per unit volume** of lead is 11.2 times greater than the mass per unit volume of water. Lead obviously packs a lot more matter into each cubic centimetre than water. It has more **mass for its size** than water. It has a greater density. Scientists call the 'mass per unit volume' of a substance its **density**.

It would be rare to find two substances with exactly the same density. Thus, a substance can be identified by its density. Density is said to be a **characteristic** property of a substance.

Calculating Density

The density of a substance is its mass per unit volume.

The formula for finding density is:

$$\text{density} = \frac{\text{mass}}{\text{volume}}$$

This formula can be rearranged to give:

mass = density × volume
and
volume = mass ÷ density

To remember all three of these formulas, it is useful to remember the density triangle, *Fig. 35.6*, page 258. Simply cover what you are looking for and you will be left with its formula. Make up your own mnemonic, like **M**odern **D**ay **V**ices, or **M**onkeys **D**rink **V**odka. This will help you to remember the positions of **M** (Mass), **D** (Density) and **V** (Volume).

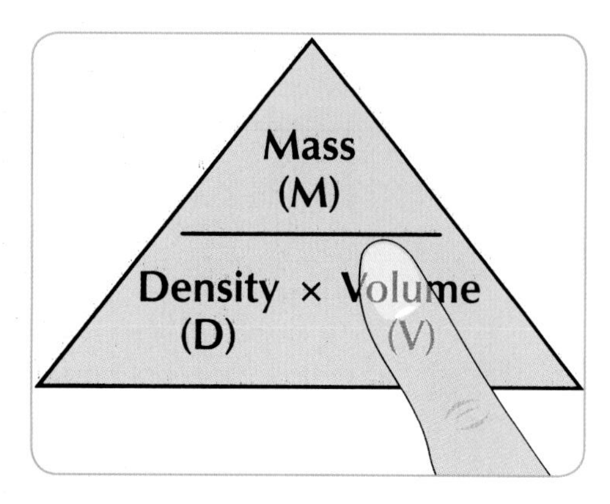

Fig. 35.6 Cover what you are looking for. If you are looking for Volume, cover Volume and you are left with what Volume is equal to, i.e. Mass ÷ Density.

Units of Density

Density is mass divided by volume. Thus, the **unit of density is the unit of mass divided by the unit of volume.** The S.I. unit of density is the **kg/m³** (or kg m^{-3}). However, the more practical unit is the **g/cm³** (or g cm^{-3}).

Substances with Densities Lower than Water

SUBSTANCE	DENSITY (g/cm³)
Hydrogen	0.00008
Air	0.0012
Oxygen	0.0013
Polystyrene	0.02
Cork	0.3
Wood	0.5 to 1.1
Paraffin Oil	0.8
Ice	0.9
Water	1.0

Substances with Densities Higher than Water

SUBSTANCE	DENSITY (g/cm³)
Perspex	1.2
Coal	1.6
Glass	2.5
Aluminium	2.7
Iron	7.9
Copper	8.9
Lead	11.2
Mercury	13.6
Gold	19.3

Table 35.1 Densities of some common substances.

Mandatory experiment 35.2

To determine the densities of solids and liquids

To find the density of a substance, the following three steps should be carried out:

(i) Find the mass of the substance.

(ii) Find the volume of the substance.

(iii) Divide the mass of the substance by its volume.

(a) To find the density of a regular solid

Apparatus required: a rectangular block (metal or wood); a balance; a ruler

Method

1. Place the rectangular block on a balance or scales. Note the mass of the block.
2. Measure the length, width and height of the block with a ruler. Record each result.

Results

length (l) = ______ cm

width (w) = ______ cm

height (h) = ______ cm

mass of block (M) = ______ g

volume ($l \times w \times h$) = ______ cm³

$$\text{density} = \frac{\text{mass } (M)}{\text{volume } (l \times w \times h)} = \_\_\_\_\_\_ = \_\_\_\_\_\_ \text{ g/cm}^3$$

(b) To find the density of an irregular solid

Apparatus required: an irregular solid like a stone (or a potato); a balance; an overflow can; a piece of thread; a measuring cylinder

Method

1. Find the mass of the stone by placing it on a balance. Note the reading.
2. Find the volume of the stone. Using a measuring cylinder, measure the volume

of the water it pushes out of an overflow can. This method was described in *Experiment 35.1*, page 257.

Results

mass of stone (M) = ______ g
volume of stone (V) = ______ cm^3

$$\text{density} = \frac{\text{mass } (M)}{\text{volume } (V)} = \_\_\_\_\_\_ = \_\_\_\_\_\_ \text{ g/cm}^3$$

(c) To find the density of a liquid

Apparatus required: a liquid (e.g. water or methylated spirits), a balance, a burette (or pipette)

Method

1. Find the mass of an empty dry beaker on a balance, *Fig. 35.7*.
2. Using a burette (or a pipette), transfer a known volume (say, 50 cm^3) of the liquid into the beaker. Note the mass of the 'beaker + liquid' on the balance. Then, by subtraction, calculate the mass of the liquid on its own. (***Note:*** The mass of the liquid can also be got by re-zeroing the balance.)

Fig. 35.7

Results

mass of liquid (M) = ______ g
volume of liquid (V) = ______ cm^3

$$\text{density} = \frac{\text{mass } (M)}{\text{volume } (V)} = \_\_\_\_\_\_ = \_\_\_\_\_\_ \text{ g/cm}^3$$

35.4 Flotation

Fig. 35.8 A peach floats in water.

Experiment 35.3

Investigating flotation for a variety of solids in water and other liquids

Some objects float in water, some do not. A peach, for instance, floats in water (***Fig. 35.8***), but lead sinks. Is it simply because the lead is heavier than the peach? But, a large heavy ship floats in water. A tiny pebble sinks. Why? An egg floats in salty water. But the same egg sinks in ordinary water. Why is that?

Use this investigation to attempt to answer the above questions.

Apparatus required: a wide variety of objects to be tested; a large container of water; assorted cubes; a second liquid (e.g. methylated spirits or water with dissolved salt)

Investigation (a)

Method

1. Test a large variety of objects to see whether they sink or float in water as in *Fig. 35.9*, page 260. First write down your prediction for each item. Then in a separate column, write down the actual result.
2. Repeat the investigation, but, use a liquid other than water this time.

Conclusions

What conclusions can you make? Did the large items sink and the small items float? Did the heavy objects sink and the light objects float? Did the items that are heavy for their size sink and the items that are light for their size float? Are the results exactly the same in a different liquid?

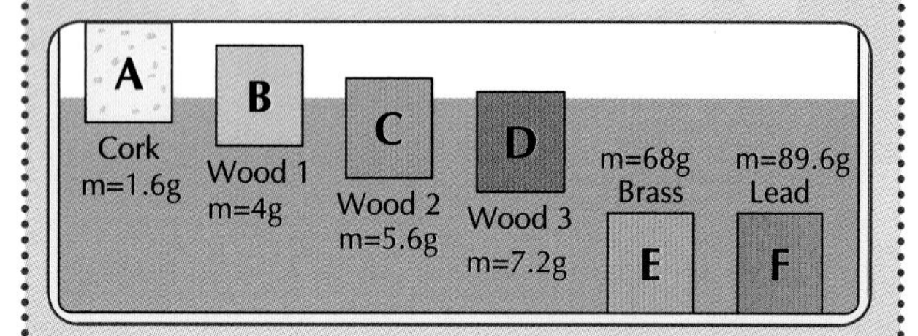

Fig. 35.9 All of these cubes are made of different materials. They are each of side 2 cm. Some float. Some don't.

Investigation (b)

Method

1. Find the mass of each of the assorted cubes. Label each of them *A*, *B*, *C*, etc.
2. All these cubes are the same size. (Their sides are of length 2 cm usually.) Find the volume of each cube.
3. Find the density of each cube by dividing its mass by its volume.
4. Place the assorted cubes in water as in *Fig. 35.9*. Note which cubes sink and which cubes float.
5. Place the assorted cubes in another liquid. Find the density of this liquid as in *Experiment 35.2(c)*. Note again which cubes sink and which cubes float.

Conclusion

Compare the densities of each of these cubes with the liquids they are in. What do you notice about the densities of the cubes that sink? What conclusion can you make?

How it Works

An object will always **float** in a liquid if the object is **less dense** than the liquid.

Examples of Flotation

- Liquids, like water, try to support solid objects. In *Fig. 35.9*, the water could support the **cork** and the three different kinds of **wood**. But, the brass and the lead were too heavy for their size. The water could not support them.
- A **ship** floats in water. This is because its overall density (including all the air inside it) is **less** than that of the water. Even though it is very heavy, it is not that heavy for its size. In contrast, a small pebble is heavy for its size. It sinks in water.
- Water can also support other liquids like **oil** that are less dense. Milk can similarly support **cream**.
- A **balloon filled with helium** will float in air. This is because helium is less dense than air. Also, **hot air** rises because it is less dense than cold air.

TEST YOURSELF:
Now attempt questions *35.4–35.6* and *W35.6–W35.11.*

35.5 Speed

Speed is a measure of how quickly a body can travel from one point to another. Speed is a quantity that comes from the basic quantities of distance and time.

Calculating Speed

The speed of an object is the distance it travels per unit time.

or

$$\text{speed} = \frac{\text{distance travelled}}{\text{time taken}}$$

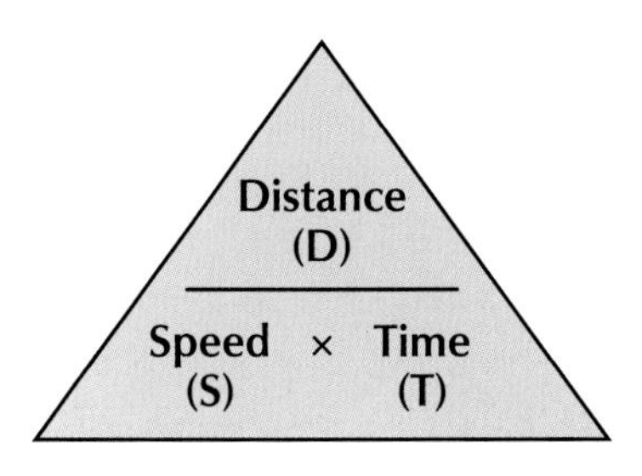

Fig. 35.10 The "D Street" (or D St.) method of finding distance, speed or time. Cover the one you are looking for and you will be left with the formula for finding it. For example, Time = Distance ÷ Speed.

The **unit of speed** is the unit of distance divided by the unit of time.

- It is the **metre per second**. The symbol for this unit is **m s^{-1}** or **m/s**.
- Another unit commonly used is the **kilometre per hour**. Its symbol is **km h^{-1}** or **km/h**.

Walking speed, for example, is about 5 km/h or 1.4 m/s. But, a space shuttle travels at 30 000 km/h or 8333 m/s.

Example 4

An athlete ran a 1500 m race in exactly 4 minutes. Calculate, in metres per second, her average speed.

Minutes are not the S.I. unit of time. Minutes must be converted to seconds. 4 minutes = 240 s.

$$\text{Average speed} = \frac{\text{distance travelled}}{\text{time taken}}$$

$$= \frac{1500 \text{ metres}}{240 \text{ seconds}} = 6.25 \text{ m/s}$$

Example 5

A train travelling at an average speed of 120 km/h covers a certain journey in 2.5 hours. Calculate the length of the journey.

Here, we are looking for the *distance* travelled. According to the 'D St.' triangle, *Fig. 35.10*, D = S × T = 120 × 2.5 = 300 km

Distance-Time Graphs

In physics, **graphs** are often used to represent information at a glance. *Table 35.2* shows the distances travelled by a walker in given time intervals.

DISTANCE (m)	0	2	4	6	8	10	12	14
TIME ELAPSED (s)	0	1	2	3	4	5	6	7

Table 35.2 Table showing how the distance travelled by a walker varied with time.

Fig. 35.11 shows how this data is represented on a distance-time graph. A distance-time graph can be used to answer many questions (see below).

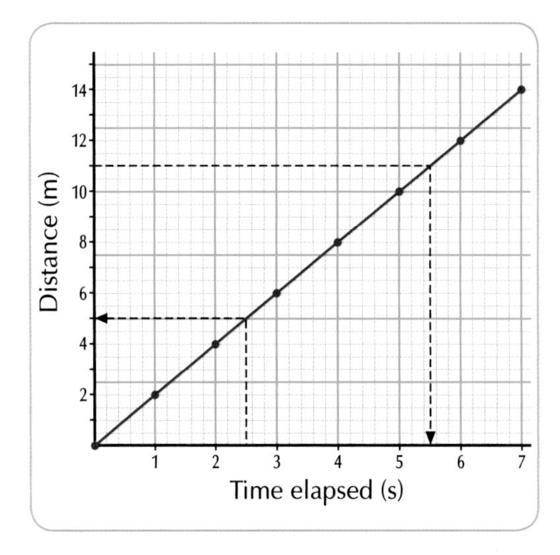

Fig. 35.11 Distance-time graph showing how distance varies with time.

(i) *How long does it take for the walker to travel 11 m?*
We can see from the broken line drawn across from the 11 m mark, that it takes 5.5 s for the walker to reach this point.

(ii) *How far has the walker walked after 2.5 s?*
We can see from the graph (see broken line) that, after 2.5 seconds, the walker has travelled 5 m.

(iii) *Is the speed of the walker constant?*
Yes, the graph is a straight line. This shows that the walker is travelling equal distances in equal time intervals. Therefore the speed is constant.

(iv) *What is the speed of the walker?*
The constant speed can be found by dividing any chosen distance travelled by the corresponding time interval.

$$\text{speed of walker is} = \frac{\text{distance}}{\text{time}} = \frac{14}{7} \text{ or } \frac{8}{4}$$

$$= 2 \text{ m s}^{-1}$$

PHYSICS

Examine the distance-time graphs shown in *Fig. 35.12*. Describe and compare the movement indicated by each of these graphs.

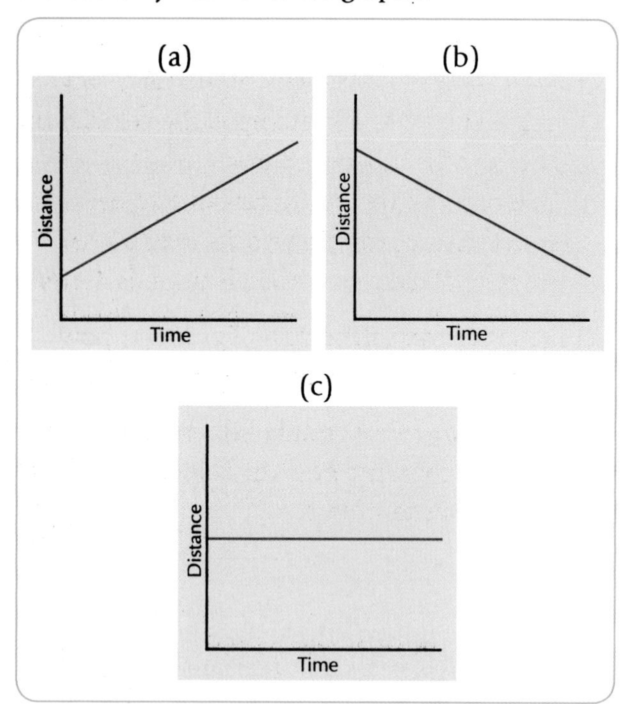

Fig. 35.12 In which of these three distance-time graphs is there no movement at all?

Velocity

The velocity of an object is its speed in a certain direction.

A pilot uses the plane's instruments to measure the wind that is blowing. It is not just the **speed** of the wind that must be noted. Its **direction** is also very important.

Velocity is the term used when **both** the **speed** and **direction** of a particular motion are in question.

- The speed of a plane, flying from Dublin to London might be 700 km/h. But, its **velocity** is **700 km/h, south-east**, *Fig. 35.13*.
- Another plane travelling in the opposite direction might have exactly the same speed. But, its velocity would be different, i.e. **700 km/h, north-west**.

Velocity, like speed, is measured in **m/s** or **km/h**, but it is followed by a direction.

Fig. 35.13 You must always show direction for Velo City!

35.6 Acceleration

You may be familiar with the 'accelerator' pedal on a car. If this pedal is pressed, the car will move faster. The acceleration of a car is a measure of how quickly it can change its speed or velocity. Formula One racing cars can change their speeds by a huge amount in 1 second. Thus, they have a huge **acceleration**.

Calculating Acceleration

Acceleration is the rate of change of velocity (or speed).

or $$\text{acceleration} = \frac{\text{change in velocity}}{\text{time}}$$

The **unit** of acceleration is the unit of speed divided by the unit of time (i.e. m/s ÷ s). The unit is the **metre per second per second** or **metre per second squared**. This is written as **m/s^2** or **$m\ s^{-2}$**.

If a cheetah accelerates from rest to a speed of 30 m/s in 6 seconds, its acceleration is 5 m/s^2. This means that its speed increases by 5 m/s every second.

Example 6

Along a straight section of track, a train changed its speed from 12 m/s to 27 m/s in 30 s. Calculate the acceleration of the train.

$$\text{acceleration} = \frac{\text{change in speed}}{\text{time taken}} = \frac{(\text{final speed} - \text{initial speed})}{\text{time}}$$

$$= \frac{(27-12)}{30} = \frac{15}{30} = 0.5\ m/s^2$$

Example 7

A bullet enters a thick tree travelling at 300 m/s and comes to rest in the tree 0.2 seconds later. Find the acceleration of the bullet.

$$\text{acceleration} = \frac{\left(\begin{array}{c}\text{final velocity}-\\ \text{initial velocity}\end{array}\right)}{\text{time taken}} = \frac{(0-300)}{0.2}$$

$$= \frac{-300}{0.2} = -1500 \text{ m/s}^2$$

Note: In *Example 7*, it is obvious that the bullet is *losing* velocity each second in the tree. This is called **deceleration**. It can also be regarded as **negative acceleration.** The velocity is **reducing** by 1500 m/s each second. The negative sign in the answer indicates that the bullet is decelerating.

Velocity–Time Graphs

Earlier we examined motion using distance-time graphs. **Velocity-time** graphs (or speed-time graphs) are also used. They give us different information. *Table 35.3* shows how the velocity of a racing car varies with time as it starts off a race along a straight road.

VELOCITY (m/s)	0	4	8	12	16	20	24
TIME ELAPSED (s)	0	1	2	3	4	5	6

Table 35.3 Velocity of a racing car changing with time

This data can be represented by a velocity-time graph as shown in *Fig. 35.14*.

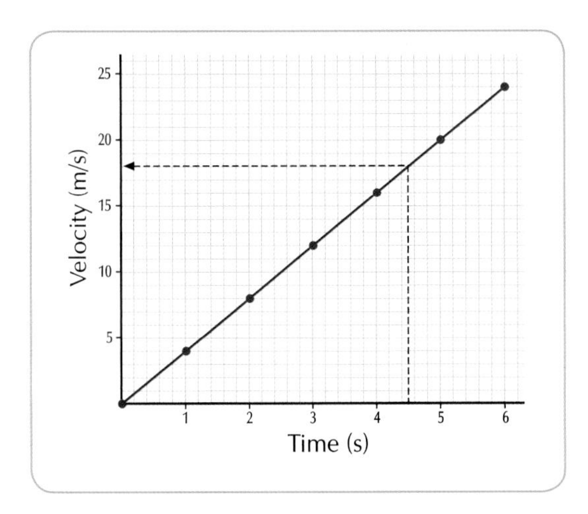

Fig. 35.14 Graph of velocity against time for a racing car.

Again, lots of information can be taken from a graph like this. We see immediately that the velocity of the car is **increasing steadily. It is** accelerating.

- We can find the **velocity** of the car at any particular time. For example, the velocity of the car after **4.5 s is 18 m/s.**
- The **acceleration** of the car can also be found. The change in velocity over **6 s**, is 24 m/s. Therefore, the acceleration is **4 m/s²**.

Can you find from the graph, Fig. 35.14, how long it takes the car to reach a velocity of 10 m/s?

Fig. 35.15 shows four simple velocity-time graphs. They compare how the velocities of four cars vary in a certain time.

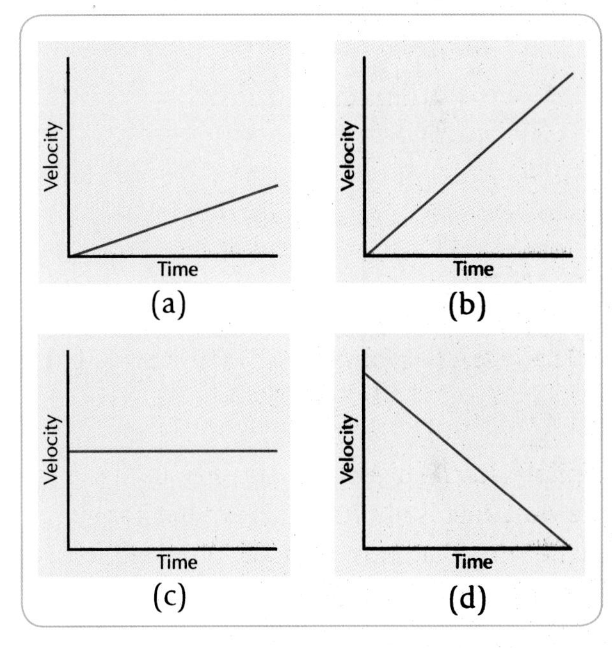

Fig. 35.15 Velocity-time graphs tell us how the velocity of an object varies with time.

(i) Which diagram shows a car travelling at a constant velocity?
(ii) Which diagram shows a car slowing down to rest?
(iii) Is the acceleration of the car in diagram (a) less than or greater than the acceleration of the car in diagram (b)?

TEST YOURSELF:
Now attempt questions *35.7–35.10* and *W35.12–W35.16.*

PHYSICS

What I should know

- The area of any shape is the amount of surface enclosed within its boundary lines.
- The area of a rectangle is its length multiplied by its width. The area of a triangle is half the base multiplied by the perpendicular height. The area of a circle is $\pi \times (\text{radius})^2$.
- The volume of an object is the amount of space it takes up.
- The volume of a rectangular block is found by multiplying its length by its width by its height.
- The volume of a liquid is found using a measuring cylinder. Measuring cylinders should be read at eye level. The level of liquids like water should be read from the bottom of the meniscus.
- Burettes and pipettes are used to transfer definite amounts of liquid.
- The volume of an irregular shaped object is found by measuring the volume of the liquid that it pushes out.
- Volume is measured in cubic metres (m^3) or cubic centimetres (cm^3). Another commonly used unit is the litre, which is equal to 1000 cm^3.
- The density of a substance is the mass of a unit volume of it.

or $$\text{density} = \frac{\text{mass}}{\text{volume}}$$

- The unit of density is g/cm^3 or kg/m^3. (These can also be written as $g\ cm^{-3}$ or $kg\ m^{-3}$.)
- To find the density of a substance you must (i) find the mass of the substance, (ii) find the volume of the substance, and (iii) divide the mass of the substance by its volume.
- The density of a substance is a characteristic property of that substance.
- An object will float in a liquid if the object is less dense than that liquid.
- The speed of an object is the distance it will travel per unit time.

$$\text{average speed} = \frac{\text{distance travelled}}{\text{time taken}}$$

- Velocity is speed in a given direction.
- Acceleration is the rate of change of speed (or velocity).

$$\text{acceleration} = \frac{\text{final speed} - \text{initial speed}}{\text{time taken}}$$

- Speed and velocity are measured in m/s or $m\ s^{-1}$. Acceleration is measured in m/s^2 or $m\ s^{-2}$.

QUESTIONS

Write the answers to the following questions into your copybook.

35.1 (a) The ..... of a shape is the amount of surface enclosed within its boundary lines.

(b) The unit in which area is measured is the ..... . It may also be measured in ..... .

(c) A rectangle of area 63 cm^2 has a length of 9 cm. The width of this rectangle is ..... .

(d) The radius of disc *A* is three times bigger than the radius of disc *B*. Therefore, the area of disc *A* is ..... times bigger than the area of disc *B*.

35.2 (a) The ..... of an object is the amount of space it takes up.

(b) The volume of a rectangular block is found by multiplying its ..... by its width by its ..... .

(c) The unit in which volume is measured is the ..... . It may also be measured in ..... .

(d) Burettes and ..... are used to transfer definite amounts of ..... .

(e) The volume of an irregular shaped object like a stone is found by measuring the volume of the liquid that it ..... .

35.3 A rectangular box is of length 13 cm, width 2.5 cm and height 2 cm. A cube has a side of length 4 cm. Which has the greater volume, the rectangular box or the cube? Justify your answer.

35.4 (a) The standard S.I. unit of mass is the ..... , whose symbol is ..... , but mass is also often measured in ..... , which has ..... as its symbol.

(b) The density of a glass stopper is found by dividing its ..... by its ....

(c) The unit in which density is measured is the ..... or the ..... .

(d) The density of a substance is a ..... property of that substance.

35.5 A student used a balance and a measuring cylinder containing water to find the density of a small stone. His results were as follows:

mass of the stone = 20 g

original volume of water in cylinder = 30 cm^3

volume of stone and water = 38 cm^3

What was the density of the stone?

35.6 *Fig. 35.16* shows three liquids: oil, water and syrup. It also shows three solids: a cork, a plastic block and a grape at various depths in the liquid.

(a) Which of the three liquids has the greatest density?

(b) Which solid has the lowest density?

(c) Would the plastic block float in the syrup?

(d) Would the grape sink in the oil?

(e) Why does the plastic block sink in the oil but float in the water?

Fig. 35.16 Flotation depends on densities.

35.7 The speed of an object is the ..... it travels per unit ..... .

QUESTIONS

35.8 The graph, *Fig. 35.17*, shows how the distance travelled by a car increased with time.

(a) How far away was the car at the beginning?

(b) At what time was the car 5 m away?

(c) What was the distance of the car after 3 s?

(d) What was the speed of the car?

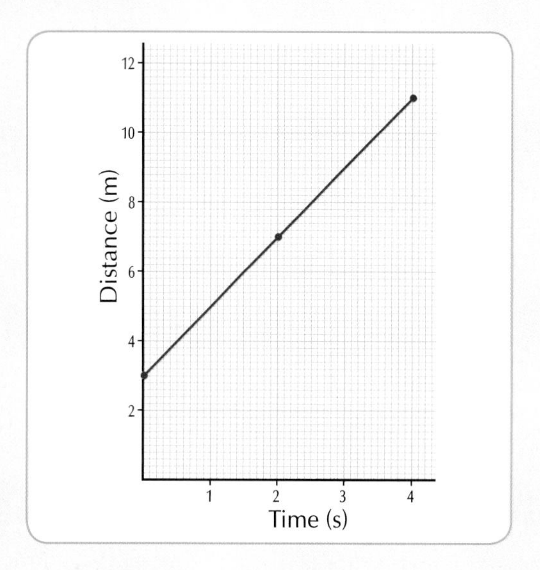

Fig. 35.17 How distance varied with time for a car over 4 s.

35.9 A 400 m runner, starting from rest, took 5 s to reach a maximum speed of 10 m s^{-1}. What was her acceleration?

35.10 A bullet enters an apple as shown in *Fig. 35.18* at a velocity of 200 m/s. It emerges from the other side of the apple at a velocity of 180 m/s just 0.00005 s later. Find the deceleration of the bullet on passing through the apple.

Fig. 35.18 The passing of a bullet through an apple.

Chapter 36 Force and Work

36.1 What is Force?

Objects do not **move** by themselves. They may be pushed. They may be pulled. Something **causes** them to move. This something is what we call a **force**. A force is something that causes an object to **accelerate** (or change velocity). This acceleration can be positive or negative. This means that a force may cause an object to **move faster**, **slow down**, **stop** or **change direction**.

A force is something that causes an object to accelerate.

Fig. 36.1 Two strong forces balancing each other. What would happen if one team suddenly let go?

Sometimes forces don't **seem** to cause acceleration at all. This is because forces don't always act on their own. *Fig. 36.1* shows two tug-of-war teams. Both teams are exerting considerable force; yet, neither is moving. The force applied in one direction is being **balanced** by an equal force in the opposite direction. But, if either force was acting alone, it would cause acceleration.

Measuring Force

The **unit** of force is the **newton** (symbol: **N**). This was named to honour the great scientist Sir Isaac Newton. **One newton (1 N)** is approximately the force needed to support the weight of an apple, *Fig. 36.2*. Force can be **measured** using a spring balance or **newton-meter** (or force-meter),

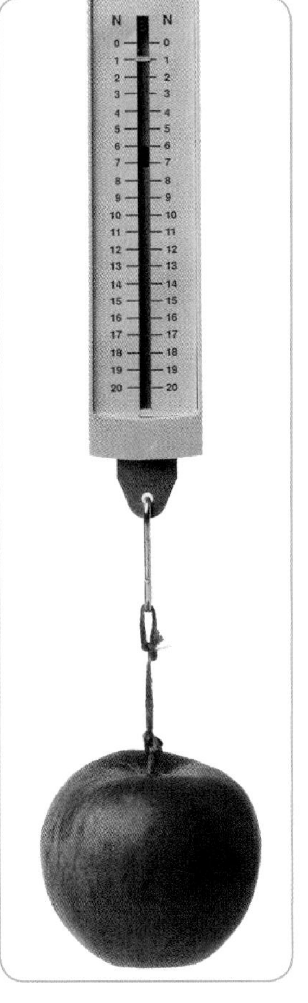

If possible, use a newton-meter now to measure the force needed to, say, open a drawer or to pull a door.

Fig. 36.2 A newton-meter (or spring balance) showing that the force needed to support an apple is about 1 N. (So, as the apple was falling from the tree onto Newton's head, it could be said: "There heads a force of 1 newton to one Newton's head!")

Isaac Newton (1642–1727) was from England. He contributed hugely to our knowledge of physics in areas such as forces and motion, gravity, light, colour, etc.

36.2 Types of Force and their Effects

There are many different **types** of force. *Table 36.1* gives examples of various types of force and the **different effects** these forces can have.

FORCE	EFFECT	EXAMPLE
Push	Moves or increases speed or changes direction	Pushing a child on a swing
Pull	Causes movement (as above)	Pulling off a boot
Weight	Moves objects towards the centre of the earth	A ball falling
Friction	Slows objects moving over each other	Brakes on a bicycle
Tension	Moves objects on pulleys or stops a body falling	A light suspended from a ceiling
Reaction	Supports or balances weight	A book resting on a table
Stretching	Changes shape of object	Stretching an elastic band
Twisting	Causes rotation	Opening a screw cap on a bottle
Compression	Changes shape of object	Sitting on a cushion
Magnetic	Moves magnetic materials	Picking up iron in a scrap yard
Electric	Moves charged particles	Positive charge attracting negative

Table 36.1 Types of forces and their effects.

Fig. 36.3 Try to match up each of the forces shown in the sketches with the force that matches it best in Table 36.1.

36.3 Weight

The **force of gravity** is what causes **weight**. If this book was not supported right now, it would be pulled rapidly towards the centre of the earth.

> **The weight of an object is the force of gravity on it.**

Weight on the Earth, on the Moon and in Space

- The **earth** is not a perfect sphere. The poles are closer to the centre of the earth than places at the equator. Thus, the pull of the earth is **slightly stronger at the poles**. So, an object's weight at the North Pole is a little bit more than its weight at the equator.
- On the **moon**, objects weigh much less. The pull of the moon is about six times less than the pull of the earth. So, the weight of an object on the moon would be about **one sixth** of what it is on the earth.
- Deep **in space**, far from any planet, an object experiences no pull. It is **weightless**.

Weight and Mass

In English we mix up the words '**weight**' and

'**mass**' all the time. We often say that we weigh something or find its weight. Usually we mean that we find its **mass**. A **weighing** scales normally gives us the **mass** of an object!

- Remember that **mass** is measured in **kilograms** (kg).
- But, **weight** is a force. It is measured in **newtons** (N).

Look at the multi-flash photograph in *Fig. 36.4*. Do you notice the effect of gravity on the ball? As it falls, the ball travels **larger distances** in the same time intervals. Its speed is **increasing**. In fact, its speed increases by **10 m/s** each second. The **acceleration** due to gravity is approximately 10 m/s^2.

Fig. 36.4 A ball travels larger distances in the same time intervals, increasing its speed by 10 m/s each second.

The **weight** of an object is **calculated** by multiplying its mass (in kilograms) by the acceleration due to gravity (approximately 10 m/s^2). In other words, the weight of a **1 kg** mass is **10 N** approx.

weight (in newtons)
= mass (in kilograms) × 10

Example 1

(a) What is the weight of a packet of sugar, which has a mass of 400 g?
(b) What is the weight of a block of mass 7.5 kg?
(c) What is the weight of the same block on the moon, if the pull of gravity on the moon is one sixth of the pull of gravity on earth?

(a) Mass of packet of sugar = 400 g = 0.4 kg
∴ Weight of packet of sugar
= mass × 10 = 0.4 × 10 = 4 N
(b) Weight of block
= mass × 10 = 7.5 × 10 = 75 N
(c) Weight of block on earth = 75 N
Weight of block on moon
= ($^1/_6$) (weight on earth)
= ($^1/_6$) (75)
= 12.5 N

Example 2

What is the **mass** *of a book that has a weight of 15 N?*

Weight of book = 15 N

$$\text{Mass of book} = \frac{\left(\text{weight of book}\right)}{10} = \frac{15}{10} = 1.5 \text{ kg}$$

TEST YOURSELF:
Now attempt questions *36.1–36.7* and *W36.1–W36.9.*

36.4 Friction

The small rubber disc used in ice hockey is called a puck.

- If hit gently, it will move a long distance over the **ice**.
- On a **road**, the puck would stop very quickly with a similar hit.

What is the difference? Something obviously slows the puck down on the road. That something must be a force. We call this force **friction**. The surface of the road may seem flat. But, on a microscopic scale, it is full of tiny humps and hollows. These catch and stick to small humps and hollows on the puck's surface, *Fig. 36.5*.

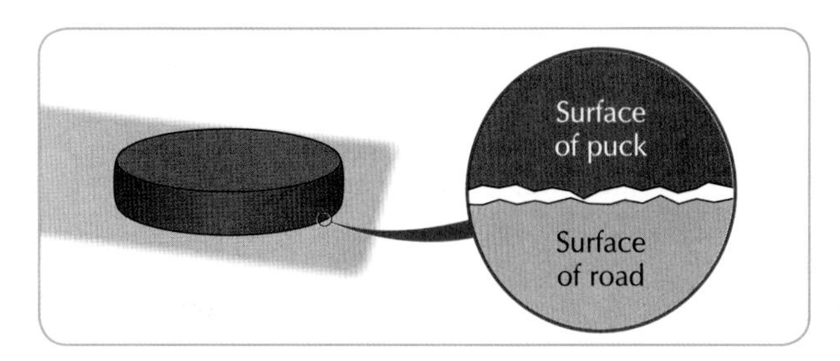

Fig. 36.5 Friction caused by one object trying to move over another.

PHYSICS

Friction is a force which opposes motion between two objects in contact.

Disadvantages of Friction

(1) Friction often **slows** things down. Greater friction is the main reason why ships travel much **slower** than planes.
(2) Friction often results in **waste of energy**. In food mixers and car engines, for example, a lot of unwanted **heat** and **sound** energy are produced.
(3) Friction **wears away** surfaces in contact with each other. Tyres lose their thread and brake pads on bicycles wear away due to constant contact.

Advantages of Friction

(1) We use friction **to stop** or **slow things down**. The brakes of a car or bicycle, for example, use friction.
(2) Friction between the tyres of a car and the road allow a car **to move**. The wheels of a car simply spin around without a grip on mucky ground! Likewise, we need friction **to walk**.
(3) Without friction you would not be able to **pick things up** with your fingers. **Writing** would be almost impossible.

Lubrication Reduces Friction

Oil (or grease) is a **lubricant** used in engines to **reduce friction** between moving parts. Otherwise,very high temperatures would be created. **Synovial fluid** acts as a lubricant in our joints, see *Section 10.5*, page 58.

Fig. 36.6 What are these cyclists doing to reduce friction between themselves and the air?

Experiment 36.1

Investigating friction and the effects of lubrication

Apparatus required: newton-meter; block of wood (with a hook attached); large weight
Also required: oil; sandpaper

Method

1. Attach the newton-meter to the block on a bench or surface. Pull the block.
2. Note the reading just as the block is about to move. You may need to do this a few times for accuracy. Use the same part of the surface or bench each time.
3. Place a large weight on top of the block. Measure again the force needed to move the block, *Fig. 36.7*.
4. Attach sandpaper onto the bottom of the block. Repeat steps 1, 2 and 3 again.
5. Remove the sandpaper. Put some oil on the surface beneath the block. Repeat steps 1, 2 and 3 again.

Note: The force required to move the block in each case is equal to the force of friction. The block will not move unless the force you apply to it matches the frictional force at least.

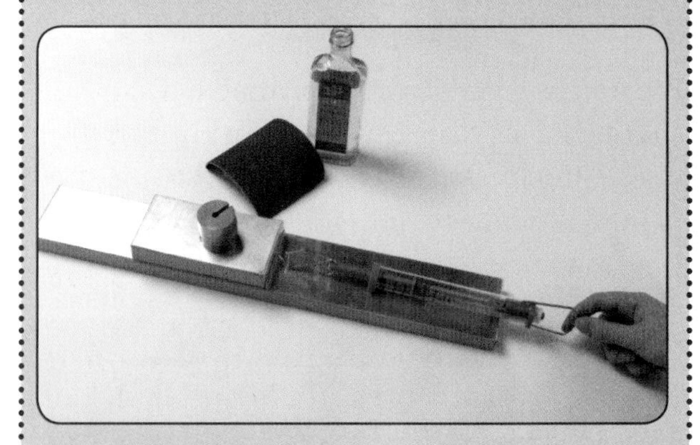

Fig. 36.7 Measuring force needed to move wood.

Results

Record your measurements on a table as shown, *Table 36.2*.

SURFACE USED	FORCE REQUIRED TO MOVE BLOCK (N)	FORCE REQUIRED TO MOVE BLOCK AND LARGE WEIGHT (N)
Untreated surface		
Sandpaper		
Oiled surface		

Table 36.2 Investigating friction and the effects of lubrication.

Conclusions

- Does the frictional force increase or decrease when the weight is added to the block?
- Which surface required the least force to move the block? What was the reason for this?
- Can you explain how two people, using the same block, but on different parts of a bench, may get different friction measurements.

Note: Lubricants should be cleaned off the equipment and surfaces at the end of this investigation.

36.5 Stretching Forces

When you lie on a bed, the springs beneath you change shape. When you get up from the bed, the springs return to their original shape. Objects that behave like this are said to be **elastic**.

- When a force is put on them, they **change shape.**
- When the force is removed, they **return to their original shape.**

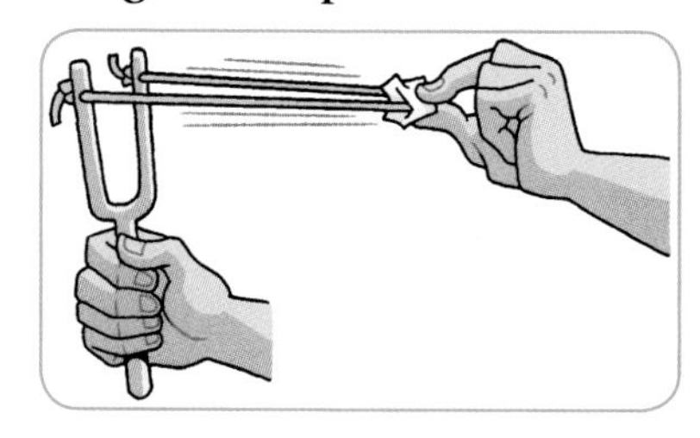

Fig. 36.8 The elastic band will return to its original shape when let go.

Mandatory experiment 36.2

Investigating the relationship between the extension of a spiral spring and the force applied to it

Apparatus required: retort stand; spiral spring with a pointer attached; metre stick; slotted weights hanger with several equal weights (it helps if the weight of the hanger is equal to the weight of each of the slotted weights)

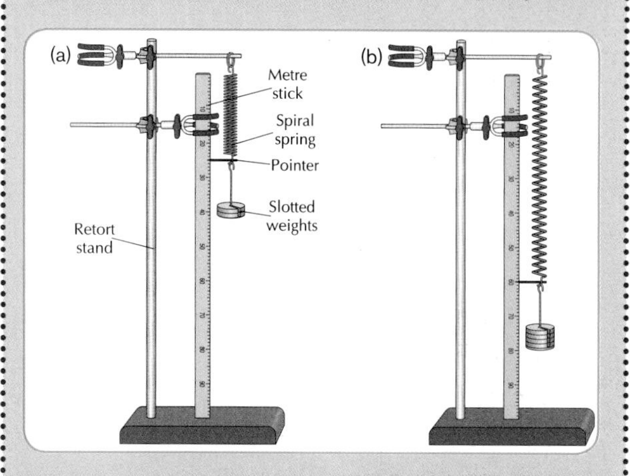

Fig. 36.9 Investigating how extension changes when extra forces are applied.

Method

1. Measure the **natural or unstretched length** of the spiral spring. This is given by the position of the pointer on the scale. (This length must be subtracted from all other lengths to calculate each **extension** below.)
2. Attach a hanger of known weight. This stretches the spring. Measure and note the **new length** of the spring. Calculate the **extension** of the spring.
3. Add a weight to the hanger. Measure the new length. Calculate the extension.
4. Add more weights, one by one. Note the new length of the spring each time. Calculate each corresponding extension.

   Take care **not** to stretch the spring beyond what is called its **elastic limit**. It will not return to its original shape.

5. Record your results in a table similar to *Table 36.3* below.
6. Draw a graph of **extension** (vertical axis) against the stretching **force** (horizontal axis).

	TRIAL						
	1	2	3	4	5	6	7
FORCE OR WEIGHT ADDED (N)							
POSITION OF POINTER ON SCALE (mm)							
EXTENSION (mm)							

Table 36.3 Investigating how extension varies with the applied force.

Result and Conclusion

The graph results in a straight line going through (0, 0). Thus, the extension is **directly proportional** to the force which caused it.

HOOKE'S LAW:
The extension of a spring is directly proportional to the force that is stretching it.

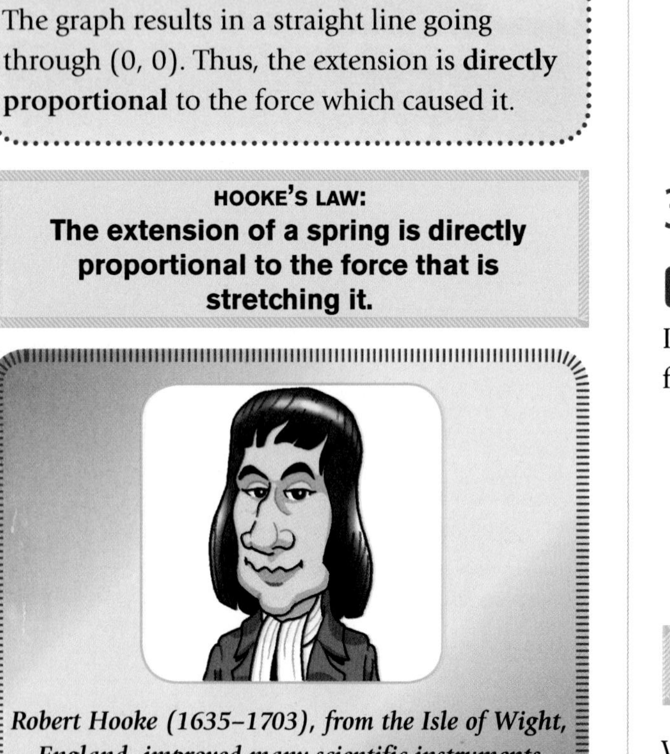

Robert Hooke (1635–1703), from the Isle of Wight, England, improved many scientific instruments including barometers and telescopes and was one of the first scientists to study elasticity.

Fig. 36.10 shows the graph obtained by a student who was trying to verify Hooke's Law.

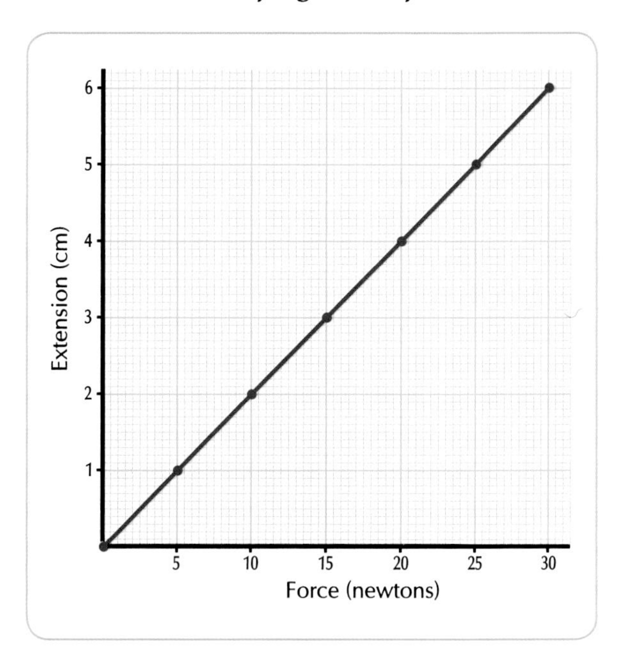

Fig. 36.10 Extension against force applied.

Using this graph, estimate:

(i) The extension caused by a force of 12 N.

(ii) The force that would cause an extension of 5.5 cm.

TEST YOURSELF:
Now attempt questions *36.8–36.10* and *W36.10–W36.15*.

36.6 Work, Energy and Power

(1) Work

In physics, **work** is only said to be done when a force **moves** an object.

- **Swimming**, for instance, may be fun but it is work too. You keep **pushing** water out of your way.
- When you **write**, you are forcing a pen to **move** over paper.

work done = force × distance moved

We **calculate** the amount of work done by multiplying the applied **force** by the **distance** that the object moves.

Force is measured in newtons. The distance moved is measured in metres. Work done could be measured in newtons × metres. But, we call this unit the **joule** (symbol: **J**).

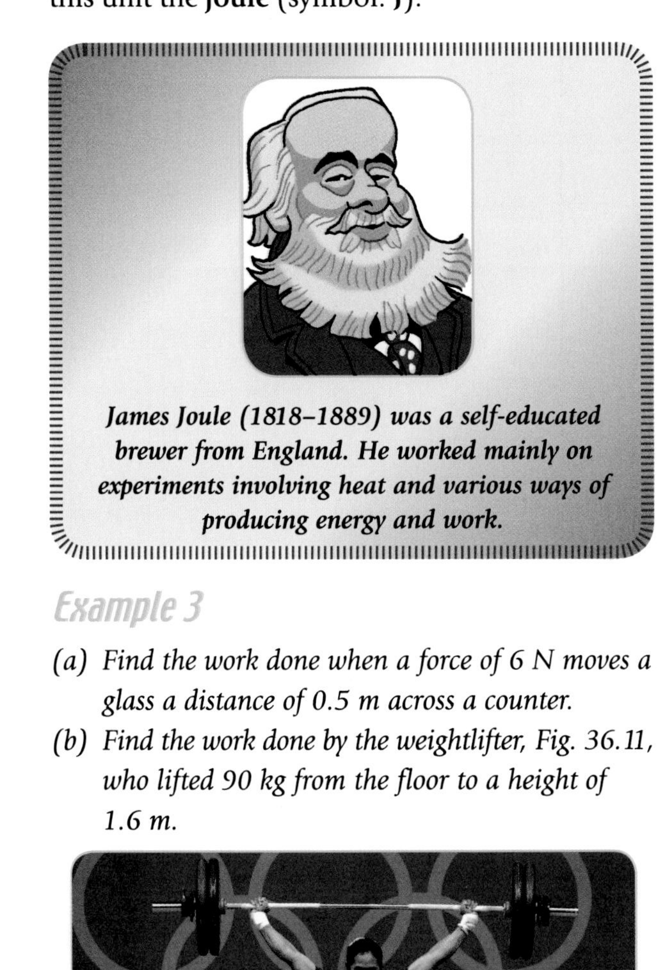

James Joule (1818–1889) was a self-educated brewer from England. He worked mainly on experiments involving heat and various ways of producing energy and work.

Example 3

(a) Find the work done when a force of 6 N moves a glass a distance of 0.5 m across a counter.

(b) Find the work done by the weightlifter, Fig. 36.11, who lifted 90 kg from the floor to a height of 1.6 m.

Fig. 36.11 Olympic weightlifter, who has done some work.

(a) Work done = force × distance moved

= 6 × 0.5

= 3 J

(b) Mass lifted = 90 kg

⇒ weight lifted = 90 × 10

= 900 N

∴ work done = force × distance moved

= 900 × 1.6

= 1440 J

(2) Energy

When you do work, you use up **energy**. The work you do is, in fact, equal to the energy you use. Thus, the unit of energy is the same as the unit of work, i.e. the **joule**. We will see more about energy again in *Chapter 39*.

Energy is the ability to do work.

(3) Power

Power is a measure of how **quickly** work is done. It is the **rate** of doing **work** or the work done per unit time. The unit of power is the **watt** (symbol: **W**).

Power is also the **energy used per second**. A **100 W bulb**, for example, uses 100 J of energy each second.

Power is the work done per unit time.

or $$\text{power} = \frac{\text{work done}}{\text{time taken}}$$

James Watt (1736–1819) was a Scottish engineer, who developed the steam engine. His inventions contributed hugely to the Industrial Revolution in Britain.

Example 4

A boy used a force of 25 N to push a lawnmower a distance of 20 m across a lawn. If this task took 8 seconds, what was the average power developed by the boy?

PHYSICS

Power developed by boy:

$$= \frac{\text{work done}}{\text{time taken}} = \frac{\text{force} \times \text{distance}}{\text{time taken}}$$

$$= \frac{25 \times 20}{8} = 62.5 \text{ W}$$

TEST YOURSELF:
Now attempt questions *36.11–36.18* and *W36.16–W36.22.*

What I should know

- There are many different types of force: pulling, pushing, weight, friction, tension, stretching, compression, electric and magnetic forces, etc.
- A force is something which causes an object to accelerate or change velocity.
- Force is measured in units called newtons (symbol: N).
- A force can have many effects. It can change the speed, direction, shape or size of objects.
- Forces can be measured using a spring balance or newton-meter (or force-meter).
- Gravity exerts a downward pull on all objects.
- The weight of an object is the force (or gravitational pull) that the earth exerts on it.
- Weight can vary at different parts of the earth. The weight of an object on the moon is about one sixth of its weight on earth.
- Weight (in newtons) = mass (in kilograms) × 10.
- The earth exerts a force of approximately 10 N on every 1 kg mass at its surface.
- Friction is a force which always tries to resist movement between any two surfaces in contact.
- Friction can be a nuisance force slowing things down and wasting energy.
- Friction can also be a very useful force. It allows us to walk easily and drive. It is also useful to stop objects moving, as in the brakes of a car or bicycle.
- One way of reducing the effects of friction is to use a lubricant, like oil.
- Objects that return to their original shape when the altering force is removed are said to be elastic.
- Hooke's law: The extension of a spring is directly proportional to the force that is stretching it.
- A graph of extension against stretching force for a spring gives a straight line through the origin.
- Work is done only when a force moves an object.
- Work done = force × distance moved.
- Energy is the ability to do work.
- Work done is measured in joules (symbol: J). This is also the unit of energy.
- Power is the rate of doing work. It is the work done per unit time. It is also the energy used per unit time.
- Power is measured in watts (symbol: W).

QUESTIONS

Write the answers to the following questions into your copybook.

36.1 (a) A force is something which causes an object to ….. .

(b) Give an everyday example of each of the following: a pushing force, a twisting force, a pulling force, a stretching force, a force of tension and a force of compression.

(c) State two possible effects of a pushing force.

(d) The unit of force is the ….. . The symbol for this unit is ….. .

(e) In which direction does gravity exert a force on all objects?

36.2 Helium is less dense than air. A child is holding a helium balloon by a string. Name the force that stops the balloon from drifting upwards.

36.3 Examine again the multi-flash photograph of the falling ball in *Fig. 36.4* on page 269.

(a) What force is causing the ball to move?

(b) How do we know from this photograph that the ball is accelerating or changing speed?

36.4 What force does the earth exert on a boy of mass 50 kg?

36.5 A 100 kg mass weighs approximately 1000 N when weighed at sea level. On top of Mount Everest, the same mass weighs less (about 2.5 N less, in fact). Can you explain why?

36.6 Match each of the following masses on the left to their corresponding weights on the right:

(a) a 3.5 kg new baby	(i) 4 N
(b) a 400 g jar of marmalade	(ii) 3.5 N
(c) a 40 kg bag of coal	(iii) 35 N
(d) a 350 g bag of oranges	(iv) 400 N

36.7 Match each of the following weights on the left to their corresponding masses on the right:

(a) a book of weight 10N	(i) 12 000 kg
(b) a heavyweight boxer of weight 1000 N	(ii) 175 g
(c) a satellite that weighs 120 000 N on earth	(iii) 1 kg
(d) a piece of gold that weighs 1.75 N	(iv) 100 kg

36.8 (a) Friction is a ….. which opposes ….. between any two objects that are in contact.

(b) State two disadvantages of friction in everyday life.

(c) Give one method of reducing the effects of friction between two surfaces.

(d) Mention two examples of where the force of friction is useful.

36.9 Explain briefly the principle involved in each of the following:

(a) People, long ago, were able to start fires by rubbing two sticks together.

(b) It is hard to hold a wet bar of soap.

(c) A children's slide in a park is polished to make it more effective.

(d) A squeaky hinge on a door can be quietened by adding some oil.

QUESTIONS

36.10 (a) State Hooke's law.

(b) The graph below, *Fig. 36.12*, shows how the extension of a spiral spring varied with the weights attached to it.

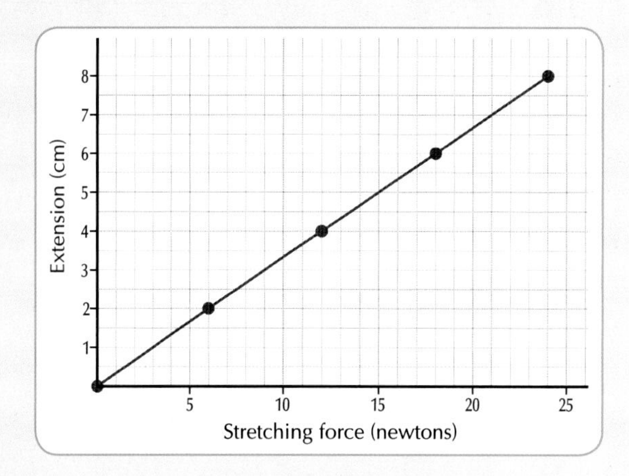

Fig. 36.12 Extension against stretching force for a spiral spring.

(i) What equipment might have been used to obtain the information in this graph?
(ii) How can you say from this graph that Hooke's law has been obeyed?
(iii) If a weight of 9 N was placed on this spring, estimate from the graph what extension it would cause.
(iv) An unknown weight was put on the spring. It caused an extension of 6.5 cm. Estimate from the graph the value of this unknown weight.

36.11 In physics, work is said to be done only when an object is ….. by a force.

36.12 Work done is calculated by multiplying the applied ….. by the ….. moved.

36.13 Joules are the units used to measure both ….. and energy.

36.14 ….. is defined as the ability to do work.

36.15 Power is the ….. of doing work or the work done per unit ….. .

36.16 Power is measured in ….. (symbol: …..).

36.17 What is the work done by a shopper, who uses a force of 120 N to push a trolley a distance of 15 m? If the shopper takes 9 s to carry out this work, what power has been used?

36.18 Anne and Emer wanted to compare their powers. They devised a competition. Each of them lifted bricks of mass 10 kg from the ground onto a wall 1.5 m high. In one minute, Anne managed to lift 24 bricks. Emer lifted 25.

(a) What was the weight of each brick?

(b) How much work was done in lifting one brick?

(c) How much work was done altogether by each girl?

(d) What power was used by each girl?

(e) Why did they stop after the same amount of time?

Chapter 37 Moments and Centre of Gravity

37.1 Turning Effect of a Force

A door swings about its **hinges**. In physics, we call the fixed point or line through which something swings, the **fulcrum**. In *Fig. 37.1(a)*, a large force is needed to close the door. In *Fig. 37.1(b)*, the clever baby uses a much smaller force. This small force has as big a **turning effect** as the larger force.

(a)

(b)

Fig. 37.1 A small force can have the same turning effect as a larger force.

The **turning effect** of a force can be increased in the following two ways:

(i) By increasing the size of the force itself, or;
(ii) By increasing the distance from the applied force to the fulcrum.

The measure of the turning effect of a force is called the **moment** (or torque) of the force. It is calculated as follows:

moment of a force

= (force) × (perpendicular distance from the force to the fulcrum)

Example 1

Fig. 37.2 shows a spanner trying to loosen a bolt.

(i) What is the moment of the force being applied?
(ii) If the same force was used on a spanner double this length, what would the moment be?
(iii) Which of these two spanners would have a better chance of loosening a very tight bolt?

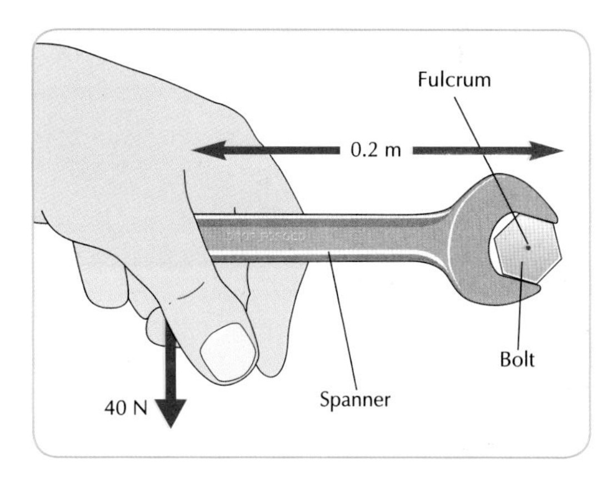

Fig. 37.2 Is the moment of the force applied here acting in a clockwise or an anti-clockwise direction?

(i) moment of a force

= (force) × (perpendicular distance from the force to the fulcrum)

= 40 × 0.2

= 8 N m

(ii) moment of a force

= (force) × (perpendicular distance from the force to the fulcrum)

= 40 × 0.4

= 16 N m

(iii) The second spanner has a greater moment and has a better chance of loosening a tight bolt.

37.2 Levers

The door handle in *Fig. 37.1* and the spanner in *Fig. 37.2* are examples of **levers**. Can you identify other examples shown in *Fig. 37.3*?

Fig. 37.3 Everyday levers.

- All of these levers are **rigid**. They do not easily bend.
- Each of them turns about a fixed point or axis, called the **fulcrum**. Levers allow us to move or turn things more easily.

A lever is any rigid body that is free to turn about a fixed point or axis called a fulcrum.

Experiment 37.1

To investigate the law of the lever

Apparatus required: metre stick, retort stand, strings (or threads), different known weights (e.g. a 1 N and a 2 N weight)

Method

1. Tie a **string** around the **centre** of the metre stick. Hang the metre stick from the retort stand. Make sure the metre stick is **balanced**, i.e. horizontal.
2. Hang a 1 N weight on the left side of the metre stick. Hang a 2 N weight on the other side, as shown in *Fig. 37.4*. Move these weights in and out until the lever is again balanced. Note the distances d_1 and d_2, *Fig. 37.4*.
3. Calculate the **moment** of each of these weights.
4. **Repeat** the experiment several times with the same weights in different positions.

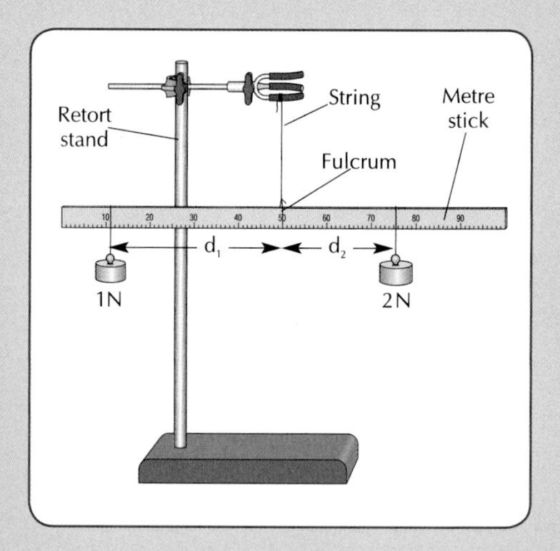

Fig. 37.4 Verifying the law of the lever.

Results

Results should be laid out in a table as follows:

	TRIAL					
	1	2	3	4	5	6
ANTI-CLOCKWISE MOMENT ($1 \times d_1$)						
CLOCKWISE MOMENT ($2 \times d_2$)						

Table 37.1

What do you notice about the clockwise moments and the anti-clockwise moments each time? Can you draw any conclusion?

Note: Repeat the experiment again. Put **two weights** on one side of the metre stick. Balance this by adjusting **one weight** on the other side. Calculate the moment for **each** of these weights. Verify that the **sum** of the clockwise moments (i.e. on the right) is equal to the **sum** of the anti-clockwise moments (i.e. on the left).

LAW OF THE LEVER:
When a lever is balanced by any number of forces, the sum of the clockwise moments acting on it is equal to the sum of the anti-clockwise moments.

Example 2

A metre stick is balanced when suspended at the 50 cm mark. A weight of 6 N is hung at the 22 cm mark. This is balanced by an unknown weight, W, hanging at the 74 cm mark. Find the unknown weight.

(*Note:* When calculating moments, it is the **distance from the fulcrum** that is important. Also, it is a good idea to draw a quick sketch of the information given.)

anti-clockwise moment = 6 × (50 – 22)
= 6 × 28
= 168

clockwise moment = W × (74 – 50)
= W × 24
= 24W

But, because the lever is balanced

anti-clockwise moment = clockwise moment

∴ 168 = 24 W

⇒ (168) ÷ (24) = W

⇒ W = 7 N

This example shows how a metre stick and one known weight can be used to find an unknown weight.

TEST YOURSELF:
Now attempt questions *37.1–37.7* and *W37.1–W37.8.*

37.3 Centre of Gravity

Fig. 37.5 shows a boy carrying a ladder. He is struggling! It would be easier if he supported the ladder at its **centre**. The entire weight of the ladder seems to be concentrated here. It appears as if gravity acts only at this point. This is why this point is known as the **centre of gravity**. The centres of gravity of regular symmetrical shapes like rectangles or discs, etc., are right at their centres.

The centre of gravity of an object is the point through which the weight of the object appears to act.

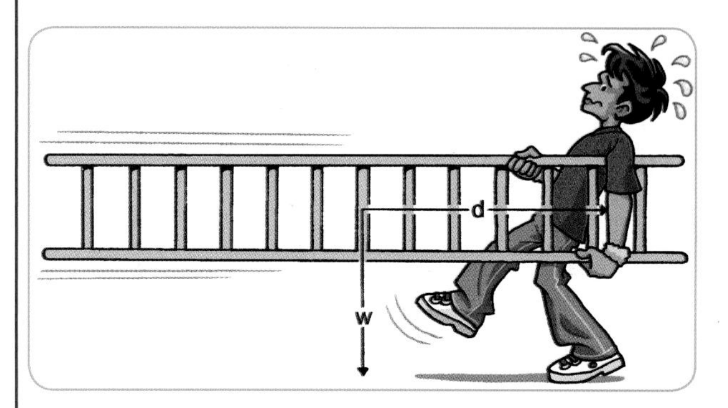

Fig. 37.5 Why would the ladder be much easier to carry if the boy supported it at its centre of gravity?

In *Fig. 37.5*, the **weight** of the ladder has a large **turning effect**. The fulcrum is at the boy's point of support. The moment of the weight is large because the **distance, *d*,** is large. The boy struggles because he has to balance this large turning effect himself.

Experiment 37.2

To find (accurately) the centre of gravity of a thin lamina, e.g. an irregular-shaped piece of cardboard

Apparatus required: irregular-shaped piece of cardboard; retort stand; pencil; cork; pin and plumbline. (A **plumbline** is a weight tied to the end of a string. It is used to indicate the vertical direction.)

Method

1. Make three or four small **holes** in the cardboard near its edges.
2. Hold a **cork** securely in the clamp of a retort stand. Stick the pin through one of the holes in the cardboard and into the cork as shown in *Fig. 37.6(a)*, page 280.

3. Hang the **plumbline** from the pin. Make sure that the string of the plumbline and the cardboard are hanging freely. Allow the whole system to come to rest.
4. Mark with **×s** (as shown) two points on the cardboard directly behind the plumbline.
5. **Repeat** at least twice more, hanging the cardboard from different corners each time as shown in *Fig. 37.6(b)*.
6. Remove the cardboard and **join the lines** between each pair of **×s**. The **point** at which these lines cross is the centre of gravity.
7. Check your result by trying to balance the cardboard at the point you have found.

Explanation

If an object is pivoted and free to move, *Fig. 37.6(a)*, it will swing **until** its centre of gravity is **directly** below the fulcrum. The weight acts through the centre of gravity. The weight will cause **turning** (or have a moment) **unless** it is exactly in line with the fulcrum.

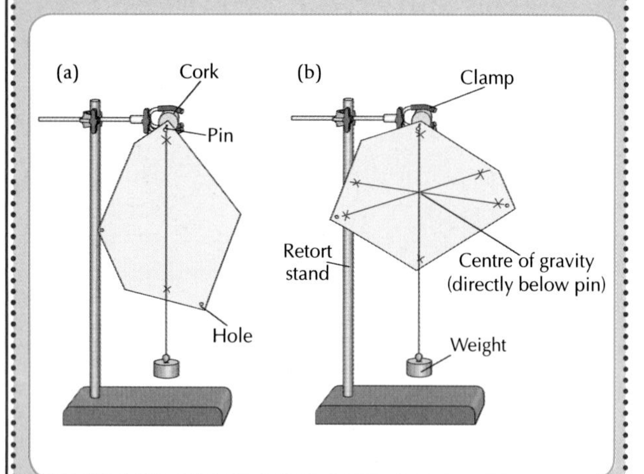

Fig. 37.6 Finding the centre of gravity of an irregular shape.

TEST YOURSELF:
Now attempt questions *37.8–37.10* and *W37.9–W37.11.*

37.4 Equilibrium

An object that is balanced is said to be 'in equilibrium'. The **stability** of an object is closely linked to the position of its centre of gravity.

- The yellow card in *Fig. 37.7(a)* is in **stable equilibrium**. The fulcrum is **above** the centre of gravity. If the card is slightly moved, its centre of gravity **rises**. When let go, the card returns to its original position.
- The red card in *Fig. 37.7(b)* is in **unstable equilibrium**. The fulcrum is **below** the centre of gravity. If this card is slightly moved, its centre of gravity **falls**. When let go, the card will not return to its original position.
- The green card in *Fig. 37.7(c)* is in **neutral equilibrium**. The fulcrum is at the **centre** of gravity. If this card is moved, it will just stay in its new position. Its centre of gravity **doesn't rise** or fall.

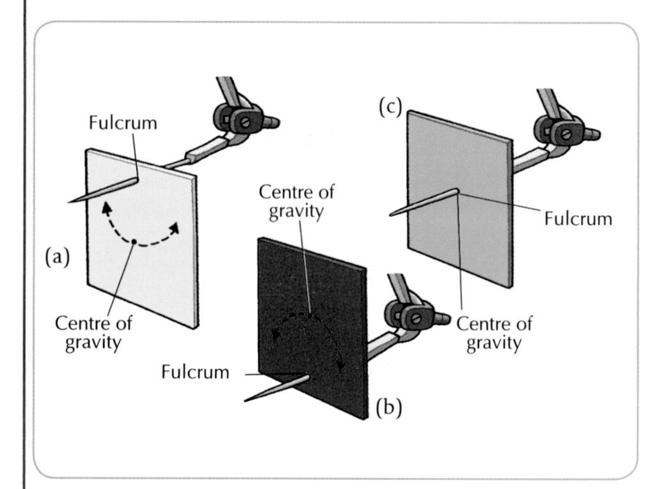

Fig. 37.7(a) Stable equilibrium. (b) Unstable equilibrium. (c) Neutral equilibrium. The centre of gravity of each of these regular shapes is at their centres.

37.5 Designing for Stability

Centre of gravity is very important when designing anything that can **topple** over. This would include cars, buses, boats, toys, televisions, buildings, chairs, benches, tables, bottles, glasses, vases, lamps, etc.

Experiment 37.3

To investigate how the centre of gravity affects stability and equilibrium

Apparatus required: Three plastic bottles; liquid (e.g. water); two Bunsen burners; some books of different thicknesses; a meter stick and a trough (or basin)

Method

1. *Fig. 37.8(a)* has three identical plastic bottles. Each bottle contains a different amount of liquid. Estimate roughly the position of the centre of gravity of each of the objects.
2. Predict which object is the easiest to topple. Gently try to topple each of the objects and test your predictions.
3. Repeat *Step 2* for two Bunsen burners as in *Fig. 37.8(b)*.
4. Repeat *Step 2* for some upright textbooks of different thicknesses.
5. Place a metre stick on an upturned trough as in *Fig. 37.8(c)*. Now slowly slide the metre stick along the top of the trough? When exactly does the metre stick topple off?

Conclusions

Now comes the hard part. Can you draw any conclusions from the investigations above? Why were some objects easier to topple? What has the stability of the object got to do with the centre of gravity of the object? What is the effect of having a wide base? Why does an object like the metre stick topple over at a certain point?

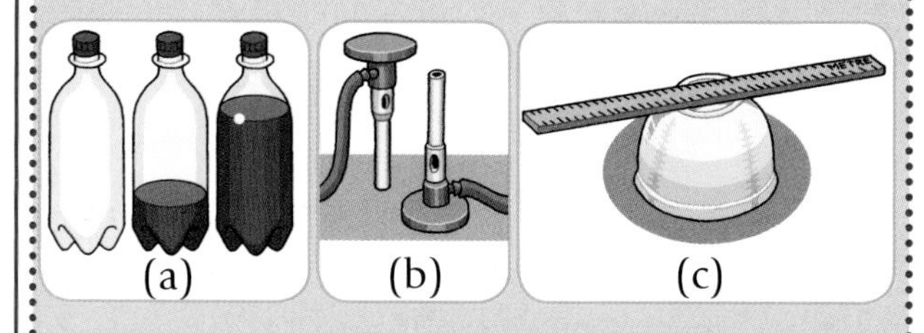

Fig. 37.8 Why is one object more stable than another?

Factors which Affect Stability

Step 5 of the investigation above shows that a body will topple when a **vertical line** through its centre of gravity falls **outside** its supporting **base**. Thus, there are two main factors which help the stability of an object:

(1) A Low Centre of Gravity Helps Stability

- In *Fig. 37.9(a)*, the bus has a **low** centre of gravity at *A*. It can tilt over a long way. The vertical line from its centre of gravity is still **inside its base**.
- If the centre of gravity of the bus was at *B*, *Fig. 37.9(b)*, it would be in trouble. The vertical line from its centre of gravity would now be **outside its base**. It would topple over.
- In *Fig. 37.9(c)*, designers test the angle at which the bus topples. The centre of gravity of all buses is deliberately made low. Heavier parts like the engine are kept down low. Notice too how the bus gets narrower as it goes upwards. Passengers in a bus are not allowed to stand upstairs. This makes sure the centre of gravity of the bus is not raised too high.

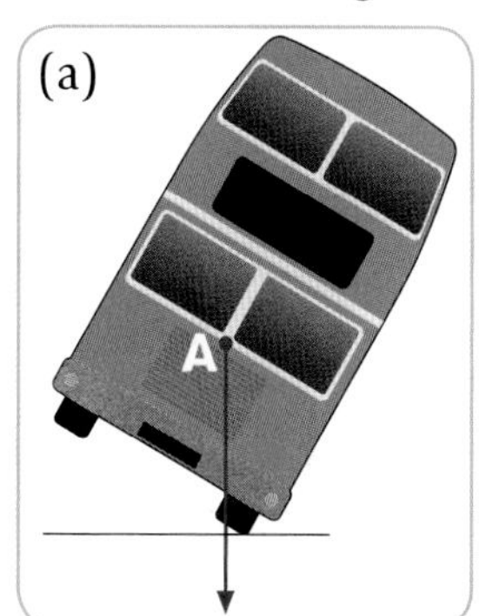

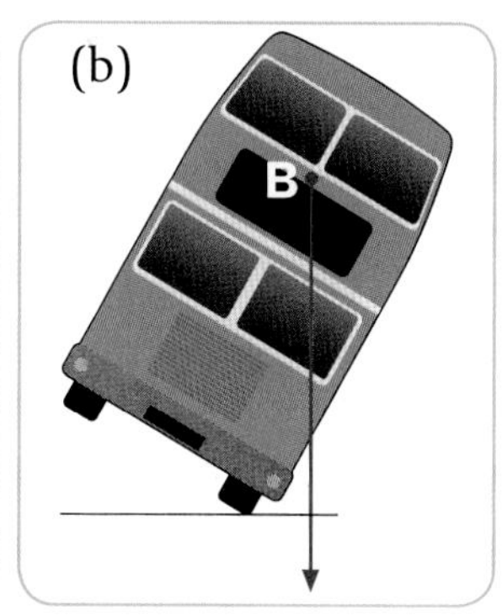

Fig. 37.9 A low centre of gravity makes a bus more stable and less likely to topple.

(2) A Wide Base Helps Stability

The centre of gravity will have a better chance of staying directly over the base if the base is **wide**. Racing cars, for example, have to be very stable. They must not flip over even at very high speeds. For stability, their wheels are spread wide apart, *Fig. 37.10*. This gives them a **large base**.

Fig. 37.10 A wide base and a low centre of gravity help keep this car stable.

TEST YOURSELF:

Now attempt questions *37.11–37.15* and *W37.12–W37.17*.

What I should know

- The moment of a force is a measure of its turning effect.
- The turning effect of a force can be increased by (i) increasing the size of the force itself, or (ii) increasing the distance from the applied force to the fulcrum.
- Moment of a force = (force) × (perpendicular distance from the force to the fulcrum).
- Levers are simple machines that allow us to move or turn things more easily.
- A lever is any rigid body that is free to turn about a fixed point or line called a fulcrum.
- The law of the lever states that when a lever is balanced, the sum of the clockwise moments is equal to the sum of the anti-clockwise moments.
- The centre of gravity of an object is the point through which all the weight of the object appears to act.
- The centres of gravity of regularly shaped symmetrical objects are at their centres.
- An object can be balanced if supported either at its centre of gravity or at a point vertically above or below its centre of gravity.
- The stability of an object is linked to the position of its centre of gravity.
- A body is in stable equilibrium if, when slightly moved or tilted, its centre of gravity rises.
- A body is in unstable equilibrium if, when slightly moved or tilted, its centre of gravity falls.
- A body is in neutral equilibrium if, when moved or tilted, its centre of gravity neither rises nor falls.
- A body will topple if a vertical line through its centre of gravity falls outside its base.
- Objects can be made stable by giving them (i) a low centre of gravity, (ii) a wide base.

QUESTIONS

Write the answers to the following questions into your copybook.

37.1 (a) The measure of the turning effect of a force is called the ..... of the force.

(b) The point or line about which an object turns is referred to as the ..... .

(c) The moment of a force = the ..... multiplied by the perpendicular ..... from the force to the ..... .

37.2 A boy is challenged to hold a book of weight 50 N in his outstretched hand for five minutes. The length of his arm is 55 cm. (He only lasts three minutes!)

(a) What is the turning effect of the weight of the book in his outstretched hand?

(b) What would be the turning effect of the weight of the book if it were held 20 cm from his shoulder?

(c) What (roughly) would the turning effect be if the book was held right at his side?

(d) Why is it more difficult to hold the book in his outstretched hand than at his side?

37.3 (a) What is the turning effect of the 70 N force in *Fig. 37.11*?

(b) Is this moment enough if the bolt needs a moment of 18 N m to turn it?

(c) What is the minimum force which must be applied (at the same point) to turn the nut?

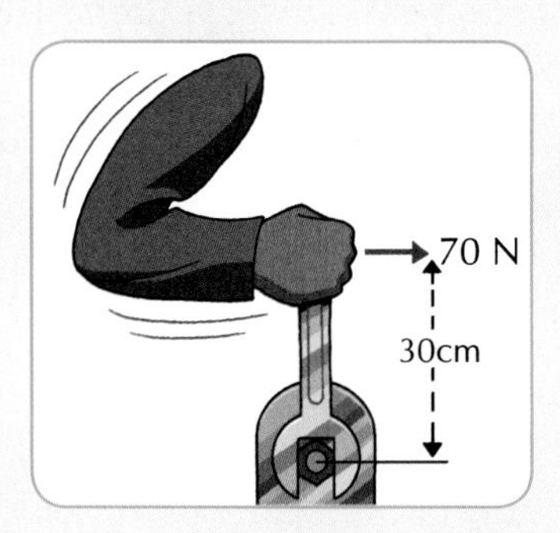

Fig. 37.11 Momentous effort!

37.4 (a) A ..... is any rigid body which is free to turn about a fulcrum.

(b) The *law of the lever* states that when a lever is ....., the sum of the clockwise ..... is equal to the sum of the ..... moments.

(c) List five everyday examples of levers.

37.5 A metre stick is suspended at its centre of gravity at the 50 cm mark. A weight of 6 N is hung at the 18 cm mark and this is balanced by a weight, *W*, which is hung at the 98 cm mark.

(a) Which of the two weights has an anti-clockwise moment about the fulcrum?

(b) What is the value of the unknown weight, W?

37.6 The lever in *Fig. 37.12* is balanced with the weights hanging from the marks, as shown. Calculate the weight, *W*, that is suspended from the 19 cm mark to maintain balance?

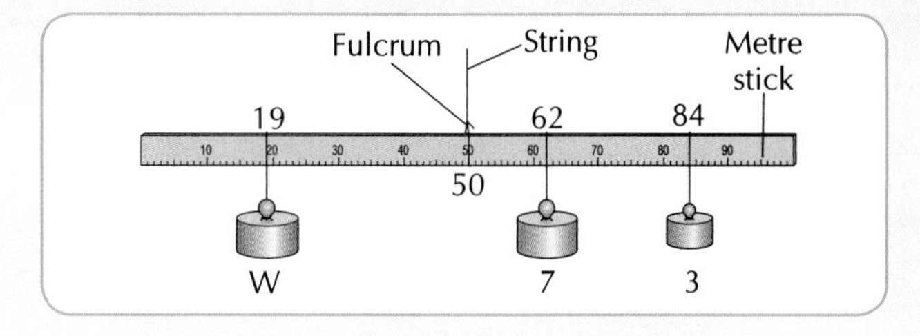

Fig. 37.12 Using the law of the lever, find W.

QUESTIONS

37.7 A see-saw is supported at its centre of gravity. A girl of weight 500 N sits 3 m from the fulcrum. How far from the fulcrum on the other side should a force of 600 N be applied to balance the girl?

37.8 What is meant by the centre of gravity of an object?

37.9 An object can be balanced if supported either at its ..... or at a point vertically ..... or ..... this point.

37.10 Julie was taking part in a treasure hunt. She was given a large irregular-shaped card. On the card was a map. Julie was told that the 'treasure' could be found precisely at the centre of gravity of the card. Give Julie instructions as to how she might accurately find the centre of gravity of the card. Explain also how she can easily check whether the point she finds is, in fact, the centre of gravity.

37.11 If a person stands up in a small rowing boat, any slight disturbance can cause the boat to tip over. Yet, there is no problem when the person is sitting down. Can you explain why the stability of the boat changes when the person stands up?

37.12 Which of the two gymnasts on the beams in *Fig. 37.13* is in trouble and why?

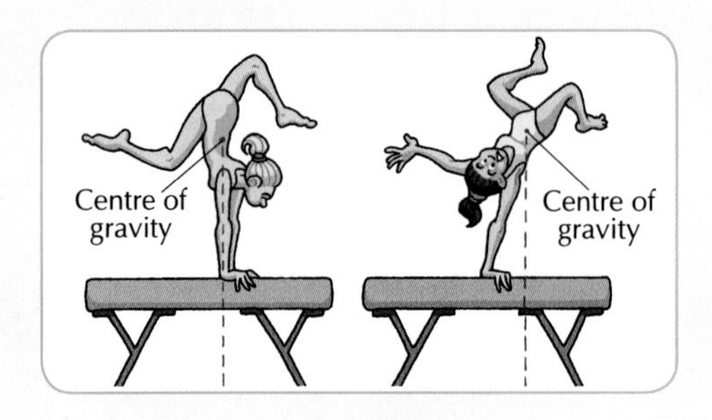

Fig. 37.13 Past the point of no return!

37.13 (a) A body will topple if a vertical line through its ..... falls ..... its base.

(b) Objects can be made more stable by giving them a ..... base or a ..... centre of gravity.

37.14 Mention *two* features of the common traffic cone, *Fig. 37.14*, that makes it a good stable design.

Fig. 37.14 A good stable design.

37.15 *Fig. 37.15* shows four lamps with different designs. In each case the centre of gravity is indicated by the letter C. Arrange these lamps in order of their stability, starting with the most stable one.

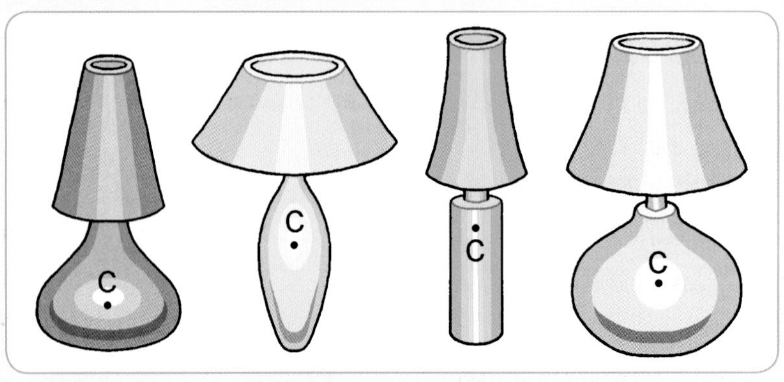

Fig. 37.15 Which design is the most stable?

Chapter 38 Pressure

38.1 What is Pressure?

Pressure is caused by **force**. *Fig. 38.1(a)* and *(b)* show an **equal force** pressing down on two coins. Yet, the coin sinks further into the plasticine in *Fig. 38.1(b)*. This is because the second coin is on its side and the pressing force is concentrated on a **smaller area**.

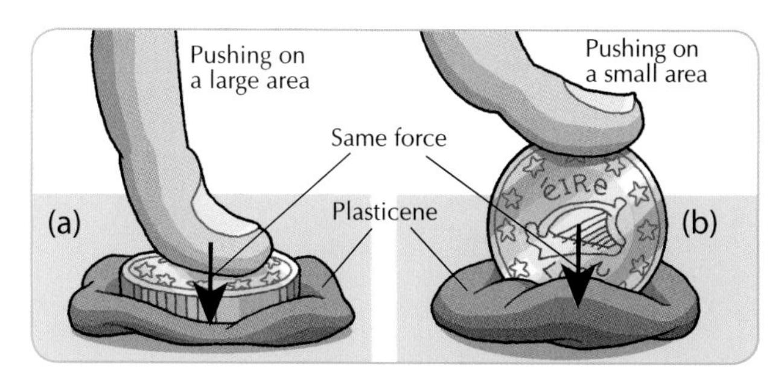

Fig. 38.1 Pressure is greater when the area being pressed is smaller.

The **pressure** on an object can be **increased** in the following two ways:

(1) Increase the **size** of the force.
(2) Decrease the **area** that the force is acting on.

From these two facts we get the definition of pressure.

pressure is force per unit area

or $$\text{pressure} = \frac{\text{force}}{\text{area}}$$

The **unit of pressure** is the unit of force divided by the unit of area, i.e. newton per square metre (N/m^2 *or* $N\ m^{-2}$). This unit is also called the **pascal (Pa)**. (However, for some measurements, the N/cm^2 is a more convenient unit to use.)

Pressure gauges are often used to measure pressure. You may have seen them being used at a garage to measure tyre pressure.

The blunt knife in *Fig. 38.2* does a bad job at slicing the tomato. The sharp knife slices it easily. Can you explain why?

Fig. 38.2 Putting it bluntly, there is a sharp contrast in the cutting ability of these two knives!

Example 1

A stone block of weight 225 000 N is to be used as a base for a sculpture. The block has dimensions 3 m × 2 m × 1.5 m as shown in Fig. 38.3. Any face of the block can be placed on the ground.

(i) What is the greatest possible pressure the block could exert on the ground?

(ii) What is the least possible pressure the block could exert on the ground?

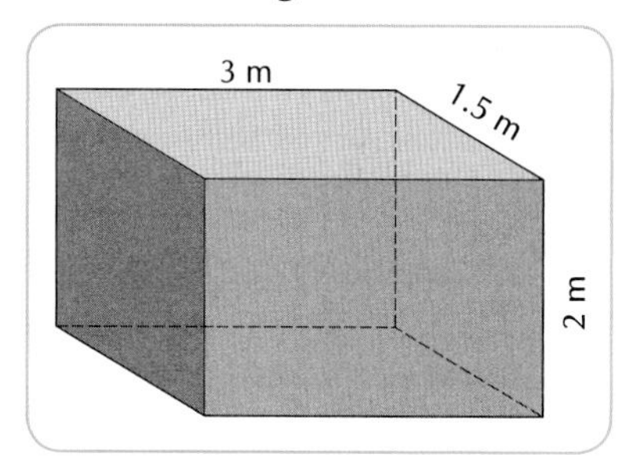

Fig. 38.3 Large stone exerting pressure on the ground.

(i) The pressure of the block will be greatest when the area that it is resting on is smallest.

$$\textbf{pressure} = \frac{\textbf{force}}{\textbf{area}} = \frac{\textbf{225 000}}{\textbf{2} \times \textbf{1.5}} = \textbf{75 000 N/m}^2 \textbf{ (or 75 kPa)}$$

(ii) The pressure of the block will be smallest when the area that it is resting on is greatest.

$$\textbf{pressure} = \frac{\textbf{force}}{\textbf{area}} = \frac{\textbf{225 000}}{\textbf{3} \times \textbf{2}} = \textbf{37 500 N/m}^2 \textbf{ (or 37.5 kPa)}$$

Example 2

Which of the following exerts more pressure on the ground underneath them:

(i) A circus elephant of weight 27 000 N standing on two hind legs, with a total area of 1800 cm² in contact with the ground.

OR

(ii) A lady of weight 600 N standing on two stiletto heels, each of area 1.5 cm²?

(i) pressure $= \frac{\text{force}}{\text{area}} = \frac{27\,000}{1800} = 15\ \text{N/cm}^2$

(ii) pressure $= \frac{\text{force}}{\text{area}} = \frac{600}{2 \times 1.5} = 200\ \text{N/cm}^2$

The lady exerts more pressure on the ground than the elephant. (This is why high heels often mark soft floors.)

Note: $1\ \text{m}^2$ is equal to $10\,000\ \text{cm}^2$. Therefore, the answers to the above questions in **N/m²** or **pascals** would be: (a) 150 000 Pa (or 150 kPa) and (b) 2 000 000 Pa (or 2000 kPa).

38.2 Pressure in Liquids

Deep-sea divers wear special protective suits when they dive to large depths, *Fig. 38.4*. This is because the **weight** of the water above them produces a very large **pressure**. This pressure is exerted on them **equally in all directions**. The further they go down, the greater the weight of the column of water pressing on them. Pressure increases, in fact, by **100 000 Pa** for every **10 m** they go down.

Fig. 38.4 The special equipment that divers wear stops the pressure of the water crushing them.

Experiment 38.1

To investigate if the pressure in a liquid depends on depth

Apparatus required: plastic container (e.g. 2-litre soft drink bottle); dish or sink; water; compass (from a mathematical set)

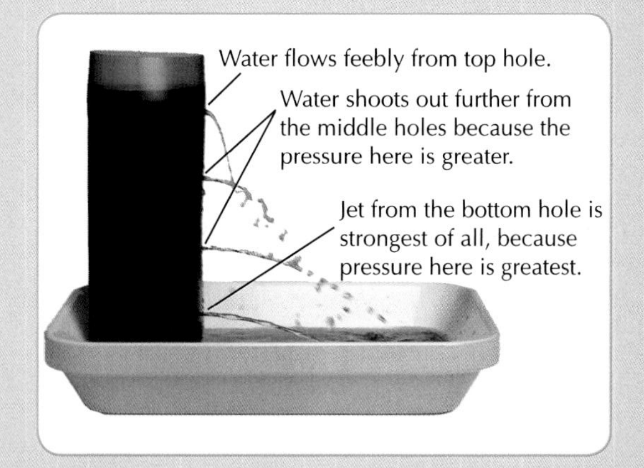

Fig. 38.5 Investigating how the pressure of a liquid varies with depth.

Method

1. Set up the apparatus as shown in *Fig. 38.5*.
2. Make four equal-sized holes in the container at different levels with a compass. Make sure the holes are facing towards the dish or sink. (Take care when using the compass.)
3. Fill the container with water. Note what happens.

Result

The lower the hole, the further the jet of water shoots out.

Conclusion

The pressure in the water depends on its depth. The greater the depth, the greater the pressure.

TEST YOURSELF:

Now attempt questions *38.1–38.3* and *W38.1–W38.15*.

38.3 Atmospheric Pressure

Air is made up of particles of nitrogen and oxygen mainly. These particles have mass. Therefore, **air** has **weight**. An experiment to verify that air has mass (or weight) is shown in *Experiment 20.2*, page 129, in the chemistry section. The pressure due to the weight of air pressing down on the earth is called **atmospheric pressure**.

Atmospheric Pressure at Sea Level

We live at the bottom of a 'sea' of air, called the **atmosphere**. This 'sea' is so deep that the air exerts a force of about 100 000 N on every square metre of the earth. This pressure acts on our bodies also. It acts **equally** in **all directions**. Luckily, our bodies exert an equal pressure outwards to balance this.

Fig. 38.6 As you go higher, the weight of the column of air above your head reduces.

Atmospheric Pressure above Sea Level

As we **increase** our **height above sea level**, atmospheric pressure **decreases**, *Fig. 38.6*. There is **less** air above us.

- Our ears often click when we go up a **steep hill**. This is our body's way of trying to equalise the pressure on either side of our eardrums.
- **Aeroplanes** fly at heights of about 10 km above sea level where the atmospheric pressure is quite low. To compensate for this, the cabins of aeroplanes have to be specially 'pressurised'.
- In **space**, there is no atmospheric pressure at all.

Experiment 38.2

To demonstrate atmospheric pressure

(a) Using a glass and a card

Apparatus required: glass tumbler (or bottle or gas jar or jam jar, without a spout); postcard or playing card

Method

1. Fill the glass tumbler right to the brim with water. (A few drops of dye added to the water make it more visible.)
2. Slide a card across the top of the water. There must be no air gaps between the card and the water.
3. Carefully turn the glass over with your hand still supporting the card. (Make sure you try this over a sink!)
4. Take away your supporting hand.

Result

The card (hopefully!) stays in place.

Conclusion

The atmospheric pressure pushing up on the card is greater than the pressure of the water pushing down, *Fig. 38.7*. This experiment also shows that the pressure of the atmosphere acts equally in all directions.

Fig. 38.7 Atmospheric pressure acting upwards on the bottom of the card is enough to support the weight of the liquid.

PHYSICS

(b) Using a soft drink's can (teacher demonstration only)

Apparatus required: Bunsen burner; tongs.

Also required: soft drink's can; water; basin of water.

Method

1. Pour water into a soft drink's can until the bottom of the can is just covered.
2. Hold the can with a tongs over a Bunsen burner and boil the water.
3. When the steam is gushing out of the can, quickly turn the can upside-down into a basin of cold water.

Result

The can collapses, *Fig. 38.8*.

Conclusion

The air inside the can is removed by boiling the water in the can. The steam produced drives out the air before it. When the remaining steam is suddenly cooled again, it condenses. The can collapses because there is now no air inside to counteract the atmospheric pressure acting from the outside.

Fig. 38.8 Collapsing can.

38.4 Measuring Atmospheric Pressure

A barometer is an instrument which is used to measure atmospheric pressure.

The two main types of barometers used are the **mercury barometer** and the **aneroid barometer**.

Uses of Barometers

- Aneroid barometers contain no liquid and are often found hanging in the halls of houses, where they are used to indicate **weather changes**.
- Altimeters are aneroid barometers that are adapted to **measure height above sea level**. They use the fact that atmospheric pressure **decreases** as the height above sea level **increases**.

Units of Atmospheric Pressure

The units used for atmospheric pressure can be confusing. Normal atmospheric pressure is taken as **76 cm** of mercury (Hg). This means that the atmosphere can normally support a column of mercury 76 cm high. The proper SI unit for pressure, however, is the **pascal** (Pa).

76 cm of mercury corresponds to a pressure of **1.013×10^5 Pa** or **1013 hectopascal (hPa)**.

The hectopascal ('hecto' = 100) is also equal to 1 millibar, an older unit of pressure.

TEST YOURSELF:

Now attempt questions *38.4–38.6* and *W38.16–W38.18*

38.5 Atmospheric Pressure and Weather

The atmosphere has a huge effect on our weather. It acts like an **insulating blanket** around the earth. Without the atmosphere, the earth would suffer extremes of temperature (from −140 °C in places at night to +80 °C at the equator).

Meteorologists are scientists who study and forecast the weather. They build up weather maps with lines called isobars, *Fig. 38.9*. These isobars join places of **equal** pressure. If the isobars are **close** together, the winds will be **strong**. If they are far apart, the winds will be light.

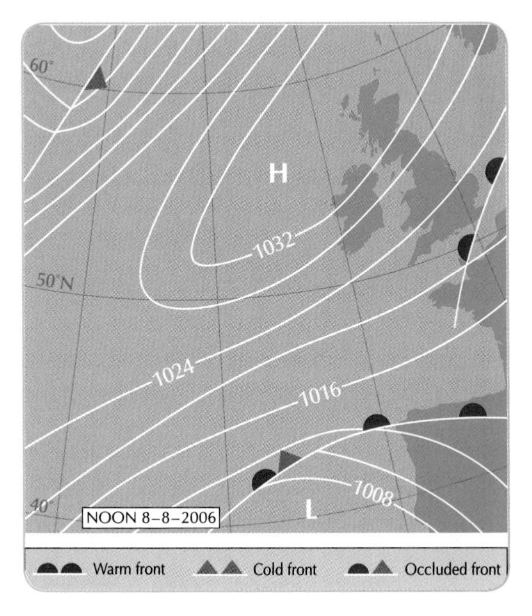

Fig. 38.9 High pressure over Ireland. What weather is this likely to bring?

- **Low** pressure gives **wet** and **windy unsettled** weather. The winds raise the air from the ground and sea. Water vapour also rises and forms clouds. The cloud cover does, however, give extra insulation. This prevents heat escaping, especially at night.
- **High** pressure means the weather will be **dry and settled**. The high pressure does not allow water vapour to rise; so there are no clouds. In summer, this will bring bright clear skies, sunny days and light winds. In winter, high pressure results in bright days. But, the days may be cold and the nights frosty without the insulating clouds.

38.6 Everyday Applications of Pressure

Our knowledge of pressure and how it can be used has led to many everyday practical applications. (We have already seen how it helps with forecasting the weather.)

(1) Flight. An aeroplane flies because of the **difference in pressure** above and below its wings. Air is pushed rapidly over the top of the wings. The pressure above the wing is reduced. This creates a pressure difference. The pressure difference creates the lift, *Fig. 38.10*.

Fig. 38.10 Why does the paper rise when you blow over the top of it like this?

(2) Hydraulics. Liquids can easily **transfer pressure** from one place to another. In *Fig. 38.11*, the pressure down on the small piston is transferred through the liquid. It is then used to lift a large force. Car brakes, power steering, garage lifts, etc., are hydraulic machines that use liquid pressure like this.

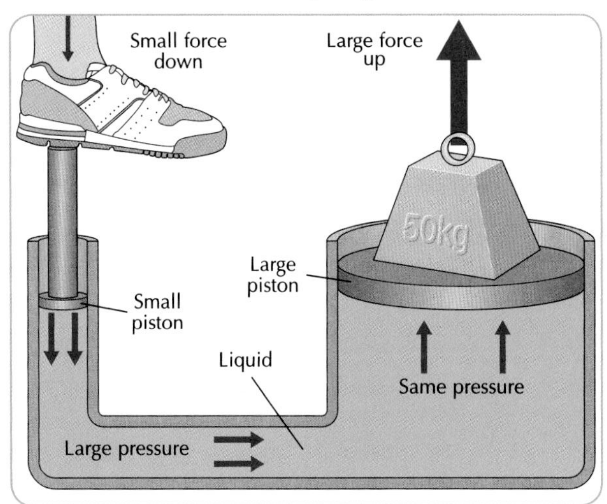

Fig. 38.11 Hydraulic jack uses a small force to lift a big force.

(3) **Water Supply System**. The greater the height of a water reservoir above a house, the greater the **pressure of the water** in the pipes to the house. Similarly, if there is a large drop in height from the water tank in a house to the shower and taps, there will be strong pressure in the shower and taps.

(4) **Medicine**. The state of a person's health is often checked by finding their **blood pressure**. Blood pressure that is too high or too low can indicate that something may not be right.

TEST YOURSELF:
Now attempt questions *38.7–38.9* and *W38.19–W38.22.*

What I should know

- Pressure is caused by a force that presses or pushes against an object.
- Pressure increases when the force applied increases or when the area it is acting on decreases.
- Pressure is force per unit area; or pressure = (force) ÷ (area).
- The unit of pressure is the pascal (Pa) or newton per metre squared (N/m^2).
- The pressure in a fluid increases with depth.
- The atmosphere, because of its weight, exerts enormous pressure on the earth.
- Atmospheric pressure decreases as the height above sea level increases.
- A barometer is an instrument used to measure atmospheric pressure.
- A barometer can be used as an altimeter, which measures heights above sea level.
- Normal atmospheric pressure is 1013 hectopascals (or 76 cm of mercury).
- Areas of high atmospheric pressure will normally have clear, dry, settled weather.
- Low pressure normally results in windy, cloudy, wet weather.
- Some everyday applications of pressure include flight, hydraulics, water supply systems, medicine, etc.

QUESTIONS

Write the answers to the following questions into your copybook.

38.1 (a) Pressure is caused by a ..... that ..... against an object.

(b) Pressure increases if the ..... applied increases or if the area it is acting on ..... .

(c) The unit of pressure is the ..... or ..... . The symbols for each of these units of pressure respectively are ..... and ..... .

(d) Pressure ..... are used to measure pressure, like for example, measuring tyre pressure at a garage.

38.2 A boy is sitting on a chair. The weight of the boy and chair is 720 N. The area of the chair touching the floor is 18 cm^2. Find the pressure exerted on the floor. The boy then tilts back on the chair. Now, only a part of two legs touch the floor. If the area in contact with the floor is now 2 cm^2, what is the pressure exerted on the floor? Apart from the danger of over-balancing, what effect might this situation have on a soft floor or floor covering?

QUESTIONS

38.3 Why is hammer *(a)* in *Fig. 38.12* having more success at driving the nail into the wood than hammer *(b)*, even though the same force is applied to both?

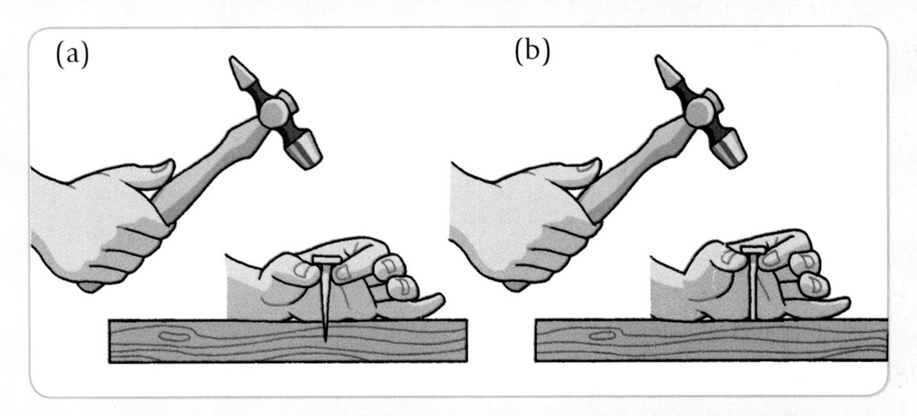

Fig. 38.12 Why is hammer (a) having more success at driving the nail?

38.4 (a) The pull of gravity on the 'sea' of air above us gives rise to what we call ..... .

(b) Atmospheric pressure can be measured using an instrument called a ..... .

(c) The pressure of the atmosphere is capable of supporting, on average, a column of mercury ..... cm in height.

38.5 Why does atmospheric pressure get less and less as you go higher and higher?

38.6 Describe an experiment to demonstrate that the atmosphere exerts pressure.

38.7 (a) Water reservoirs are normally at a ..... level than the houses they supply.

(b) If the local water supply is high above your house, the ..... of water in the pipes to your house will be ..... .

(c) The ..... in a house acts as a reservoir for the water for the whole house.

38.8 Look at the weather map in *Fig. 38.9*, page 289.

(a) Is the pressure over Ireland above or below normal?

(b) What type of weather is likely with this pressure?

(c) Where, roughly, is the nearest centre of low pressure?

38.9 Explain *each* of the following:

(a) A deep-sea diver has to wear a special protective suit when diving deep.

(b) It is uncomfortable to hold a heavy bag or suitcase by a string.

(c) When in space, astronauts have to wear spacesuits.

(d) It is quicker to fill a bucket from a downstairs tap rather than from an upstairs one.

(e) Low pressure areas tend to have unsettled wet weather.

Chapter 39 Energy

39.1 What is Energy?

When you are **very tired**, you seem to have **no energy** at all. You feel unable to do any work or even move. So, what is energy?

Energy is the ability to do work or move something.

The world is made up of **matter** and **energy**. Matter has **mass** and takes up **space**. Energy does **not** have mass. It does **not** occupy space. We cannot see energy. We cannot touch it. Yet, it is all around us. In *Fig. 39.1*, the energy got from burning fuel is enough to send the space shuttle into orbit.

Energy is measured in **joules** (*symbol*: **J**).

Fig. 39.1 Energy causing movement.

39.2 Forms of Energy

Energy can exist in many different forms. Sometimes these forms are broken up into two main groups, **stored energy** and **energy in action**.

(1) Stored Energy

Stored energy is energy that is not 'in action' right now, but is ready for action.

- **Potential energy** is the energy a body has due to its position or shape. Potential energy is the stored energy that water has behind a dam. **Stretched elastic bands**, **springs**, **wound-up clocks** also have potential energy.
- The **nucleus** at the centre of an atom has an enormous supply of stored energy. This **nuclear** energy can be released in nuclear reactors or in nuclear bombs.
- **Fuels** like petrol, timber, coal, oil and gas have stored energy. This is stored **chemical** energy. This energy is released by burning.
- **Food** is also stored **chemical** energy. The energy is released by combining the food with oxygen in a process called respiration (see *Chapter 6*). This is how our bodies get the energy to move and to keep warm (i.e. at 37 °C).

Fig. 39.2 Examples of stored energy.

(2) Energy in Action

- **Kinetic energy** is the energy associated with all moving objects. Speeding trains or falling rocks have kinetic energy.
- **Heat energy** is also energy in action. The hotter the water particles get in a kettle, the more they move around.

- **Light energy** can turn the vanes of a Crooke's radiometer (see page 318, *Chapter 41*). Light also provides the energy for plants to grow and make food and oxygen in a process called **photosynthesis** (see *Chapter 14*).
- **Sound energy** makes things vibrate. The fact that sound energy causes the air and our eardrums to vibrate allows us to hear.
- **Electrical energy** comes from moving charges. It is one of the most convenient forms of energy. It can very easily be changed into other forms of energy. It can cause a motor to move, an electric fire to give out heat, a radio to give out sound, a light bulb to give out light, etc.
- **Magnetic energy** is the ability of certain substances to attract or repel other substances. Magnetic energy is used, for example, in a compass, to show direction.

Fig. 39.3 Examples of energy in action.

39.3 Conservation of Energy

Often it might *appear* that energy is lost or destroyed. This is not so. It simply changes from one form to another. Other times, energy just spreads out until it is not noticed anymore.

PRINCIPLE OF CONSERVATION OF ENERGY:
Energy cannot be created or destroyed, but it can change from one form to another.

In a light bulb, for instance, only about 5 per cent of the **electrical** energy going in appears as **light**. The remaining energy is turned into **heat** energy. This heat energy then **spreads out**. It is not useful. Heat energy, in fact, is produced as 'waste' in many energy transfers.

TEST YOURSELF:
Now attempt questions *39.1–39.5* and *W39.1–W39.8.*

39.4 Sources of Energy

(1) The Sun

The sun is our **primary source** of energy. It provides almost 1 000 000 J of energy each day to every square metre of the earth's atmosphere.

- All green plants trap sunlight energy. They use it in **photosynthesis** to make **food**.
- **Light** from the sun is used so that you can see the world around you.
- **Heat** from the sun keeps us **warm**.
- The sun also heats up the land and the sea. This produces winds. Winds produce waves. **Winds and waves** can be used to produce electricity.
- Some of the sun's energy evaporates the water in the sea. This eventually causes rain. Rain forms rivers. From rivers we get **hydroelectricity**.
- **Solar panels**, **solar cells** and solar power stations capture a tiny amount of the sun's energy. They convert it into heat and electricity.

(2) Fossil Fuels

Plants have been making food for **millions** of years. Some of this food, or chemical energy, was eaten by animals. Over the ages the remains of these plants and animals got squashed. They eventually turned into **peat**, **coal**, **oil** and **gas**. This is why these are called **fossil** fuels.

- In Ireland, we mainly use coal, gas, oil and peat to produce **electricity**.

- But fossil fuels are **non-renewable sources**. They get used up. They will come to an end some day. Also, they will become more expensive as they get scarcer.

(3) Renewable Energy Sources

Some sources of energy do not run out when they are used. They are called **renewable energy sources.**

- Renewable sources of energy create **no harmful waste products**.
- But, they are **expensive** to set up. It is difficult to trap the energy and transfer it to electricity in some cases.

Examples of renewable energy include: solar, wind, hydroelectric, wave, tidal, biomass and geothermal energy.

Renewable sources of energy are those that will never be used up.

(4) Nuclear Energy

Nuclear Fission

In **nuclear reactors**, enormous amounts of energy can be produced in a process called nuclear fission. This energy is released by **splitting up a large nucleus** into two almost equal parts, *Fig. 39.4*. In this process, 1 kg of the fuel uranium or plutonium can yield as much energy as 3 000 000 kg of coal.

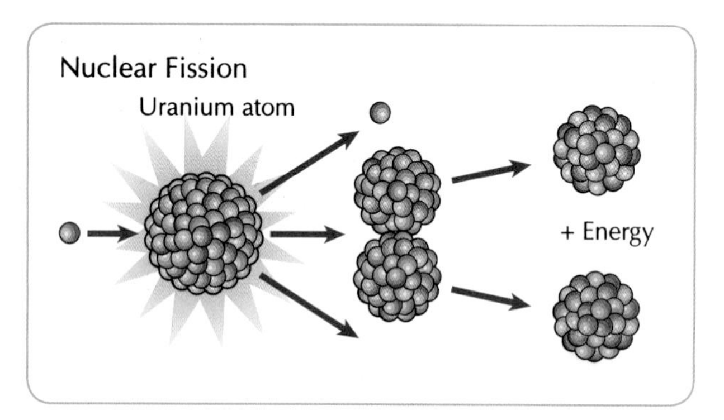

Fig. 39.4 Fission is the splitting of a large nucleus.

Problems with Nuclear Fission

Radioactive substances are also produced in this reaction. The disposal of these dangerous wastes is a constant and expensive problem. Some can continue emitting radiation for thousands of years.

Radiation can damage or **kill cells**. It can also induce forms of **cancer**. Yet, the entire nuclear power industry is responsible for less than 1 per cent of all the nuclear radiation in the world today.

Nuclear Fusion

Nuclear fusion is another method of producing energy from the nucleus. Here **light nuclei are forced to join together**, *Fig. 39.5*. This is the process by which the **sun** produces enormous amounts of energy. If the nuclear fusion reaction can eventually be controlled on earth, it may be the answer to most of our energy problems.

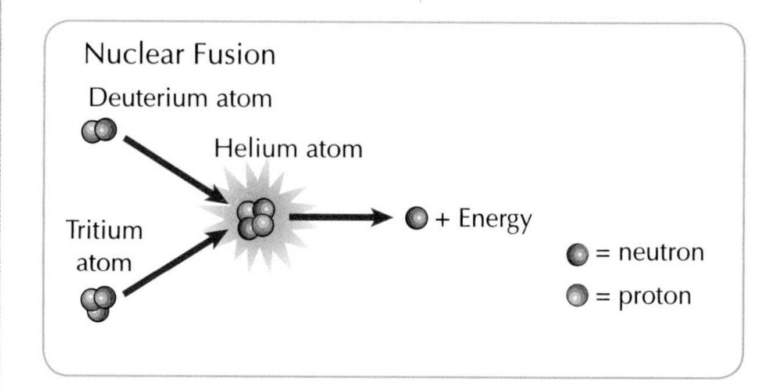

Fig. 39.5 Fusion is the joining together of two light nuclei. The 'fuel' for nuclear fusion is deuterium, a form of hydrogen. There would be an almost endless supply of cheap 'fuel' for the reaction and it would be relatively clean.

(5) Other Sources

Other sources of energy are still being developed by scientists.

Hydrogen

Fuel cells which turn **hydrogen** into electricity are well established for specialised uses like space travel. The only emission is water. Many **car manufacturers** are experimenting with hydrogen as an alternative power source for their vehicles, *Fig. 39.6*. Hydrogen-fuelled buses have been introduced on a trial basis in some countries.

Fig. 39.6 Hydrogen-fuelled cars like this may replace the internal combustion engine when oil runs out.

39.5 National Energy Needs and Strengths of Different Energy Sources

The earth is like a large travelling spaceship. It has an ever-increasing crew. This crew needs to be fed. Fuel or energy is needed during the journey. But, the supply of food and fuel is limited. Population increases and greater demands for energy cause terrific strains on our resources. Careful planning is needed.

Ireland's Energy Needs

Ireland too has an energy problem. In the last 15 years the demand for electricity has more than doubled. At present more than 90 per cent of the electricity we produce comes from **fossil fuels**. This is a bad position to be in. **Alternate sources** need to be considered.

Tables 39.1 to *39.9* show the advantages and disadvantages of various energy sources.

ENERGY SOURCE	Fossil fuels (coal, oil, gas, turf, peat, etc.)
ORIGINAL SOURCE	Sun
ADVANTAGES	Electricity supply is guaranteed (once the fuel is there). Fuel can be stored, extra demand can be catered for. Relatively cheap.
DISADVANTAGES	Limited supply of fuel available. Will get much dearer as fuel supplies dwindle. It pollutes the air, produces acid rain (see *Section 28.5*, page 200 *and Section 18.2*, page 116) and increases the greenhouse effect (also see *Section 18.2*).

Fig. 39.7 Turf being collected.

Table 39.1 Fossil fuels.

ENERGY SOURCE	Solar energy
ORIGINAL SOURCE	Sun
ADVANTAGES	Free supply will never run out. No pollution. Solar panels or cells on roofs do not interfere with anything.
DISADVANTAGES	Initial costs can be very high. Least energy is produced in winter when it is needed most and no energy can be produced at night. Solar energy is spread over a wide area and is difficult to collect.

Fig. 39.8 Over 1800 computer controlled movable concave mirrors, which follow the sun in the Mojave desert solar power plant.

Table 39.2 Solar Energy.

ENERGY SOURCE	Wind energy
ORIGINAL SOURCE	Sun
ADVANTAGES	Free supply will never run out. No pollution. Land with wind turbines can still be used for farming. Low level technology costs about 1/5 of what it cost 20 years ago.
DISADVANTAGES	The wind supply is variable and unreliable, windy sites are essential. Some people think turbines are noisy and unsightly. Many turbines are needed to provide as much power as one modern power station.

Fig. 39.9 Wind farm in the Arklow Bank, off the Wicklow coast.

Table 39.3 Wind Energy.

ENERGY SOURCE	Hydroelectric energy
ORIGINAL SOURCE	Sun (sun → evaporation of water → rivers)
ADVANTAGES	Free supply will never run out. No pollution and can store energy from other sources. Can be switched on in a matter of seconds.
DISADVANTAGES	Only a small number of suitable sites (wet and hilly). Initial costs are high. Lands have to be flooded to make reservoirs, lakes get silted up.
Note: At Turlough hill in Wicklow, off-peak or otherwise unused electricity is used to pump water up to an artificial lake. This water can later be released when electricity demand is high.	

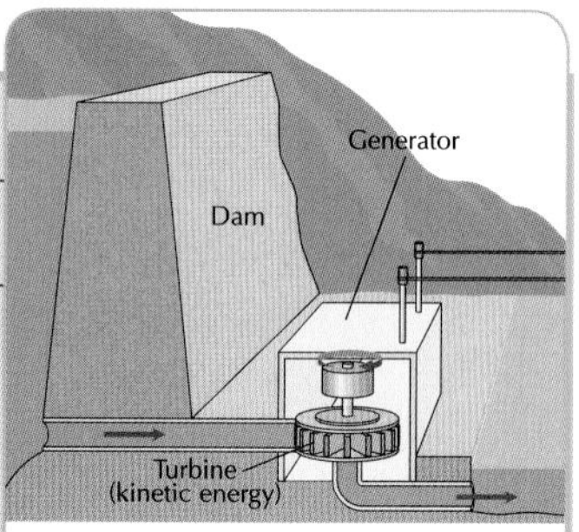

Fig. 39.10 Kinetic energy of falling water is used to generate electricity.

Table 39.4 Hydroelectric Energy.

ENERGY SOURCE	Wave energy
ORIGINAL SOURCE	Sun (sun → winds across sea → waves)
ADVANTAGES	Supply will never run out. No pollution. Vast amounts of energy possible.
DISADVANTAGES	Supply can vary with the winds. Some people think turbines at sea are unsightly. About 25 km of large floats would be needed to produce as much energy as an average generating station. High maintenance costs.
Note: The potential of wave energy is vast. But there are still a lot of engineering problems to be overcome.	

Fig. 39.11 The awesome energy of a wave.

Table 39.5 Wave Energy.

ENERGY SOURCE	Tidal energy
ORIGINAL SOURCE	Moon and sun (gravitational attraction)
ADVANTAGES	Very reliable and supply will never run out. No pollution, the barrier can also act as a bridge across the estuary. Cost of each unit of electricity produced is slightly less than average.
DISADVANTAGES	Damages habitats of birds and other creatures that live in river estuaries. Electricity is generated only at certain times during the day. Considering the cost of construction and size, it does not deliver a large amount of energy.
Note: At high tide the gates of the dam trap the water. At low tide the water is let out again.	

Fig. 39.12 Tidal generating station in Brittany, France.

Table 39.6 Tidal Energy.

ENERGY SOURCE	Biomass energy (wood, waste from plants, energy crops)
ORIGINAL SOURCE	Sun (through photosynthesis)
ADVANTAGES	Supply will be secure and will never run out. Reduction in greenhouse gas emissions. Create employment in rural areas and particularly good for developing countries.
DISADVANTAGES	Large areas of land are needed to produce enough biomass. Can cause some pollution when burned. People do not like the smell of rotting plants, etc.
Note: This is the chemical energy stored in fast-growing plants or in the waste from animals that eat the plants. Solid biofuel like firewood can be used to heat our homes. Liquid biofuel like the oil from oil-seed rape can be used as a vehicle fuel. Biogas (e.g. methane) can be produced from rotting plant and animal matter, *Fig. 39.13*.	

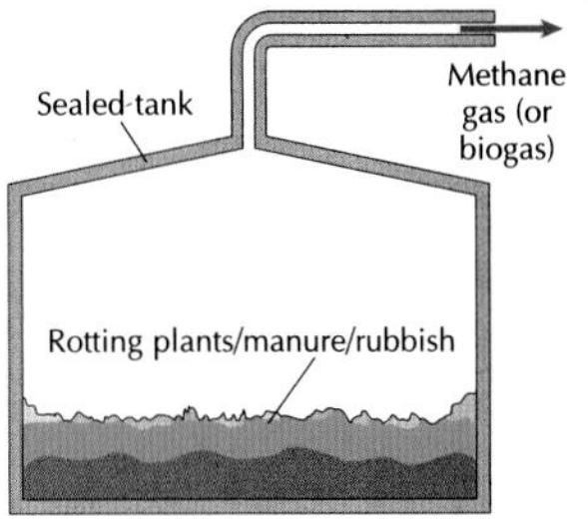

Fig. 39.13 The methane gas produced can be used for cooking and heating.

Table 39.7 Biomass Energy.

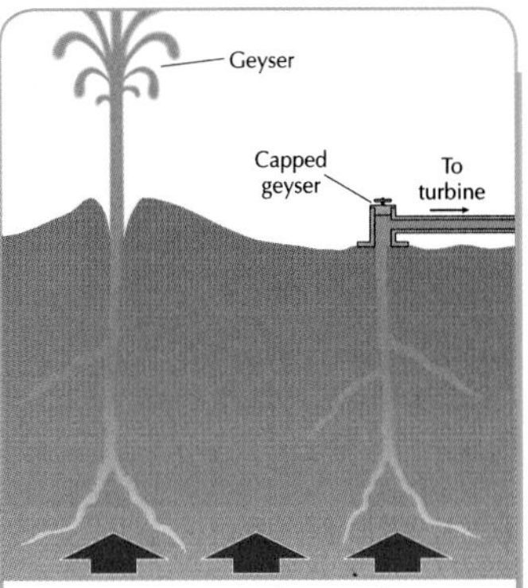

ENERGY SOURCE	Geothermal energy
ORIGINAL SOURCE	Earth (the decay of radioactive material in the earth's core)
ADVANTAGES	Supply is free and unlikely to run out. No pollution. Extremely low-cost energy where it happens naturally.
DISADVANTAGES	Limited number of suitable sites. Expensive to drill so deep into the earth. Possible damage caused to the earth's crust.
Note: Hot water springs up naturally in 'geysers'. In some places, two deep holes are drilled in the ground. Cold water is pumped down one hole. It returns as hot water or steam from the other hole.	

Fig. 39.14 Geothermal energy can be used to drive turbines and produce electricity.

Table 39.8 Geothermal Energy.

ENERGY SOURCE	Nuclear energy
ORIGINAL SOURCE	Nuclei in atoms (splitting in fission, joining in fusion)
ADVANTAGES	Huge energy from small amount of fuel. Produces no acidic gases or carbon dioxide. Nuclear radiation produced in reactors has many benefits, including medical, industrial and food preservation.
DISADVANTAGES	Fission is non-renewable, the uranium or plutonium will run out. Radioactive waste is difficult to store safely. Small chance of a very serious accident.

Fig. 39.15 Nuclear power plant.

Table 39.9 Nuclear Energy.

Which Energy Source is Best?

As we have seen, each energy source has its problems. Different sources of energy suit different areas in the world.

- In hot countries, **solar** power stations are worth building.
- **Biomass** might be the best answer for a country like Brazil, with its climate and rain forests.
- Iceland or New Zealand might avail of the **geothermal** energy inside the earth.
- In Ireland, our climate and location would suggest that **wind** energy could be one of our best options.

TEST YOURSELF:
Now attempt questions *39.6–39.11* and *W39.9–W39.18.*

39.6 Conserving Energy

'What do you do if the bath water is escaping? Do you (i) get a better bath plug or (ii) turn on the taps faster?'

This is a famous quote from an energy expert. It sums up our energy situation well. Many times we opt for the second solution with energy. In our homes and schools we often **turn the heat up** when we are cold rather than trying to **stop the heat escaping**. Saving energy is called **energy conservation**. Conserving energy means not wasting it.

Fig. 39.16 shows how heat is lost in our houses. We can reduce the amount of heat flowing from our warm rooms by using substances called **insulators**. These don't allow heat to flow through them.

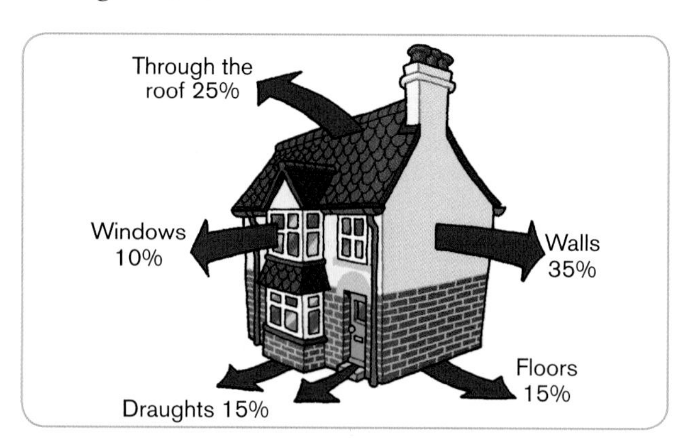

Fig. 39.16 Joules (of heat) constantly go missing in our houses!

Methods of Saving Energy in Our Homes and Schools

- Put lagging jackets on **hot water tanks**.
- **Insulate attics** with polystyrene or fibreglass (good insulators).
- Use **draught excluders** on doors and windows (heat is lost through spaces that cause draughts).
- **Walls** should be fitted with air cavities (trapped air is a great insulator).
- Have **double-glazed** windows (trapped air again between two glass panes).
- Use **thick carpets** and curtains.

- Do not leave doors or **windows open** when it is cold outside.
- Put thermostats on all **radiators** to control heat in each room.
- **Lower the settings** when the weather warms up.
- Choose **'A' energy rated** appliances. They use less energy.
- If it is not in use, **turn it off**! TVs, computers, printers, music systems, etc., use up to 20% of power in standby mode.

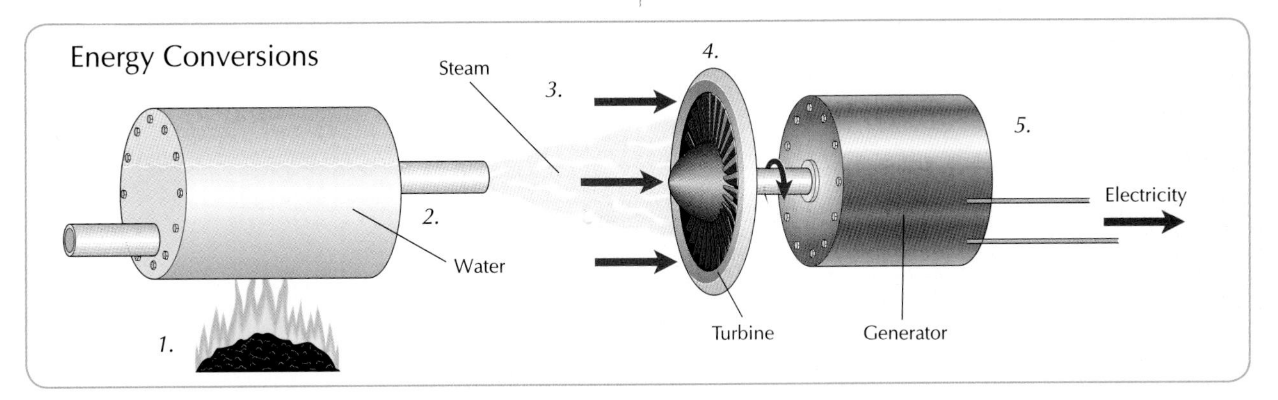

Fig. 39.17 1. Burning coal changes chemical energy into heat. 2. This heat is used to give energy to water. 3. This turns into the kinetic energy in the steam. 4. The moving steam is then used to turn a turbine (more kinetic energy). 5. The generator then converts this kinetic energy into electricity (also using magnetic energy).

39.7 Energy Conversions

We have already seen that energy changes from one form to another. This is called **energy conversion** or **energy transfer**. Look, for example, at the energy conversions involved in generating electricity in *Fig. 39.17*. But you could still ask the following:

(i) Where did the fossil fuel get its energy?
(ii) What will happen next to the electricity generated?

No matter what energy you think about, there are similar chains of **energy conversions** linked to it. The following experiments demonstrate some more energy conversions.

Mandatory experiment 39.1

(a) To show chemical energy changing to electrical energy to heat energy

Apparatus required: 6 V battery; beaker (or polystyrene cup); coil of resistance wire (e.g. nichrome or constantan); connecting wires; switch; thermometer (or temperature sensor and data logging equipment)

Method

1. Place the resistance wire (or coil) under the cold water in the beaker.
2. Connect the resistance coil through a switch and to the battery as shown in *Fig. 39.18*.
3. Place the thermometer in the cold water. After a few minutes, note the temperature.
4. Close the switch. Note the temperature of the water every two minutes. (If using data logging equipment, a graph of temperature against time will be plotted.)

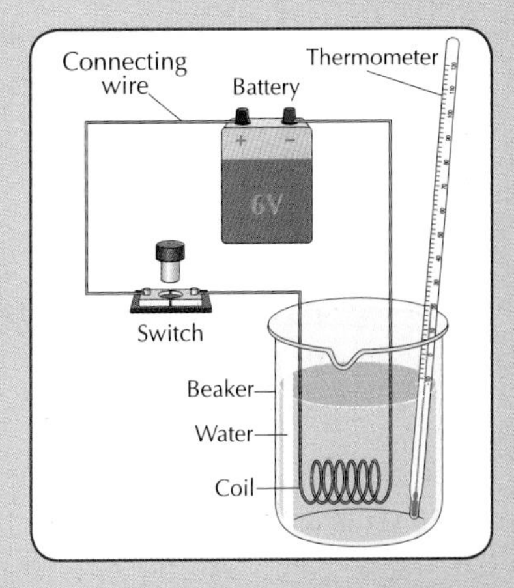

Fig. 39.18 Chemical energy → electrical energy → heat energy.

Result

The temperature of the water rises as long as the battery is supplying energy.

Conclusion

The chemical energy of the battery was converted to electrical energy in the circuit. The electrical energy in the resistance coil was converted to heat. The heat caused the temperature of the water to rise.

(b) To show electrical energy changing to magnetic energy to kinetic energy

Apparatus required: a long nail; a length of insulated wire (60 cm approx.); a 6 V battery or a power supply; some paper clips

Method

1. Wind the insulated wire closely around the nail.
2. Connect the two ends of the wire to a power supply or a 6 V battery.
3. Place some loose paper clips under the nail, *Fig. 39.19*.
4. Switch on the current. Note what happens.
5. Switch off the current. Note again what happens.

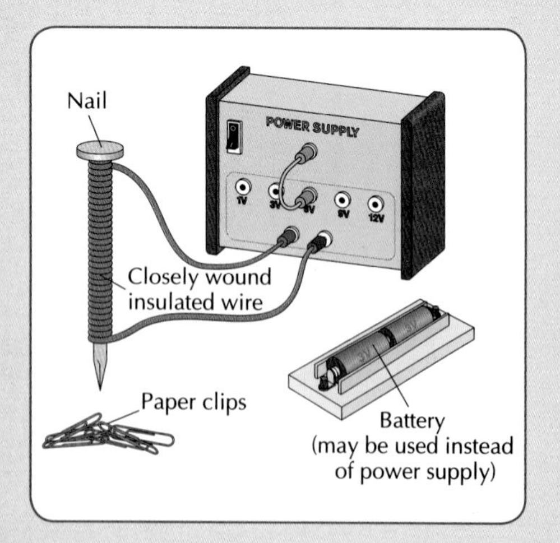

Fig. 39.19 Electrical energy → magnetic energy → kinetic energy.

Conclusion

The electrical energy from the power supply (or battery) changes to magnetic energy in the nail. This magnetic energy in the nail causes the paper clips to move towards the nail, i.e. kinetic energy. (The kinetic energy of the paper clips then changes to potential energy.)

(c) To show light energy changing to electric energy to kinetic energy

Apparatus required: solar cell; miniature electric motor; strong light source (e.g. strong sunlight or light from a projector); connecting wires

Method

1. Connect the solar cell to a miniature electric motor, *Fig. 39.20*.
2. Place the solar cell in the path of strong sunlight or in the beam of a projector. Note what happens.
3. Block the light arriving to the solar cell. Again note what happens.

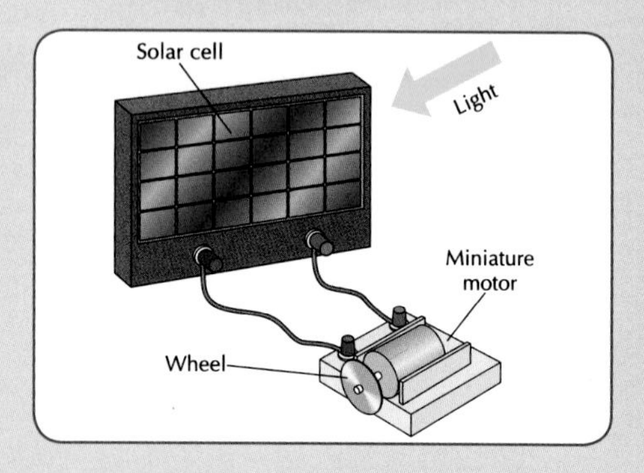

Fig. 39.20 Light energy → electrical energy → kinetic energy.

Conclusion

Light energy is converted into electrical energy by the solar cell. Electrical energy is changed to kinetic energy when the wheel of the motor rotates.

TEST YOURSELF:
Now attempt questions *39.12–39.18* and *W39.19–W39.22.*

What I should know

- Energy is the ability to do work or to move something.
- The unit in which energy is measured is called the joule (symbol: J).
- Energy has no mass and it does not take up space.
- The principle of conservation of energy states that energy can neither be created nor destroyed but that it can change from one form to another.
- Types of stored energy include potential energy (due to height or shape), chemical energy (as in food, fuel and a battery) and nuclear energy.
- Types of energy in action include kinetic energy (moving objects), heat, light, sound, magnetic and electrical energy.
- The sun is the original source of most of our energy supplies. The sun's energy is used by plants to produce food.
- Sources of energy can be renewable or non-renewable.
- Non-renewable sources get used up. They include fossil fuels like coal, oil, gas and peat.
- The burning of fossil fuels pollutes the air, water and soil. It also increases the greenhouse effect.
- Renewable sources of energy are those that will never be used up. They include: solar, hydroelectric, wind, tidal, wave, biomass and geothermal.
- Biomass is the chemical energy stored in fast-growing plants like trees and oil-seed rape. Methane can be produced as a biogas.
- Geothermal energy comes from beneath the surface of the earth. This is due to radioactive material heating the rocks far below the surface.
- Nuclear energy is the energy released from the nuclei of atoms.
- Nuclear fission is a process used in nuclear power stations in which the nuclei of large atoms are split. Enormous energy is released.
- Nuclear fusion occurs in the sun. Nuclei of light atoms combine to produce enormous amounts of energy.
- Ireland needs to look for alternative sources of energy to satisfy its increasing energy demands.
- Most of the energy produced in the world ends up as wasted heat.
- A good practical way to avoid an energy crisis is to conserve a lot of the energy that we waste at the moment. Good insulation can save a lot of energy.
- An energy conversion takes place when energy changes from one form to another. This is usually done to produce a more useful form of energy.

QUESTIONS

Write the answers to the following questions into your copybook.

39.1 Energy is defined as the ability to do ..... or the ability to ..... something.

39.2 The unit in which energy is measured is called the ..... . The symbol for this is ..... .

39.3 The chemical energy stored in coal and oil and gas is released by ..... .

39.4 Humans get the energy to move and to heat their bodies from the ..... energy in their food. This energy is released when ....., that we breathe in, combines with the food.

39.5 The principle of conservation of energy states that energy can neither be ..... nor ..... but that it can ..... from one ..... to another.

QUESTIONS

39.6 (a) Name three fossil fuels that are used in Ireland to generate electricity.

(b) Give two harmful side effects that the burning of fossil fuels may have on the environment.

(c) Name five renewable sources of energy.

39.7 (a) The principal source of energy available to us is the ….. . Plants use light energy from this source to produce ….. .

(b) Natural gas, coal and ….. were formed by energy from the ….. over millions of years.

39.8 Which of the following energy sources does not depend on the sun?

(i) wind (ii) hydroelectric (iii) solar (iv) geothermal (v) wave

39.9 (a) Biomass energy is the chemical energy stored in ….. which can be used to produce useful substances like ….. or ….. .

(b) Give one advantage of using geothermal energy to produce electricity.

39.10 (a) ….. energy is the energy released from the nuclei of atoms.

(b) The 'splitting' of the nucleus at nuclear power stations, in a process called nuclear ….., releases ….. amounts of energy.

39.11 Examine the pie-chart shown in *Fig. 39.21*. It shows where the world presently gets its energy supplies to produce electricity. How do you think the pie-chart for the year 2050 will differ from this? Explain your answer.

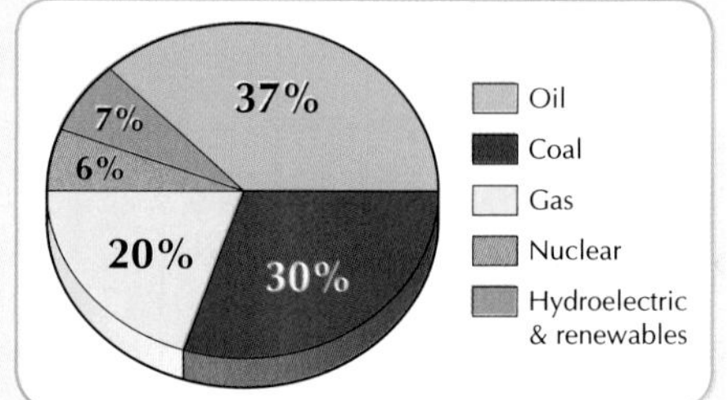

Fig. 39.21 Energy sources used to produce the world's electricity.

39.12 We can reduce the amount of heat escaping from our warm houses by using substances called ….., which do not allow heat to ….. them. Two examples of such substances would be (i) ….. and (ii) ….. .

39.13 (a) Give two reasons why is it in everyone's interest to try to conserve energy.

(b) Mention any two things a teenager might do that would help to conserve energy.

39.14 Give any five examples of energy either doing work or converting to other forms of energy.

39.15 Why does a bouncing ball not bounce back to its original height after hitting the ground?

39.16 (a) An energy ….. takes place when energy changes from one form to another to produce a form of energy which is usually more ….. .

(b) The ….. energy of water is used in hydroelectric power stations to produce ….. energy.

39.17 Name one place where sunlight energy is changed into chemical energy?

39.18 Name the energy changes that are taking place in each sketch in *Fig. 39.22*.

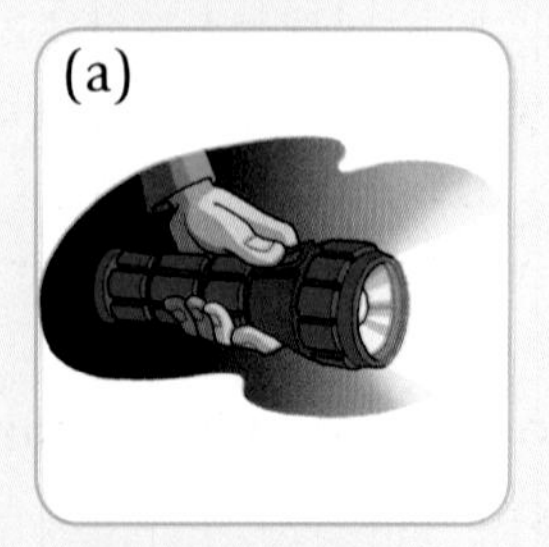

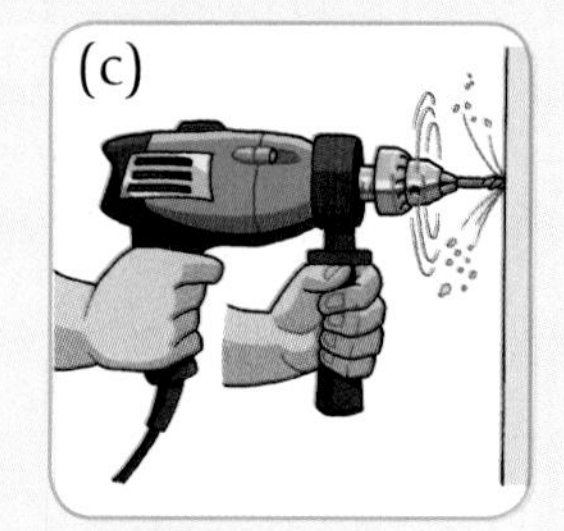

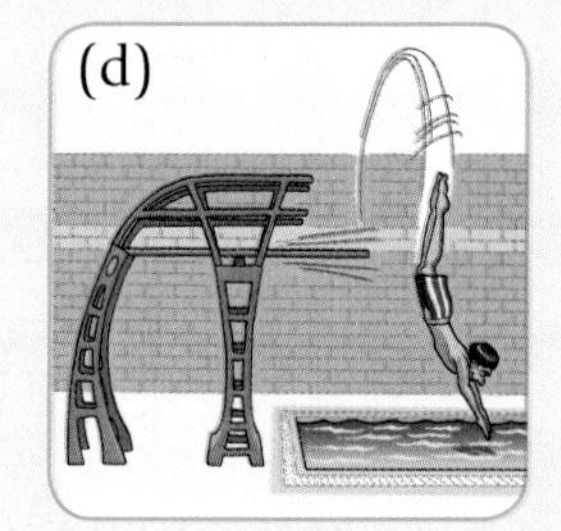

Fig. 39.22 Energy conversions.

Chapter 40 Heat

40.1 Heat is a Form of Energy

Heat has the ability to do work or move something, *Fig. 40.1*. Thus, heat is a form of **energy**. Further proof comes from the fact that heat can change into other forms of energy and vice versa.

- *Fig. 40.1* shows that **heat** can be converted into **light** and **sound**.
- **Kinetic energy** turns into **heat** energy when you rub your hands together.
- *Fig. 39.17*, page 299, shows that **heat** can be changed into **kinetic energy** and **electricity**. It also shows that **chemical** energy in coal can turn into **heat** energy.

Fig. 40.1 Lava is rock that has got so hot (about 600 °C) that it has melted. It now flows like a thick liquid. Heat caused this movement.

Because heat is a form of energy, it is measured in the unit of energy, the **joule** (symbol: **J**).

40.2 Expansion and Contraction

One common **effect of heat** is that it causes substances to **expand**. The opposite happens when objects lose heat: they **contract**.

Disadvantages of Expansion and Contraction

Expansion and contraction can sometimes be a nuisance.

- Long lengths of **railway track** need tapered joints to allow for expansion in hot weather.
- Large **slabs of concrete** can buckle in the heat if not given room to expand.
- A 700 m **bridge** can be 25 cm longer in summer than in winter, *Fig. 40.2*.

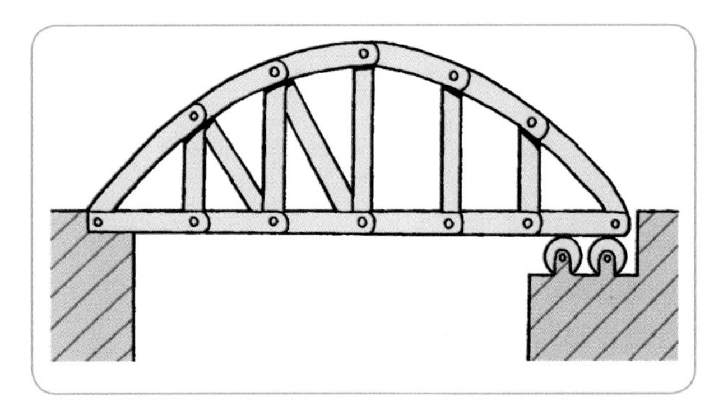

Fig. 40.2 Steel bridge on rollers to allow for expansion and contraction.

Advantages of Expansion and Contraction

But expansion and contraction can sometimes be put to good use.

- **Thermometers** work on the basis that liquids expand when heated, and contract when cooled.
- **Bimetallic strips** are used to trigger alarms or switch off heating. They are made from two metals joined together. With heat, one metal expands faster than the other. This causes the strips to bend and complete or break a circuit.
- Expansion is used to great effect in the **car engine** and in **rockets**.

Mandatory experiment 40.1

(a) To show that a solid expands when heated and contracts when cooled

Apparatus required: metal ball and ring apparatus; Bunsen burner; tongs

Method

1. Set up the apparatus as shown in *Fig. 40.3(a)*. Check that the ball just fits through the ring when cool.
2. Using tongs to hold the chain, heat the ball over a Bunsen flame for a few minutes.
3. Using the tongs again, try to pass the ball through the ring as in *Fig. 40.3(b)*.
4. Allow the ball to cool.
5. Check again to see if the ball will now pass through the ring.

Warning!: Do not touch the ball or the chain when they are hot!

Result

At normal temperature, the metal ball just passed through the ring. When heated, the ball did not fit through the ring. When cooled, the ball did pass through the ring again.

Conclusion

The ball expanded when heated and contracted when cooled.

Fig. 40.3 At normal temperature, the ball passes through the ring. When heated the ball cannot fit through the ring.

(b) To show that a liquid expands when heated and contracts when cooled

Apparatus required: round-bottomed flask with stopper and narrow glass tube at the top; water bath; Bunsen burner; tripod; wire gauze

Also required: liquid (e.g. water with a dye added)

Method

1. Fill the flask and narrow glass tube with the liquid to the level shown, *Fig. 40.4*. Mark this level with an elastic band or a pen.
2. Heat the flask in a water bath for several minutes (as shown).
3. Note the new level of the liquid in the tube.
4. Turn off the Bunsen burner. Allow the liquid to cool. Observe the level of the liquid after several minutes.

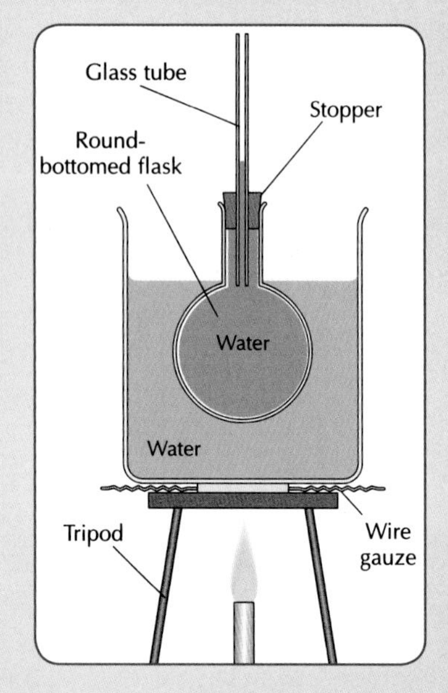

Fig. 40.4 A liquid expands when heated and contracts when cooled.

Result

When heated, the level of the liquid in the glass tube went up. When the heat was removed, the liquid level went back down.

Conclusion

Liquids expand when heated and contract when cooled.

(c) To show that a gas expands when heated and contracts when cooled

Apparatus required: round-bottomed flask with a stopper and long narrow glass tube at the top; retort stand; large beaker or basin of water; a hair dryer

Method

1. Set up the apparatus as shown in *Fig. 40.5*.
2. Heat the flask very gently with the hair dryer or by warming it with your hands.
3. Note that bubbles appear at the end of the glass tubing in the large beaker.
4. Now, take away the hair dryer. Allow the flask to cool. Observe what happens.

Result

On heating, the air in the long glass tube expands. When the air cools, the water rises up through the sloped glass tube.

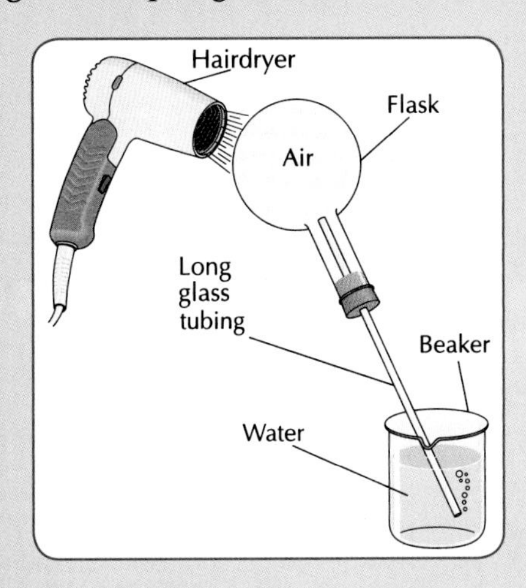

Fig. 40.5 Gases expand when heated and contract when cooled.

Conclusion

The gas (air) expanded when heated. Air is pushed from the long glass tube in the form of bubbles. When the air cools, it contracts. But some of the air particles have escaped and there is less air now than there was at the beginning. This creates a partial vacuum. Water is pushed or sucked up the long tube into this empty space.

Water Breaks its Contract!

We saw that as liquids cool down they contract. There is one major exception to this rule. **Water** does contract on cooling, but only until it reaches **4 °C**. Here it stops contracting ('breaks its contract'!) and begins to expand again as it cools further. This is one reason why water is not used as a liquid in thermometers.

Fig. 40.6 This water expanded on freezing. The lunchbox ensures that you don't have to touch the sharp pieces of the bottle. It also stops them from falling into your freezer.

Fig. 40.6 shows how to **demonstrate that water expands on freezing**. This soft drink bottle was filled to the very top with water. Its top was tightly secured. It was then enclosed in a lunchbox in the freezer for a couple of days. The result shows that the frozen water expanded and shattered the bottle.

Disadvantage and Advantage of Water Expanding at Low Temperatures

- In cold weather, exposed **water pipes** often burst. This is because the water, which was filling the pipe, freezes. It then takes up more space and causes the pipe to burst.
- One **advantage** of this peculiar behaviour of water is that **fish can survive** in lakes in very cold weather, *Fig. 40.7*, page 306. Water is at its **most dense** at 4 °C. (Its volume is smallest at this temperature.) Being dense, this 'warm' water **sinks** to the bottom. 'Colder' water is less dense and floats to the top.

Fig. 40.7 Even when the top of the lake is frozen over, fish can survive in the warmer denser water at the bottom.

TEST YOURSELF:
Now attempt questions *40.1–40.4* and *W40.1–W40.5*.

40.3 Heat Transfer

Heat moves. It travels from a warm object to a less warm object.

- If you touch a **warm radiator**, you feel that it is hot. Heat leaves the radiator and moves to your colder fingers.
- If you touch a **cold radiator**, you feel that it is cold. This time heat leaves your fingers and moves to the colder radiator.

There are three ways that heat can be transferred from one place to another. These are called **conduction**, **convection** and **radiation**.

(1) Conduction

Solids pass heat on from particle to particle by **conduction**. This is similar to the way in which the balls, which represent heat, are being transferred from *A* to *B* to *C* to *D*, etc., along a line, *Fig. 40.8*.

Fig. 40.8 The balls represents the heat being transferred.

Metals are good **conductors** of heat. If the bottom of a metal spoon is placed in hot water, the top of the spoon will very soon become hot as well.

Conduction is the method by which heat travels from particle to particle through a solid, e.g. along a metal spoon.

Insulators

But, a **wooden** spoon does **not** allow heat energy to flow through it easily. It is said to be a good **insulator** of heat.

An insulator is a substance which does not allow heat to flow through it easily.

Gases are also good **insulators** of heat. We saw already in *Chapter 39* that **trapped air** is a good insulator (or a poor conductor). Clothes, quilts, sleeping bags, feathers on birds, double-glazed windows, etc., all use trapped air to provide insulation. Other good insulators are: wool, fibreglass, cork, plastic, polystyrene, carpets, curtains, liquids (except mercury), etc.

Experiment 40.2

To compare the insulating abilities of different materials

Apparatus required: calorimeter (or can or beaker or boiling tube) with cover (or bung); thermometer; beaker

Also required: hot water; some insulating materials such as cotton wool, polystyrene beads, aeroboard, crumpled paper, cloth, foam, wool, etc.

Method

1. Arrange the apparatus as shown in *Fig. 40.9*. Place the calorimeter inside a beaker. Fill the area between the calorimeter and the beaker with one of the available insulators.
2. Pour hot water into the calorimeter. Place the lid back on the calorimeter. (Take care using hot water!)

3. When the temperature of the water cools to, say, 70 °C, start a timer or stopclock and record the temperature of the water every two minutes.
4. Draw a graph of temperature (vertical axis) against time (horizontal axis).
5. Repeat the experiment again. This time fill the area around the calorimeter with a different insulator. Record your results as above.
6. Repeat again, this time using no insulator at all. All the results can be plotted on the same graph.

Result

The graphs will show which of the insulators work best. The best insulator will be the slowest to lose the heat. The calorimeter with no insulation at all loses heat the quickest.

Note: This experiment can also be carried out using a temperature sensor and data logging equipment. The apparatus is set up in the same way. The data logging equipment will give graphs of how the temperatures change with time.

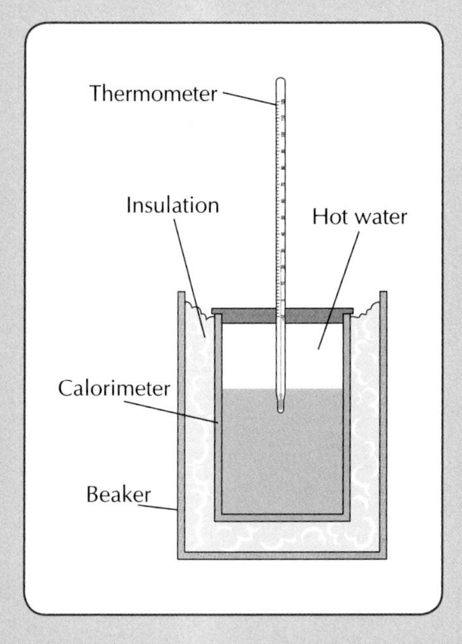

Fig. 40.9 Insulation slows down the loss of heat.

Mandatory experiment 40.3

To show the transfer of heat by conduction and to compare the conductivity of various substances

Apparatus required: rods of various substances (selection of metals, glass, wood or plastic could be used); metal container with holes and stoppers for inserting rods

Also required: vaseline (or wax); small nails (or drawing pins)

Method

1. Set up the apparatus as shown in *Fig. 40.10*. Each rod has a similar length and diameter.
2. Vaseline (or wax) is put at the end of each rod. Attach a small nail (or drawing pin or matchstick) to each rod using Vaseline.
3. Pour boiling water into the metal container, covering the inside ends of the rods.

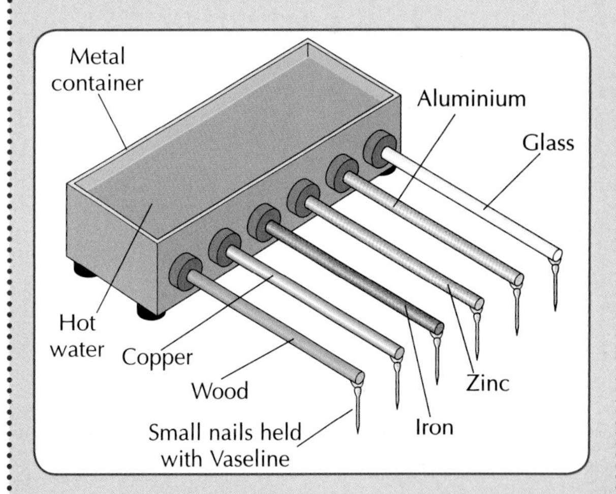

Fig. 40.10 Each substance conducts heat but some are better conductors than others.

Result

The nails drop from the rods when the heat arrives. Different materials conduct at different rates. The rod which drops its nail first is the best conductor. All the rods can then be placed in order of conductivity as their nails drop off.

Mandatory experiment 40.4

To investigate conduction in water

Apparatus required: boiling tube; wire gauze; metal tongs; Bunsen burner

Also required: water, ice

Method

1. Trap a lump of ice with a wire gauze at the bottom of the boiling tube as shown in *Fig. 40.11*. (The ice could also be wrapped in the wire gauze.)
2. Add ordinary tap water to the boiling tube until it is about two-thirds full.
3. Hold the boiling tube at an angle with metal tongs. Heat the water near the top of the boiling tube until it begins to boil. Observe what happens to the ice.

Warning!: When heating the water, don't point the boiling tube towards anyone, including yourself.

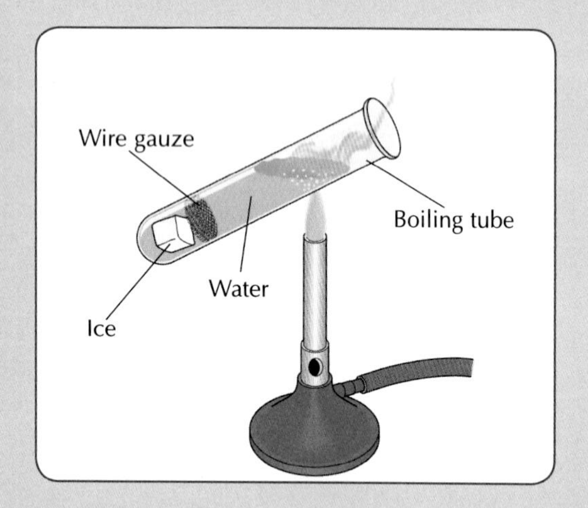

Fig. 40.11 Water is a poor conductor of heat.

Result

The water at the top of the boiling tube boils. The ice at the bottom does not even melt.

Conclusion

Water is a poor conductor of heat. Heat is unable to pass from the top of the liquid to the ice at the bottom.

(2) Convection

When liquids or gases are heated, the particles near the heat source rise through the fluid bringing the heat with them. This is similar to what happens in *Fig. 40.12*.

Fig. 40.12 The person at the end 'A' was first to receive the ball (or heat). He swims through the water and brings the ball (or heat) with him. The person at 'B', then moves into the space left at 'A'. She collects another ball (or more heat). She will then swim to the other end too.

Convection is the transfer of heat through a liquid or gas when the particles of the liquid or gas move and carry the heat with them.

Fig. 40.13 shows a **smoke box**. This can be used to show convection currents in air. A taper is lit and then blown out. It is then placed above the hole or chimney on the right.

1. The air above the candle on the left is heated. It expands. It becomes less dense and **rises**.
2. The **colder air** and smoke nearby move in to take its place.

The whole process is repeated again. Convection currents are set up. **Fires** and **chimneys** ventilate ordinary rooms in houses in a method similar to this.

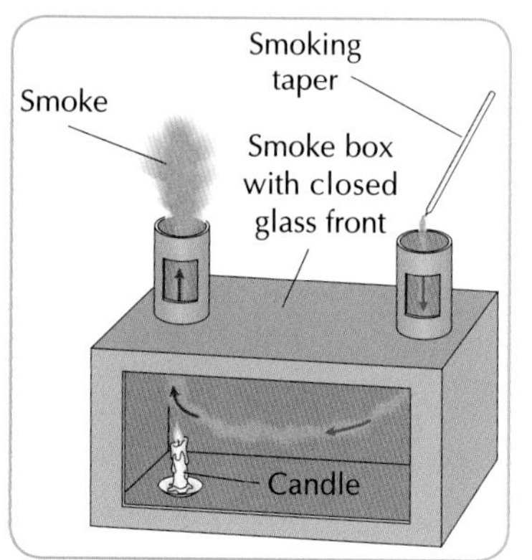

Fig. 40.13 Demonstrating convection currents in air.

Mandatory experiment 40.5

To investigate convection in water

Apparatus required: large glass beaker; forceps; tripod; glass tubing

Chemical required: potassium permanganate or potassium manganate (V11) crystal

Also required: water; candle

Method

1. Set up the apparatus as shown in *Fig. 40.14*.
2. Using forceps, drop a crystal of potassium manganate (V11) to a corner of the beaker directly above the candle. (The crystal is best added by putting it down through a glass tubing, which touches the bottom of the beaker. Then, to remove any coloured water, place your finger over the top of the glass tube and lift it out to the sink.)
3. Heat the beaker gently with the candle (or a low-level Bunsen flame).
4. Observe what happens.

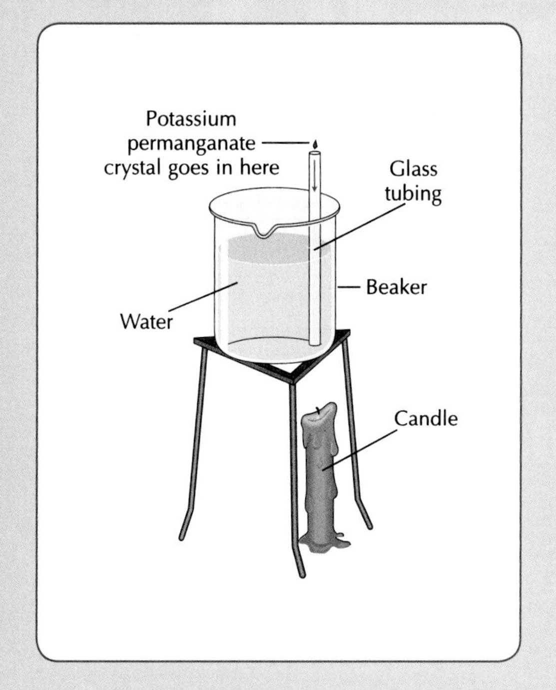

Fig. 40.14 Convection currents in water.

Result

The crystal colours the water near it and 'currents' are seen spreading throughout the water.

Conclusion

The water molecules near the crystal are the first to receive heat. This causes them to expand. They become less dense than the molecules around them. Thus, they 'float' to the top. Their path is shown by the purple dye. Their place is then taken by denser colder water. This, in turn, is heated. This process is continued until all of the liquid is heated. The moving currents of water are referred to as '**convection currents**'.

Breezes

- On warm **sunny days** the land absorbs more heat from the sun than the sea. The warmer air over the land rises. Cooler air over the sea rushes in to take its place, *Fig. 40.15*. A convection current, which we call a **sea breeze**, is the result.
- **At night** the reverse happens. The air over the sea is warmer. Colder air moves from the land out to sea. A **land breeze** results.

Fig. 40.15 A sea breeze with colder air above the sea replacing the rising warm air over the land.

(3) Radiation

On warm days we can directly feel the heat from the sun. This heat travels almost 150 000 000 km from the sun to Earth in just over eight minutes. Space is empty so no medium is required. This method of transferring heat is called **radiation**.

Fig. 40.16 Radiation is similar to the way in which the person at 'A' simply throws the ball (or heat) quickly to the person at the other end, 'F', without affecting the people in between.

Radiation is the rapid transfer of heat from a hot object without needing a medium.

Fig. 40.17 How do the sausages receive heat? It is not conduction as air is a very poor conductor. It is not convection as the heat travels downwards, not upwards. Therefore, the sausages receive heat by radiation.

All hot objects, even our own bodies, radiate heat. The hotter an object, the more heat it radiates. The rate at which a body radiates heat depends on the type of surface it has. This will be shown by the following experiment.

Mandatory experiment 40.6

To show the transfer of heat by radiation and that a dull black surface radiates heat better than a bright shiny surface

Apparatus required: two cans (or calorimeters) of equal size (one polished and shiny, the other black or dull); two temperature sensors and data logging equipment (or two thermometers), retort stand

Also required: hot water

Method

1. Pour equal amounts of hot water into each can.
2. Place a temperature sensor, linked to data logging equipment (or a thermometer) into each can as shown in *Fig. 40.18*. Make sure that both cans are placed on a non-conducting surface.
3. Allow the hot water to stand in each can, stirring the water from time to time.
4. Compare the graphs of temperature dropping with time for each can, *Fig. 40.18*.

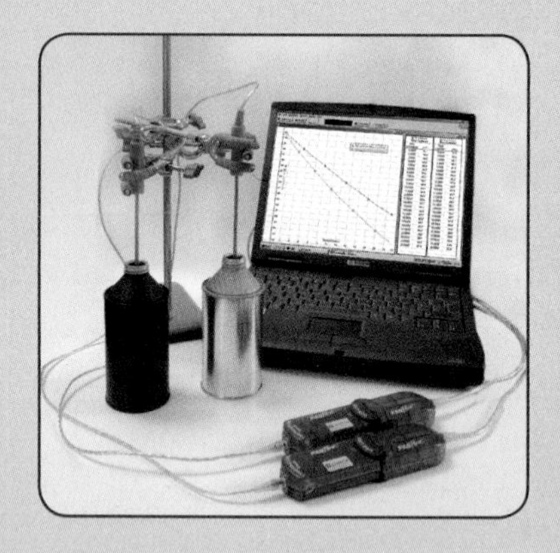

Fig. 40.18 Comparing the rates at which dark and bright surfaces radiate heat.

Result and Conclusion

The temperature falls faster in the container with the dark surface. Therefore, a dark surface radiates heat better than a bright surface.

TEST YOURSELF: Now attempt questions *40.5–40.8* and *W40.6–W40.11*

40.4 Temperature and Heat

Heat is a form of **energy**. It is measured in **joules (J)**. Temperature is a measure of the **level** of heat in an object. It is measured in **degrees Celsius** (°C). When two objects are in contact, heat flows from the hotter object to the colder object. If object *A* is at a **higher temperature** than object *B*, heat will flow from *A* to *B*.

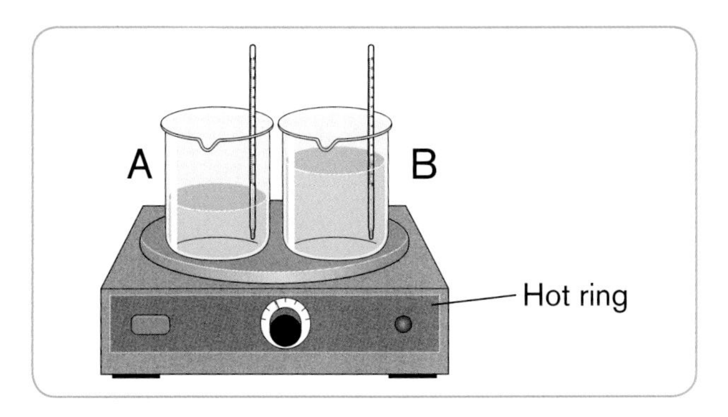

Fig. 40.19 Showing that heat and temperature are different.

Beaker *A* in *Fig. 40.19* holds 50 g of water. Beaker *B* holds 150 g of water. If both receive **equal heat** for two minutes, the temperature of the water in beaker *A* will rise more. They both receive the same amount of heat, but their temperatures do not increase by the same amount. This shows that temperature and heat are different.

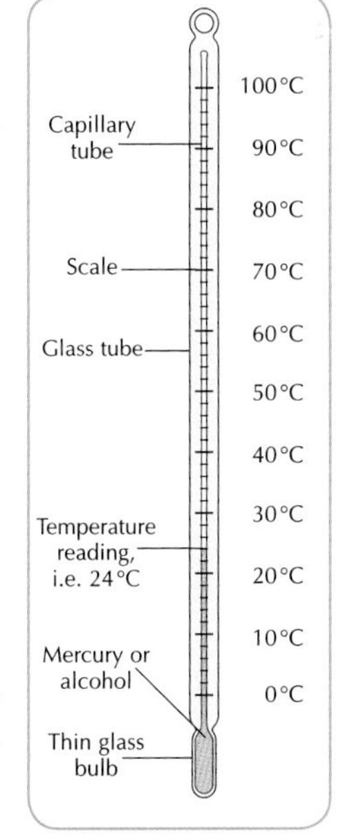

Fig. 40.20 Common laboratory thermometer using the Celsius scale.

Fig. 40.20 shows a typical laboratory thermometer. Other thermometers were shown in *Fig. 34.11*, page 251. Important temperature values can be compared in *Fig. 40.21*. The temperature at which ice melts, or the **melting point of ice,** is included. So is the temperature at which water turns into steam, or the **boiling point of water**. Anders Celsius used these as fixed points in his temperature scale, *Fig. 40.20*.

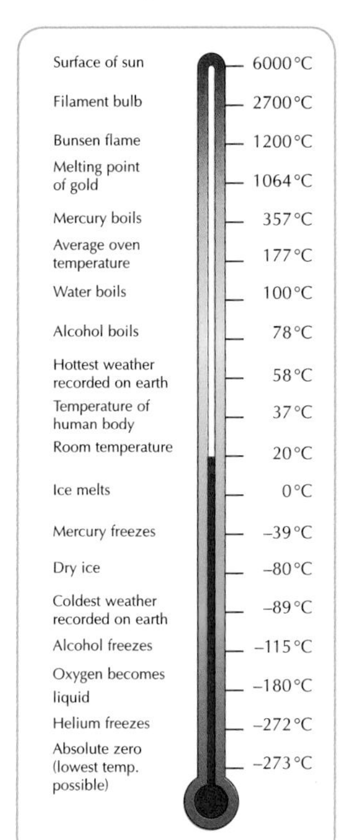

Fig. 40.21 Some important temperature values.

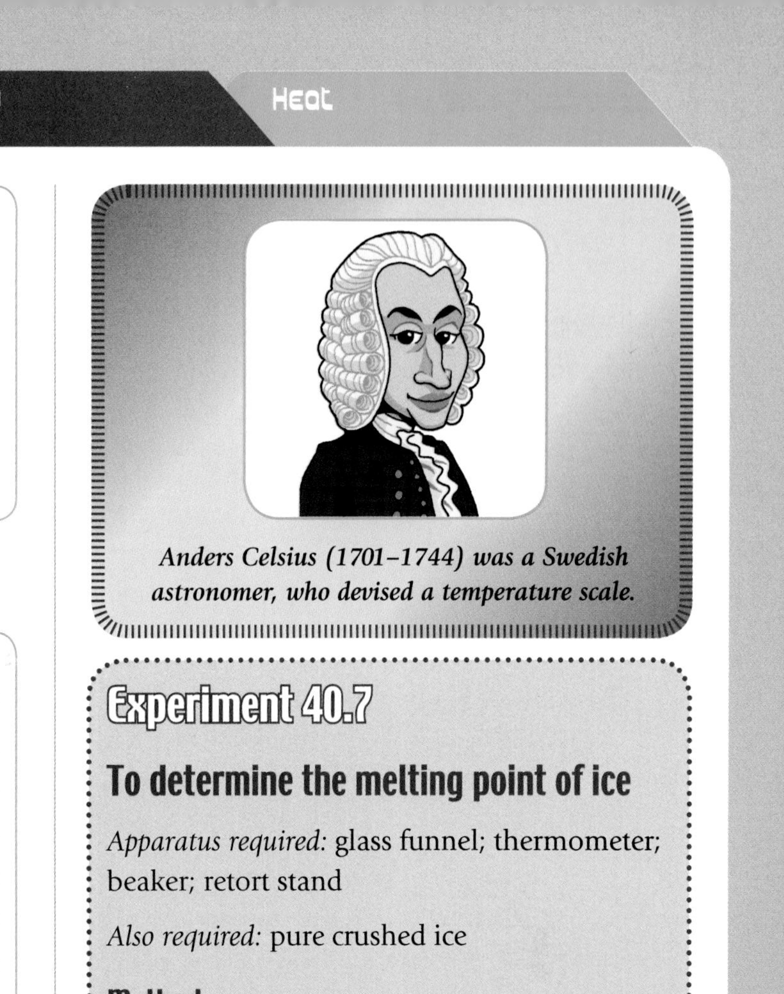

Anders Celsius (1701–1744) was a Swedish astronomer, who devised a temperature scale.

Experiment 40.7

To determine the melting point of ice

Apparatus required: glass funnel; thermometer; beaker; retort stand

Also required: pure crushed ice

Method

1. Set up the apparatus as shown in the diagram, *Fig. 40.22*.
2. Place the bulb of the thermometer into the crushed ice.
3. Wait until the ice begins to melt and you can see the water (or melted ice).
4. Note the reading on the thermometer. The temperature should stay at this reading until all the ice melts. This temperature is the melting point of ice.

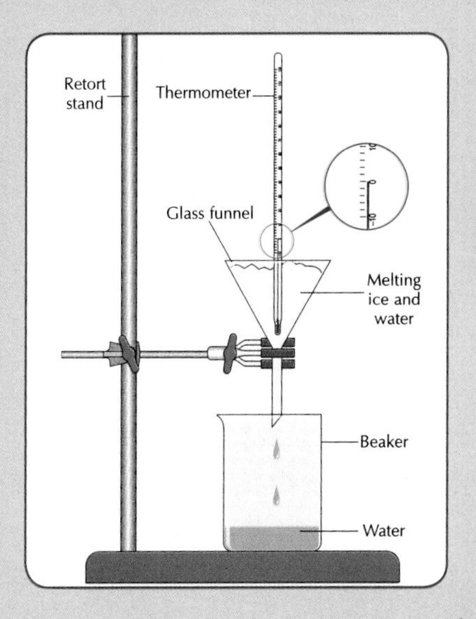

Fig. 40.22 Determining the melting point of ice.

Result

The ice melts at 0 °C. (Impurities in the ice could change its melting point.)

Note: A temperature sensor and data logging equipment can be used in this experiment in place of the mercury thermometer.

Experiment 40.8

To determine the boiling point of water

Apparatus required: round-bottomed flask; tripod; wire gauze; thermometer; glass vent; two-holed cork or bung; retort stand; Bunsen burner

Also required: water

Method

1. Set up the apparatus as shown in *Fig. 40.23*.
2. Adjust the thermometer so that the bottom of it is above the water level.
3. Heat the water until it is clearly boiling throughout. Steam at this stage will be flowing freely from the glass vent.
4. Note the reading on the thermometer. The temperature stays at this reading (the boiling point) until all the water boils.

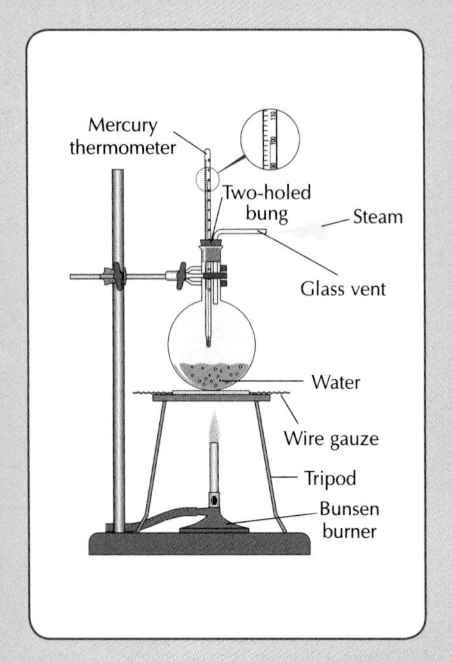

Fig. 40.23 To measure the boiling point of water.

Warning!: Great care must be taken with the **boiling water** and **steam** in this experiment. Also, the tripod and wire gauze are likely to get very hot. They should **not** be touched directly until they have cooled properly.

Result

If the water is pure, the temperature at which it boils is 100 °C. (Impurities in the water change its boiling point.)

Note: An alternative method for this experiment, using a temperature sensor and data logging equipment was shown in *Fig. 34.11(d)*, page 251.

The Effect of Pressure on the Boiling Point of Water

Pure water does not always boil at 100°C.

- On **Mount Everest** water boils at 70 °C. The atmospheric pressure is very low at that height.
- In a **pressure cooker**, water boils at about 120 °C. Here, the pressure acting down on the water is very high.

The following experiments investigate the effects of pressure on the boiling point of water.

Experiment 40.9

To investigate the effect of (a) increased pressure and (b) decreased pressure on the boiling point of water

(a) Increased pressure

Note: Teacher demonstration only. A protective screen should be used.

Apparatus required: strong round-bottomed flask with two-holed bung; retort stand; thermometer; glass tubing with attached rubber tubing and clip; wire gauze; tripod; Bunsen burner

Also required: pure water (with some anti-bumping granules added)

Method

1. Half fill the round-bottomed flask with pure water. Leave the clip on the rubber tubing open. Heat the water to boiling point.
2. Close the clip. Continue heating for no more than a further 10 to 15 seconds, *Fig. 40.24(a)*.

Result

The thermometer will show that the temperature of the boiling water has gone up above 100 °C.

Conclusion

Increasing the pressure raises the boiling point of water.

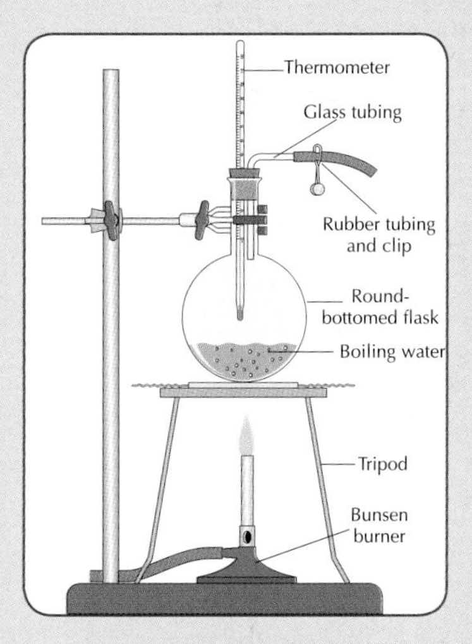

Fig. 40.24(a) Increased pressure increases the boiling point of water.

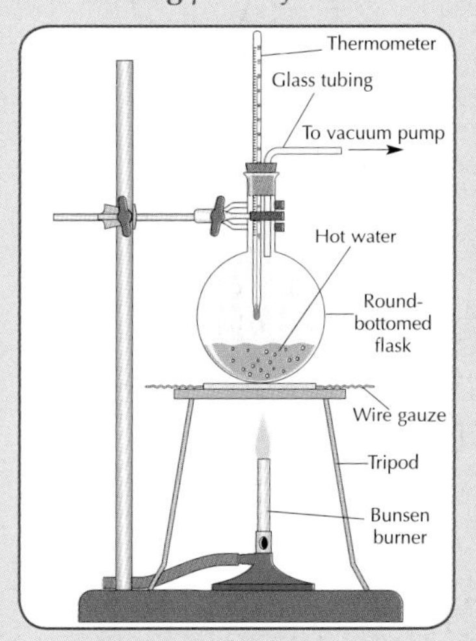

Fig. 40.24(b) Decreased pressure decreases the boiling point of water.

(b) Reduced pressure

Note: Teacher demonstration only. A protective screen should be used.

Apparatus required: strong round-bottomed flask with two-holed bung; retort stand; thermometer; glass tubing; wire gauze; tripod; Bunsen burner; vacuum pump

Also required: pure water

Method

1. Half fill the round-bottomed flask with pure water. Heat the water until it is close to boiling point.
2. Turn on the pump to reduce the pressure above the water, *Fig. 40.24(b)*.

Result

The thermometer will show that the temperature is well below 100 °C, but the water will still be seen to be boiling.

Conclusion

Reducing the pressure lowers the boiling point of water.

Note: A temperature sensor and data logging equipment would also work well instead of the thermometer in **both** of these experiments.

40.5 Changes of State

We are familiar with water in its three states, i.e. ice, water and steam. All substances (in theory) can change state. Each substance has its own melting and boiling point.

Experiment 40.10

To demonstrate changes of state

(a) To demonstrate melting and freezing

Apparatus required: test tube; thermometer; beaker; tripod; wire gauze; Bunsen burner; stirrer

Also required: butter or lard or candle wax

Method

1. Set up the apparatus as shown in the diagram, *Fig. 40.25(a)*.
2. Heat the beaker of water, using a gentle flame from a Bunsen burner. Use a stirrer to ensure an even temperature. Record the temperature every 30 seconds.
3. When all the butter has melted, allow the temperature to increase a little bit more. Keep recording the temperature every 30 s. Plot a graph of temperature against time, *Fig. 40.25(b)*.
4. Then, take away the Bunsen burner. Allow the butter to cool. (This is the whole process in reverse.)
5. The butter freezes after a short while. It turns back into a solid again.
6. Again plot a graph of temperature against time. This is called a **cooling curve**, *Fig. 40.25(c)*.

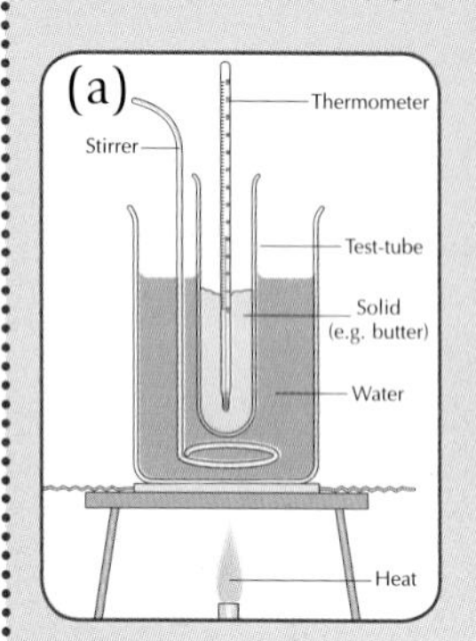

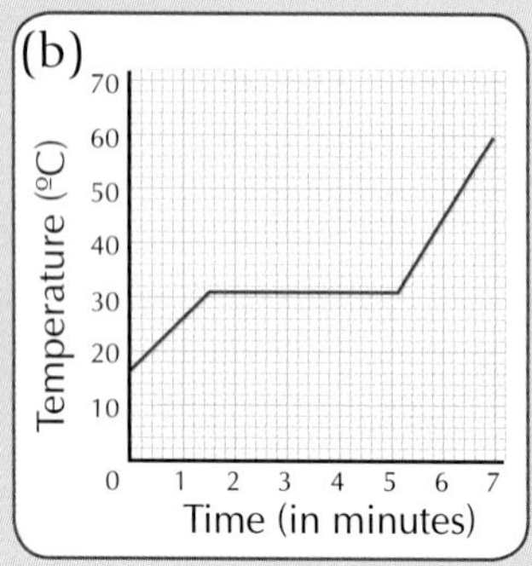

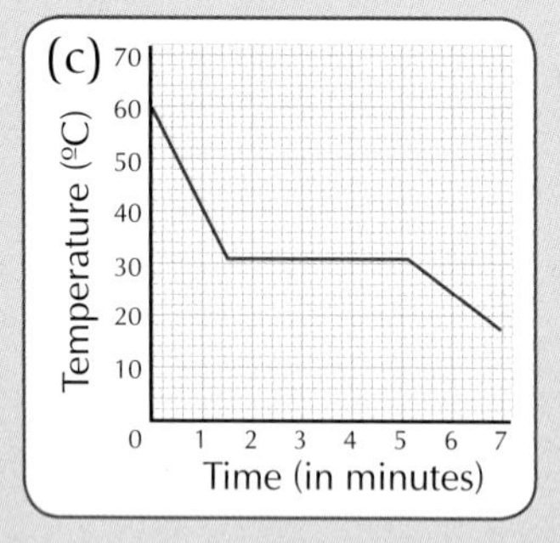

Fig. 40.25(a) Melting. (b) Temperature changing with time as heat is supplied. (c) Cooling curve.

Result

When heating, the temperature of the butter rises until it reaches a constant temperature. Then it stays fixed at this temperature for a while. This constant temperature is the **melting point**. When all the butter has melted, the temperature increases again, *Fig. 40.25(b)*. The reverse happens when cooling. The constant temperature on the cooling curve is the **freezing point**, *Fig. 40.25(c)*.

(***Note:*** The melting/freezing point of butter is approximately 31 °C.)

(b) To demonstrate boiling and condensing

Apparatus required: pear-shaped flask; conical flask; tripod; wire gauze; Quickfit apparatus; Liebig condenser; thermometer; Bunsen burner

Also required: water

Method

1. Set up the apparatus as shown in *Fig. 23.7*, page 158. Half fill the pear-shaped flask with water. Connect this flask to the Liebig condenser.
2. Let cold water flow through the outside tube of the Liebig condenser.
3. Heat the water in the pear-shaped flask using a Bunsen burner.

Result

The water changes from liquid to gas (boils) in the pear-shaped container. The steam produced condenses or changes from gas to liquid in the condenser. The water dropping into the conical flask proves this.

40.6 Latent Heat

Heating Ice

Heat crushed ice in a beaker. What happens? Its temperature rises with time until it reaches 0 °C, *Fig. 40.26*. Then, even though heat is still being supplied, its temperature remains at 0 °C until all the ice has melted. Then the temperature begins to rise again.

Heating Water

With further heat, the temperature of the water rises to 100°C. But then, the same thing happens again. The temperature remains at 100 °C until all the water has turned into steam. It is only when all the water has turned into steam that the temperature rises again, *Fig. 40.26*.

What is Latent Heat?

The effect of the heat supplied during both changes of state seems to be hidden. The temperature does not change. This hidden heat is referred to as **latent heat**. This heat is used to break down the attractive forces between the particles. Similarly, this heat is given out when a gas cools to become a liquid and a liquid cools to become a solid.

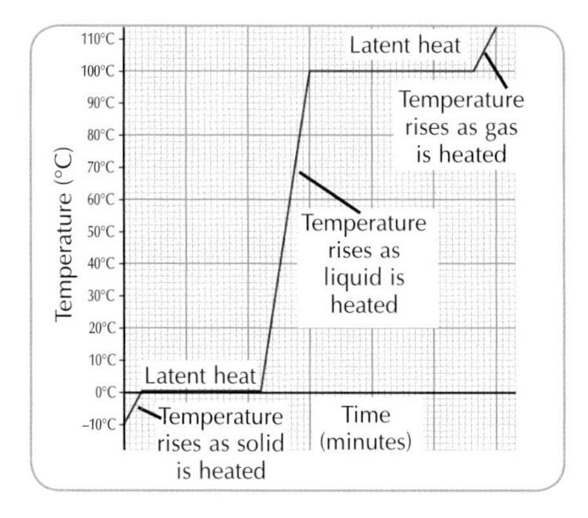

Fig. 40.26 Heating crushed ice all the way beyond melting point and boiling point.

Latent heat is the heat taken in or given out when a substance is changing state without changing temperature.

Amount of Latent Heat

Quite a lot of energy is lost or gained when a substance changes state. In fact, it takes **five** times more energy to change water at 100 °C into steam at 100 °C, than to heat water at 0 °C up to water at 100 °C. Thus, steam at 100 °C has a lot more heat in it than water at 100 °C. This is why a scald from steam is much worse than a scald from boiling water.

Similarly, ice cubes are good at cooling down a drink because the ice takes a lot of latent heat from the drink to melt it.

TEST YOURSELF:

Now attempt questions *40.9–40.12* and *W40.12–W40.17.*

What I should know

- Heat is a form of energy. It can do work and change into other forms of energy.
- One of the effects of heat is that it can cause solids, liquids and gases to expand. They contract again when cooled.
- Expansion can be a nuisance but it also can be put to good use.
- Water is exceptional in that when cooled below 4 °C, it expands. This is why fish can survive in lakes even in extreme winter conditions.
- Conduction is the transfer of heat through a solid from particle to particle.
- Metals are good conductors. Water is a poor conductor of heat.
- Insulators, like wool, polystyrene, plastics, trapped air, gases, etc., are very poor conductors of heat.
- Convection is the transfer of heat through a liquid or gas when the particles of the liquid or gas move and carry the heat.
- Warm liquids or gases rise because they expand and become less dense.
- Convection currents cause sea and land breezes and wind.
- Radiation is the rapid transfer of heat from a hot object without needing a medium.
- A dull black surface radiates heat better than a bright shiny surface.
- Heat will travel from a warm region to a colder region. The unit of heat is the joule (J).
- Temperature is a measure of how hot an object is. It is measured in degrees Celsius (°C).
- On the Celsius scale the lower fixed point is the freezing point of water (0 °C). The upper fixed point is the boiling point of water (100 °C).

What I should know

- The boiling point of water is increased when the pressure on it is increased. The boiling point of water is decreased when the pressure on it is decreased.
- Latent heat is the heat taken in or given out when a substance is changing state without changing temperature.

QUESTIONS

Write the answers to the following questions into your copybook.

40.1 (a) Heat is a form of ..... which travels from warm objects to ..... objects.

(b) Give one example of heat energy changing to any other form of energy.

40.2 (a) Why must roads that are made of large concrete slabs have gaps put between these slabs?

(b) State an everyday example of an allowance being made for a metal to expand when it is heated.

40.3 What apparatus can be used to show that solids expand when heated and contract when cooled.

40.4 When water is cooled below ..... °C, it starts to expand. This explains why ice is less ..... than water. It is also one of the reasons why water is not used as a liquid in a ..... .

40.5 (a) Convection is the transfer of ..... through a ..... or a ....., where the particles of the substance itself ..... and carry the heat with them.

(b) On a warm day the air over the land is ..... than the air over the sea. The hot air ..... and cooler air over the sea rushes in to take its place. This sets up ..... currents or sea breezes.

40.6 (a) ..... is the rapid transfer of heat from a ..... object without needing a medium.

(b) A dull black surface ..... heat ..... than a bright shiny surface.

40.7 Why is a saucepan often made of a material like aluminium or iron but the handle is made of wood or plastic?

40.8 Explain, with the aid of a diagram, how, at night, a breeze often blows from the land out to the sea.

40.9 (a) Temperature is a measure of how ..... an object is and it is measured in degrees ..... . The symbol for this unit is ..... .

(b) The lower fixed point on the Celsius scale is the ..... point of water or ..... °C. The upper fixed point is the ..... point of water or ..... °C.

(c) The two liquids most commonly used in thermometers are ..... and ..... .

(d) The normal temperature of humans is ..... °C.

(e) Increased pressure on water ..... its boiling point .

(f) Latent heat is the heat ..... in or given out when a substance is changing ..... without changing ..... .

QUESTIONS

40.10 *Fig. 40.27* shows two thermometers, one measuring the temperature of water, the other measuring the temperature of the air in a room.

(a) What is the temperature of the water?

(b) What is the temperature of the air?

(c) What is the temperature difference between the water and the air?

(d) Will heat flow from the air to the water or from the water to the air?

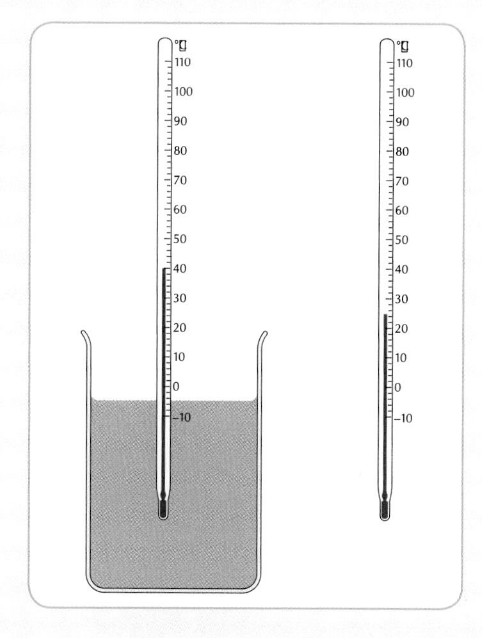

Fig. 40.27 The temperature of the water is different to the temperature of the air.

40.11 Describe a simple experiment to show the difference between heat and temperature.

40.12 In an experiment to find the melting point of a substance, its solid form was first heated and melted in a boiling tube inside a water bath. Then the boiling tube was removed from the hot water and allowed to cool. The following table shows how the temperature of the substance dropped with time.

TIME (MINUTES)	0	1	2	3	4	5	6	7	8	9
TEMPERATURE (°C)	78	69	60	56	56	56	56	54	48	42

(a) Why was the substance (in solid form) heated in a water bath?

(b) Plot a graph of temperature against time for this substance.

(c) What was happening to the substance between the 3rd and 6th minute?

(d) What is (i) the freezing point and (ii) the melting point of this substance?

PHYSICS

Chapter 41 Light

41.1 Light is a Form of Energy

The following evidence shows that light is a form of energy:

(1) Light can do **work** or cause movement. It causes the vanes of a Crooke's radiometer to move, *Fig. 41.1(a)*.

(2) Light can be converted **to other forms** of energy.
- **Solar cells** convert light to electrical energy. They are often used to power calculators, space ships, satellites, *Fig. 41.1(b)*, etc.
- **Solar panels** convert light to heat. They are often found on the roofs of buildings, *Fig. 41.1(c)*.
- A **green plant** converts light energy to chemical energy (or food) during photosynthesis.

(3) Other forms of energy can be converted **to light**.
- **Nuclear energy** is converted to light on the sun.
- **Chemical energy** changes to light in oil lamps, candles, fires, fireworks, etc.
- **Electrical energy** is converted to light in a light bulb.

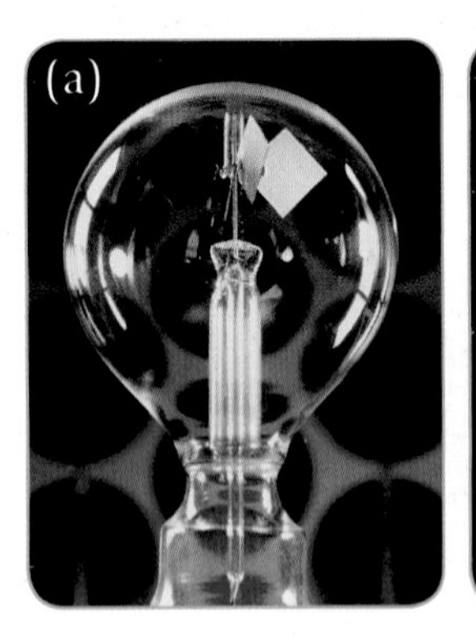
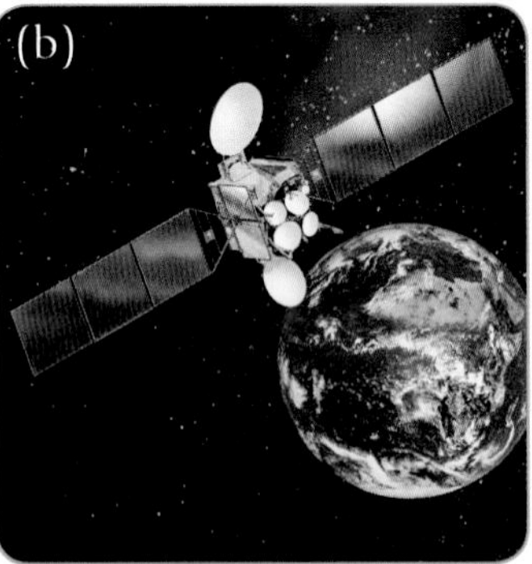

Fig. 41.1(a) Crooke's radiometer: Light energy → kinetic energy (or light doing work).
(b) Solar cell: Light energy → electrical energy.

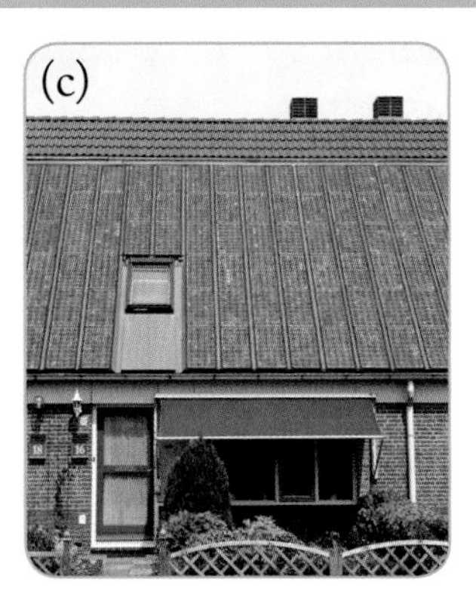

Fig. 41.1(c) Solar panel: Light energy → heat energy.

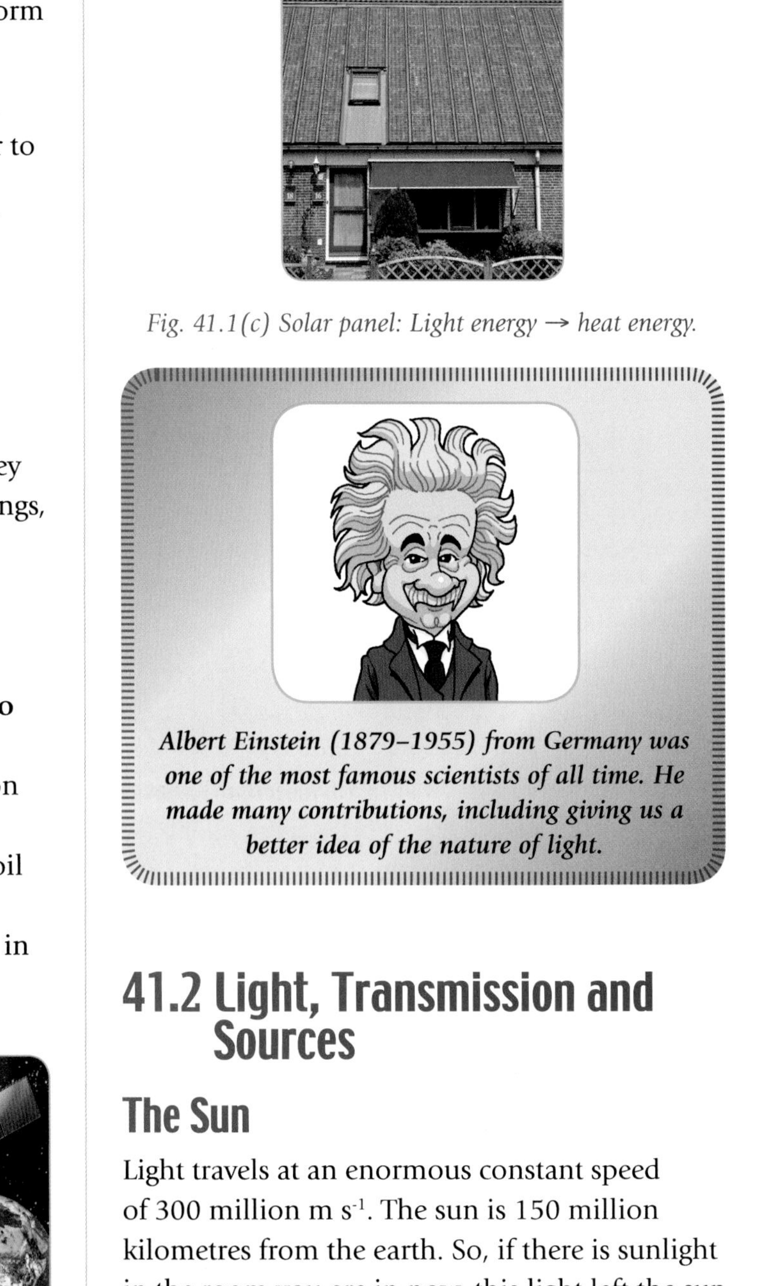

Albert Einstein (1879–1955) from Germany was one of the most famous scientists of all time. He made many contributions, including giving us a better idea of the nature of light.

41.2 Light, Transmission and Sources

The Sun

Light travels at an enormous constant speed of 300 million m s^{-1}. The sun is 150 million kilometres from the earth. So, if there is sunlight in the room you are in now, this light left the sun less than 8.5 minutes ago. Light from the sun passes mainly through space, which is a vacuum. It also passes through the atmosphere, which is **transparent**, i.e. it allows light through.

Luminous Objects

Sources that give out their own light are said to be **luminous**. The **sun** is our main luminous source of light, *Fig. 41.2*. Other luminous sources would include **light bulbs**, **televisions**, **candles**, **fires**, **fireworks**, etc.

Fig. 41.2 The sun is our bright star in the sky, providing us with many joules! We could not survive without it.

Non-luminous Objects

Most objects we see do **not** give out their own light. They are said to be **non-luminous**. We see them because they **reflect** light into our eyes. The eye can be thought of as a **light detector**. Right now, light from some luminous source like the sun or a bulb is arriving at this page. The page is reflecting some of this light to your eyes. This is why you can see the page (see *Fig. 41.7*, page 320).

TEST YOURSELF:
Now attempt questions *41.1–41.3* and *W41.1–W41.4*.

41.3 Light Travels in Straight Lines

In a dusty cinema, you will see the beam of light from the projector heading towards the screen. It travels in a straight line. The beam of a powerful torch, lighthouse, spotlight or car headlight also shows the straight line path taken by light. The path along which light travels from a source is called a **ray**.

Fig. 41.3 We cannot see around corners because light travels in straight lines.

Mandatory experiment 41.1

To show that light travels in straight lines

Apparatus required: light source (candle, bulb or small torch); three identical cards with a small hole in the middle of each

Also required: plasticine; a length of string

Method

1. Place the three cards about 10 cm apart in a line in front of the candle. Use the plasticine to hold them to the bench, *Fig. 41.4*.
2. Thread a length of string through the holes. Pull the string tight. This makes sure the holes in the cards are in a straight line.
3. The light from the candle should now be clearly visible through the holes.
4. Move one of the cards slightly to one side. Observe the result.

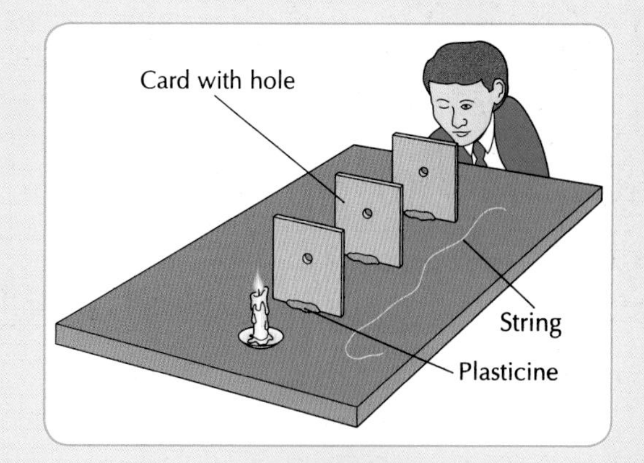

Fig. 41.4 Light is seen only when the three holes are in line with each other.

Result

The light cannot be seen when the holes in the cards are out of line.

Conclusion

The light from the candle travels in straight lines.

Shadows

We saw that light can pass through **transparent** substances like air, water, glass, etc. Other substances such as wood, metal, etc., stop light passing through. These are said to be **opaque**. Shadows are formed when opaque objects are placed in the path of light. The existence of shadows is further evidence that light travels in straight lines.

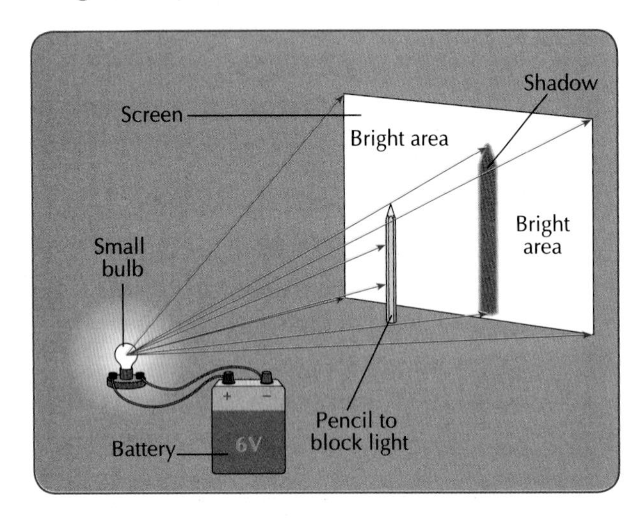

Fig. 41.5 Explaining how shadows are formed.

Rays heading in straight lines towards the pencil in *Fig. 41.5* are blocked by the pencil. This creates a dark space or **shadow**. The shape of this shadow will be the same as the shape of the blocking pencil. When the pencil is close to the bulb, the shadow is large. When the pencil is far from the bulb and close to the screen, the shadow is smaller.

Eclipses

One of the most spectacular shadows known is caused by a **solar eclipse**, *Fig. 41.6*. It occurs when the moon passes between the sun and the earth. One part of the earth goes **completely dark**. No light from the sun reaches this area. The sun literally disappears for a few minutes. An area that has **partial darkness** surrounds this central dark area. Here, only some of the light is blocked off. A person in this region will see a partial eclipse.

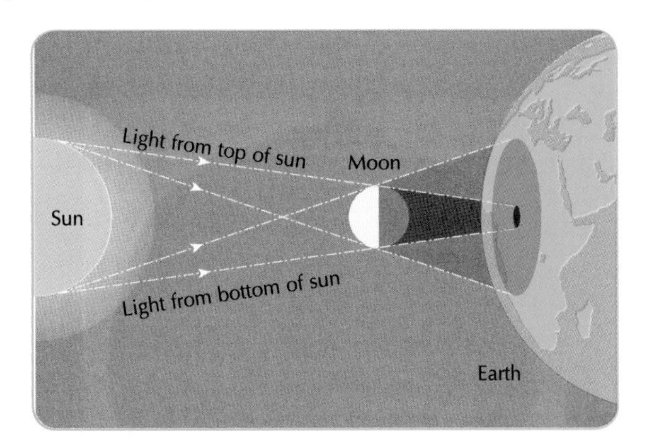

Fig. 41.6 Solar eclipse. There is complete darkness in the dark central circle with no light reaching.

41.4 Reflection of Light

When light hits a surface some of the light bounces or **reflects** from the surface.

Reflection is the bouncing of light from a surface.

Most surfaces, on a microscopic scale, are full of humps and hollows. They reflect light in many directions. You do not see a clear image of the source of light. The page of this book, for example, is not smooth and reflects light as in *Fig. 41.7*.

Fig. 41.7 Light is reflected in many directions from a non-smooth or bumpy surface.

Mirrors

A mirror, however, reflects light in a special way. This is because the surface of the mirror is **smooth, flat and polished**. The light is reflected in a regular, orderly pattern. If the surface of water is calm and smooth, it too can reflect regularly, like a mirror, *Fig. 41.8*. But if it is bumpy with ripples or waves, it does not reflect clearly.

Fig. 41.8 Smooth calm water reflects clearly. When the surface is rough, there is no clear reflection.

Periscopes

Periscopes use the reflection of light.

- They are used in **submarines** to enable the crew to see above the surface of the water.
- They are sometimes used by people **watching sports**, for example a golf tournament, *Fig. 41.9*. People at the back of a crowd can use them to see over the heads of those in front.

Fig. 41.9 Those at the back can still get an 'open' view!

Mandatory experiment 41.2

(a) To investigate the reflection of light by plane mirrors, and illustrate this using a ray diagram

Apparatus required: ray box with single slit; power supply; smooth plane mirror (with mirror holders)

Also required: white paper; pencil

Method

1. Set up the apparatus as shown in *Fig. 41.10*. A sheet of white paper is placed on the bench underneath everything.
2. Darken the room. Shine a ray of light from the ray box at the plane mirror.
3. Mark, with a pencil, at least two Xs on the path of the ray going from the ray box to the mirror (the incident ray). Do the same for the ray coming out from the mirror (the reflected ray), *Fig. 41.10*.
4. Move the ray box. Increase the angle, *A*, but keep the ray of light aimed at the same point on the mirror. What happens the corresponding angle, *B*?
5. Repeat for different angles.

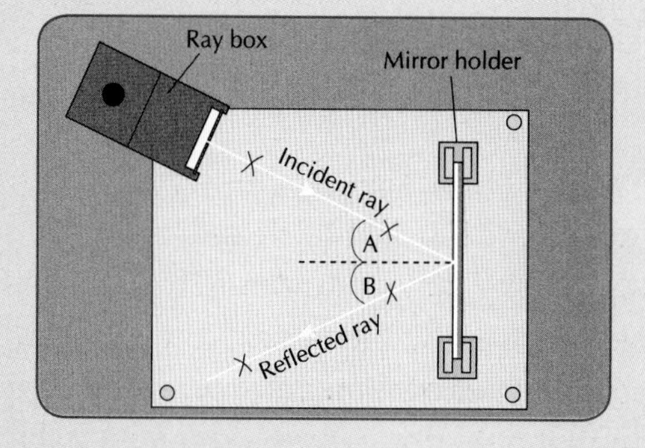

Fig. 41.10 Investigating reflection in a darkened room.

Result

The light is reflected from the mirror in straight lines. There is a regular pattern. When the angle, *A*, is increased on one side, the angle, *B*, increases on the other side in a similar way.

Warning!: Ray boxes can get quite hot, even after a short time.

PHYSICS

(b) To demonstrate and explain the operation of a simple periscope

Apparatus required: Two plane mirrors; retort stand; large obstacle (e.g. a block or a box); object to view; protractor

Method

1. Set up two plane mirrors on a retort stand as shown, *Fig. 41.11*. Place the top mirror vertically above the lower mirror. Tilt the top mirror downwards at an angle of 45° to the horizontal. Tilt the lower mirror upwards at an angle of 45° to the horizontal.
2. Now, keep your eye level with the lower mirror. You should be able to see the image of the object on top of the block.

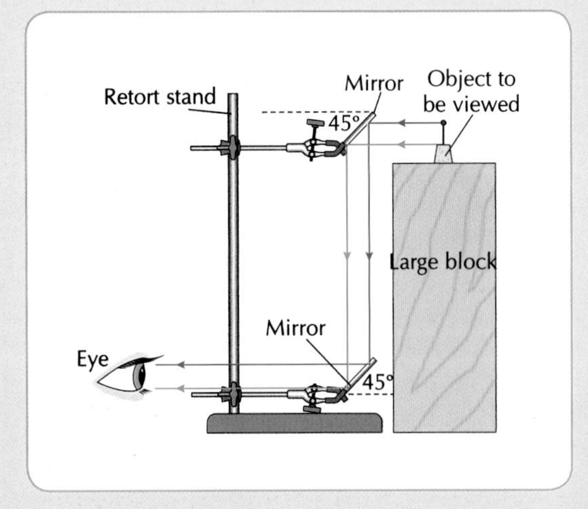

Fig. 41.11 How a periscope works.

Explanation

The top mirror reflects the light from the object. It changes or deviates its path by 90°. The bottom mirror also changes the path of the light through 90° by reflection. Thus, you can see the object on top of the block.

Applications of Reflection

(1) We can **see** non-luminous objects because they reflect light.

(2) **Periscopes** use reflection to see objects that otherwise would be difficult to see.

(3) Reflective mirrors in cars are used for **safety**. Drivers can see what is happening behind them. Bicycle reflectors and cat's-eyes on roads also use reflection to make our roads safer.

(4) **Make-up** or **shaving mirrors** use reflection so that we can see our own faces. These are often slightly curved mirrors, which give a magnified image when the object is up close.

(5) **Microscopes** often use mirrors to direct light on to the slide or viewing stage.

(6) Clever use of mirrors in rooms will often give **the impression of space**.

(7) **Kaleidoscopes** are toys that use reflection by mirrors to create beautiful symmetric patterns.

(8) Curved mirrors are placed behind bulbs in car **headlights** and **spotlights**. The reflected light then emerges as a parallel beam and little light is wasted.

(9) **Security mirrors** in stores, train stations, ends of blind driveways, etc., are curved mirrors that give a wide field of vision in a small space.

(10) Movable mirrors are used in the 'flashing' lights of a **lighthouse** and on top of **police cars**. They are also used sometimes to send Morse code signals.

TEST YOURSELF:

Now attempt questions *41.4–41.8* and *W41.5–W41.11*.

41.5 Refraction of Light

We saw earlier that light travels in straight lines. This is not true, however, when light passes from one substance into another of **different density**. When light goes from air into glass or water, for example, it bends. It also bends when passing from glass or water back into air. This bending of light is known as **refraction**.

Refraction is the bending of light as it passes from one medium to another.

Experiment 41.3

To demonstrate the refraction of light

(a) Refraction from air to glass and from glass to air

Apparatus required: ray box; power supply; glass block, darkened room

Method

Direct a single ray from a ray box onto a rectangular block of glass as shown in *Fig. 41.12*.

Fig. 41.12 Refraction of light going from air to glass and from glass to air.

Result

The path of the light ray bends as it goes from air into glass. It also bends as it passes from glass back into air. (The fainter line shows that some reflection occurs also as the light hits the glass from air.)

Warning!: Ray boxes can get very hot, even after quite a short time.

(b) From air to water

Apparatus required: laser; power supply

Also required: chalk dust; basin of water with tiny amount of milk added (or calcium hydroxide powder or Dettol)

Method

Shine light from a laser from air into water. If the laser beam is difficult to see in the air, blow some smoke or chalk dust onto its path. To see the beam in water, a few drops of milk are added.

Result

The beam bends on going from air into the water, *Fig. 41.13*.

Fig. 41.13 Light bending (refraction) as it goes from air into water.

(c) Refraction from water to air

Apparatus required: straight rod (or ruler); large beaker of water

Method

Set up the apparatus as shown in *Fig. 41.14*, holding the rod at an angle in the water.

Result

The straight rod seems to be broken or bent. The rays of light coming from the bottom of the rod bend as they emerge from the water. Refraction takes place.

Fig. 41.14 Light bends or refracts as it goes from water into air.

Lenses

A lens is a piece of glass or other transparent material that has at least one curved surface.

Lenses refract or bend light passing through them. They are used for this purpose in many optical instruments.

Experiment 41.4

To demonstrate the converging and diverging of light by lenses

Apparatus required: ray box; power supply; converging lens; diverging lens

Fig. 41.15(a) Converging (convex) lens converging or bringing together some of the rays coming from a ray box.

Fig. 41.15(b) Diverging (concave) lens diverging or spreading apart the rays coming from a ray box.

Method

1. In a darkened room, direct a number of rays from the ray box onto a converging lens, *Fig. 41.15(a)*.
2. Observe what happens.
3. Repeat the experiment, replacing the converging lens with a diverging lens, *Fig. 41.15(b)*.
4. Observe the result again.

Result

The converging (or convex) lens refracts the light as shown in *Fig. 41.15(a)*.

The diverging (or concave) lens refracts the light as shown in *Fig. 41.15(b)*.

Warning!: Ray boxes can get very hot, even after quite a short time.

Magnifying Glass

A magnifying glass is simply a (convex) **lens**. It uses **refraction** to give an enlarged view of an object placed near it, *Fig. 41.16*.

Fig. 41.16 A magnifying glass is simply a refracting lens.

Is White Light Really all White?

Nature answers the above question with the beauty of a rainbow, *Fig. 41.17(b)*. Rainbows appear when there is rain and sunshine together. When the sunlight passes through the raindrops, the colours that make up white light are separated. The colours in the rainbow are **red**, **orange**, **yellow**, **green**, **blue**, **indigo** and **violet**. These seven colours are known as the **spectrum** of white light.

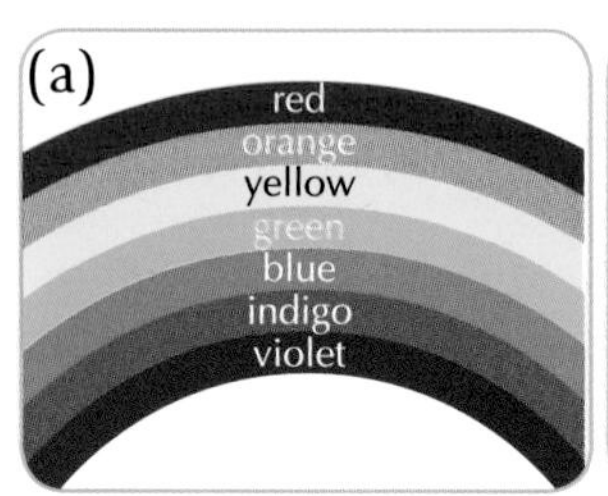

Fig. 41.17 White light is separated into different colours in a rainbow.

Note: To **remember** the colours of the spectrum of white light in the right order, you might use a mnemonic to help your memory. An example would be a sentence like: 'Ring Out Your Great Bell In Victory' or a person's name like ROY G BIV. You may prefer to make one up yourself.

The spectrum of white light is sometimes referred to as the **visible spectrum**. This distinguishes it from other types of radiation that our eyes cannot detect. These would include radio waves, microwaves, infra-red, ultra-violet, X-rays and gamma rays.

Experiment 41.5

To produce a spectrum of white light

Apparatus required: ray box (or a slide projector) with a single narrow slit; power supply; prism; screen

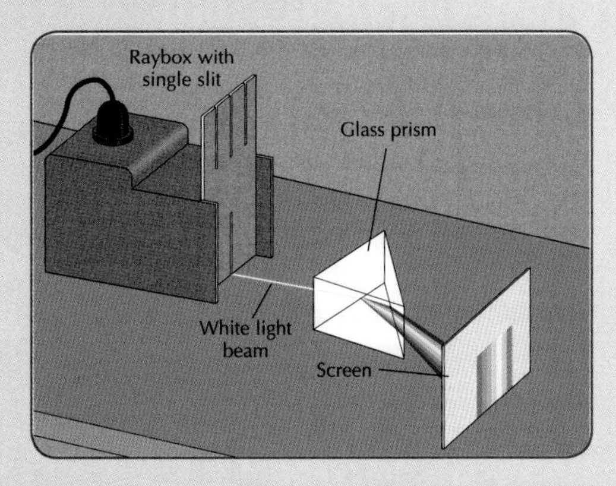

Fig. 41.18 Producing a white light spectrum.

Method

1. Shine a single narrow beam of white light on to the prism from a ray box, *Fig. 41.18*.
2. Move the ray box or screen until the spectrum of white light appears on the screen.

Result

The colours that result on the screen are the colours of the rainbow. Therefore, white light is made up of several colours.

Warning!: Ray boxes can get very hot, even after quite a short time.

When white light is broken up into its constituent colours, it is said to be **dispersed**. We saw that a prism disperses white light.

Dispersion is the breaking up of white light into the colours that make it up.

Applications of Refraction

(1) Refraction is used in **lenses**. Lenses are used in contact lenses and spectacles to correct eyes that are short-sighted and long-sighted. Magnifying glasses, microscopes, telescopes, cameras, etc., also use lenses.

(2) **Forensic scientists** often measure how much a particular glass bends light. Each glass is different. Any traces of glass found on a suspect can be compared to the glass broken at the scene of the crime.

(3) Refraction sometimes, on a hot day, causes a **mirage** of water to appear on the road ahead.

(4) The colours seen in diamonds and other jewellery and in a **rainbow** are caused by the refraction and dispersion of light.

(5) We 'see' the sun for up to four minutes before it actually appears over the horizon in the morning. Thus, we get **'extra' daylight** every day due to the refraction of light in the atmosphere.

TEST YOURSELF:
Now attempt questions *41.9–41.13* and *W41.12–W41.20.*

What I should know

- Light is a form of energy. It can do work (as in Crooke's radiometer) and change into other forms of energy.
- Some objects are luminous, e.g. the sun, a light bulb and a fire. They give out their own light.
- The sun is our main source of light. The sun converts nuclear energy to heat and light.
- Most things we see do not give out light. They are non-luminous. But they do reflect light. We see them because of the light that they reflect.
- Light can travel through a vacuum like space or through transparent substances like air and glass and water. It travels at the enormous speed of 300 000 000 m s^{-1}.
- Light travels in straight lines. Evidence of this is given by the existence of shadows.
- A solar eclipse occurs when the moon passes between the sun and the earth.
- Reflection is the bouncing of light from a surface.
- Light is reflected in a regular manner from a surface like a mirror, which is smooth and polished.
- Reflection allows us to see objects that reflect light. Reflection is used in periscopes and kaleidoscopes. It is used for safety in car mirrors and for security in store mirrors, etc.
- Light reflected from a rough surface is scattered in many directions.
- Periscopes use reflection to allow people to see things that would otherwise be difficult to see.
- Refraction is the bending of light as it passes from one medium to another.
- Lenses refract light. They can bring rays of light together or spread them out.
- A magnifying glass refracts light to give an enlarged view of an object.
- Refraction is used in lenses to correct eye defects and by forensic scientists to gather evidence. It is responsible for giving us extra daylight minutes each day and causes rainbows and mirages, etc.
- White light is made up of a mixture of colours, called a spectrum. These colours are red, orange, yellow, green, blue, indigo and violet.
- The breaking up of white light into the colours that make it up is called dispersion.
- A spectrum can be seen using white light and a prism. A spectrum is also seen when light from the sun passes through water drops or rain. This spectrum is called a rainbow.

QUESTIONS

Write the answers to the following questions into your copybook.

41.1 Objects which give off their own light are said to be ..... . Other objects, like this page, can only be seen when they ..... light.

41.2 How can you verify that light is a form of energy?

41.3 In photosynthesis, light ..... is converted to ..... energy. In a coal fire, ... energy is converted to ..... and ..... energy. Light energy is converted to electrical energy in a ..... .

41.4 Light travels in ..... lines. The paths along which light travels from a source are called ..... .

41.5 State two things that are scientifically incorrect in the sketch drawn in *Fig. 41.19*.

Fig. 41.19 Spot the two deliberate errors in this sketch.

41.6 If given a torch, three similar cards with holes in the middle of each one, a string and some plasticine, describe an experiment to verify that light travels in straight lines.

41.7 If you look at a mirror hanging on a wall, you will see your image. If you look at a landscape painting hanging on a wall, you will not see your image. Yet, you know that the painting reflects light because, otherwise, it would not be visible. So, why can you not see your image when you look at the painting?

41.8 What type of vehicle sometimes goes around with the sign shown in *Fig. 41.20* written on the front of it? Why do you think that the sign is written like this? If you saw the words '**POLICE**' or '**FIRE**' in the mirror, write down what is actually written on the front of the police car or fire engine?

AMBULANCE (printed mirror-reversed)

Fig. 41.20

41.9 ..... is the bending of light as it passes from one medium to another.

41.10 Describe an experiment which demonstrates the refraction of light.

41.11 A ..... is a piece of glass or other transparent material which has at least one curved surface.

41.12 White light is a mixture of the colours **red**, ... , yellow, ... , **blue**, ... and **violet**. Together, these colours form the visible ..... of white light.

41.13 ..... is the breaking up of white light into its component colours. A ..... is an example of an object that breaks up white light like this.

Chapter 42 Sound

42.1 Vibrations and Sound Energy

What is sound? If we think about how various sounds are produced, we get a clue to the answer. The sound sources listed in *Table 42.1* are all produced by **vibrations**. When the vibrations stop, there is no sound.

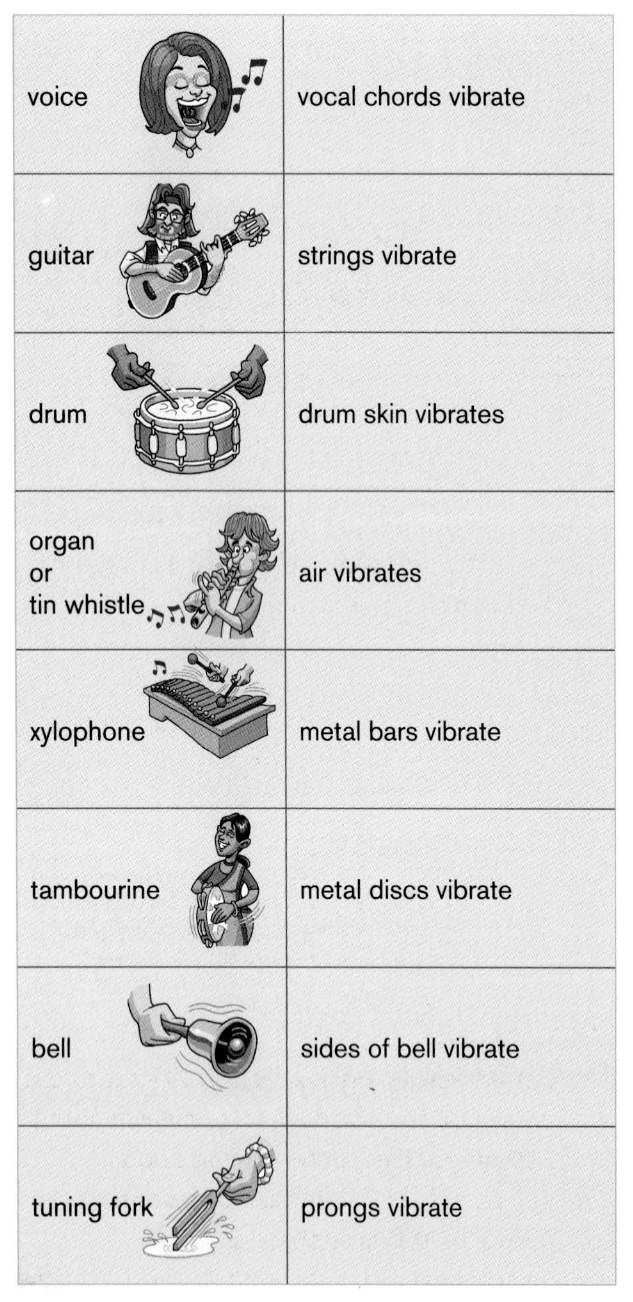

voice	vocal chords vibrate
guitar	strings vibrate
drum	drum skin vibrates
organ or tin whistle	air vibrates
xylophone	metal bars vibrate
tambourine	metal discs vibrate
bell	sides of bell vibrate
tuning fork	prongs vibrate

Table 42.1 All sources of sound are produced by vibrations.

A tuning fork vibrates so fast you may not **see** it moving. However, if you touch the end of the tuning fork to the surface of some water, you will know that it is vibrating, *Table 42.1*. Vibrations are a form of **kinetic energy**. Vibrations cause sound. Thus, sound is also a form of **energy**.

Sound is a form of energy, which is caused by vibrations.

Experiment 42.1

To verify that sound is a form of energy

Apparatus required: signal generator and loudspeaker (or loud music system); light ball (e.g. pith ball or table-tennis ball) taped to a string; split cork; retort stand

Method

1. Hang the light ball from a split cork clamped in a retort stand.
2. Place the light ball directly in front of the loudspeaker, *Fig. 42.1*.
3. Switch on a loud sound from the signal generator, which is connected to the loudspeaker.

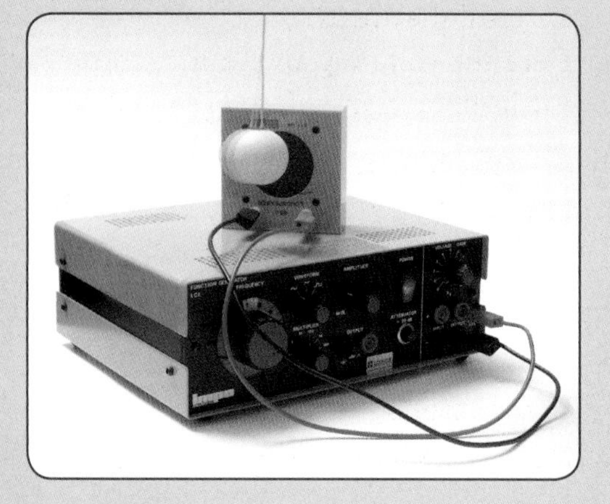

Fig. 42.1 To prove that sound is a form of energy.

Result and Conclusion

The ball moves when the sound is emitted. This shows that sound can do work. It causes movement. It changes to kinetic energy. Therefore sound is a form of energy.

42.2 Transmission of Sound

We saw in *Chapter 41* that **light** can travel through a vacuum. It does not need a substance or **medium** to travel through. But, **sound does need a medium**. In space there is no medium. This is why astronauts have to use radios to communicate with each other.

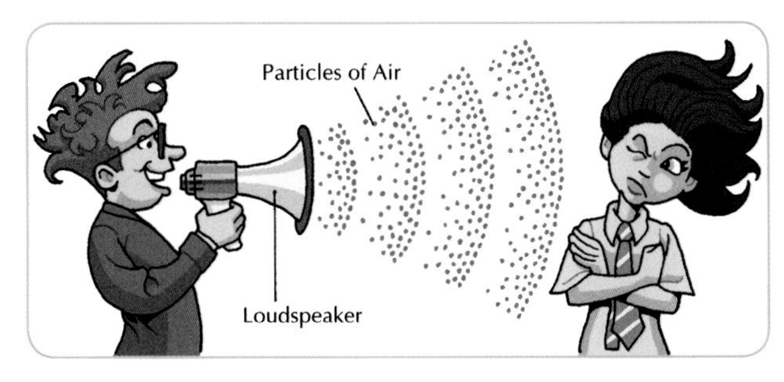

Fig. 42.2 Sound is carried through the air by vibrations of air particles.

How it Works

1. The vibrating cones of a loudspeaker **vibrate the air particles** near them.
2. These vibrating air particles, in turn, make the **air particles beside them vibrate** also, *Fig. 42.2*.
3. These vibrations are carried on **through the air** until they reach your ear.
4. Here, the vibrating air particles cause your **eardrum to vibrate** (see *Fig. 42.6*, page 332).

If there were no medium, then the sound could not travel, as there would be nothing to pass on the vibrations.

Experiment 42.2

(teacher demonstration only)

To show that sound cannot travel through a vacuum

Apparatus required: bell jar; vacuum pump; electric bell; battery (or low-voltage supply)

Method

1. Set up the apparatus as shown in the diagram, *Fig. 42.3*.
2. Connect the bell to a battery and switch it on. You should be able to hear the bell clearly.
3. While the bell is ringing, turn on the vacuum pump. This gradually removes the air from the bell jar.

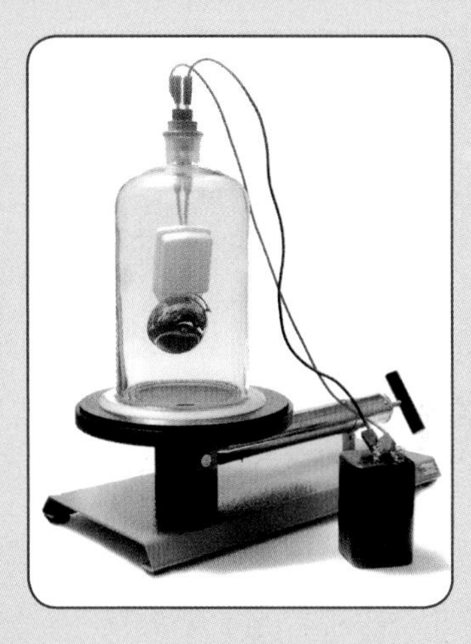

Fig. 42.3 Sound cannot travel through a vacuum.

Result

The sound of the bell fades out as the air is removed. You can still see that the hammer of the bell is striking the gong.

Conclusion

Sound requires a medium to travel through. It cannot travel through a vacuum.

TEST YOURSELF:
Now attempt questions *42.1–42.6 and W42.1–W42.3.*

42.3 Speed of Sound

- Sound travels very fast. In air, vibrations are passed on from particle to particle at about **340 m s^{-1}**. (This is the same as about 1224 km/h or 760 mph.) Over short distances, this is almost instantly.
- Over longer distances you may notice a **time lag**. In a fireworks display, for instance, you

PHYSICS

will see the explosion first. Then, after a short time, you will hear the bang.

- The speed of sound is **faster in denser materials**. For this reason, sound generally travels faster in solids than in liquids; and sound travels faster in liquids than in gases. Iron or steel transmits sound 15 times faster than air. Water transmits sound about 4.5 times faster than air.
- Sound may be fast but it is no match for light! The **speed of light** in air is nearly a million times faster than the speed of sound in air.

Thunder and Lightning

Evidence that light travels faster than sound is provided by **thunder and lightning**. These both originate together at the same point. The lightning flash travels at the speed of light. The thunder arrives a little later; it travels at the speed of sound. We can calculate how far away the lightning is by measuring the **time lag** between the flash of lightning and the clap of thunder. Each second in this time lag will represent a distance of about 340 m.

Example 1

A flash of lightning was observed from a house. Exactly 6 seconds later, the clap of thunder was heard at the house. If the speed of sound is 340 m/s, calculate how far the lightning was from the house.

We are looking for the distance travelled by the thunder. According to the 'D St.' triangle (*Fig. 35.10*, page 261)

D (distance) = S (speed) × T (time)

= 340 × 6

= 2040 m or 2.04 km

TEST YOURSELF:

Now attempt questions *42.7–42.12* and *W42.4–W42.6*.

42.4 Reflection of Sound

Echoes

Just like light, sound is reflected when it meets a barrier. While shiny polished surfaces are best at reflecting light, hard solid surfaces are best at reflecting sound.

Echoes are sounds that are reflected from a surface.

If you stand at a distance from a large wall (or cliff) and shout, the sound you emit will bounce off the wall and back to you again. If you are more than about 17 m from the wall, you will hear the echo shortly after the original sound. The further you are from the reflecting wall, the longer the gap between the emitted sound and its echo. The existence of echoes is evidence that sound can be reflected.

Experiment 42.3

To demonstrate the reflection of sound

Apparatus required: two long cardboard tubes; metal (or glass) sheet backed by cotton wool; ticking watch; cardboard (or wooden) screen; large sheet of paper

Method

1. Set up the apparatus as shown in the diagram, *Fig. 42.4*.
2. Arrange the cardboard screen at 90° to the metal sheet.
3. Move the tube, *B*, until you can best hear the ticking watch.
4. Repeat by rotating tube *A*, giving different values of the angle, *i*.

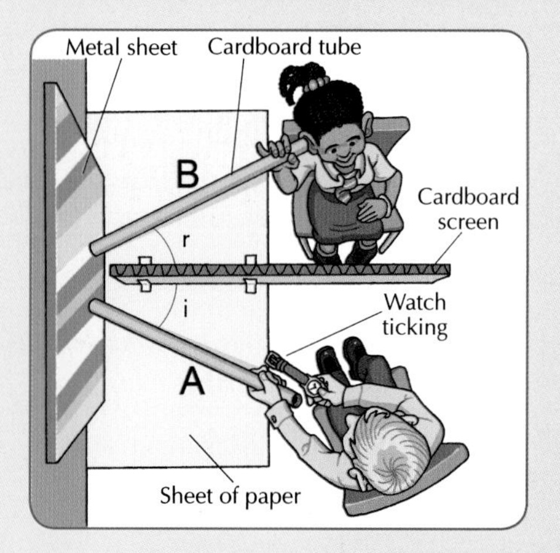

Fig. 42.4 Sound waves can be reflected.

Result

The metal sheet reflects the sound. As the angle i increases, so does the angle, r.

Note: A sound sensor and datalogger could very effectively replace the listening ear in this experiment.

Disadvantages of Echoes

Echoes can often be a nuisance in large halls, cinemas or theatres. Reflected sounds can get mixed up with new sounds and arrive together to the audience. Soft padded walls and thick carpets are used to absorb unwanted echoes. Soft surfaces are bad reflectors or good absorbers of sound.

Advantages of Echoes

Echoes can also be very **useful**. Ships use **echo-sounding** (or **echo-location**) to find out the **depth of the sea**, *Fig. 42.5*. The ship sends out a sound. The seabed reflects the sound back to the receiver on the ship. The speed of sound in water is 1500 m/s. Knowing the time taken by the sound signal to return, the depth can be calculated. The depth of a shoal of fish that pass under the ship can also be found in the same way.

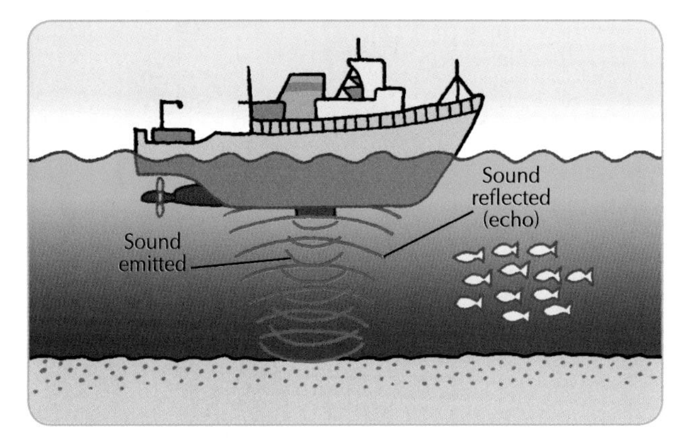

Fig. 42.5 Echo-sounding can measure the depth of the seabed and locate shoals of fish.

Example 2

(i) *A sound is sent down to the seabed by a ship. A receiver on the ship picks up its echo 1.2 seconds later. If the speed of sound in water is 1500 m/s, calculate the depth of the water beneath the ship.*

(ii) *Suddenly a shoal of fish swims under the ship at a depth of 250 m below the surface. How long does it take for the receiver to pick up the signal now?*

(i) According to the 'D St.' formula (*Fig. 35.10*, page 261)

distance = speed × time

∴ distance travelled by sound = 1500 × 1.2

= 1800 m

But, depth of water = half of the distance travelled by the sound (i.e. down and up)

= 1800 ÷ 2

= 900 m

(ii) This time the sound travels 250 m down and 250 m back, a total distance of 500 m

time = (distance) ÷ (speed)

= (500) ÷ (1500)

= 1/3 s or 0.33 s

42.5 Sound Detection in the Ear

How it Works

Our ears are our very own sound vibration detectors.

1. The **outer ear** is shaped to collect sound. It directs the vibrating air particles to your eardrum, *Fig. 42.6*, page 332.
2. The **eardrum** is like a thin drum skin. Incoming sound makes it vibrate back and forth.
3. The vibration is carried through your middle ear to your **inner ear**.
4. Here, **nerves** pick up the vibrations and send them as electrical signals to the brain.

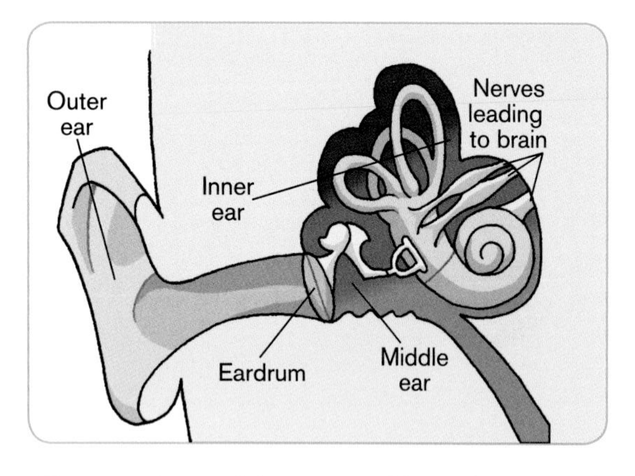

Fig. 42.6 A sound detector used by early man and modern man!

Noise Level

Our ears are extremely sensitive.

- They can hear very soft or **quiet sounds**. These are low energy sound vibrations.
- They can also hear **loud sounds**. These carry much more energy. A very loud sound like an explosion can break a person's eardrum. This is very painful. After it heals, the eardrum is not as sensitive as before. Loud sounds can also damage the nerves in our ears. Nerves cannot recover once they are damaged.

The Decibel Scale

But when is a sound loud? A scale of different **sound levels** exists so that we can compare the loudness of different sounds. It is often referred to as the decibel scale. The sound level of various sounds can be measured using a **sound level meter**, *Fig. 42.7*.

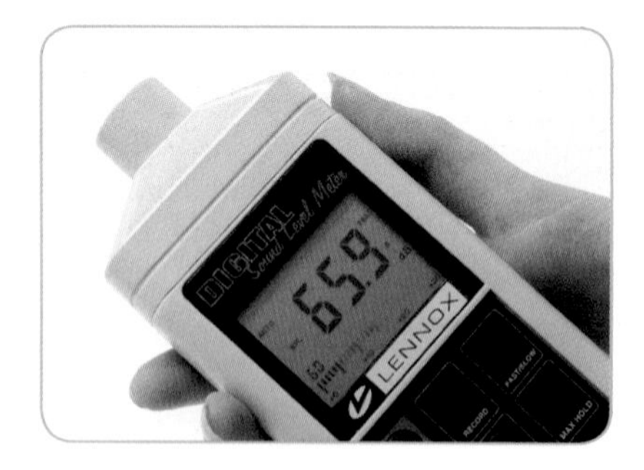

Fig. 42.7 Sound level meter.

Fig. 42.8 shows the approximate sound levels of various sounds on the decibel scale. The zero on this scale (called the threshold of hearing) is the lowest sound that the human ear can hear.

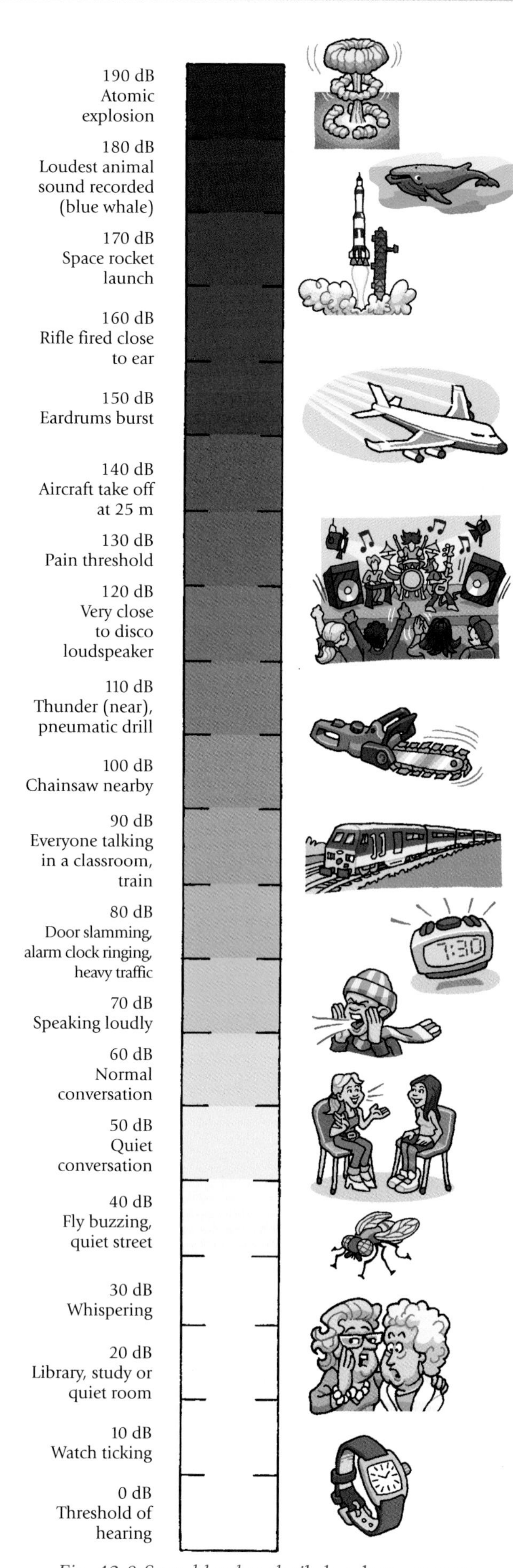

Fig. 42.8 Sound level or decibel scale to compare different sounds.

Noise Level and Damage

- Sounds above **85 dB** can be harmful to your ears. Sometimes the damage is temporary. But the possibility of permanent harm exists, especially if exposed to it over a long time.
- Sounds above **115 dB** will almost certainly do damage to your ears. It may cause headache, earache and loss of hearing sensitivity. Exposure to this level of sound for more than two minutes may result in permanent hearing loss.
- The threshold of pain is at about **130 dB**.
- Sounds above **150 dB** will very likely burst your eardrums.

The **simple test** is as follows: If you are in a place where you have to shout to be heard by a friend who is an arm's length away, you should leave that place. The place is too noisy. The risk of doing damage to your ears is far too great.

Hearing Protection

We should have more **respect** for our sensitive ears. We will need them for a long time. We should always take simple precautions. Some of these precautions are listed across:

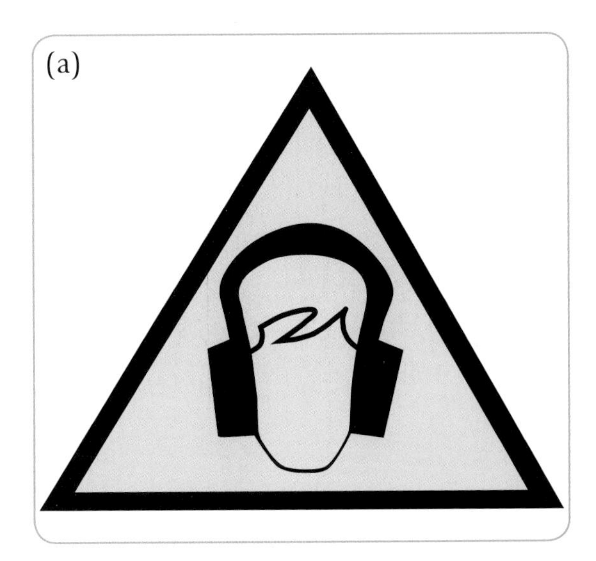

Fig. 42.9(a) Warning of high sound levels.

Fig. 42.9(b) Ear protection is legally required for some workers.

(1) Heed **warning signs** like *Fig. 42.9(a)*. This means that loud noises are close. Your ears may be damaged if you don't protect them.
(2) Heed **early warning signs** of problems, like ringing in your ears. You should avoid situations where this arises.
(3) Wear **ear muffs** where necessary. Workers are now obliged by law to wear hearing protection when exposed to noises over 90 dB, *Fig. 42.9(b)*.
(4) **Move away** from sources of high-level sound. When you double the distance between you and a source, the loudness of the sound at your ears is quartered.
(5) **Double-glazed windows** or thick heavy curtains or shutters will reduce the noise from outside.
(6) A **high wall** or trees will reduce the noise from a busy road.

TEST YOURSELF:
Now attempt questions *42.13–42.18* and *W42.7–W42.13.*

What I should know

- Sound is a form of energy, which is caused by vibrations.
- Sound needs a substance or medium to travel through. Sound cannot travel through a vacuum.
- The speed of sound in air is approximately 340 m/s. While this is fast, it is very slow when compared to the speed of light at 300 000 000 m/s.
- Thunder and lightning give evidence that light travels faster than sound.
- Sound travels faster through some materials than others. Denser materials are better at carrying sound generally. Solids are better than liquids. Liquids are better than gases.
- Sound may be reflected from certain surfaces.
- Echoes are sounds that are reflected from a surface.
- Echoes can be a nuisance in large halls, cinemas or theatres.
- Echoes can also be very useful. Echo-sounding or echo-location is used by animals, ships, submarines, etc., to provide information about what is near them.
- Our ears detect sound vibrations. The outer ear collects the sound. The sound vibrates our eardrum.
- Sound levels give us an idea of how different sounds compare.
- Exposure to very loud sound damages our hearing.
- A simple test of noise level is as follows: If you are in a place where you have to shout to be heard by a friend who is an arm's length away, you are in a dangerously noisy place.
- Hearing protection must be worn near high-level sounds.

Questions

Write the answers to the following questions into your copybook.

42.1 Sound is a form of ..... which is produced by ..... .

42.2 Sound needs a substance or ..... to travel through. Sound cannot travel through a ..... .

42.3 Describe an experiment to verify that sound is a form of energy.

42.4 Why is the moon sometimes referred to as the 'silent planet' by astronauts that have been there?

42.5 There are many objects in the solar system and universe that make noises. Why can we never hear any of them?

42.6 Why do sounds get fainter as you move away from their source?

42.7 (a) Thunder and lightning give evidence that ..... can travel ..... than sound.

(b) The speed of sound in a solid is ..... than the speed of sound in air.

42.8 Sometimes we feel the vibrations of a big sound source like a truck passing nearby outside the house before we hear the truck itself. Why is this?

42.9 (a) What is the approximate speed of sound in air?

(b) If the time lag between seeing the flash of lightning and hearing the thunder is 4 seconds, how far away is the lightning?

42.10 When a girl was a certain distance from a high cliff she shouted loudly. One and a half seconds later the echo returned. How far was the girl from the cliff? (Take the speed of sound to be 340 m/s.)

42.11 What does it mean if you see a flash of lightning and hear a clap of thunder at almost exactly the same time?

QUESTIONS

42.12 The foghorn in the lighthouse shown emitted a short blast, *Fig. 42.10*. The distance between the lighthouse and the cliff was 1340 m. The echo was heard at the lighthouse 8 seconds later. What was the speed of the sound?

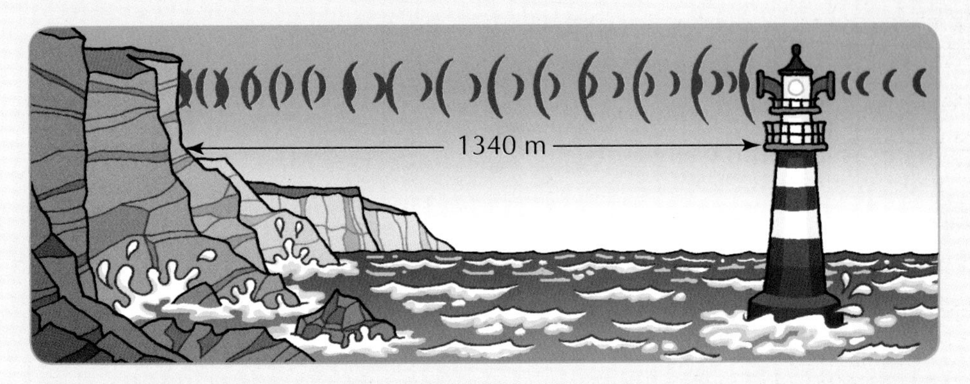

Fig. 42.10 The echo of the sound of the foghorn returns to the lighthouse from the cliff.

42.13 An echo is evidence that ..... can be ..... .

42.14 Why do noises seem louder in a tunnel than in open air?

42.15 Explain why a bat's method of hunting to find a moth is called echo-location.

42.16 A ship sends a pulse of sound to the seabed, which is 3000 m beneath the ship. If the speed of sound in water is 1500 m/s, after how long will the ship receive the echo?

42.17 A guitar string is plucked. The sound is heard. It seems simple, but there are a few stages in between. Can you put the following stages in the correct order?

vibrating eardrum, brain, vibrating source, vibrating air, nerve

42.18 If you were visiting a factory and came across a sign as in *Fig. 42.11*, what precautions do you think you should take and why?

Fig. 42.11

Chapter 43 Magnetism

43.1 Magnetism

If you hold a pin or a paper clip close to a magnet, you will feel a pull. This pulling force is called **magnetism**. You cannot see magnetism but you can feel its effects.

A magnet is a piece of metal that can attract other substances to it.

Magnets were first discovered about 2,500 years ago. A rock called **lodestone** was found to attract iron. This was the first **natural** magnet. It occurred freely in nature.

Magnetic Materials

Today, even the most powerful magnets can attract only certain materials. These materials are called **magnetic materials**. Magnets can only be made from these magnetic materials.

Experiment 43.1

To test a variety of materials for magnetism

Apparatus required: a strong magnet (e.g. a horseshoe magnet)

Also required: a variety of everyday materials. (Items might include some of the following: paper clips, pencil, needle, variety of coins, paper, pen, ruler, protractor, comb, scissors, pin, eraser, stapler, calculator, nail, bunch of keys, copper wire, etc. Use your own initiative to expand this list.)

Method

1. Make a list of the materials you are testing. Write down what each material is made of.
2. Move the strong magnet slowly over each of these materials, *Fig. 43.1*.
3. Note whether each material can be picked up or not.

Fig. 43.1 Testing materials to see if they are magnetic or not.

Result and Conclusion

The magnet attracted certain materials (**magnetic materials**). It had no effect on others (**non-magnetic materials**). The magnetic materials all had **iron** or steel in their makeup.

Note: Steel is added to the centre of some coins. These coins are magnetic. Others, without steel, are not.

Man-made Magnets

Nowadays, **magnets** come in many shapes and sizes. Almost all are artificial or man-made. Only three of the elements on the Periodic Table can be magnetised. These are **iron**, **nickel** and **cobalt**. However, strong magnets can also be made by mixing these elements with other metals to form alloys (see *Section 33.3*, page 236).

- **Steel**, for example, which is made by adding a small amount of carbon to iron, can be made into a strong magnet.
- **Alnico**, another strong magnetic material, is a mixture of aluminium, nickel, iron and cobalt.

43.2 Attraction and Repulsion

If a bar magnet is dipped into a jar of iron filings, the result will be as in *Fig. 43.2*. Most of the iron filings cling to each end. The magnet seems to be strongest here. These ends are called **poles**. Poles always occur in pairs on a magnet. They are referred to as the **north pole** and the **south pole**.

Fig. 43.2 Magnets are strongest at both ends or 'poles'. (Blue end = north pole, red end = south pole.)

Experiment 43.2

To demonstrate the attraction and repulsion of magnets

Apparatus required: two bar magnets (with north and south poles clearly marked); paper stirrup; string; non-metal retort stand

Method

1. Hang one bar magnet in a paper stirrup from a retort stand.
2. Bring the south pole of a second magnet close to the north pole of the hanging magnet, as shown in *Fig. 43.3*. Note whether the magnets attract or repel each other.
3. Now bring the north pole of the second magnet near the north pole of the hanging magnet. Record the result on a table as shown in *Table 43.1*. Then try the other combinations mentioned in the table.

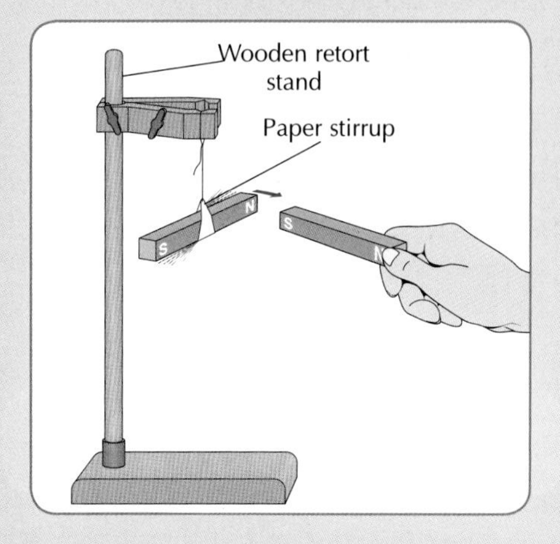

Fig. 43.3 Opposite poles attract each other.

Result

HANGING MAGNET	SECOND MAGNET	RESULT
north pole	south pole	attract
north pole	north pole	
south pole	south pole	
south pole	north pole	

Table 43.1

The results of *Experiment 43.2* lead to a very important conclusion:

Like poles repel each other; unlike poles attract.

Fig. 43.4 Magnetic attraction could turn to repulsion at the change of a hat!

In summary:

- A magnet has **no effect** on non-magnetic materials.
- A magnet can (only) **attract** magnetic materials.
- A magnet can **attract or repel** another magnet.
- The true test for determining if a piece of metal is a magnet or not is **repulsion**. Only a magnet can **repel** another magnet.

TEST YOURSELF:
Now attempt questions *43.1–43.4* and *W43.1–W43.5*.

43.3 Magnetic Fields

If a steel paper clip is placed far enough away from a magnet, the magnet has no effect on it. The effect of a magnet gets smaller (or weaker) as you go further from it. The area around a magnet, over which the magnetic force exerts its power, is called the magnetic field.

A magnetic field is the space around a magnet over which it has a magnetic effect.

Mandatory experiment 43.3

To plot the magnetic field of a bar magnet using iron filings

Apparatus required: bar magnet; iron filings; card

Method

1. Place a bar magnet on the bench. Place a card on top of it. Draw the outline of the position of the magnet on the card.
2. Sprinkle iron filings evenly all over the card.
3. Tap the paper gently with your finger, *Fig. 43.5.*

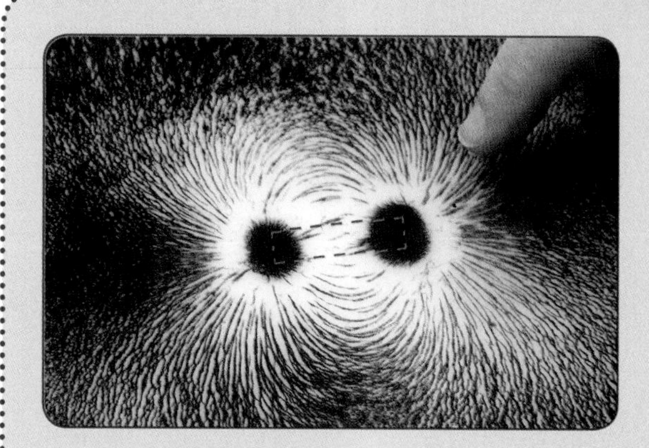

Fig. 43.5 The red broken lines show the outline of the magnet underneath the card.

Result

The iron filings are influenced by the strong magnetic forces from the poles of the bar magnet beneath. They are forced to point themselves along the direction of the magnetic field. Thus, they show the magnetic field around the bar magnet as in *Fig. 43.5.* Magnetic field lines always go **from north pole to south pole**.

Note: You can also use **plotting compasses** to show the direction of magnetic field lines. Simply place the compasses near the bar magnet as shown in *Fig. 43.6.* Put dots on a paper at both ends of each compass needle. Join the dots to trace out the magnetic field line. Various magnetic field lines can then be got by moving the compasses to different positions.

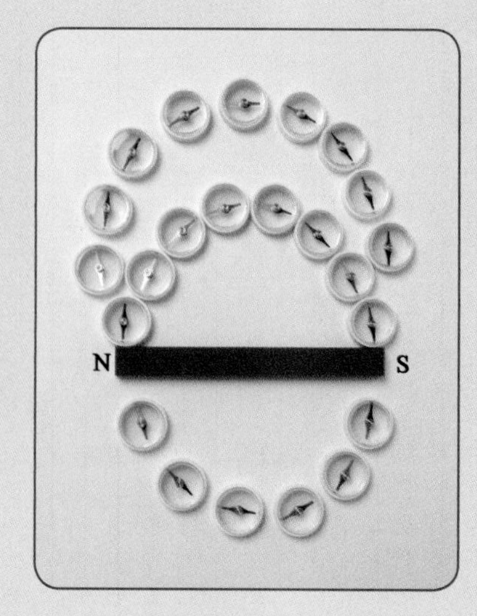

Fig. 43.6 Magnetic field lines (curved) go from the north pole around to the south pole.

43.4 The Magnetic Field of the Earth

The earth has a magnetic field. This magnetic field forces all free magnets to line up in a north-south direction. This is why we can use a free magnet to locate north and south. We call this magnet a **compass**, *Fig. 43.7*.

Fig. 43.7 Here, a magnetised needle is free to rotate and act as a compass.

A compass is a magnet, which is free to rotate and indicate direction.

The earth's magnetic field behaves **as if** there is a huge bar magnet in its centre. (We know that the centre of the earth is far too hot for such a magnet to really exist.) In *Fig. 43.8*, you will notice that the south (S) pole of this imaginary bar magnet is near the North geographical Pole. Thus, the earth's **magnetic** south pole attracts the north pole of all freely suspended bar magnets.

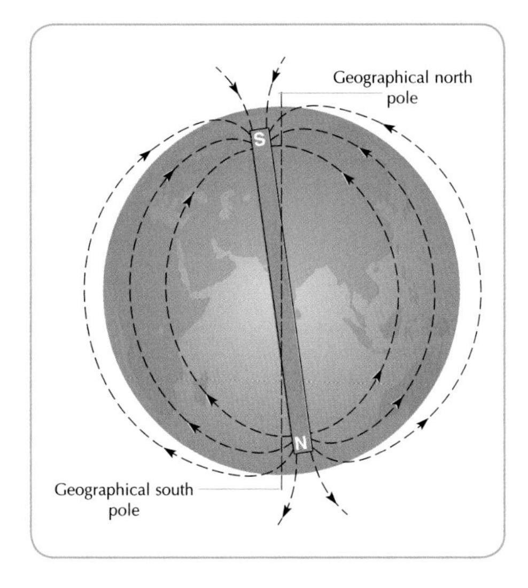

Fig. 43.8 The magnetic field of the earth.

Experiment 43.4

To demonstrate that the earth has a magnetic field and to locate north/south direction

Apparatus required: bar magnets; paper stirrup; string; non-metal retort stand

Method

1. Hang a bar magnet in a paper stirrup from a retort stand as shown in *Fig. 43.9*. The magnet should be horizontal and free to swing.
2. When the magnet comes to rest, note the position in which it is pointing. Then, give it a slight twirl. Test to see if it comes to rest in the same position again. Do this a few times.
3. Repeat the experiment for other bar magnets.

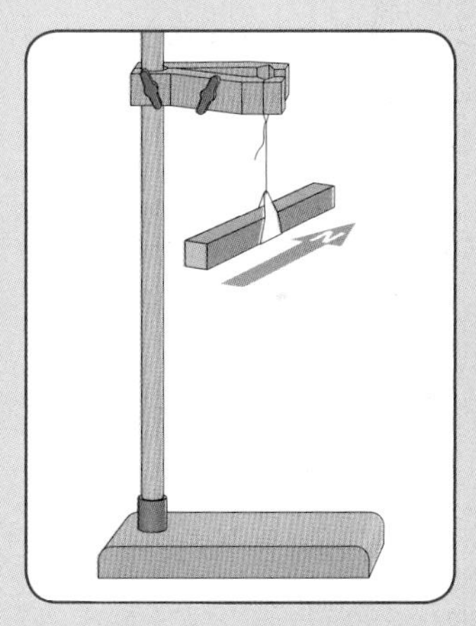

Fig. 43.9 The bar magnet always comes to rest in a north-south direction.

Result and Conclusion

Each bar magnet **always** comes to rest with one end pointing in a **northerly** direction. This end is the **north pole** of the magnet. The opposite end, or the 'south-seeking' pole, is referred to as the **south pole**. The magnets are responding to the magnetic field of the earth.

TEST YOURSELF:
Now attempt questions *43.5–43.8* and *W43.6–W43.11.*

43.5 The Magnetic Effect of a Current

Hans Christian Oersted, a Danish scientist, discovered in 1819 that an **electric current** flowing in a wire had a magnetic effect. He showed that a **magnetic field** surrounds a wire when a current flows through it.

Experiment 43.5

To show the magnetic effect of a current

Apparatus required: battery or power pack; switch; plotting compass; wire

Method

1. Allow the compass needle to come to rest in its usual north-south direction.
2. Hold a straight wire **parallel** to the compass needle over the compass, *Fig. 43.10*. Connect the wire in a circuit to a battery and switch, as shown.
3. Close the switch for a few seconds. This allows current to flow in the wire.

Result

The compass needle moves at right angles to the wire.

Conclusion

This movement is caused by the magnetic effect of the current in the wire. Because it is very near, the magnetic effect of the current is greater than the magnetic effect of the earth.

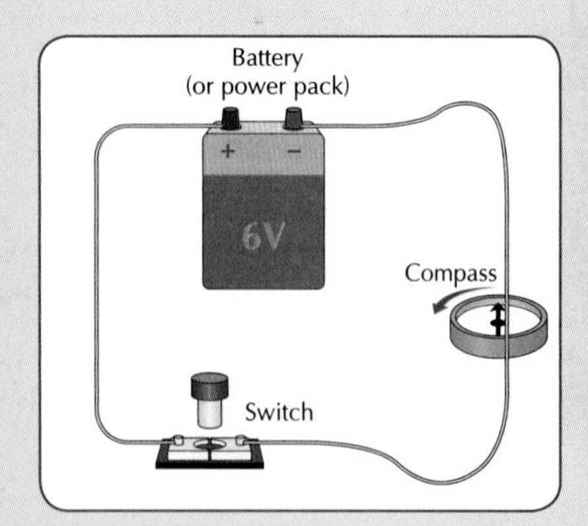

Fig. 43.10 To show the magnetic effect of an electric current.

The Electromagnet

An electromagnet consists of an iron rod (or core) surrounded by an insulated coil of wire. An electromagnet also demonstrates the **magnetic effect** of an electric current.

The simple electromagnet in *Fig. 43.11* consists of an insulated wire wrapped closely around an iron nail. A power pack or a battery is used to drive a current through this wire.

When the power supply is switched **on**, the iron nail **attracts** the paper clips. When the current is switched **off**, the nail **loses its magnetism**. The paper clips fall off. While a current flows through the wire, the iron nail is a strong magnet.

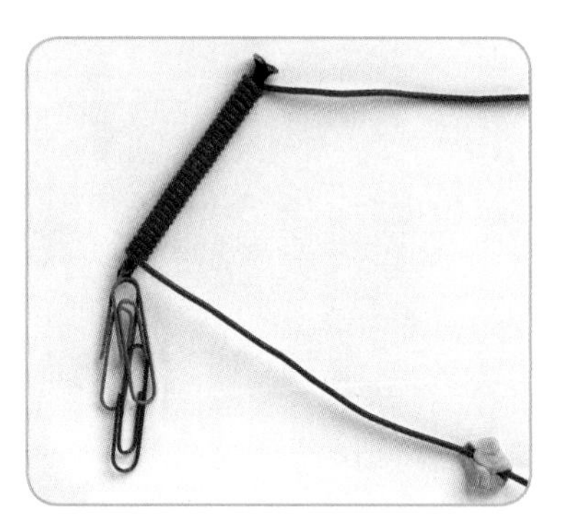

Fig. 43.11 A simple electromagnet.

Uses of Magnets

- Magnets can be used as **compasses**.
- Magnets are used in hospitals to do **body scans** (NMR scans).
- Magnets are also used in **electric motors**, electric bells, bicycle dynamos, transformers, ammeters, loudspeakers, telephones, televisions, microwave ovens, computers, tape recorders, circuit breakers, metal detectors, etc.
- They are used to keep the **doors of fridges** and presses closed.
- Magnetic ink on automatic **cash-cards** and on cheques can store information.
- Electromagnets are used to **move scrap iron** and steel in scrap yards.

TEST YOURSELF:

Now attempt questions *43.9–43.12* and *W43.12–W43.16*.

What I should know

- Natural magnets, like lodestone, are found in nature. Artificial or man-made magnets are made by magnetising iron, nickel, cobalt or their alloys.
- Magnets are strongest at their two ends, which are called the 'north pole' and the 'south pole'.
- Like poles repel each other but unlike poles attract.
- A true test for determining if a metal is magnetised is repulsion.
- A magnetic field is the space around a magnet over which it has a magnetic effect.
- The earth behaves as if there is a huge bar magnet in its centre giving it a magnetic field.
- A freely suspended magnet always comes to rest with one end pointing in a northerly direction.
- A compass contains a magnet, which is free to rotate, and it reacts to the magnetic field of the earth.
- A wire, which has a current flowing through it, has a magnetic field around it.
- An electromagnet becomes a magnet only when a current flows through its coil.
- Magnets have various uses. They are used in electric motors, electric bells, bicycle dynamos, transformers, ammeters, loudspeakers, telephones, computers, tape recorders, etc.

QUESTIONS

Write the answers to the following questions into your copybook.

43.1 Name the three elements that are attracted to magnets and can also be magnetised themselves.

43.2 Magnets are strongest at their two ends which are referred to as the ..... and the ..... .

43.3 One of the properties of magnets is that like poles .... each other and unlike poles ..... Thus, a north pole will ..... a north pole and will attract a ..... pole.

43.4 You are given a bar magnet and two pieces of steel. You are told that only one piece of steel is a magnet. Describe how you could find out which of the two pieces of steel is a magnet.

43.5 (a) A ..... is the space around a magnet over which it has a magnetic effect.

(b) A small suspended magnet which is free to rotate and indicate direction is called a ..... .

43.6 (a) The earth behaves as if there is a huge ..... in its centre giving it a magnetic field.

(b) A freely suspended magnet always come to rest with one end pointing ..... .

43.7 If you were given an unmarked bar magnet, describe how you would determine which side was the north pole.

43.8 (a) A bar magnet sitting on a table does not act as a compass. Why?

(b) Describe how such a bar magnet could be made into a magnetic compass.

43.9 An electromagnet consists of an ..... surrounded by an insulated coil of wire.

43.10 When a current flows through an electromagnet, it is a strong ....., but when the current is switched off, it ..... its magnetism.

43.11 Describe a simple experiment to show that a current flowing in a wire has a magnetic effect.

43.12 List *five* everyday uses of magnets.

Chapter 44 Static Electricity

44.1 Introduction

Electricity was first discovered in 600 B.C. by Thales de Miletus, a Greek philosopher. He noticed that a hard dry yellow substance called **amber** had a mysterious property. When he rubbed it with wool or fur, it attracted light materials such as hair, dust, etc.

Many other substances have similar properties to amber.

- A **plastic comb** will pick up small pieces of paper, *Fig. 44.1(a)*. But, this only happens if the comb is first rubbed with wool or in your hair.
- *Fig. 44.1(b)* shows how a **plastic biro** can affect a thin stream of water. Again, the biro must first be rubbed in your hair or on your sleeve.

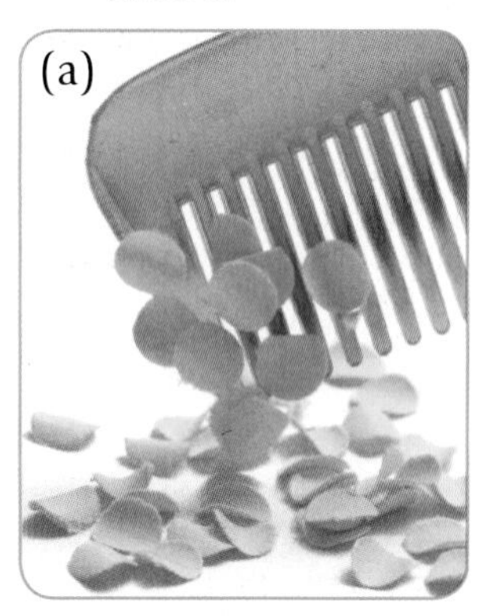

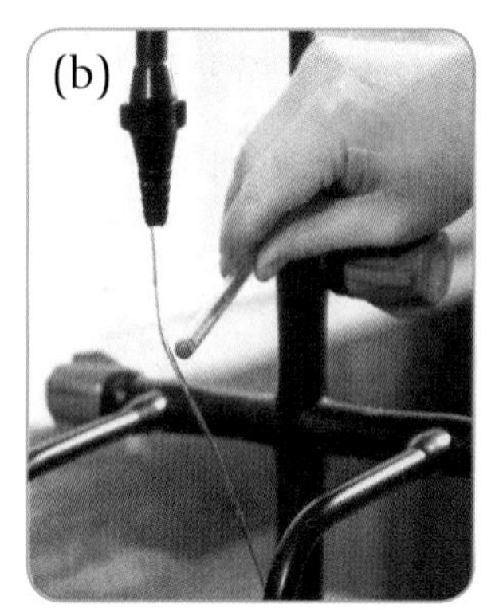

Fig. 44.1 Light papers and a thin stream of water from a tap being attracted to a plastic comb and biro.

Static and Current Electricity

The type of electricity described above is called static electricity. It involves **electric charges** that collect in one place and remain **stationary**. Static electricity was the only form of electricity known until the end of the eighteenth century. In 1798, an Italian, called Allesandro Volta, produced electricity chemically using a cell or a battery. This was the first source of '**electricity that moved**' or **current electricity** (see *Chapter 45*).

44.2 Electric Charge

To understand electricity, you must first know something about the atom. (You may have already studied the atom in *Chapters 29* and *30*.) The atoms of all substances have smaller particles inside them.

- The centre or 'nucleus' of an atom contains **neutrons**, with a neutral charge.
- It also contains **protons**. These protons have a positive electrical charge (+).
- In orbit around the nucleus are **electrons**. These are tiny particles with a negative charge (–).

Neutral Atoms

Atoms are generally uncharged or electrically **neutral**. This means that they have the **same** number of protons (+) as electrons (–). **Lithium**, for example is number three on the Periodic Table of Elements. Normally, it has three protons and three electrons. Thus, its total charge is $(+3) + (-3) = 0$, i.e. it has a charge of zero.

Charged Atoms

A lithium atom can sometimes lose an electron. It will then have three protons and two electrons. Its charge will be $(+3) + (-2) = +1$, i.e. it will be positively charged, *Fig. 44.2*. Thus, substances that **lose electrons** become **positively** charged. Other substances have atoms that **gain** electrons. When they do, they become **negatively** charged.

Fig. 44.2 A positively charged lithium atom has three protons (+ 3), but only two electrons (– 2).

How Substances Become Charged

Polythene

Polythene is a grey/white plastic. When a rod of polythene is rubbed with a woollen or fur cloth, **a transfer of electrons** takes place. Friction causes some of the outside electrons of the wool to escape. These get transferred to the polythene rod. The rod **gains electrons**; thus, it is **negatively charged**. (The wool has lost electrons. It now has more protons than electrons; therefore, it is positively charged.)

Perspex

A perspex (or cellulose acetate) rod, on the other hand, tends to **lose** electrons. When it is rubbed on your sleeve, the energy produced by the friction sets some of its electrons free. These electrons are transferred to your sleeve. But the perspex rod ends up with a **positive** charge.

Experiment 44.1

To demonstrate the force between charged objects

Apparatus required: two polythene (or ebonite) rods; two perspex (or cellulose acetate or glass) rods; woollen cloth; non-metallic retort stand; paper stirrup; dry thread

Method

Note: In **all** experiments involving static electricity, the apparatus must always be very **dry**. This can be achieved with the help of a hair dryer and by avoiding damp weather conditions.

1. Charge the first rod, *rod 1*, by rubbing it with a woollen cloth. Then suspend it in a paper stirrup as shown in *Fig. 44.3*.
2. Charge a second rod, *rod 2*, by rubbing it with a woollen cloth. Hold it in your hand. Bring it towards the suspended rod. Note what happens.
3. Repeat the experiment for different combinations of rods as shown in *Table 44.1*. The result should be noted each time.

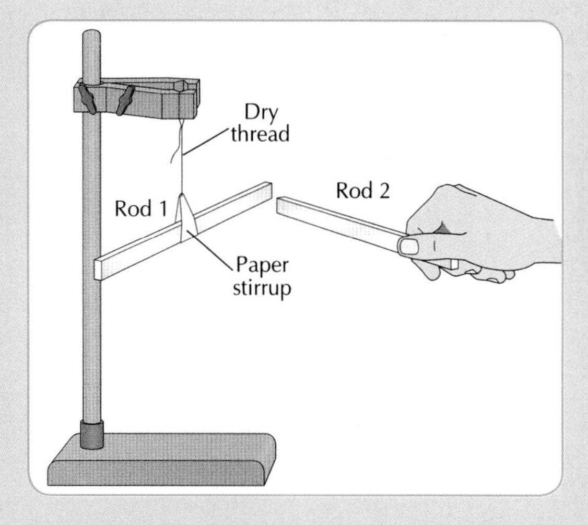

Fig. 44.3 Demonstrating the force between charged rods.

Result

ROD 1 (SUSPENDED)	ROD 2 (IN HAND)	RESULT
polythene (negative)	polythene (negative)	repulsion
polythene (negative)	perspex (positive)	attraction
perspex (positive)	perspex (positive)	repulsion
perspex (positive)	polythene (negative)	attraction

Table 44.1 Different combinations of positively and negatively charged rods.

The result of the above experiment leads to an important conclusion. Charges in static electricity act in a similar way to the poles of magnets, *Chapter 43*.

Like charges repel each other; unlike charges attract.

TEST YOURSELF:
Now attempt questions ***44.1–44.7* and *W44.1–W44.9*.**

PHYSICS

44.3 Effects of Static Electricity

A dramatic effect of static electricity occurs in the **Van de Graaff generator**. The metal dome can store a huge charge. *Fig. 44.4* shows what happens when a person with soft dry hair is touching the dome.

Fig. 44.4 A hair-raising moment, but definitely not to be tried at home.

Static electricity can have many other effects. Some are useful. Some are a nuisance. Some can be very dangerous.

(1) Useful Effects of Static Electricity

- It is used to **remove soot** from chimneys, so that it does not pollute the atmosphere.
- In **photocopiers**, it is used when the toner (or ink) is attracted to a charged rotating drum.
- In **spray painting**, droplets of paint are charged. They are then attracted to the surface that is to be painted.
- When a farmer is spraying, charged **pesticide** droplets are attracted to leaves.

(2) Nuisance Effects of Static Electricity

- Sometimes you may pick up a **small electric shock** when you touch a car door after a journey. You may also get a shock when removing nylon clothes from a tumble dryer.
- When you brush your **hair**, especially if it is dry, some hairs may repel each other and refuse to stay together or lie flat. This is because they pick up a similar electric charge.
- Plastics and car windscreens and other objects that have been rubbed with a cloth, often **attract dust**. Television screens also attract dust.

(3) Dangerous Effects of Static Electricity

- Static electricity can be very dangerous when there are **inflammable liquids** or gases around. One spark could cause tremendous damage if it ignited the fuel. When **aeroplanes** cut through the air and clouds, they pick up a large charge. This charge must be removed before the aircraft is re-fuelled.
- Great care must also be taken also on board **oil tankers** to avoid making sparks. The crew wear special 'anti-static' clothes and shoes.
- **Lightning** can be a dangerous result of static electricity. Inside a storm cloud, water and ice particles bump off each other. The friction creates a build-up of static electricity. Positive charges can build up at the top of the cloud. Negative charges may build up at the bottom of the cloud. When this charge gets very large, it is released by a stroke of lightning. The charge may be released to the earth (fork lightning), *Fig. 44.5*, or to a nearby cloud (sheet lightning).

Fig. 44.5 The awesome power of lightning. (We see lightning before we hear the thunder, proving that light travels faster than sound.)

TEST YOURSELF:
Now attempt questions *44.8–44.11* and *W44.10–W44.14*.

44.4 Earthing

Conductors and Insulators

- We have seen that static electricity is due to electric charges that remain at rest on an object. These charges cannot move. This is because the material on which they stand are **insulators**. Insulators, like polythene, perspex, glass, etc., do not allow charges to flow.
- In current electricity, which we will study in the next chapter, the charges move along **conductors**. A conductor is a substance, like copper, that allows charge to flow through it freely.

Insulators are substances which do not normally allow charge to flow through them.

Conductors are substances which allow charge to flow through them freely.

What is Earthing?

A plastic rod is a good example of an **insulator**. It can build up and store quite a large electric charge. This charge has no route to get away. However, if a **conductor touches** the charged plastic rod, things change. If you, for instance, touch this charged plastic rod, you may get a small shock. You become the escape route for the built-up charge. The charge will pass through you to the earth because you are a conductor. An escape route like this for electricity is called an **earth** connection. An object connected to the earth cannot really build up a charge.

Earthing means connecting an object to the earth by means of a conductor, so that the object shares its charge with the earth.

Examples of Earthing

(1) **Static electricity** – In a large heated shop, a person often picks up charges as their shoes brush off a dry carpet. This charge has no easy way to get to the earth. If the person then touches a **metal** handrail or doorknob, they get a **small shock**. They may feel a slight tingle or see a spark as the charge jumps to the metal. The metal is acting as an **earth**. It provides a route for the charge that has built up on the person to escape to the earth.

(2) **Lightning conductor** – We saw earlier that a **thunder cloud** builds up a huge charge. An American, called Benjamin Franklin, designed a **lightning conductor** to protect buildings. It consists of a sharp metal spike, which is erected at the top of the building. A copper strip connects this to a metal plate buried in the ground, *Fig. 44.6*. The lightning conductor provides an easy route for the built-up charge to get to the earth. Thus, it protects the building from being damaged.

Fig. 44.6 The lightning earths itself through a lightning conductor, which provides an easy escape route for the huge charge to get to earth.

Exam Question

Junior Cert Science Ordinary Level Examination Question

Fig. 44.7

The diagram shows a car. In dry weather you can sometimes get an electric shock when you get out of (or touch) a car. What name is given to the type of electricity which gives rise to this problem?

A: Static electricity.

How can this problem be prevented?

A: By earthing. [Hang on to a metal part of the door or firmly grasp the roof of the car as you exit. Keep holding on until you are fully out of the car. Alternatively, grip your keys (or metal pen) and touch the car door with the tip of the key (the spark will still jump but will not be as painful).]

Experiment 44.2

To demonstrate the effect of earthing

Apparatus required: two polythene rods; woollen cloth; metal rod; two non-metal retort stands; two paper stirrups; dry thread

Method

1. Charge two polythene rods by rubbing them with a woollen cloth, *Fig. 44.8(a)*.
2. Suspend each of them in paper stirrups from a retort stand. Bring the two rods close to each other. Because like charges repel, the two rods repel each other, *Fig. 44.8(b)*.
3. Now touch each of the rods in several places with a metal rod. Note what happens when the two rods are brought close again.

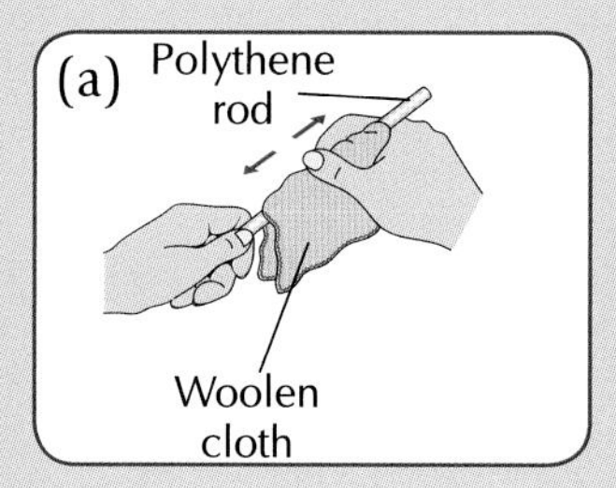

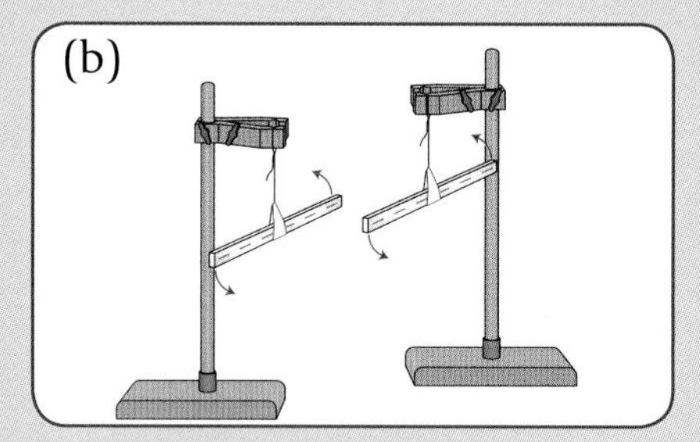

Fig. 44.8

Result

Nothing happens when the rods are brought close now. They neither attract nor repel.

Conclusion

This shows that the two rods lost their charge. They were earthed. The charge that was on each rod moved off through the metal rod and through the person holding it to the earth.

TEST YOURSELF: Now attempt questions *44.12–44.17* and *W44.15–W44.21.*

What I should know

- Electricity was discovered when it was noticed that when amber was rubbed, it attracted light objects like dust and leaves to it.
- Objects like a plastic biro, a polythene rod, a perspex rod, etc., become charged when rubbed with a woollen cloth or other materials. They can then attract small pieces of paper.
- Static electricity involves charges that remain at rest on an object.
- When two objects are rubbed together, electrons are transferred from one to the other.
- An object becomes positively charged if it loses electrons. An object becomes negatively charged if it gains electrons.
- Like charges repel each other; unlike charges attract.
- Substances like plastic and rubber are electrical insulators. This means that they do not allow charges to flow through them. Thus, quite a lot of charge can build up on them.
- A conductor is a substance that allows charge to flow through it freely, e.g. a metal.
- Earthing means connecting an object to the earth by means of a conductor, so that the object shares its charge with the earth. While a body is earthed, a charge cannot build up on it.

QUESTIONS

Write the answers to the following questions into your copybook.

44.1 A comb becomes ….. when brushed a few times in your hair. It can then be used to ….. small pieces of paper.

44.2 Normally, atoms are uncharged or electrically ….. . That means that they have the same number of protons as ….. . Protons have a ….. charge, whereas ….. have a negative charge. The particle in the atom that has no charge at all is called the ….. .

44.3 When a polythene rod is rubbed with a woollen cloth, a ….. of electrons takes place. The polythene rod gains electrons and thus, will have a ….. charge. The wool ….. electrons and therefore, it will have a ….. charge.

44.4 Similar charges ….. each other but opposite charges ….. .

44.5 Explain, in terms of electrons, how a lithium atom becomes positive when it loses an electron.

44.6 Two light charged metal spheres, hanging from a dry nylon thread have come to rest as shown in *Fig. 44.9.* Which of the following statements might be used to explain how this has happened?

(i) Sphere 1 has a negative charge and sphere 2 has an equal positive charge.
(ii) Both spheres have an equal positive charge.
(iii) Sphere 1 has a positive charge and sphere 2 has an equal negative charge.
(iv) Both spheres have an equal negative charge.

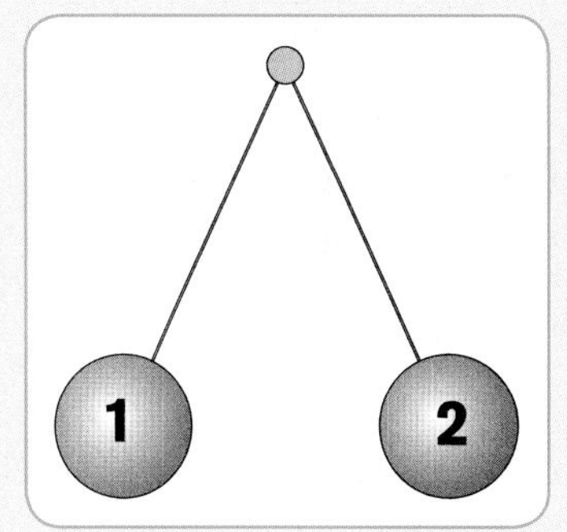

Fig. 44.9 Charged spheres.

44.7 When you try to remove some sticky tape (or clingfilm) from a roll, it often tries to go back on the roll again. Why?

44.8 Air is normally an insulator. But under certain conditions (e.g. 30 000 V across each cm) it will allow electricity to spark across it. Give an example of where air does, in fact, conduct a charge.

44.9 Mention two examples of where static electricity can be a nuisance.

44.10 A light table-tennis ball is at rest on a flat table. Describe how you could use static electricity to move the ball without touching it.

44.11 Thunder clouds pick up charges and eventually earth themselves by sparking to the ….. or to a nearby ….. . The resulting flash of light is called ….. and the resulting crashing sound is called …...

44.12 Earthing means connecting an object to the ….. by means of a ….. . While a body is earthed, a ….. cannot build up on it.

44.13 Conductors are substances which allow ….. to flow through them. Substances which do not normally allow this to happen are called ….. .

44.14 High buildings usually have a ….. to protect them from lightning damage. This consists of ….. spikes which are erected at the ….. of the building. The spikes are then connected by a ….. strip to a ….. plate buried in the ground.

44.15 Sometimes people stepping out of a car after a journey on a hot day feel a slight shock as they touch the ground. Why is this?

44.16 Oil delivery lorries usually have a metal chain or strip hanging from them, which touches the ground. Why is this needed?

44.17 To charge a perspex rod we simply hold it in our hand and rub it with a woollen cloth. Why can we not charge a copper rod in a similar way?

Chapter 45 Current Electricity

45.1 Electric Current

Conductors and Electrons

Electrons are negatively charged. They orbit the nuclei. The **outer** electrons in the atoms of some materials are only loosely attached to the nucleus. These electrons can easily escape from the atoms. Materials containing **free electrons** like these are known as **conductors**.

If a battery is connected to each end of a conductor, the positive terminal will attract these free electrons. At the other end, the negative terminal will replace the missing electrons. The result will be a constant **flow of charges** (electrons) through the conductor. A flow of charge like this is called an **electric current**.

An electric current is a flow of electric charge.

Insulators and Electrons

In some materials, there is a **strong** force of attraction between the outer electrons and the nucleus. These materials have virtually **no free electrons**. These materials are known as **insulators**.

The following experiment describes how to test whether a substance is a conductor or an insulator.

Mandatory experiment 45.1

To test if a substance is a conductor or an insulator

Apparatus required: 6 V battery; 6 V bulb; conducting wires; two crocodile clips

Also required: several items to be tested, e.g. copper wire, nichrome wire, candle, key, various coins, plastic bottle, nail, test tube, the graphite ('lead') of a pencil, comb, ruler, compass, cork, paper clip, biro, tongs, steel wool, any available metal, fruit, potato, paper, cotton wool, wet cotton wool, water (in a beaker), water with some salt added, etc.

Method

1. Make a list of all the materials you are going to test. Divide them into two columns, predicting whether they are **conductors** or **insulators**.
2. Set up the circuit as shown in *Fig. 45.1*. At the beginning there is a gap between the crocodile clips *X* and *Y*. The bulb does not light. The circuit is not complete. Air is not a conductor.
3. Now, bridge the gap with each item you are testing. Either insert each item between the clips *X* and *Y* or just press the item firmly against both clips. Note the result each time. Correct any prediction that you did not get right.

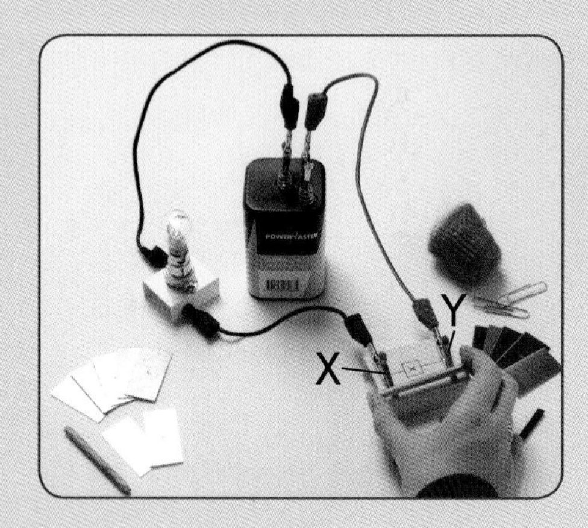

Fig. 45.1 If it is a conductor, the bulb will light.

Result

- The materials that cause the bulb to light brightly, like copper, steel and most metals, are good **conductors**.
- Other materials, like plastic, paper, etc., are **insulators**. Charge cannot travel through these materials. Thus, the light does not come on at all.

- Other materials (e.g. nichrome wire) are **not great conductors**. They offer a certain opposition or friction to the flow of charge. The bulb lights dimly.

45.2 Simple Electric Circuits

In static electricity, charges remain at rest. But, when charges move, as in current electricity, they become really useful. Normally, these charges flow around a path. This path is called an **electric circuit**. The same charges (usually electrons) keep going round and round the same path.

Points to Remember about Electric Circuits

(1) Circuits must be complete

For an electric circuit to work, the circuit must be complete. If there is a gap anywhere, the charges cannot flow. In *Fig. 45.1*, page 348 the bulb at first did not light. There was a gap. There was no conductor between *X* and *Y*.

In *Fig. 45.2*, the **switch** provides a similar gap.

- Current flows when the switch is **closed** (or 'ON').
- The current cannot flow when the switch is **open** (or 'OFF').

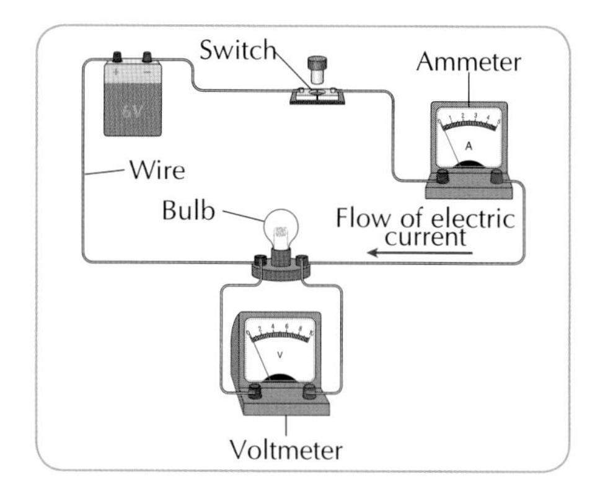

Fig. 45.2 Simple electric circuit.

(2) Potential Difference

A circuit must have a source of **potential difference** (or **p.d.**, for short). This is also referred to as **voltage** or **e.m.f.** (electron-moving-force). This provides the energy to push the electrons around the circuit. Electrons cannot move around a circuit without this energy. The battery in *Fig. 45.2* is the source of this energy.

- A **voltmeter** is used to measure the difference in potential or voltage between any two points in a circuit.

(3) Current

The potential difference or e.m.f. in a circuit causes electrons to flow, i.e. a **current**.

- To measure electric current, an instrument called an **ammeter** is used. This measures the amount of charge passing a particular point in 1 second.

(4) Resistance

Moving electrons may meet a part of a circuit that is difficult to pass through. The thin filament of the bulb in *Fig. 45.2* is made of wire that is not a very good conductor (usually, tungsten). It offers a friction or **resistance** to the electrons. It slows them down. Electrical energy is converted into heat. The bulb gets hot and glows. The bulb in this case is acting as a **resistor**. Resistors are often used in circuits to produce heat or light or to reduce the current flowing in the circuit.

Resistors, whose resistance can be changed, also exist. They are called **variable resistors** or **rheostats**. They are used, for example, in the volume control of a radio or in a dimmer switch. If you increase the resistance, less current will flow. If you decrease the resistance, more current will flow.

- An **ohmmeter** is the instrument used to measure the resistance of a resistor.

Fig. 45.3 shows a sample of resistors, fixed and variable.

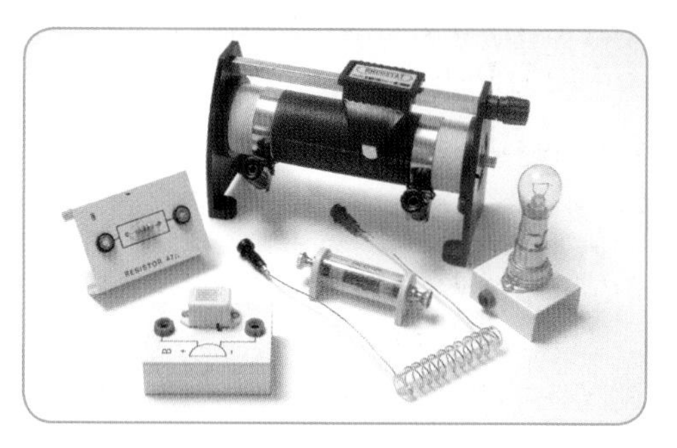

Fig. 45.3 Collection of resistors.

Units Used in Electricity

Different **units** are used in measuring various quantities of electricity.

(1) Volt (symbol: V): This is the unit of **voltage** (or **potential difference**). It is a measure of the 'push' or 'strength' of a battery or power supply to drive a current. You will have seen this unit stamped on batteries, e.g. 1.5 V or 6 V, etc.

Alessandro Volta (1745–1827) was from Como in Italy and is remembered mostly for inventing the battery.

(2) Ampere or **amp** (symbol: A): This is the unit of **current**. It gives a measure of the amount of charge flowing past a point in a wire per second. You should never forget how dangerous electricity can be. A current as small as 0.05 A going across your heart for a few seconds could be fatal.

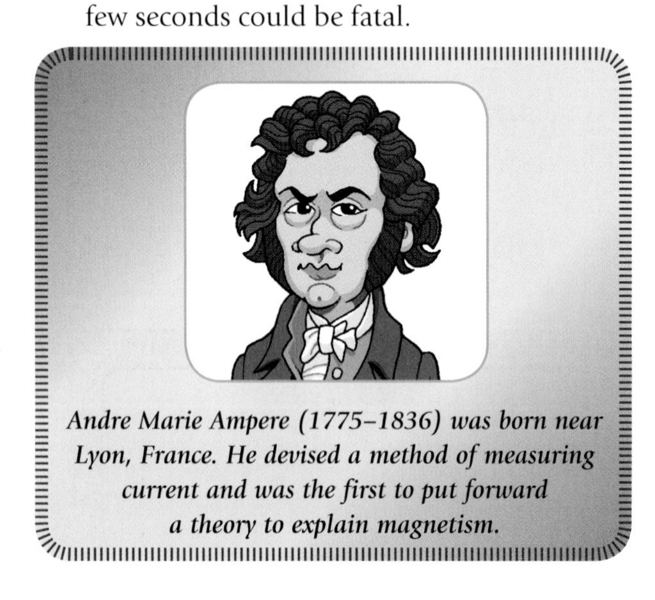

Andre Marie Ampere (1775–1836) was born near Lyon, France. He devised a method of measuring current and was the first to put forward a theory to explain magnetism.

(3) Ohms (symbol: Ω). This is the unit of **resistance**. It measures the opposition of part of a circuit to the flow of charge. The greater the resistance, the smaller the current.

Symbols for Parts in Electric Circuits

Scientists normally use symbols when describing parts of electrical circuits. *Fig. 45.4* shows some of these symbols. They will be used in some of the upcoming circuit diagrams.

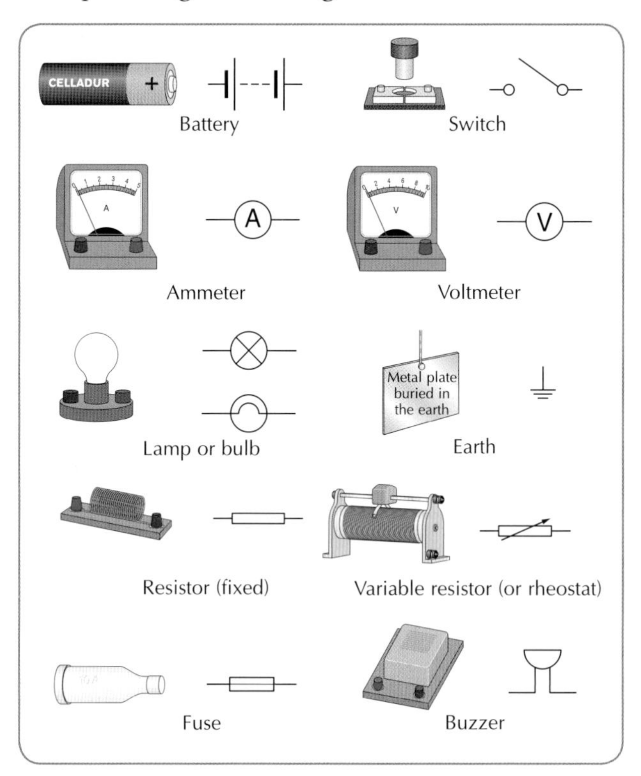

Fig. 45.4 Symbols for parts used in electric circuits.

TEST YOURSELF:
Now attempt questions *45.1–45.4* and *W45.1–W45.10*

45.3 Relationship between Voltage, Current and Resistance

The following investigation will try to find a link between the voltage, current and resistance across a coil of wire. An electric circuit will be used.

1. The **resistance** of the coil will be measured.
2. Then the **voltage** across the coil and the **current** going through the coil will be measured.
3. Finally, we will try to **establish a link** between the resistance, voltage and current.

Note:

- When **setting up a circuit**, a battery is one choice to use as a power supply. It provides a voltage that is safe to use. A low voltage power supply is also safe. It has the added advantage that the supply voltage can be easily varied when necessary.

- When **measuring voltage**, a voltmeter is not connected in the same way in a circuit as an ammeter. Its function is to measure the potential difference between two points. In *Fig. 45.6(a)*, the voltmeter measures the potential difference between *X* and *Y*. A voltmeter must always be connected to **two different** points in a circuit. It is said to be connected **in parallel** in a circuit. An ammeter is always connected **in series**.

Mandatory experiment 45.2

(a) To measure the resistance of a coil (or resistor)

Apparatus required: ohmmeter (digital); coil (resistor); test leads or connecting wires

Note: An **ohmmeter** is an instrument that measures resistance. Usually a digital ohmmeter is just one part of an instrument called a multimeter. Multimeters can also measure current, potential difference and much more.

Method

1. Set the multimeter to act as an ohmmeter by changing the dial to the lowest resistance setting. Switch it on.
2. Connect the test leads to each end of the resistor, as shown in *Fig. 45.5*. Note the reading on the ohmmeter.

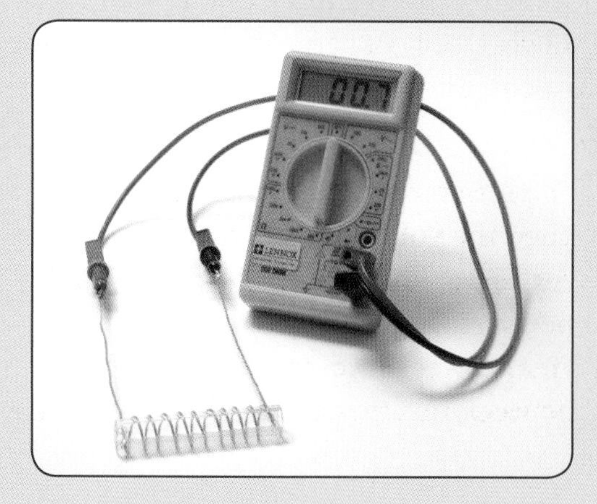

Fig. 45.5 Digital multimeter, being used as an ohmmeter to measure the resistance across a coil.

Result

The reading on the ohmmeter is the resistance of the coil.

(b) To measure potential difference (voltage) and current and establish a relationship between potential difference, current, and resistance

Apparatus required: low voltage power supply; ammeter; voltmeter; switch; coil (or resistor) in a beaker of water (same coil as used above); connecting wires

Method

1. Use connecting wires to connect up the circuit as shown in the diagram, *Fig. 45.6(a) or (b)*. (The positive terminals of the ammeter and voltmeter are connected to the positive terminal of the power supply.)

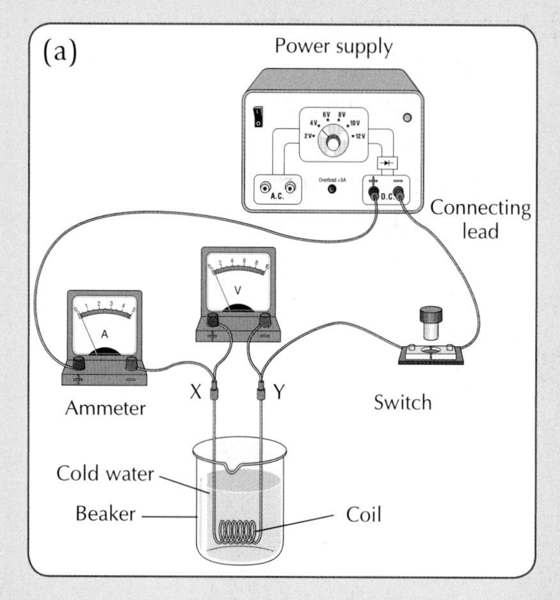

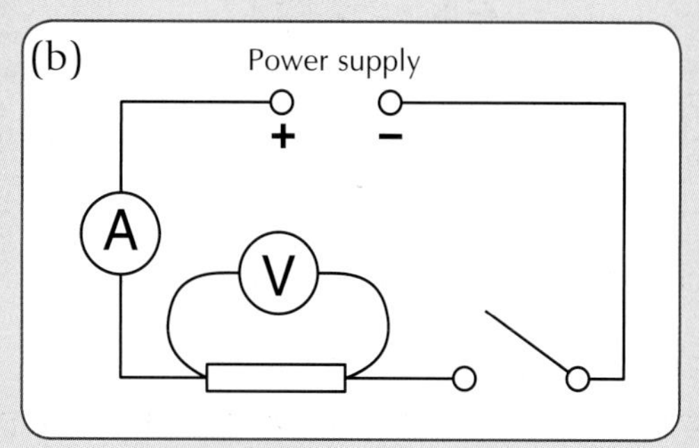

Fig. 45.6 Simple circuit to measure current and potential difference.

PHYSICS

2. Switch on the power supply at its lowest setting. Close the switch. Current will flow.
3. Note the reading on the voltmeter.
4. Note the reading on the ammeter.
5. Change to the next setting on the power supply. Take readings of voltage and current again. Repeat the process for several other settings. Record your results on a table as shown in *Table 45.1*.

VOLTAGE (V)	2	4	6	8	10	12
CURRENT (A)	...	...	...	...	...	...

Table 45.1. Current varies with applied voltage.

6. Plot a graph of voltage (vertical axis) against current (horizontal axis). What do you notice about the resulting graph? Does it give a straight line or a curve?
7. Divide each voltage in *Table 45.1* by its corresponding current. What do you notice about all of these results?
8. Now find the average of these results. Compare this average result with the resistance of the same coil, which you measured it in *part (a)*. Are the two results similar?

Further option: Find the slope of the line drawn in *step 6* above. (The slope of a line is the difference of the *y*'s over the difference of the *x*'s.) Do you get the same result again?

Results and Conclusions

- The **voltage across the resistor** (or the potential difference between *X* and *Y*) is given by the voltmeter reading.
- The reading on the **ammeter** is the current in the **circuit**.
- A straight-line graph going through the origin shows that current is **proportional** to voltage. This means that the current increases directly with the voltage. Thus, if the voltage doubles, the current also doubles.
- The **resistance of a resistor** can be found by dividing the potential difference across it by the current passing through it.
- The **resistance of the coil** can also be found by finding the slope of the graph of potential difference plotted against current.

Note: If the positions of the ammeter and switch are swapped above, the ammeter gives the same reading. This shows that the current is the same at all points in the circuit.

Calculations Based on the Relationship between Potential Difference, Current and Resistance

Georg Ohm, a German physicist, was the first to show, in 1827, that for a given conductor, at constant temperature, **the voltage and the current are proportional to each other**. This became known as Ohm's law. The unit of resistance, the **ohm** (Ω), was named in honour of Georg Ohm.

At constant temperature,

$$\frac{\textbf{voltage (V)}}{\textbf{current (I)}} = \textbf{resistance (R)}$$

This equation is often written more simply in symbols as:

V = I × R where V = voltage or p.d. in volts (V)

I = current in amps (A)

R = resistance in ohms

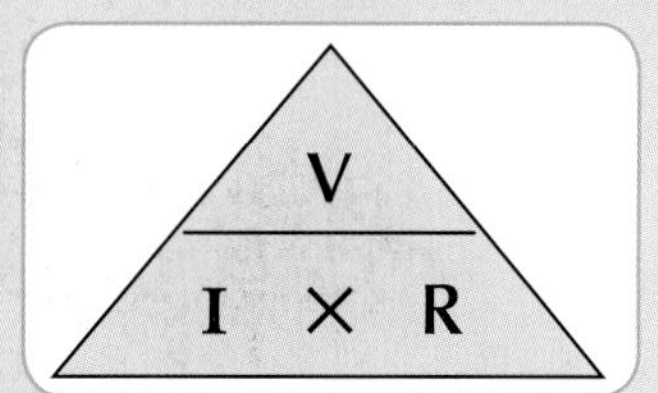

Fig. 45.7 Triangle to help you remember three formulas. Cover the quantity you are looking for and you will be left with its formula.

Example 1

A battery provides a potential difference of 9 volts across a metallic conductor of resistance 5 ohms. Calculate the current flowing.

Use the triangle in *Fig. 45.7*. We are looking for the current, *I*. We cover *I*, in the triangle. We are left with the formula for *I*,

$$I = V \div R = 9 \div 5 = 1.8\text{ A}$$

Example 2

What voltage is needed to drive a current of 1.5 A through a resistor of resistance 8 Ω?

Using the triangle in *Fig. 45.7*, cover the quantity you are looking for, *V*,

$$V = I \times R = 1.5 \times 8 = 12\text{ V}$$

Example 3

A bulb in a circuit is connected to an ammeter that reads 0.8 A. A voltmeter across the bulb reads 5 V. What is the resistance of the bulb?

Using the triangle in *Fig. 45.7*, cover the quantity you are looking for, *R*,

$$R = V \div I = 5 \div 0.8 = 6.25\ \Omega$$

TEST YOURSELF:
Now attempt questions *45.5–45.8* and *W45.11–W45.17.*

45.4 Series Circuits and Parallel Circuits

Two bulbs in a circuit can be connected in two different ways.

(1) They can be connected **in series**. Here, they are connected one after the other. There is only one path through which the current can flow, *Fig. 45.8(b)*, page 354.

(2) They can be connected **in parallel**. Here, they are connected side by side. There is a branch in the circuit and the current will have a choice of paths, *Fig. 45.8(c)*, page 354.

Experiment 45.3

To demonstrate simple series and parallel circuits, containing a switch and two bulbs

Apparatus required: a battery or low voltage power supply; two similar 6 V bulbs; switch; connecting wires

Method

1. Set up the circuit as shown in *Fig. 45.8(a)*. Close the switch and observe the brightness of the bulb.
2. Then set up the circuit shown in *Fig. 45.8(b)*. Here, an extra similar bulb is added in series with the first bulb. Close the switch. Again observe the brightness of the bulbs. What do you notice?
3. Now remove one of these two bulbs from its holder. What happens?
4. Then set up the circuit shown in *Fig. 45.8(c)*. This time, the second similar bulb is added in parallel with the first bulb. Close the switch. Again observe the brightness of the bulbs. What do you notice this time? How does the brightness compare with the brightness of the bulbs in the previous two circuits?
5. Now remove one of these two bulbs from its holder. What happens?

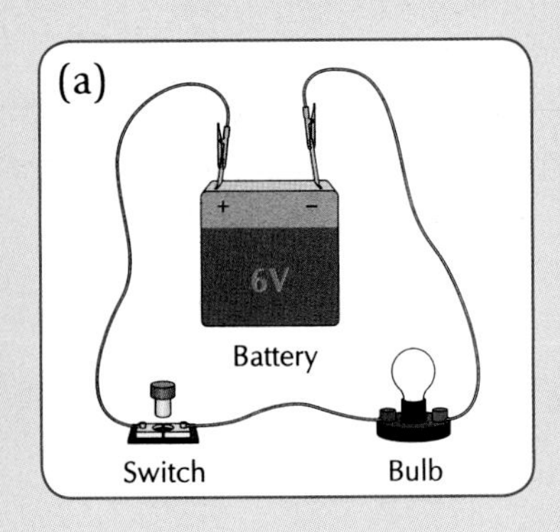

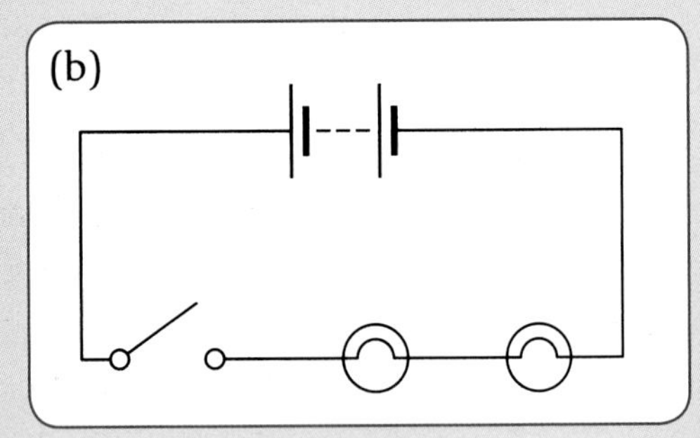

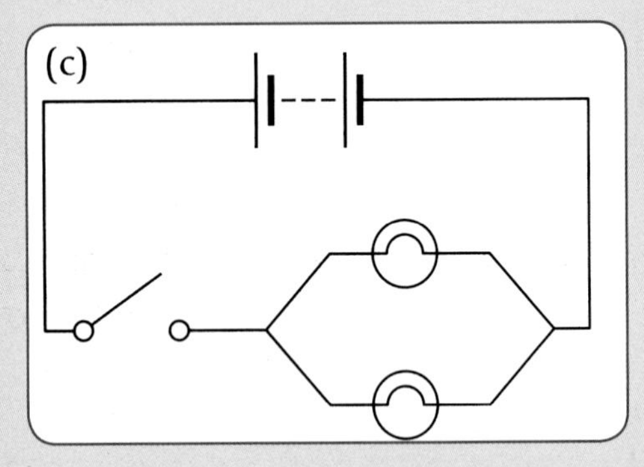

Fig. 45.8(a) Simple circuit with one bulb and switch. (b) Circuit with switch and two similar bulbs in series. (c) Circuit with switch and two similar bulbs in parallel.

Results

- Two bulbs **in series** shine **less brightly** than if one of these bulbs was on its own. In effect, the resistance doubles and thus, the current is halved. If one bulb is removed or if one bulb blows, the circuit is no longer complete. No current will flow.
- Two bulbs **in parallel** shine **as brightly** as one of these bulbs on its own. If one bulb is removed or if one bulb blows, the circuit is still complete. The current can travel by the alternative route.

Note: Old sets of Christmas tree lights are sometimes arranged **in series**. If one bulb blows, all the lights go off. Most lights are arranged **in parallel**. A huge advantage of this is that they can be individually switched on and off without affecting other lights in the circuit.

45.5 Effects of an Electric Current

We cannot **see** electric charges moving. Yet, we can still work with the 'invisible'. We know charges move because of the effects they cause. We will now look at three major effects of electricity. These are when electrical energy is converted to (1) heat energy, (2) magnetic energy and (3) chemical energy.

(1) Heating Effect

Electric current passes easily through a copper wire. Copper is a good conductor of electricity. However, current has great difficulty passing through a wire made of say, nichrome. Nichrome has a high resistance. More energy is needed to force it through. As a result, heat is given off. This is why nichrome wire is used in the element of an electric fire.

Experiment 45.4.

To demonstrate the heating effect of an electric current

Apparatus required: polystyrene cup (or beaker or calorimeter); thermometer; coil of wire (e.g. nichrome or constantan) or small 'element' (as in Joule's calorimeter); low voltage power supply (or battery)

Also required: cold water

Method

1. Set up the apparatus as shown in *Fig. 45.9*. Connect the ends of the nichrome wire to a low voltage power supply or battery.
2. Turn on the power supply. Note the temperature of the water every two minutes.

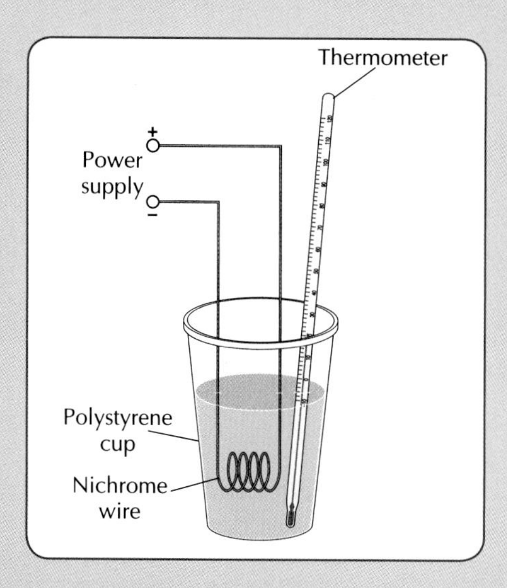

Fig. 45.9 Demonstrating the heating effect of an electric current.

Result and Conclusion

The temperature of the water rises steadily as the current passes through the coil. Therefore, an electric current has a heating effect.

Note: A temperature sensor and data-logging equipment could replace the thermometer in this experiment. The shape of the graph drawn on the screen would then show the increase in temperature.

Everyday Applications of the Heating effect of an Electric Current

- The arrangement above in *Fig. 45.9* with the coil in water is very similar to how an **electric kettle** or an **immersion heater** works.
- Electric current produces heat also in **electric fires, electric blankets, ovens, irons, toasters, hair dryers, bulbs**, etc.
- Another important application of the heating effect of an electric current is a **fuse**. A fuse is simply a thin piece of wire, enclosed usually in a porcelain case, *Fig. 45.10*. It is a **weak link**, which is deliberately put into a circuit to make it safer. If a fault occurs and a current above the expected flows, the fuse wire will get hot and melt safely in its case. This will switch off the current before the conducting wires overheat or a fire occurs.

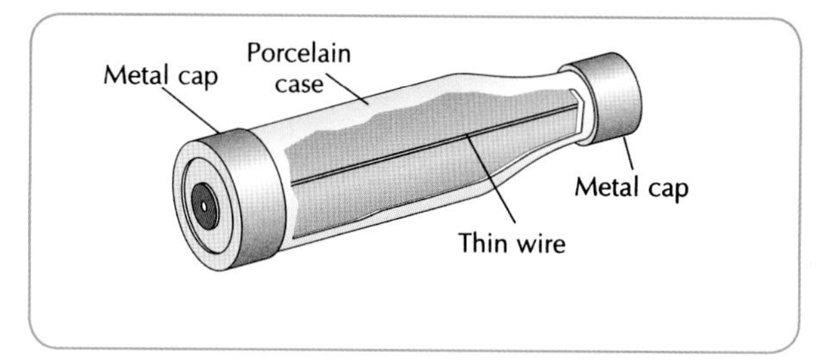

Fig. 45.10 Drawing of inside of a fuse.

(2) Magnetic Effect

We have already seen the magnetic effects of an electric current in *Chapter 43*.

- In ***Fig. 43.10***, page 340, we saw how a compass needle moved, when placed near a wire that had a current flowing through it.
- Also, in ***Fig. 43.11***, page 340, we saw how an electromagnet works. When a current flows in a wire that is wound around a nail, the nail becomes a strong magnet.

Everyday Applications of the Magnetic Effect of an Electric Current

- Electromagnets are used in **electric motors, electric bells, transformers, loudspeakers, ammeters, tape recorders, computers, scrap yards**, etc.
- Electromagnets are used in **circuit breakers**, which we will look at again in the next chapter, *Fig. 46.1*, page 358.

(3) Chemical Effect

The chemical effect of an electric current can also be put to good use. It can be used to separate water into hydrogen and oxygen as shown in *Fig. 27.14*, page 191. This is called the **electrolysis of water**.

Everyday Applications of the Chemical Effect of an Electric Current

- **Electrolysis** can also be used to extract metals from solutions. In industry it is used in refining or **purifying metals**.
- It is used in beauty treatments to **remove** unwanted **hair** from the body.
- **Electroplating** is used in making cutlery, kettles, car bumpers and many other items. A cheap metal is covered with a layer of

another metal to protect it from corrosion and to make it look better.

45.6 Direct and Alternating Current

Direct Current

A **battery** has a positive terminal and a negative terminal. It always pushes current in **one direction only** around a circuit. Current that travels in one direction only like this is referred to as **direct current** or **d.c.** for short.

Alternating Current

However, the current from the mains supply does not flow in the same direction all of the time. It changes direction, in fact, 100 times every second. So, for one hundredth of a second, the current is being pushed around in one direction. Then, in the next one hundredth of a second, it is being pushed around in the opposite direction. This type of current is called **alternating current** or **a.c.** for short.

AC to DC

For most appliances it does not matter whether the current is alternating or direct. A bulb, for instance, will work if the current is flowing through it in one direction only or if it is constantly changing direction. However, some appliances, like a television for example, need a d.c. supply. Therefore, a device called a **rectifier** is needed. A rectifier converts a.c. to d.c. You will learn more about a rectifier in the next chapter.

TEST YOURSELF:

Now attempt questions *45.9–45.12* and *W45.18–W45.27.*

What I should know

- An electric current is a flow of electric charge.
- Electrical conductors allow current to flow through them freely. Insulators do not normally allow current to flow through them.
- Electricity is a very convenient form of energy. It can easily be changed into many other forms of energy.
- Circuits must be complete. A potential difference or voltage or source of energy must also exist across the circuit to move the electrons.
- A switch is a device which either completes or breaks a circuit, allowing current to flow or not.
- Voltage is a measure of the 'push' or 'force' that a power supply has to drive a current. Voltage is measured with a voltmeter. It is measured in volts (V).
- Current is measured in amperes (A). Current is measured with an ammeter.
- The resistance of a material is a measure of its opposition to the flow of current through it. It is measured in ohms (Ω). Resistance is measured with an ohmmeter.
- A rheostat is a variable resistor, whose resistance can be changed.
- A variable resistor can control the current flowing in a circuit. When the resistance is high, the current is low and when the resistance is low, the current is high.
- Ohm's law states that the current flowing through a metallic conductor is proportional to the voltage across it, provided the temperature remains constant.
- Resistance (at constant temperature) = Voltage ÷ Current **or** $R = V / I$.
- With two bulbs in series, the same current flows through each bulb. If one bulb blows or is switched off, the other will also go off.
- With two bulbs in parallel, if one bulb blows or is switched off, the other bulb will remain lighting as there is still a complete path for the current.

What I should know

- Three major effects of an electric current are (i) the heating effect, (ii) the magnetic effect, and (iii) the chemical effect.
- Everyday electric heating appliances are based on the heating effect of an electric current.
- A fuse is a safety device. It is an enclosed thin length of wire. This wire melts when a current greater than its stated value passes through. Thus, it switches off the current in a circuit.
- The magnetic effect of a current is used in electric motors, electric bells, loudspeakers, ammeters, circuit breakers, etc.
- The chemical effect of a current is seen in the electrolysis of water and in electroplating.
- Current that travels in one direction only is called direct current (or d.c.). Current which is constantly changing direction is called alternating current (a.c.).

QUESTIONS

Write the answers to the following questions into your copybook.

45.1 An electric current is a flow of ..... .

45.2 It is a cold winter's evening as you arrive home from school. There is a power failure and electricity cannot be restored until the following morning. Mention three ways in which the absence of electricity will affect the rest of your evening and night.

45.3 The amount of current flowing in a wire is measured in a unit called the ..... . Current can be measured with an ..... .

45.4 In order that a simple electrical circuit conducts electricity, the circuit must be ..... and must have a source of ..... . A switch is a device which either ..... or ..... a circuit, either allowing current to flow or not.

45.5 Ohm's law states that the ..... flowing through a metallic conductor is proportional to the ..... across it, provided the ..... remains ..... .

45.6 Draw a circuit showing a battery, a switch, a lamp and a variable resistor. Include also instruments to measure the potential difference across the lamp and the current passing through it.

45.7 What potential difference is needed to drive a current of 0.2 A across a resistor of 1150 Ω?

45.8 Calculate the current flowing through a metallic conductor of resistance 50 Ω when a potential difference of 12 V exists across its ends.

45.9 If two similar bulbs are connected in series in a circuit, why is the brightness of each bulb less than the brightness of one bulb on its own?

45.10 The fact that water can be split up into hydrogen and ..... is an example of the ..... effect of an electric current. Another example of this effect can be seen in ..... .

45.11 A circuit breaker works because of the ..... effect of an electric current.

45.12 Current that travels in one direction only is called ..... current . Current which is constantly changing direction is called ..... current.

Chapter 46 Electricity in the Home and Electronics

46.1 Mains Supply and Safety

Cables and danger

The mains supply to the sockets in your house or school is at **230 V a.c.** This voltage could push a big enough current through your body to KILL you. This is why electrical appliances, sockets, wires, etc., must always be treated with the utmost care and respect.

There are two cables from the mains supply to your home or school. One is called the **live**. The other is the **neutral**. The live cable is at +230 V (or –230 V) and is very dangerous. The neutral remains at a voltage of zero.

The Set-up

1. The main fuse is on the live wire as it comes into your house.
2. Then, both wires pass through the **electricity meter**. This records the number of units used.
3. Both wires then pass to a distribution box or fuse box. Modern distribution boxes contain several **circuit breakers** rather than fuses. Each circuit breaker (or fuse) controls a different circuit in the house, *Fig. 46.1*. One might be for sockets in the kitchen area; another might be for bedroom lights, etc.

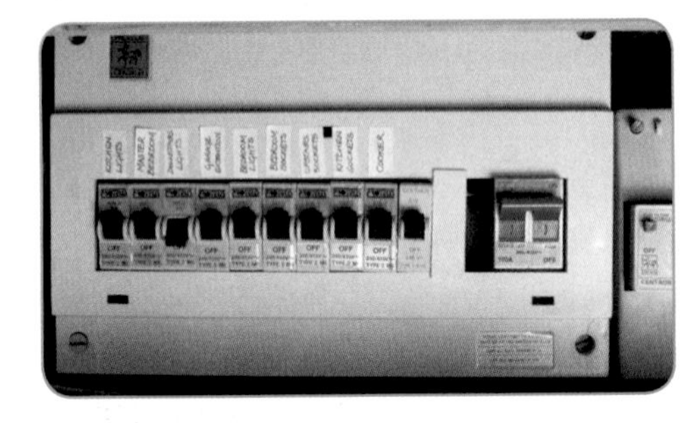

Fig. 46.1 Circuit breakers for domestic circuits in a distribution box.

Fuses and Circuit Breakers

We saw in *Chapter 45* that fuses and circuit breakers are **safety** components in electric circuits. Both cut off the current in a circuit if the current for some reason becomes too big. Fuses work using the heating effect of an electric current. Circuit breakers use the magnetic effect of an electric current.

Fuses

Fuses of various current ratings are available, e.g. 1 A, 2 A, 3A, 5A, 10 A, 13 A, etc., *Fig. 46.2.* A 5 A fuse, for example, will melt if a current of more than 5 A flows through it. When a fuse melts like this, it cannot be used again. It has to be replaced with another fuse of the same rating. For safety, both switches and fuses are always placed on the **live** wire.

Circuit Breakers

A circuit breaker has advantages over a fuse. If the current goes above a certain value, the circuit breaker simply '**trips**' off. The circuit is broken and the current stops immediately. The switch needs only to be closed again when the fault in the circuit has been corrected. The circuit breaker does not have to be replaced. A circuit breaker also acts more quickly than a fuse.

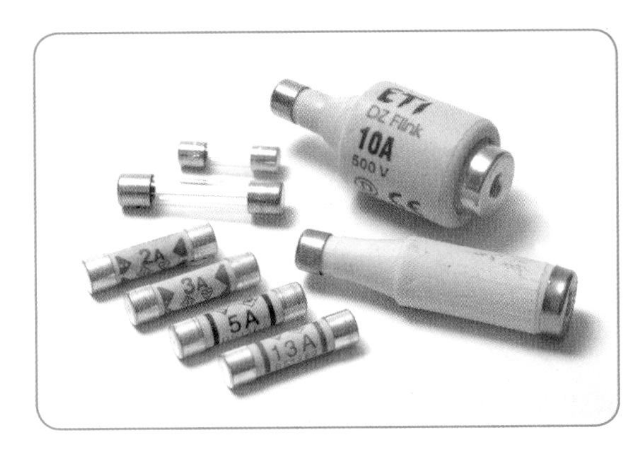

Fig. 46.2 Confused about which fuse to use? Choose the one that is just above the 'normal' current expected in the circuit.

Plugs

Most appliances are connected to an electricity supply by pushing a **plug** into a wall **socket**.

When you plug in an appliance (like a toaster), the appliance provides a path connecting the 'live' to the 'neutral'. Once an unbroken path like this exists, a current will flow. This is about the safest and simplest way of using a current from the mains supply. Most sockets and plugs have three terminals, the 'live', the 'neutral' and the 'earth', *Fig. 46.3*. As an extra protection, the 'live' and 'neutral' holes in the socket are covered with blinds. These blinds are opened only when the larger 'earth' pin is inserted.

Fig. 46.3 The 'live', 'neutral' and 'earth' terminals of the socket are matched by the 'live', 'neutral' and 'earth' pins on the plug.

The Earth Terminal

The 'earth' terminal is connected to a metal plate buried in the earth. It is there purely as a **safety** device to protect the user of an appliance. In *Fig. 46.4*, the insulation on the 'live' wire has worn away. The 'live' now touches the metal of the kettle. The kettle becomes 'live'. But, because the metal of the kettle is earthed, the current flows safely to earth through the 'earth' wire. If the kettle had not been earthed, the current would go to earth through the user who touches it.

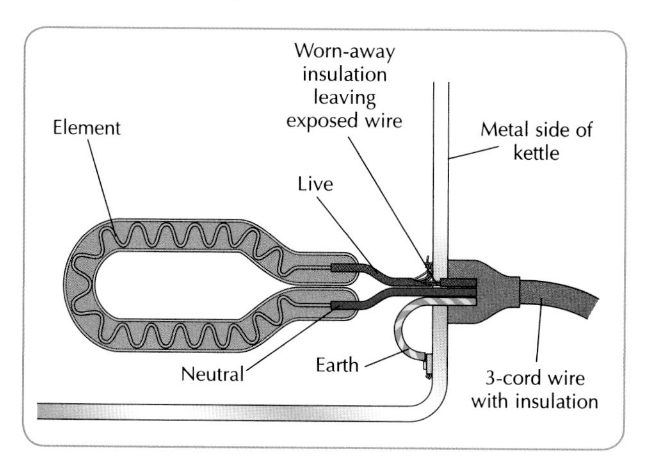

Fig. 46.4 Because the kettle is earthed, this problem will not kill anyone.

Experiment 46.1

To wire a plug correctly

Apparatus required: a three-pin plug; a three-cord flex; wire strippers (or sharp knife); a small screwdriver

Method

1. You are given a three-cord flex, which contains three wires, *Fig. 46.5(a)*.

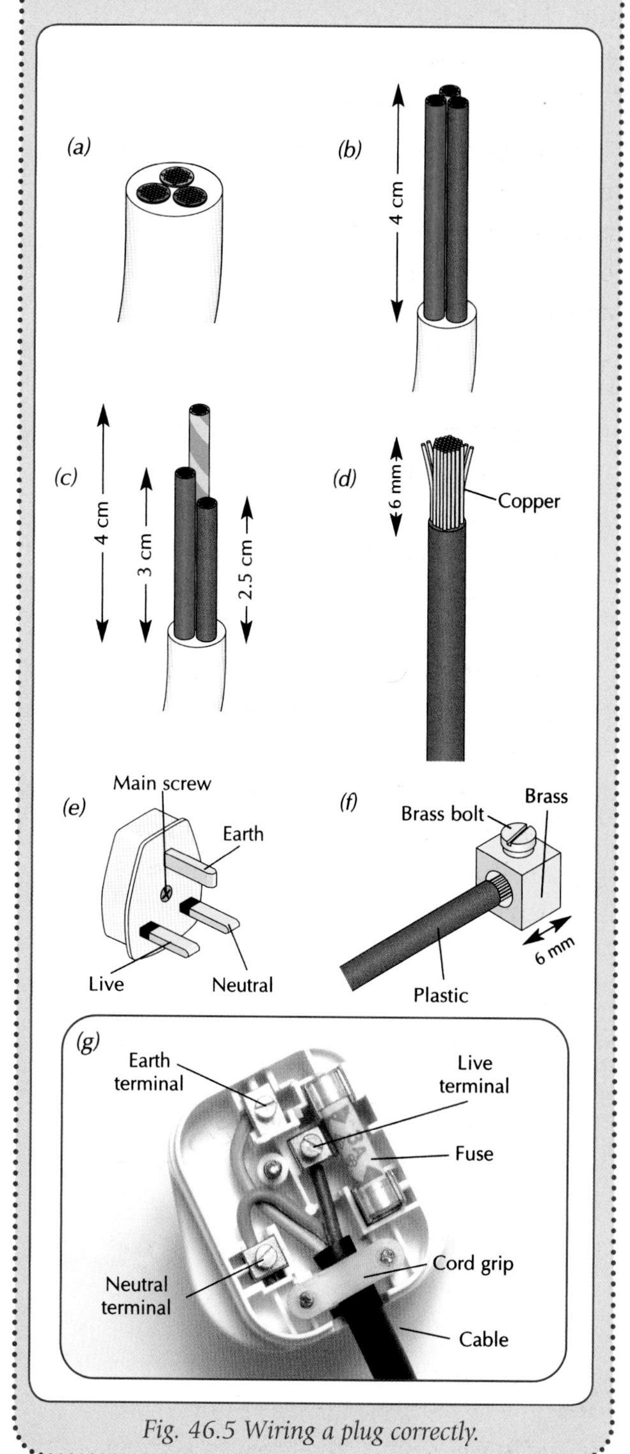

Fig. 46.5 Wiring a plug correctly.

2. Using wire strippers or a sharp knife, remove about 4 to 5 cm of the **outer** insulation on the three-cord flex, *Fig. 46.5(b)*. (Take care not to damage the insulation on the three wires by cutting too deeply.)
3. Then cut the blue wire and the brown wire roughly to the dimensions shown in *Fig. 46.5(c)*.
4. Remove approximately 6 mm of plastic insulation from the ends of each of the three wires, *Fig. 46.5(d)*.
5. Remove the cover from the plug by loosening the main screw, *Fig. 46.5(e)*.
6. Secure each of the three wires to its correct terminal by tightly fastening the brass bolt on the exposed copper wire, *Fig. 46.5(f)*. To know which coloured wire to fasten to which terminal, follow the colour guidelines shown in *Table 46.1*.

WIRE COLOUR	TERMINAL
BROWN	LIVE (L)
BLUE	NEUTRAL (N)
GREEN/YELLOW	EARTH (E)

Table 46.1

6. Tighten the cord grip, making sure it is securing the fully insulated part of the three-cord flex, *Fig. 46.5(g)*.
7. Have your work checked by your teacher before screwing back on the cover of the plug.

TEST YOURSELF:

Now attempt questions *46.1–46.4* and *W46.1–W46.4*.

46.2 The Cost of Electricity

Unit of Power

You may remember in *Chapter 36*, page 273, that power was defined as the **work done per unit time**. It can also be defined as the **energy used per unit time**. The unit of power is the **watt**. Appliances like an iron and a cooker both convert electrical energy to heat energy. However, they vary in the amount of energy that they use each second.

- A **2 kW cooker**, for example, converts 2000 J of electrical energy to heat energy each second.
- A **1 kW iron** converts 1000 J per second.

The cooker has a **higher power** rating than the iron. It uses twice the electrical energy each second. This will also make it twice as expensive to run.

Unit of Electricity

The ESB sell electricity in amounts, which they call '**units**'. The normal unit of energy is the joule but this is far too small for costing purposes. So, instead, the ESB use a unit called the **kilowatt-hour** (symbol: **kW h**).

- A **1000 W** iron running for **1 hour** uses 1 kW h or 1 'unit' of energy.
- A **2 kW** electric heater running for **3 hours** would use 6 units of electricity.

(1 kilowatt-hour is equal to 3.6 million joules.)

A kilowatt-hour is the electrical energy converted by a 1 kW appliance running for 1 hour.

One 'unit' of electricity, or 1 kW h, is the energy needed for each of the following:

(i) An 8 kW power shower for 7.5 minutes.
(ii) A 2 kW electric fire for a half-hour.
(iii) A 125 W (or 0.125 kW) television for 8 hours.
(iv) A 100 W (or 0.1 kW) bulb for 10 hours.
(v) A 25 W (or 0.025 kW) iPod for 40 hours.

It is easy to calculate the number of 'units' an appliance uses. The **power rating** of the appliance will be stamped on it. If this number is in watts (W), convert to kilowatts (kW) by dividing by 1000. Now, simply multiply the number of kilowatts by the number of hours that it is running.

number of kilowatt-hours (or 'units') used = number of kilowatts × number of hours

To work out the **cost** of the electricity, the number of 'units' used is multiplied by the price for each 'unit'.

cost = (number of kilowatt-hours used) × (price per kilowatt-hour)

OR

cost = (number of 'units' used) × (price per 'unit)

Exam Question

Junior Cert Science Ordinary Level Examination Question

The ESB charges for electricity at a rate of 12 cent per kW h. A tumble drier of power rating 2.5 kW is used for 2 hours each week for 4 weeks.

How many units of electricity are used?

A: (number of units used) = (number of kilowatts) × (number of hours)
= 2.5 × 2 × 4
= 20 kW h (or 20 'units')

What is the cost, in cent, of using the tumble drier?

A: cost = number of kW h used × price per 'kW h'
= 20 × 12
= 240 cent

Example 2

A 6 kW cooker is 'on' for half an hour. How many 'units' does it use? If each 'unit' costs 16.3 cent, how much does this electricity cost?

number of kilowatt-hours (or 'units') used
= (number of kilowatts) × (number of hours)
= 6 × ½
= 3 kW h (or 'units')

cost = (number of kW h used) × (price per 'unit')
= 3 × 16.3
= 48.9 cent

Example 3

If each 'unit' of electricity costs 18 cent, calculate the total cost of using a 1500 W electric heater for 2 hours, a 150 W freezer for 4 hours, a 1000 W hoover for 12 minutes and a 1200 W iron for 10 minutes.

Note: Make sure all units of power are in kilowatts and all units of time are in hours.

number of units used
= (number of kilowatts) × (number of hours)
= (1.5 × 2) + (0.150 × 4) + (1 × 1/5) + (1.2 × 1/6)
= 3.0 + 0.6 + 0.2 + 0.2
= 4.0 kW h (or 4 'units')

cost = (number of 'units' or kW h used) × (price per unit)
= 4.0 × 18
= 72 cent

The **electricity meter** in your home records the total number of units used. To determine the number of **units** used in any interval, simply subtract the reading taken at the start of that interval from the reading taken at the end.

Example 4

The final reading on the electricity meter two months ago was 21 185. Now, the reading is 21 935. The price per unit is 17.12 cent. The standing charge is €16.89 and VAT @ 13.5% is then added. What is the total bill for electricity over these two months?

number of units used
= ('present' reading) – ('previous' reading)
= 21 935 – 21 185
= 750

cost of electricity used	=	**750 × 17.12**
	=	**€128.40**
standing charge	=	**€16.89**
total (before VAT)	=	**€145.29**
VAT @ 13.5%	=	**€19.61**
total (including VAT)	=	**€164.90**

TEST YOURSELF:
Now attempt questions *46.5–46.10* and *W46.5–W46.12.*

PHYSICS

46.3 Electronics

Electronics has been one of the most rapidly advancing branches of science in recent history. Today we are surrounded by all sorts of electronic devices. Examples include radios, iPods, compact disc players, computers, calculators, mobile phones, automatic cameras, etc. We also have all sorts of detectors, which can indicate the presence of heat, moisture, light, sound, fire, smoke, burglars, etc. All of these devices are operated by **small carefully controlled electric currents**, the basis of electronics.

Components of a Microchip

The key to most of the developments in electronics was the 'microchip'. A tiny 'chip', made usually of silicon or germanium, can contain many complex circuits. Each of these circuits may have several electrical components or parts with various functions. We will now look at some of these components, particularly the diode, the LED and the LDR and we will see what each can do, *Fig. 46.6*.

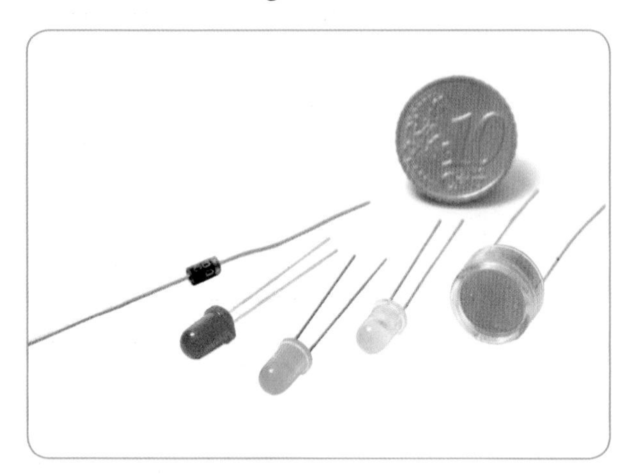

Fig. 46.6 A diode, 3 different coloured LEDs and an LDR. The 10 cent coin shows the relative size of these electronic devices.

(1) The Diode

A diode is a device which allows current to pass through it in one direction only.

You can see in *Fig. 46.6*, that the diode has a silver band at one end. This end is called the **cathode** (–). The opposite end is called the **anode** (+). The diode gets its name from the fact that it has two **(di) electrodes**, the cathode and the anode.

The symbol for the diode is shown in *Fig. 46.7(a)*. The triangle (▷) represents the anode (+). The straight vertical line (|) represents the cathode (–).

Forward Biased

If the cathode (–) is connected to the negative terminal of the battery and the anode (+) is connected to the positive terminal of the battery, the diode **will conduct** a current. When connected like this, the diode is said to be **forward biased**. *Fig. 46.7(b)* shows a diode that is forward biased in a circuit with a battery, switch and bulb. When the switch is closed, the bulb lights.

Reverse Biased

In *Fig. 46.7(c)*, the diode is connected the other way round. The cathode (–) is connected to the positive of the battery. The anode (+) is connected to the negative. The diode is now said to be **reverse biased**. When connected like this, the diode does **not conduct**. When the switch is closed the bulb does NOT light.

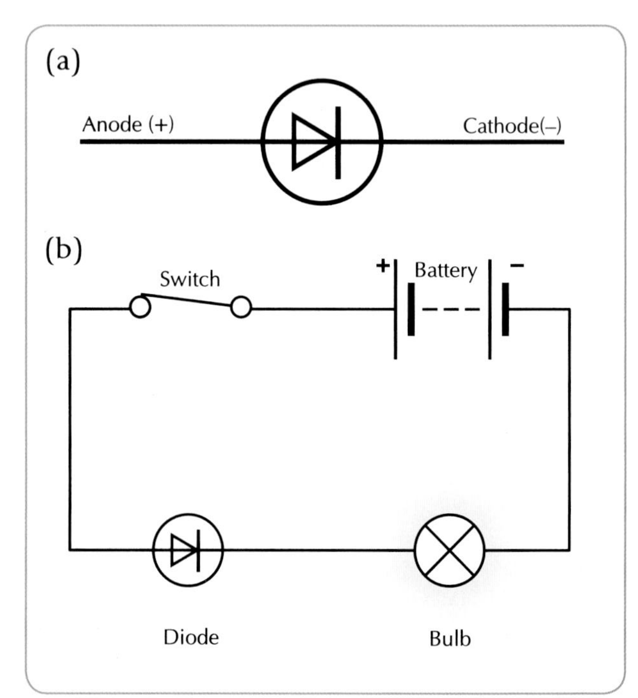

Fig. 46.7(a) Symbol for diode. (b) Forward biased diode conducts.

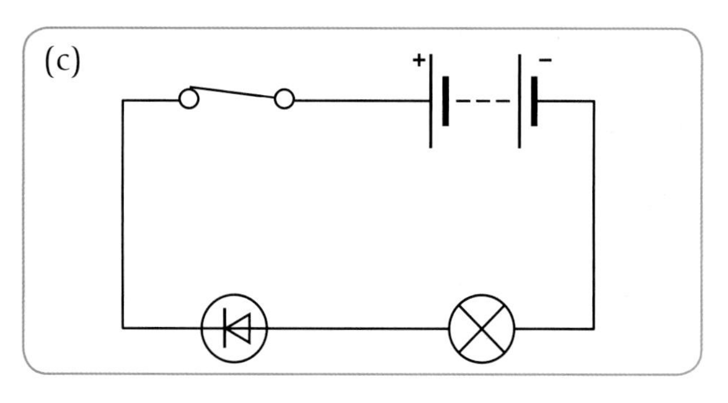

Fig. 46.7(c) Reverse biased diode does NOT conduct.

Uses of a Diode

- The main function of a diode is as a **rectifier**. This means it can be used to change **alternating current** (or **a.c.**) into **direct current** (or **d.c.**). Certain appliances need a current, which flows always in the same direction (i.e. direct current). Therefore, they need a diode to change the alternating current from the mains to direct current.
- Diodes are also used to **protect** appliances that use direct current. Computers and radios, etc., can be damaged if connected the wrong way round to a battery or power supply. But, with a diode in the circuit, no current would flow if connected wrongly and no damage would be done.

(2) The Light Emitting Diode (LED)

An LED is a special diode, which gives out **light** when current passes through it. LEDs come mainly in three colours, red, yellow and green, *Fig. 46.6*. A typical LED is shown in *Fig. 46.8(a)*.

- A **lens** at the top of the LED helps to focus the emitted light.
- The shorter **cathode** lead is near the **flat edge** at the base of the LED.

Fig. 46.8(b) shows the symbol for a light emitting diode. The arrows coming from it indicate the light being emitted.

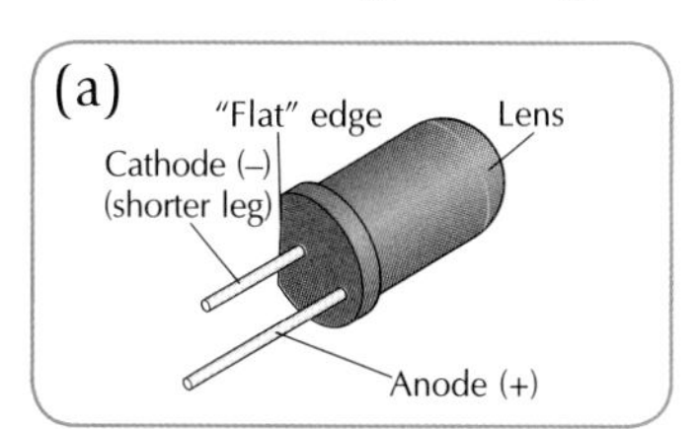

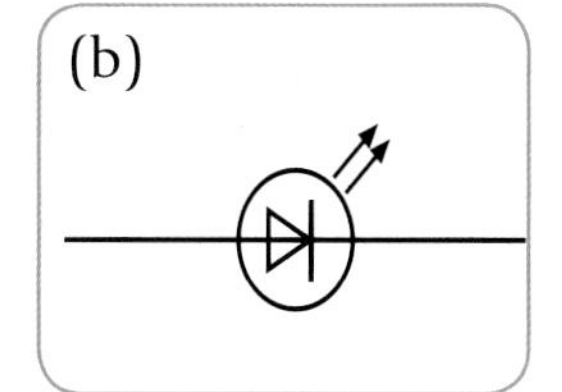

Fig. 46.8(a) A light emitting diode (LED) (b) Symbol for LED.

Two important points about LEDs:

(1) A LED is a **diode**. Like all diodes, current will only flow through it when it is forward biased.

(2) A LED requires only a **very small current**, especially when compared to an ordinary bulb. Too much current will damage it. To protect a LED, a **large resistor** should always be in series with it.

Experiment 46.2

To use two LEDs to test the polarity of a power supply or battery

Apparatus required: 6 V d.c. power supply or battery; two 330 Ω resistors; two LEDs (one red, one green); switch

Method

1. Connect up the circuit as shown in *Fig. 46.9*. The power supply or battery to be tested is placed between *A* and *B*. The red and green LEDs are connected in **opposite** directions. This means that if one of them is forward biased, the other will be reverse biased. They cannot both come 'on' together.
2. Close the switch and note which LED lights up.

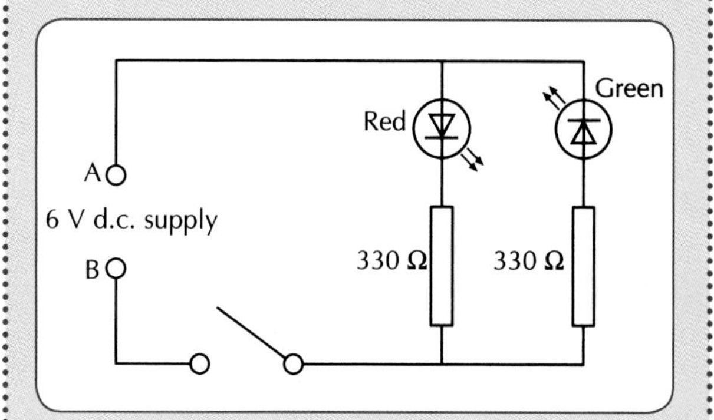

Fig. 46.9 Testing which side of the power supply is positive and which is negative using two LEDs.

Conclusion

If, when you close the switch, the red LED comes 'on', then the red LED is forward biased. Thus terminal *A* must be positive and terminal *B* negative. If the green LED comes 'on', this means that that the green LED is forward biased. Thus terminal *B* must be positive and terminal *A* negative.

Uses of LEDs

- LEDs make very useful **indicator lamps** on all sorts of electrical equipment. They show whether an appliance is switched 'on' or not. In *Fig. 46.10*, the red dot of light at the bottom right of the numbers indicates that the alarm has been switched on.
- Many clocks, calculators, cash registers, etc., use LEDs as **digital displays**. Each digit can be formed by a seven segment display. Each segment contains a LED in the shape of a bar. The number '8' in *Fig. 46.10*, for example, requires all seven LEDs to be lighting.
- LEDs use very little current compared to ordinary bulbs. They are very reliable, cheap and long lasting. They are used in bicycle lamps and in **traffic lights**, replacing ordinary bulbs.

Fig. 46.10 LED (red dot of light) indicating alarm has been set. LEDs in the shape of bars form the digits or numbers. How many LEDs are 'on' to form the digit zero?

(3) The Light Dependent Resistor (LDR)

The light dependent resistor (LDR) is a special type of **variable resistor**. Its resistance varies with the amount of light shining on it. The symbol for an LDR is shown in *Fig. 46.11*.

- In **dim** light, the light dependent resistor has a very high resistance and allows little current through.
- In **bright** light, the resistance of the LDR is low and it allows much more current through.

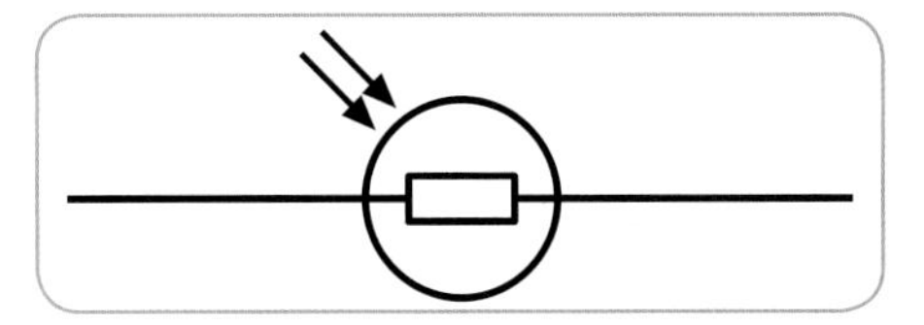

Fig. 46.11 Symbol for LDR.

Experiment 46.3

To measure the resistance of a light dependent resistor as the brightness of the light shining on it varies

Apparatus required: LDR; ohmmeter; lamp

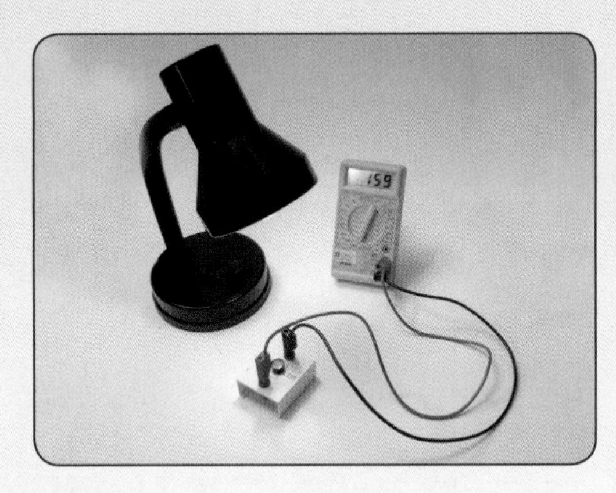

Fig. 46.12 Showing how the resistance of an LDR varies with the brightness of the incident light.

Method

1. Connect the LDR to the ohmmeter (or multimeter, set as an ohmmeter) as shown in *Fig. 46.12*. Bring the lamp very close to the LDR. Note its resistance.
2. Vary the brightness of the light shining on the LDR. This is done by gradually moving the lamp further away from the LDR. Note the resistance on the ohmmeter after each change in brightness level.

Results

The resistance of the LDR increases as the brightness of the light shining on it decreases. The resistance of some LDRs vary from 100 Ω in sunlight to 10 MΩ in the dark.

Uses of LDRs

- Light dependent resistors or LDRs are used widely in light controlled switches. They are used to **switch street lighting** on and off at the appropriate times of the day.
- They are also used in light meters for **cameras** and in **burglar alarms**.

TEST YOURSELF:
Now attempt questions *46.11–46.14* and *W46.13–W46.25*.

What I should know

- The live and the neutral are the two cables from the mains supply to your home. The live cable is at a potential difference of ±230 V. The neutral remains at a voltage of zero.
- When choosing the correct fuse or circuit breaker for a circuit, its rating should be just above the normal current that the circuit or appliance requires.
- A circuit breaker works using the magnetic effect of an electric current, whereas a fuse works using the heating effect of an electric current.
- Plugs have three terminals. Each terminal must be connected to the correct coloured wire. The live (L) is connected to the brown wire, the neutral (N) to the blue and the earth (E) to the green/yellow.
- All electric appliances with metal bodies should be earthed.
- The power of an appliance is a measure of how quickly it converts electrical energy to other forms of energy. It is measured in watts (W).
- The greater the power rating of an electric appliance, the more it will cost to run.
- The ESB sell electricity in 'units' or kilowatt-hours (kW h). One 'unit' or one kilowatt-hour is the electrical energy converted by a 1 kW appliance running for 1 hour.
- Number of kilowatt-hours (units) used = (number of kilowatts) × (number of hours).
- Cost of electricity = (number of kilowatt-hours used) × (price per kilowatt-hour or 'unit').
- The number of ESB 'units' used in a given interval is found by subtracting the previous reading from the present reading on the electricity meter or electricity bill.
- A diode is a device which allows current to flow only in one direction.
- If a diode is to conduct, the cathode (–) must be connected to the negative terminal of the battery and the anode (+) is connected to the positive terminal of the battery. The diode is now forward biased.
- If the cathode of the diode is connected to the positive of the battery and the anode to the negative, the diode is reverse biased. Current cannot now flow through it.
- A diode can act as a rectifier, which means that it changes a.c. to d.c.
- A light emitting diode (LED) is a special type of diode which gives out light when current passes through it. It uses little current and is usually protected by a large resistor in series.
- LEDs are used as indicator lamps on electrical equipment and in displays on clocks and cash registers.
- A light dependent resistor (LDR) is a variable resistor whose resistance decreases as more light falls on it.
- LDRs are used in light controlled switches, e.g. street lamps. They are also used in light meters for cameras and in burglar alarms.

QUESTIONS

Write the answers to the following questions into your copybook.

46.1 The live and the ..... are the two cables from the ..... supply to your home. The live cable is very ..... . It is at a potential difference of ..... V.

46.2 If only 1 A, 2 A, 3 A, 5 A and 13 A fuses are available, which fuse is best to put in a plug connected to each of the following appliances:

(a) A 1200 W iron that normally uses 5.2 A.
(b) A 120 W computer that normally uses 0.52 A.
(c) A 1 kW hoover that normally uses 4.3 A.
(d) A 2014 W electric kettle that normally uses 8.8 A.
(e) A 600 W microwave oven that normally uses 2.6 A.

46.3 State one advantage of a circuit breaker over a fuse in electrical circuits?

46.4 Copy the table below and fill in the missing names of the terminals and the colours of the insulating wires.

NAME OF TERMINAL	LETTER FOR TERMINAL	COLOUR(S) OF INSULATING WIRES
	L	
	N	
	E	

Table 14.2 Fill in the missing names and colours.

46.5 An electric heater produces 60 000 J of heat in 40 s. What is the power rating of the heater?

46.6 The instrument in your home that records the total number of 'units' used is called ..... . The number of 'units' used in a given two-month interval is calculated by ..... .

46.7 To calculate the number of 'units' an appliance uses, you simply multiply the number of ..... marked on the appliance by the number of ..... that it is running for.

46.8 How many units would a 1500 W immersion heater use in 2 hours?

46.9 Four 150 W bulbs are left on for five hours. What is the total number of units of electricity that they use? If each unit costs 16 cent, how much does this electricity cost?

46.10 A unit of electricity costs 18 cent. A 20 W energy saving bulb gives the same light as a 100 W ordinary bulb. The energy saving bulb, however cost €6 more than the ordinary bulb to buy. How much money was saved overall in a year when the bulb was used on average for 6 hours a day for each of the 365 days of the year?

46.11 A diode is a device which allows ..... to flow only in one ..... . The diode gets its name from the fact that it has ..... electrodes.

46.12 Electrical energy is converted to light (and heat) energy in a ..... diode.

46.13 Current cannot flow through a LED if the LED is ..... biased.

46.14 (a) Identify the components labelled *A, B, C, D, E* and *F* in *Fig. 46.13.*

(b) What is observed in (i) *C* and (ii) *F*, when *B* is closed? Explain why this is so in each case.

(c) What is the function of *E* in the circuit?

(d) Are *D* and *F* arranged in series or in parallel in this circuit?

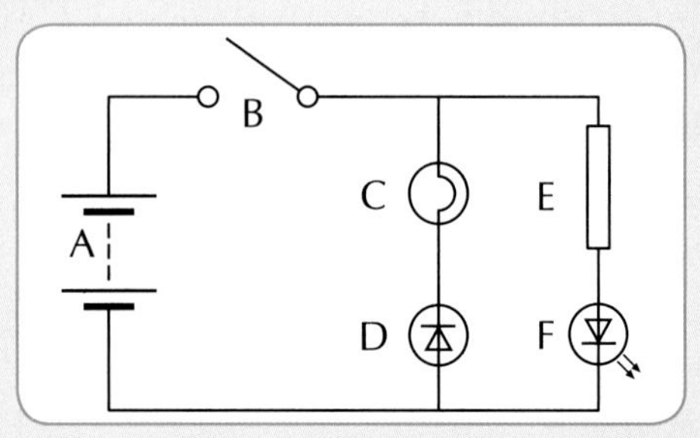

Fig. 46.13 Identifying components by their symbols.

Chapter 47 Investigations in Science

47.1 What is an Investigation?

The practical work that you have studied so far has involved mainly the mandatory experiments on the syllabus. The mandatory experiments help to give you experience of carrying out practical work in the school laboratory. They also help you to develop various laboratory skills like measuring, observing and drawing conclusions. You have been writing up these experiments as part of *Coursework A* of the syllabus. In this chapter you begin *Coursework B* of the syllabus. This involves you carrying out experiments that are called **investigations**. These investigations differ from the mandatory experiments in two important ways.

When carrying out investigations you will not have a set of instructions as you have had for the mandatory experiments. You yourself will have to plan how to carry out the experiment.

When carrying out investigations you do not usually know the expected result. For example, recall the work you did in *Mandatory experiment 33.3* To carry out an experiment to demonstrate that oxygen and water are necessary for rusting. In this mandatory experiment you knew the result before you began the experiment, i.e. that oxygen and water are necessary for rusting. Therefore, this mandatory experiment is not classified as an investigation.

Therefore, an investigation is defined as follows:

An investigation is a task for which the student cannot immediately see an answer or recall a routine method for finding it.

We will now study the various types of investigations and explain how they should be carried out.

47.2 Types of Investigations

For the Junior Certificate examination you are required to carry out **two** investigations. Some examples of Junior Cert investigations are shown in *Table 47.1*.

BIOLOGY

1. A gardener suggests that the length of time taken for marrowfat peas to germinate is decreased if they are soaked in water in advance. Carry out a quantitative investigation of this suggestion.
2. Carry out a quantitative survey of the plant species in a local habitat.

CHEMISTRY

3. Investigate a range of plant pigments to evaluate their effectiveness as acid-base indicators.
4. Investigate how the concentration of a hydrogen peroxide solution affects the speed at which it decomposes to produce oxygen gas.

PHYSICS

5. Investigate the relationship between the temperature of a rubber squash ball and the height to which it bounces.
6. Carry out an investigation of the relationship between the length of a metallic conductor (e.g. nichrome wire) and its resistance.

Table 47.1 Examples of coursework B *Investigations.*

On studying *Table 47.1* you will see that there are two types of investigations:

- **Variable Type Investigation**. The word variable means 'something that can be changed'. If you study Table 47.1 you will see that some of the investigations involve variables like length of time (Investigation 1), concentration of solution (Investigation 4) and temperature (Investigation 5). We will discuss variables in more detail in the next section.
- **Exploration Type Investigation.** This type of investigation involves you carrying out an exploratory type activity, i.e. you carry out a scientific study of a particular situation and write a report on your work. An example of an exploratory type investigation is investigation 2 in Table 47.1. In this investigation you are expected to study a local habitat and write an account of your work under various headings.

47.3 How to Carry out and Write up an Investigation

When carrying out and writing up an investigation, it is important to follow the stages listed in the *Coursework B* booklet of the State Examinations Commission. A description of each stage will now be given. Some sample answers will also be given to help you understand what is required.

1. Introduction to the Investigation

In this section you must show that you understand what is involved in the investigation. Therefore, you must include the following points:

- Give a clear statement of the problem or topic being investigated.
- List the sources used to obtain background information on the topic, e.g. books, websites and people consulted.

An example of an introduction to investigation 5 (squash ball) is shown in *Fig. 47.1*.

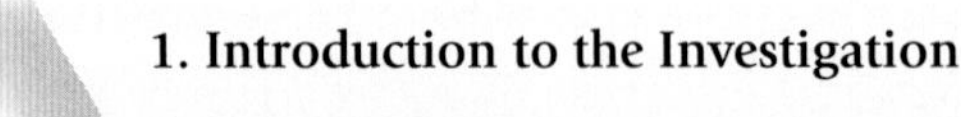

1. Introduction to the Investigation

In this experiment, I will investigate if a heated squash ball or a cooled squash ball will bounce higher than one at normal room temperature.

I searched on the internet for information about squash balls. I learned that a squash ball is a hollow rubber ball containing air. I learned that there are many different sizes and makes of squash balls.

I found some information about the bouncing properties of squash balls on the following websites: **[insert web addresses here]**

Fig. 47.1 Sample introduction to the investigation: 'Investigate the relationship between the temperature of a rubber squash ball and the height to which it bounces'.

2. Preparation and Planning

In this section you describe how you are going to plan out the experiment and what you will do during the experiment. There are three headings under which you will describe this stage:

(i) Identification of Variables and Necessary Controls

As described above, a variable is something that can be changed. In investigations of this type, you change one variable (e.g. the temperature of the ball) and measure the second variable (e.g. the height to which the ball bounces). All other variables must be controlled, e.g. the type of squash ball being used, the height from which the ball is dropped, the surface on which the ball is dropped, etc. These other variables are called **control variables** or simply **controls**. These variables must be controlled in order to have a **fair test**. *Fig. 47.2*.

A fair test means that the relationship between temperature and the height of bounce must always be studied under the same set of conditions. For example, it would not be a fair test if you used both a wooden floor and a carpet in the same investigation.

To make an experiment a fair test, just change one variable and keep the other variables constant.

An example of a sample description is given in *Fig. 47.3*.

Fig. 47.2 This is not a fair test of the speed of the runners since the conditions are not the same for all runners.

(i) Identification of Variables and Necessary Controls

The variable that I will change is the temperature of the squash ball. The variable that I will measure is the height to which the ball bounces. The control variables are the height from which it is dropped, the type of squash ball used and the surface on which it is dropped. To make sure that this investigation is a fair test, I will drop the squash ball from the same height each time. Also, I will use the same squash ball and I will drop it on to the same surface.

Fig. 47.3 Sample identification of variables and controls for the investigation: 'Investigate the relationship between the temperature of a rubber squash ball and the height to which it bounces.'

Note: In an exploration type investigation, there are no controls. Therefore, you could write something like that shown in *Fig. 47.4*.

(i) Identification of Variables and Necessary Controls

Since the quantitative survey of the plant species in a local habitat is an exploration type investigation, there are no control variables.

Fig. 47.4 Not all investigations have controls.

(ii) List of the Equipment Needed for the Investigation

In this section, simply list the equipment and chemicals (if any) that you will use in the investigation. An example of equipment used in the squash ball investigation is shown in *Fig. 47.5*.

(iii) List of the Equipment Needed for the Investigation

Squash ball, metre stick, retort stand, thermometer, beaker, water bath, tongs, ice cubes, kettle, safety goggles.

Fig. 47.5 List of all the items used in the squash ball investigation.

(iii) List of Tasks to be Carried out during the Investigation

In this section, you must write down the main stages of the investigation. You do not have to give any details at this stage. An example of the main stages for the squash ball investigation are given in *Fig. 47.6*.

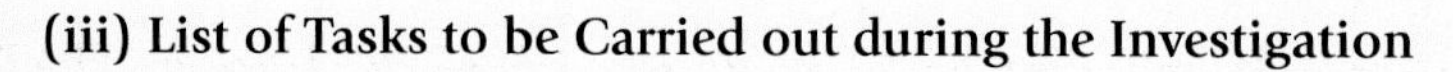

(iii) List of Tasks to be Carried out during the Investigation

I will drop the ball from a certain height above the ground. After dropping the ball, I will measure the height to which it bounces. I will change the temperature of the ball. I will then measure the new height to which it bounces. I will repeat this procedure a number of times with the ball at different temperatures. I will record my results in the form of a table. I will plot a graph to study the relationship between temperature of the ball and the height to which it bounces.

Fig. 47.6 Summary of tasks that will be carried out during the squash ball investigation.

3. Procedure, Apparatus, Safety, Data Collection/Observations

(i) Particular Safety Precautions Required by this Investigation

In this section, try to think of two safety precautions that you will follow when carrying out the investigation. As a general rule, safety glasses should be worn for all investigations. Some safety precautions for the squash ball investigation are given in *Fig. 47.7.*

(i) Particular Safety Precautions Required by this Investigation

I wore safety goggles to protect my eyes. I used tongs when handling the squash ball at various temperatures.

Fig. 47.7 Always mention at least two safety precautions.

(ii) and (iii) Procedure Followed in the Investigation

In this section, write down all the main steps that you took in carrying out the investigation. It is very important that you draw a labelled diagram. Drawing a labelled diagram helps to make it clear to the examiner what you have done in the investigation. Also, in cases where you forget to write down a step, marks can be obtained from a diagram if the step is shown in the diagram. There is plenty of space in the Reporting Booklet for you to write a detailed account of your procedure. You may also wish to draw a second diagram.

An example of a procedure written by a student for the squash ball experiment is shown in *Fig. 47.8.*

(ii) and (iii) Procedure Followed in the Investigation

1. We set up the apparatus as shown in the diagram.

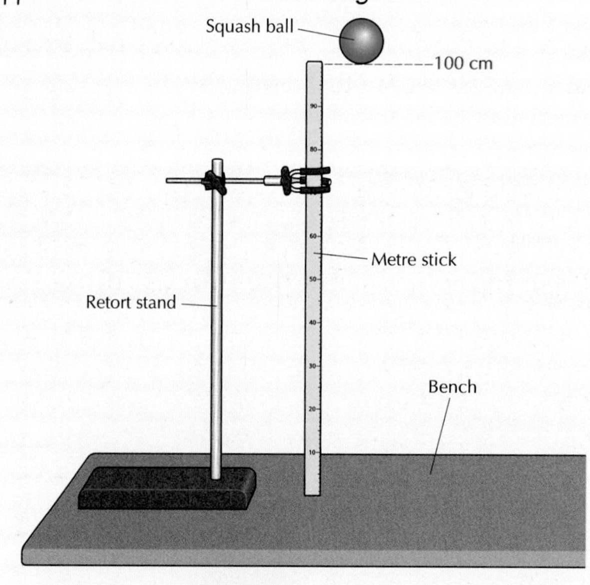

Contd.

2. We used a metre stick to measure the heights of the ball above the ground. The metre stick was held in position using a retort stand. I dropped the ball and my partner measured the height to which it bounced. All heights were measured from the bottom of the squash ball.

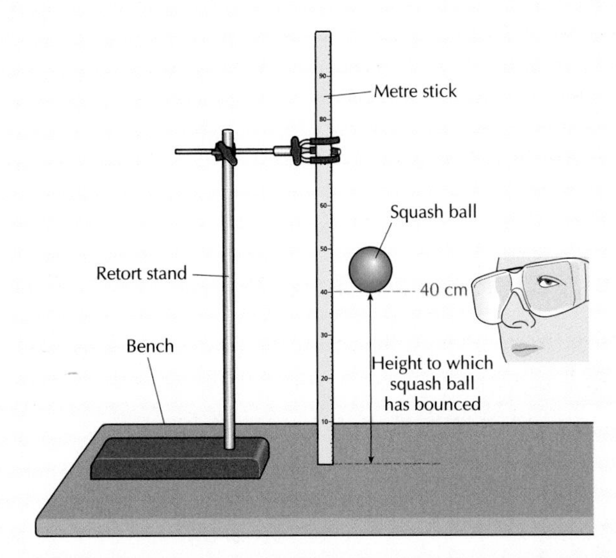

3. We bounced the ball at room temperature (18 °C) and measured the height to which it bounced. We bounced the ball five times and took the average of the heights to which the ball bounced.
4. We changed the temperature of the ball by putting it in water at different temperatures. We started off with a beaker of water containing some ice cubes. We placed the ball in the beaker for five minutes and measured the temperature of the water with a thermometer. We tried to get the water at exact temperatures like 20 °C, 40 °C, 60 °C. However, after discussing this with our teacher we realised that this was not necessary.
5. We bounced the ball five times at each temperature. In each case, we measured the height to which the ball bounced. We then calculated the average height to which the ball bounced.

Fig. 47.8 Description of procedure followed in investigation.

(iv) Recorded Data/Observations

In this section you must present the data that you have collected. You must also make a comment about the data. It is a good idea to present the data in the form of a table like that shown in *Table 47.1*.

Temperature of ball (°C)	First bounce height (cm)	Second bounce height (cm)	Third bounce height (cm)	Fourth bounce height (cm)	Fifth bounce height (cm)	Average bounce height (cm)
6	12.5	13.2	11.8	10.7	12.3	12.1
16	53.7	55.6	54.7	56.8	53.7	54.9
36	61.5	64.6	60.8	62.8	61.2	62.2
65	75.5	73.7	74.6	76.5	75.2	75.1
84	77.2	75.5	76.8	78.6	76.2	76.9
100	82.5	80.8	80.5	78.6	80.2	80.5

Table 47.1 Sample data from squash ball investigation.

If possible, also present the data in the form of a graph, *Fig 47.9*.

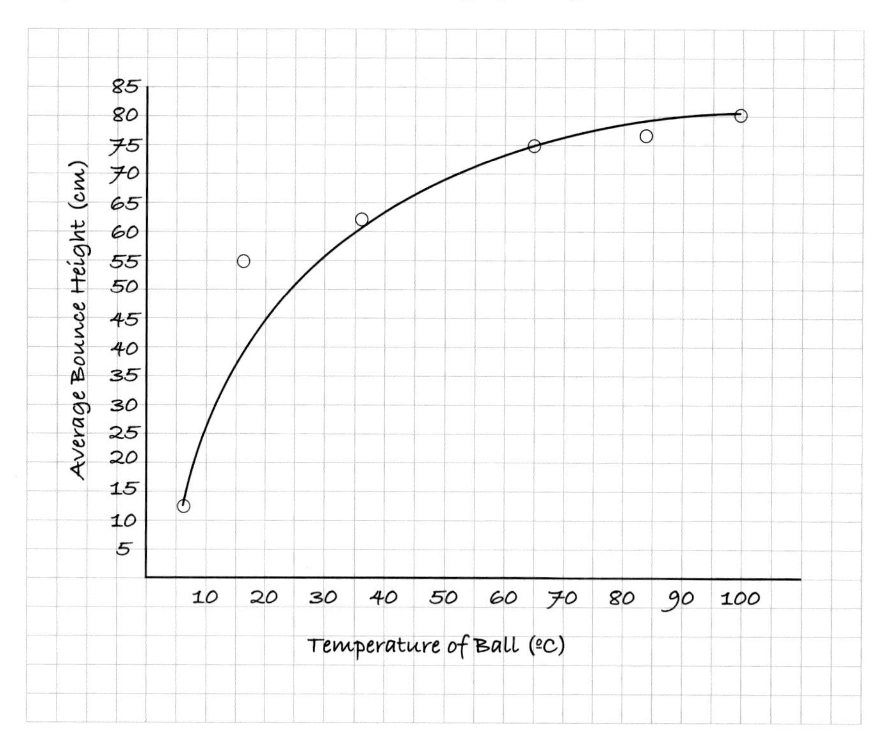

Fig. 47.9 Graph showing data from squash ball investigation.

An example of one student's comment on this data is shown in *Fig. 47.10*.

(iv) Recorded Data/Observations

Comment on data: I am surprised that the ball did not bounce to exactly the same height at the same temperature. This may be due to the fact that the temperature of the ball may not be exactly the same for each bounce. It could also be due to experimental error. It was difficult to measure the exact height of the bounce.

Fig. 47.10 An example of a student's comment on the data obtained in the squash ball investigation.

4. Analysis

(i) Calculations / Data Analysis

In this section you should carry out any calculations required and comment on the data obtained. For example, if your graph is a straight line, this indicates that the two variables are directly proportional to each other. An example of a comment for the squash ball investigation is shown in *Fig. 47.11*.

(i) Calculations / Data Analysis

A straight line graph is not obtained. Therefore, the height of the bounce is not directly proportional to the temperature. Also, it is clear from the data that as the temperature was increased, the less was the increase in the bounce height with every measurement, i.e. the increase in bounce height was most noticeable at the lower temperatures.

The graph did (almost) take the shape of a smooth curve. But one point was clearly well off the curve. This may be due to experimental error.

Fig. 47.11 A student's comments on the data obtained for the squash ball investigation.

(ii) Conclusion(s) and Evaluation of Results

In this section, state clearly what you have found out in the investigation. The conclusion should be kept short and stated as simply as possible. Also comment on the accuracy of the results. For example, the conclusion and comment on accuracy for the squash ball experiment is shown in *Fig. 47.12*.

(ii) Conclusion(s) and Evaluation of Results

The results of the investigation show that as the temperature of a squash ball is increased, the height to which the ball bounces also increases.

I feel that the results are accurate because I used the average of the five readings when graphing the data.

Fig. 47.12 The conclusion drawn from the squash ball investigation.

5. Comments (e.g. refinements, extensions, sources of error, etc.)

In this section, discuss what changes you would make if you were doing the experiment again. Also comment on possible sources of error, *Fig. 47.13*.

5. Comments (e.g. refinements, extensions, sources of error, etc.)

If I were doing this experiment again, I would take more temperature readings, e.g. around 10 °C and around 25 °C. This would help me to get more points on the graph so that I could get a clearer picture of the bounce height in the lower part of the graph. Also, I would put the squash ball in a freezer overnight and measure the bounce height at very low temperatures.

One source of error is that the temperature of the squash ball may change when it is taken out of the hot water. Therefore, the temperature may not be the same for each of the five bounces at any one particular temperature.

Another source of error is the difficulty of measuring the exact height of bounce of the squash ball. The use of video camera or a computer with datalogging equipment could give us more accurate results.

Fig. 47.13 Always comment on the changes you would make if you were carrying out the investigation again. Also, comment on the possible sources of error.

47.4 Summary

The various steps for carrying out an investigation are summarised in *Fig. 47.14*. Regardless of what investigation you are carrying out, always ask yourself the questions listed in *Fig. 47.14*. It is hoped that these questions and the sample answers given in this chapter will help you to carry out your own investigations. I wish you the very best of luck in your work!

1. Introduction to the Investigation

- Give a clear statement of the topic being investigated.
- List the sources used to obtain background information.

↓

2. Preparation and Planning

(i) Identification of Variables and Necessary Controls

For a variable type investigation:

- State what you will change.
- State what you will measure.
- State what you will keep constant ('controls').

For an exploration type investigation:

- State that there are no control variables.

(ii) List of Equipment Needed for the Investigation

- List the equipment and chemicals needed.

(iii) List of Tasks to be Carried out during the Investigation

- State what you will carry out in each stage of the investigation.

3. Procedure, Apparatus, Safety, Data Collection/Observations

(i) Particular Safety Precautions Required by this Investigation

- State two safety precautions that you will follow.

(ii) and (iii) Procedure Followed in the Investigation

- List all the main steps you took in carrying out the investigation.
- Give a number to each step in the investigation.

(iv) Recorded Data/Observations

- Present the data in a table (if possible).
- Present the data in the form of a graph (if possible).
- Make a comment on anything you notice about the data.

(ii) Conclusion(s) and Evaluation of Results

- State what you have found out in the investigation.
- Comment on the accuracy of the results.

4. Analysis

(i) Calculations / Data Analysis

- Perform any calculations needed.
- Comment on what the data is telling you.
- Comment on the shape of any graphs drawn.

(ii) Conclusion(s) and Evaluation of Results

- State what you have found out in the investigation.
- Comment on the accuracy of your results.

5. Comments (e.g. refinements, extensions, sources of error, etc.)

- State what changes you would make if you were carrying out the investigation again.
- Comment on possible sources of error.

Fig. 47.14 Stages involved in carrying out an investigation.

Appendix I

LARGE and small Numbers

Some numbers used in science are very large. Others are very small. The mass of the earth, for instance is 6 000 000 000 000 000 000 0 00 000 kg. But, the mass of a water molecule is 0.000 000 000 000 000 000 000 000 03 kg. It would be very easy to make mistakes when writing out or reading numbers like these. It would also be very tedious.

Instead, a shorthand way of writing numbers, called ***scientific notation***, is used. To write large and small numbers in scientific notation, we shift the decimal point to the left or to the right so that there is only one number (other than '0') to the left of the decimal point. We then multiply by the appropriate power of 10.

Example 1: 31 500 000 is written as 3.15×10^7, because the decimal point has shifted **seven** places to the **left**.

Example 2: 0.000 026 is written as 2.6×10^{-5}, because the decimal point has shifted **five** places to the **right**.

In this notation, the mass of the earth is written as 6×10^{24} kg and the mass of a water molecule is 3×10^{-26} kg.

Other examples of scientific notation are given in *Table A1* below. Note that for small numbers between 0 and 1, a **negative** power of 10 is used. (You will learn more about this again in your Maths course.)

Prefixes can also be used to represent large and small numbers. Some of these are also given in the table below. (A list of prefixes is given also on Page 4 or 5 of the *Mathematical Tables*.)

NUMBER	NO. OF PLACES MOVED BY DECIMAL POINT	SCIENTIFIC NOTATION	PREFIX	SYMBOL FOR PREFIX
1 000 000	Six places to the left	1×10^6	mega	M
1 000	Three places to the left	1×10^3	kilo	kg
100	Two places to the left	1×10^2	hecto	h
10	One place to the left	1×10^1	deca	da
0.1	One place to the right	1×10^{-1}	deci	d
0.01	Two places to the right	1×10^{-2}	centi	c
0.001	Three places to the right	1×10^{-3}	milli	m

Table A1 Various large and small numbers represented by different powers of 10 or by prefixes.

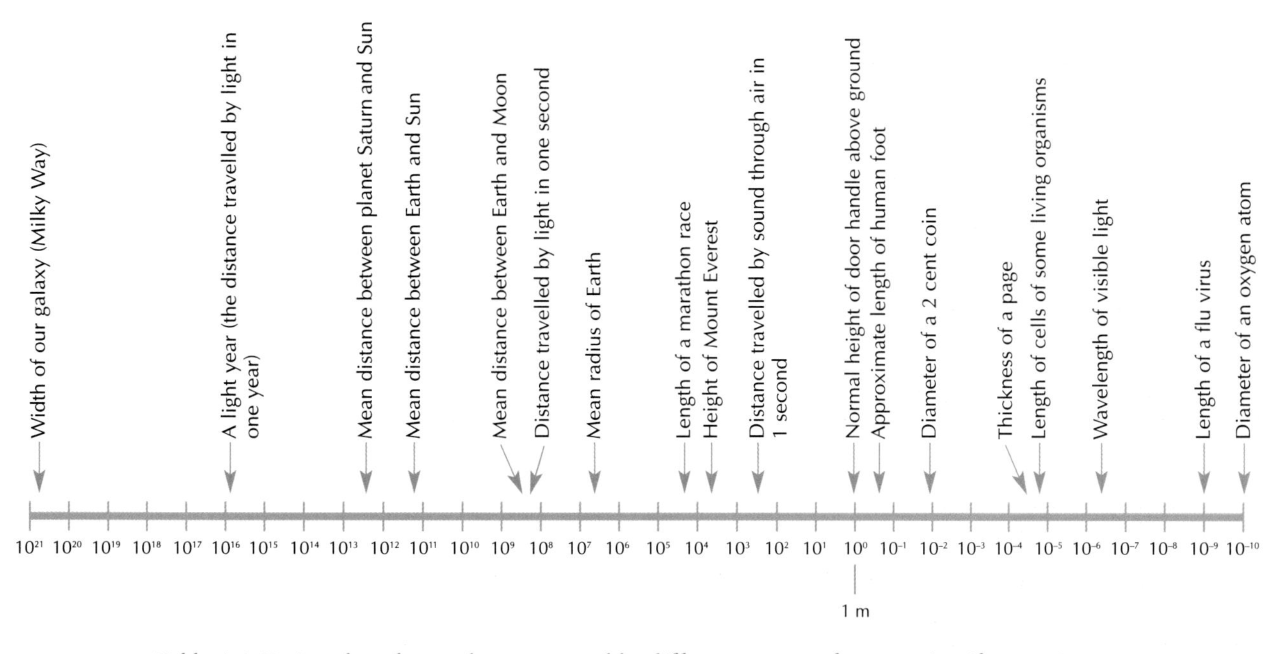

Table A.2 Various lengths can be represented by different powers of ten or scientific notation.

Appendix II

Formulas and Units in Physics

Formulas

- The area of a rectangle is its length multiplied by its width.
- The area of a circle = $\pi \times r^2$, where r is the radius of the circle and π is 3.14 approximately.
- The volume of a rectangular block is found by multiplying its length by its width by its height.
- Density $= \dfrac{\text{mass}}{\text{volume}}$
- Average speed $= \dfrac{\text{distance travelled}}{\text{time taken}}$
- Average velocity $= \dfrac{\left(\text{distance travelled in a given direction}\right)}{\text{time taken}}$
- Acceleration $= \dfrac{(\text{final velocity} - \text{initial velocity})}{\text{time taken}}$
- Weight = mass × g, where 'g' is the acceleration due to gravity (= 10 m/s^2 approx.)
- Work done = force × distance moved
- Moment of a force =

  (force) × (perpendicular distance from the force to the fulcrum)
- Pressure $= \dfrac{\text{force}}{\text{area}}$
- Number of kilowatt-hours (units) = (number of kilowatts) × (number of hours)
- Cost of electricity = (number of kilowatt-hours used) × (price per kilowatt-hour or 'unit')
- Resistance (at constant temperature) = voltage/current or R = V/I

Units

QUANTITY	UNIT	SYMBOL
Mass	kilogram	kg
Length	metre	m
Time	second	s
Area	metre squared	m^2
Volume	metre cubed	m^3
	(or) litre	L
Density	kilogram per metre cubed	kg/m^3 (or) $kg\ m^{-3}$
	(or) gram per centimetre cubed	g/cm^3 (or) $g\ cm^{-3}$
Speed and velocity	metre per second	m/s (or) $m\ s^{-1}$
Acceleration	metre per second squared	m/s^2(or) $m\ s^{-2}$
Force (and weight)	newton	N
Pressure	pascal	Pa
	(or) newton per metre squared	N/m^2 (or) $N\ m^{-2}$
Energy and work	joule	J
Temperature	degree celsius	°C
Potential difference or voltage	volt	V
Current	ampere	A
Resistance	ohm	Ω
Power	watt	W
ESB 'unit' of energy	Kilowatt-hour	kW h

Index

T

U

V

W

X

Z